Partial Least Squares Regression

Partial least squares (PLS) regression is, at its historical core, a black-box algorithmic method for dimension reduction and prediction based on an underlying linear relationship between a possibly vector-valued response and a number of predictors. Through envelopes, much more has been learned about PLS regression, resulting in a mass of information that allows an envelope bridge that takes PLS regression from a black-box algorithm to a core statistical paradigm based on objective function optimization and, more generally, connects the applied sciences and statistics in the context of PLS. This book focuses on developing this bridge. It also covers uses of PLS outside of linear regression, including discriminant analysis, non-linear regression, generalized linear models and dimension reduction generally.

Key Features:

- Showcases the first serviceable method for studying high-dimensional regressions.
- Provides necessary background on PLS and its origin.
- R and Python programs are available for nearly all methods discussed in the book.

R. Dennis Cook is Professor Emeritus, School of Statistics, University of Minnesota. His research areas include dimension reduction, linear and nonlinear regression, experimental design, statistical diagnostics, statistical graphics, and population genetics. Perhaps best known for “Cook’s Distance,” a now ubiquitous statistical method, he has authored over 250 research articles, two textbooks and three research monographs. He is a five-time recipient of the Jack Youden Prize for Best Expository Paper in Technometrics as well as the Frank Wilcoxon Award for Best Technical Paper. He received the 2005 COPSS Fisher Lecture and Award, and is a Fellow of ASA and IMS.

Liliana Forzani is Full Professor, School of Chemical Engineering, National University of Litoral and principal researcher of CONICET (National Scientific and Technical Research Council), Argentina. Her contributions are in mathematical statistics, especially sufficient dimension reduction, abundance in regression and statistics for chemometrics. She established the first research group in statistics at her university after receiving her Ph.D in Statistics at the University of Minnesota. She has authored over 75 research articles in mathematics and statistics, and was recipient of the L’Oreal-Unesco-Conicet prize for Women in science.

Partial Least Squares Regression

And Related Dimension Reduction Methods

R. Dennis Cook
Liliana Forzani

CRC Press is an imprint of the
Taylor & Francis Group, an **informa** business
A CHAPMAN & HALL BOOK

Designed cover image: © R. Dennis Cook and Liliana Forzani

First edition published 2024
by CRC Press
2385 NW Executive Center Drive, Suite 320, Boca Raton FL 33431

and by CRC Press
4 Park Square, Milton Park, Abingdon, Oxon, OX14 4RN

CRC Press is an imprint of Taylor & Francis Group, LLC

Library of Congress Cataloging-in-Publication Data

Names: Cook, R. Dennis, author. | Forzani, Liliana, author.
Title: Partial least squares regression : and related dimension reduction methods / R. Dennis Cook and Liliana Forzani.
Description: First edition. | Boca Raton, FL : CRC Press, 2024. | Includes bibliographical references and index.
Identifiers: LCCN 2024000157 (print) | LCCN 2024000158 (ebook) | ISBN 9781032773186 (hbk) | ISBN 9781032773230 (pbk) | ISBN 9781003482475 (ebk)
Subjects: LCSH: Least squares. | Regression analysis. | Multivariate analysis.
Classification: LCC QA275 .C76 2024 (print) | LCC QA275 (ebook) | DDC 511/.422--dc23/eng/20240326
LC record available at https://lccn.loc.gov/2024000157
LC ebook record available at https://lccn.loc.gov/2024000158

ISBN: 978-1-032-77318-6 (hbk)
ISBN: 978-1-032-77323-0 (pbk)
ISBN: 978-1-003-48247-5 (ebk)

DOI: 10.1201/9781003482475

Typeset in CMR10 font
by KnowledgeWorks Global Ltd.

Publisher's note: This book has been prepared from camera-ready copy provided by the authors.

For Violette and Rose

R. D. C.

For Bera, Edu and Marco

L. F.

Contents

Preface

Partial least squares (PLS) regression is, at its historical core, a black-box algorithmic method for dimension reduction and prediction based on an underlying linear relationship between a possibly vector-valued response and a number of predictors. Its origins trace back to the work of Herman Wold in the 1960s, and it is generally recognized that Wold's non-linear iterative partial least squares (NIPALS) algorithm is a critical point in its evolution. PLS regression made its first appearances in the chemometrics literature around 1980, subsequently spreading across the applied sciences. It has attracted considerable attention because, with highly collinear predictors, its performance is typically better than that of standard methods like ordinary least squares, and it has particularly good properties in the class of abundant high-dimensional regressions where many predictors contribute information about the response and the sample size n is not sufficient for standard methods to yield unambiguous answers. In some regressions, its estimators can converge at the root-n rate regardless of the asymptotic relationship between the n and the number of predictors. It is perhaps the first serviceable method for studying high-dimensional linear regressions.

The statistics community has been grappling for over three decades with the regressions in which n is not sufficient to allow crisp application of standard methods. In the early days, they fixed on the idea of sparsity, wherein only a few predictors contribute information about the response, to drive their fitting and prediction methods, resulting in a focus that likely hindered the pursuit of other plausible methods. Not being based on optimizing an objective function, PLS regression fell on the fringes of statistical methodology and consequently did not receive much attention from the statistics community. This picture began to change in 2013 with the findings that PLS regression is a special case of the then nascent envelope theory of dimension reduction (Cook et al., 2013). Through envelopes, we have learned much about PLS regression in the past decade, resulting in a critical mass of information that allows us to provide an envelope bridge that takes PLS regression from a black-box algorithm to a core statistical paradigm based on objective function optimization and, more generally, connects the applied sciences and statistics in the context of PLS. Developing that bridge is a goal of this book. We hope that our arguments will stimulate interest in PLS within the statistics community. A second goal of this book is to check out uses of PLS outside of linear regression, including discriminant analysis, non-linear regression, generalized linear models, and dimension reduction generally.

Outline

Chapter 1 contains background on the multivariate linear model and on envelopes. This is not intended as primary instruction, but may be sufficient to establish basic ideas and notation. In Section 1.5, we introduce five algebraic methods for envelope construction, some of which are generalizations of common PLS algorithms from the literature. In Chapter 2, we review envelopes for response and predictor reduction in regression, and discuss first connections with PLS regression. In Chapter 3, we describe the common PLS regression algorithms for predictor reduction, NIPALS and SIMPLS, and prove their connections to envelopes. We also discuss PLS for response reduction, PLS1 v. PLS2 and various other topics. These first three chapters provide a foundation for various extensions and adaptations of PLS that come next. Chapters 4 – 11 do not need to be read in order. For instance, readers who are not particularly interested in asymptotic considerations may wish to skip Chapter 4 and proceed with Chapter 5.

Various asymptotic topics are covered in Chapter 4, including convergence rate and abundance v. sparsity. Simultaneous PLS reduction of responses and predictors in multivariate linear regression is discussed in Chapter 5, and methods for reducing only a subset of the predictors are described in Chapter 6. We turn to adaptations for linear and quadratic discriminant analysis in Chapters 7 and 8. In Chapter 9 we argue that there are settings in which the dimension reduction arm of a PLS algorithm is serviceable in non-linear as well as linear regression.

The versions of PLS used for path analysis in the social sciences are notably different from the PLS regressions used in other areas like chemometrics. PLS for path analysis is discussed in Chapter 10. Ancillary topics are discussed in Chapter 11, including bilinear models, the relationship between PLS regression and conjugate gradient methods, sparse PLS, and PLS for generalized linear models. Most proofs are given in an appendix, but some that we feel may be particularly informative are given in the main text.

Computing

R or Python computer programs are available for nearly all of the methods discussed in this book and many have been implemented in both languages. Most of the methods are available in integrated packages, but some are standalone programs. These programs and packages are not discussed in this book, but descriptions of and links to them can be found at https://lforzani.github.io/PLSR-book/. This format will allow for updates and links to developments following this book. The web page also gives errata, links to recent developments, color versions of some grayscale plots as they appear in the book, and commentary on parts of the book as necessary for clarity.

Acknowledgments

Earlier versions of this book were used as lecture notes for a one-semester course at the University of Minnesota. Students in this course and our collaborators contributed to the ideas and flavor of the book. In particular, we would like to thank Shanshan Ding, Inga Helland, Zhihua Su, and Xin (Henry) Zhang for their helpful discussions. Much of this book of course reflects the many stimulating conversations we had with Bing Li and Francesca Chiaromonte during the genesis of envelopes. The tecator dataset of Section 9.7 is available at http://lib.stat.cmu.edu/datasets/tecator.

We extend our gratitude to

Rodrigo García Arancibia for providing the economic growth data and analysis of Section 6.3.

Fabricio Chiappini for providing the etanercept data of Section 9.8 and for sharing with us a lot of discussion and background on chemometric data. He together with Alejandro Olivieri were very generous in providing us with necessary data.

Pedro Morin for helping us comprehend the subtle link between PLS algorithms and conjugate gradient methods.

Marilina Carena for help drawing Figure 10.1 and Jerónimo Basa for drawing the mussels pictures for Chapter 4.

Special thanks to Marco Tabacman for making our codes more readable and to Eduardo Tabacman for his guidance in creating numerous graphics in R.

R. Dennis Cook	Liliana Forzani
St. Paul, MN, USA	Santa Fe, Argentina
	January, 2024

Notation and Definitions

Reminders of the following notation may be included in the book from time to time.

- r number of responses.
- p number of predictors.
- q number of predictor components and dimension of predictor envelope.
- u number of response components and dimension of response envelope.
- X an $p \times 1$ vector of predictors. $\mathbb{X}$ is the $n \times p$ matrix with rows $(X_i - \bar{X})^T$ and $\mathbb{X}_0$ is the $n \times p$ matrix with rows X_i^T, $i = 1, \dots, n$.
- Y an $r \times 1$ vector of responses. $\mathbb{Y}$ is the $n \times r$ matrix with rows $(Y_i - \bar{Y})^T$ and $\mathbb{Y}_0$ is the $n \times r$ matrix with rows Y_i^T, $i = 1, \dots, n$.
- Observable data matrices will be generally represented in mathbb font like $\mathbb{X}$, $\mathbb{Y}$, and $\mathbb{Z}$.
- $C = (X^T, Y^T)^T$, the concatenation of X and Y.
- $v_1(\cdot)$ largest eigenvalue of the argument matrix with corresponding eigenvector $\ell_1(\cdot)$ normalized to have length 1.
- $\mathrm{diag}(d_1, \dots, d_m)$ denotes an $m \times m$ diagonal matrix with diagonal elements d_j. $j = 1, \dots, m$.
- $\mathrm{tr}(A)$ denotes the trace of a square matrix A.
- $W_q(M)$ denotes the Wishart distribution with q degrees of freedom and scale matrix M.

- $X_n = O_p(b_n)$ if $X_n/b_n = O_p(1)$. That is, if X_n/b_n is bounded stochastically: for every $\epsilon > 0$ there is a finite constant $K(\epsilon) > 0$ and a finite integer $N(\epsilon) > 0$ so that, for all $n \geq N$,

$$\Pr(|X_n/b_n| > K) \leq \epsilon.$$

- $a_k \asymp b_k$ means that, as $k \to \infty$, $a_k = O(b_k)$ and $b_k = O(a_k)$, and we then describe a_k and b_k as being asymptotically equivalent.

- $a \gg b$ informal comparison indicating that the real number a is large relative to the real b.

- $(M)_{ij}$ denotes the i,j-th element of the matrix M. $(V)_i$ denotes the i-th element of the vector V.

- $A^\dagger$ denotes the Moore-Penrose inverse of the matrix A.

- $\mathbb{R}^{m\times n}$ is the set of all real $m \times n$ matrices and $\mathbb{S}^{k\times k}$ is the set of all real symmetric $k \times k$ matrices.

- For a positive definite $V \in \mathbb{R}^{t\times t}$, $P_{A(V)}$ denotes the projection onto $\text{span}(A) \subseteq \mathbb{R}^t$ if A is a matrix, and then we have the matrix representation

$$P_{A(V)} = A(A^TVA)^{-1}A^TV$$

when A^TVA is non-singular. The same notation $P_{A(V)}$ denotes the projection onto A if A is a subspace. In both cases projection is with respect to the V inner product, and P_A denotes projection with the identity inner product. Let $Q_{A(V)} = I - P_{A(V)}$.

- For a vector V, $\|V\| = (V^TV)^{1/2}$. For a real number a, $|a|$ indicates the absolute value of a. For a square matrix M, $|M|$ denotes the determinant of M.

- We indicate that a matrix A is positive definite by $A > 0$ and positive semi-definite by $A \geq 0$.

- $\text{vec}(\cdot)\colon \mathbb{R}^{a\times b} \to \mathbb{R}^{ab}$ vectorizes an arbitrary matrix by stacking its columns, and $\text{vech}(\cdot)\colon \mathbb{R}^{a\times a} \to \mathbb{R}^{a(a+1)/2}$, vectorizes a symmetric matrix by extracting its columns of elements on and below the diagonal.

- The population expectation and variance of a random vector A are denoted as $\mathrm{E}(A)$ and $\mathrm{var}(A)$. The expectation operator may occasionally be subscripted with the random vector for clarity. For instance, $\mathrm{E}_{A|B}(A)$ denotes the expectation of A with respect to the conditional distribution of A given B.

- For random vectors $A \in \mathbb{R}^a$ and $C \in \mathbb{R}^c$

$$\begin{aligned}\Sigma_{A,C} &= \mathrm{E}\left\{(A - \mathrm{E}(A))(C - \mathrm{E}(C))^T\right\} \\ \Sigma_A &= \mathrm{E}\left\{(A - \mathrm{E}(A))(A - \mathrm{E}(A))^T\right\}.\end{aligned}$$

  When the vectors are enclosed in parentheses,

$$\Sigma_{(A,C)} = \Sigma_B,$$

  where B is the vector constructed by concatenating A and C:

$$B = \begin{pmatrix} A \\ C \end{pmatrix}.$$

  This extends to multiple vectors: $\Sigma_{(A_1,A_2,\ldots,A_k)} = \Sigma_B$, where B is now constructed by concatenating $A_1, \ldots, A_k$.

- For a square nonsingular matrix A partitioned in diagonal blocks as

$$A = \begin{pmatrix} A_{1,1} & A_{1,2} \\ A_{2,1} & A_{2,2} \end{pmatrix},$$

  we set

$$A_{1|2} = A_{1,1} - A_{1,2}A_{2,2}^{-1}A_{2,1}.$$

- For stochastic vectors A and B, $\beta_{A|B} = \Sigma_{A,B}\Sigma_B^{-1}$. The subscripts may be dropped if clear from context; for instance, $\beta = \Sigma_X^{-1}\Sigma_{X,Y}$.

- Subspaces are indicated in a math calligraphy font; e.g. $\mathcal{S}$, $\mathcal{R}$, and $\mathcal{A}$. Subscripts may be added for specificity.

  - If $M \in \mathbb{R}^{m \times n}$ then $\mathrm{span}(M) \subseteq \mathbb{R}^m$ is the subspace spanned by columns of M.

  - The sum of two subspaces $\mathcal{S}$ and $\mathcal{R}$ of $\mathbb{R}^m$ is defined as $\mathcal{S} + \mathcal{R} = \{s + r \mid s \in \mathcal{S}, r \in \mathcal{R}\}$.

- $\mathcal{C}_{X,Y} = \text{span}(\Sigma_{X,Y})$, $\mathcal{C}_{Y,X} = \text{span}(\Sigma_{Y,X})$, $\mathcal{B} = \text{span}(\beta)$ and $\mathcal{B}' = \text{span}(\beta^T)$
- Called an envelope, $\mathcal{E}_M(\mathcal{S})$ is the intersection of all reducing subspaces of M that contain $\mathcal{S}$, Definition 1.2.
- $\mathcal{S}_{Y|X}$ is the central subspace for the regression of Y on X, Definition 2.1

- For an $m \times n$ matrix A and a $p \times q$ matrix B, their direct sum is defined as the $(m+p) \times (n+q)$ block diagonal matrix $A \oplus B = \text{diag}(A, B)$. We will also use the $\oplus$ operator for two subspaces. If $\mathcal{S} \subseteq \mathbb{R}^p$ and $\mathcal{R} \subseteq \mathbb{R}^q$ then $\mathcal{S} \oplus \mathcal{R} = \text{span}(S \oplus R)$ where S and R are basis matrices for $\mathcal{S}$ and $\mathcal{R}$.

- The Kronecker product $\otimes$ between two matrices $A \in \mathbb{R}^{r\times s}$ and $B \in \mathbb{R}^{t\times u}$ is the $rt \times su$ matrix defined in blocks as

$$A \otimes B = (A)_{ij}B, \quad i = 1, \ldots, r, \ j = 1, \ldots s,$$

  where $(A)_{ij}$ denotes the ij-th element of A.

- For stochastic quantities A, B, C with a joint distribution, $A \sim B$ indicates that A has the same distribution as B, $A \perp\!\!\!\perp B$ indicates that A is independent of B and $A \perp\!\!\!\perp B \mid C$ indicates that A is conditionally independent of B given any value for C. See Cook (1998, Section 4.6) for background on independence relationships.

- Partial least squares notation: w, s, l weights, scores, and loadings. w_d and W_d weight vector and weight matrix for NIPALS. v_d and V_d weight vector and weight matrix for SIMPLS. *The same notation is used for sample and population weights, as seems common in the PLS literature.* The context will be clear from the setting.

- Common acronyms
  - CB|SEM, covariance-based structural equation modeling
  - CGA, conjugate gradient algorithm
  - CMS, central mean subspace
  - MLE, maximum likelihood estimation/estimator
  - NIPALS, non-linear iterative partial least squares

- OLS, ordinary least squares
- PLS, partial least squares
- PLS|SEM, PLS-based structural equation modeling or analysis
- RRR, reduced rank regression
- SEM, structural equation model
- SIMPLS, **s**traightforward **impl**ememtation of a **s**tatistically **i**nspired **m**odification of **PLS** de Jong (1993)
- WLS, weighted least squares.

Authors

R. Dennis Cook

Dennis Cook is Professor Emeritus, School of Statistics, University of Minnesota. His research areas include dimension reduction, linear and non-linear regression, experimental design, statistical diagnostics, statistical graphics and population genetics. Perhaps best known for "Cook's Distance," a now ubiquitous statistical method, he has authored over 250 research articles, two textbooks and three research monographs. He is a five-time recipient of the Jack Youden Prize for Best Expository Paper in *Technometrics* as well as the Frank Wilcoxon Award for Best Technical Paper. He received the 2005 COPSS Fisher Lecture and Award, and is a Fellow of ASA and IMS.

School of Statistics
University of Minnesota
Minneapolis, MN 55455, U.S.A.
Email: rdcook@umn.edu

Liliana Forzani

Liliana Forzani is Full Professor, School of Chemical Engineering, National University of Litoral and principal researcher of CONICET (National Scientific and Technical Research Council), Argentina. Her contributions are in mathematical statistics, especially sufficient dimension reduction, abundance on regression and statistics for chemometrics. She established the first research group in statistics at her university after receiving her Ph.D in Statistics at the University of Minnesota. She has authored over 60 research articles in mathematics and statistics, and was recipient of the L'Oreal-Unesco-Conicet prize for Women in science. https://sites.google.com/site/lilianaforzani/english

Facultad de Ingeniería Química, UNL
Santiago del Estero 2819
Santa Fe, Argentina
Researcher of CONICET.
Email: liliana.forzani@gmail.com

List of Figures

List of Tables

1

Introduction

Partial least squares (PLS) regression is at its historical core an algorithmic method for prediction based on an underlying linear relationship between a possibly vector-valued response Y and a number p of predictors $x_j, j = 1, \ldots, p$. With highly collinear predictors, its performance is typically better than that of standard methods like ordinary least squares (OLS), and it is serviceable in the class of abundant high-dimensional regressions where many predictors contribute information about the response and the sample size n is not sufficient for standard methods to yield unambiguous answers.

The origins of PLS regression trace back to the work of Herman Wold in the 1960s, and it is generally recognized that Wold's non-linear iterative partial least squares (NIPALS) algorithm (Wold, 1966, 1975b) is a critical point in the evolution of PLS regression. It has since garnered considerable interest, making its first appearance in the chemometrics literature around 1980 (Geladi, 1988; Wold, Martens, and Wold, 1983) and subsequently spreading across the applied sciences. PLS regression algorithms stand today as central methods for prediction in the applied sciences, boasting hundreds of articles extending their capabilities and extolling their virtues. We use "PLS regression" as a collective referring to the class of PLS regression algorithms. Specific algorithms will be referenced when necessary for clarity.

There have certainly been useful papers on PLS regression in the mainline statistics journals (e.g. Frank and Frideman, 1993; Garthwaite, 1994; Helland, 1990, 1992; Næs and Helland, 1993; Nguyen and Rocke, 2004), but the development of PLS regression algorithms nearly all originates outside of statistics. This may be explained partly by differences in culture. Estimators in statistics are historically based on given stochastic models, since these approaches allow for relatively straightforward evaluation on standard criteria and typically provide useful comprehensible paradigms for data analyses. In contrast, PLS

DOI: 10.1201/9781003482475-1

regression is defined by data-based algorithms – NIPALS, de Jong's SIMPLS method (de Jong, 1993), or one of the many variations (e.g. Lindgren, Geladi, and Wold, 1993; Martin, 2009; Stocchero, 2019) – that have been historically viewed as enigmatic forms of estimation by statisticians. Helland et al. (2018) framed the cultural differences in terms of creativity and rigor. An article by Breiman (2001) is widely cited for his portrayal of the model vs. algorithmic cultures.

The relative paucity of PLS regression studies in the mainline statistics journals might also be partly explained by a contemporary principle governing high-dimensional regression. The statistics community has for over two decades embraced sparsity as a natural characterization of high-dimensional regressions, the inertia of which seems to have inhibited consideration of other plausible paradigms. Indeed, some seem to view sparsity as akin to a natural law: If you are faced with a high-dimensional regression then naturally it must be sparse. Others have seen sparsity as the only recourse. In the logic of Frideman et al. (2004), the bet-on-sparsity principle arose because, to continue the metaphor, there is otherwise little chance of a reasonable payoff. In contrast, standard PLS regression methodology works best in abundant (anti-sparse) regression where many predictors contribute information about the response, as in spectroscopy applications in chemometrics. This sets up a dichotomy that distinguishes PLS regression methods from sparse methods. If many predictors contribute information about the response in a high-dimensional regression, then sparse methods will be unserviceable generally because the regression is not sparse, but abundant methods like PLS regressions can do well, as described in later chapters. If few predictors contribute information about the response, then sparse methods may do well because the regression is sparse, but we would not expect abundant methods to be serviceable, generally. Collinearity is less of a problem for PLS regression methods than it is for sparse regression methods.

A close connection exists between PLS regression and relatively recent envelope methods for dimension reduction (Cook, 2018). There are several types of envelope methods, depending on the goals of the analysis. The two most common are response envelopes (Cook, Li, and Chiaromonte, 2010) for reducing the response vector Y and predictor envelopes for reducing the vector of predictors X (Cook, Helland, and Su, 2013). These envelope methods are based on dimension reduction perspectives that set them apart from other statistical methods, the overarching goal being to separate with clarity the information in the data that is material to the study goals from that which is immaterial, which is in the spirit of Fisher's notion of sufficient statistics (Fisher, 1922). This is the same as the general objective of PLS regression, and it is this connection of purpose that promises to open a new chapter in PLS regression by enhancing its current capabilities, extending its scope and bringing applied scientists and statisticians closer to a common understanding of PLS regression (Cook and Forzani, 2020).

Following brief expository examples in the next section, this chapter sets forth the multivariate linear model that forms the basis for much of this book and describes the algebra that underpins envelopes, which are reviewed in Chapter 2, and our treatment of PLS regression, which begins in earnest in Chapter 3. This and the next chapters are intended to set the stage for our treatment of PLS regression in Chapter 3 and beyond. *Although we indicate how results in this chapter link to PLS algorithms, it nevertheless can be read independently or used as a reference for the developments in later chapters.*

Throughout this book we use 5-fold or 10-fold cross validation to pick subspace dimensions and the number of compressed predictors in PLS. These algorithms require a seed to initiate the partitioning of the sample, and the results typically depend on the seed. However, in our experience, the seed does not affect results materially.

1.1 Examples

In this section we describe three experiments where PLS regression has been used effectively. This may give indications about the value of PLS and about the topics addressed later in this book. All three examples are from chemometrics, where PLS has been used extensively since the early 1980s. Lasso (Tibshirani, 1996) fits were included to represent sparse methodology. PLS and lasso fits were determined by using *library{pls}* and *library{glmnet}* in *R* (R Core Team, 2022).

1.1.1 Corn moisture

In our first example, the goal is to predict the moisture content of corn using spectral measurements at 2 nm intervals in the range 1100–2498 nm, for a total of 700 predictors (e.g. Allegrini and Olivieri, 2013). The model was trained on a total of $n_{\text{train}} = 50$ corn samples and then tested on a separate sample of size $n_{\text{test}} = 30$. Let Y denote the moisture content of a corn sample, and let X denote the corresponding 700×1 vector of spectral predictors. We assume that the mean function is linear, $\text{E}(Y \mid X) = \alpha + \beta^T X$. The overarching goal is then to use the training data (Y_i, X_i), $i = 1, \ldots, 50$, to produce an estimate $\widehat{\alpha}$ of the scalar α and an estimate $\widehat{\beta}$ of the 700×1 vector β to give a linear rule, $\widehat{E}(Y \mid X) = \widehat{\alpha} + \widehat{\beta}^T X$, for predicting the moisture content of corn.

Since $n_{\text{train}} \ll p$, traditional methods like OLS are not serviceable. It has become common in such settings to assume sparsity, which reflects the view that only a relatively few predictors, $p^* < n_{\text{train}}$, furnish information about the response. The lasso represents methods that pursue sparse fits, producing estimates of β with most coefficients equal to zero. Another option is to use PLS regression, which includes a dimension reduction method that compresses the predictors onto $q < n_{\text{train}}$ linear combinations traditionally called components, $X \mapsto W^T X$ where $W \in \mathbb{R}^{p \times q}$, and then bases prediction on the OLS linear regression of Y on $W^T X$. The number of components q is typically determined by using cross validation or a holdout sample. It is known that PLS can produce effective predictions when $n_{\text{train}} \ll p$ and the predictors are highly collinear, as they often are in chemometrics applications. A third option is to use OLS, replacing inverses with generalized inverses. This method is known to produce relatively poor results; we use it here as an understood reference point.

The second row of Table 1.1, labeled "Moisture in corn," summarizes the results. Columns 2–6 give characteristics of the fits. With a 10-fold cross validation, PLS selected 23 compressed predictors or components, while the lasso picked 21 of the 700 predictors as relevant. The final three columns give the root mean squared error (RMSE) of prediction for the $n_{\text{test}} = 30$ testing observations,

$$\text{RMSE} = \left[n_{\text{test}}^{-1} \sum_{i=1}^{n_{\text{test}}} (Y_i - \widehat{E}(Y \mid X_i))^2 \right]^{1/2}.$$

Clearly, PLS did best over the test dataset. This conclusion is supported by Figure 1.1, which shows plots of the PLS and lasso fitted values versus the observed response for the test data.

TABLE 1.1
Root mean squared prediction error (RMSE), from three examples that illustrate the prediction potential of PLS regression.

						RMSE		
Dataset	p	n_{train}	n_{test}	PLS q	Lasso p^*	Lasso	PLS	OLS
Moisture in corn	700	50	30	23	21	0.114	0.013	6.52
Protein in meat	100	170	70	13	19	1.232	0.787	3.09
Tetracycline in serum	101	50	57	4	10	0.101	0.070	1.17

Columns 2–6 give the number of predictors p, the size of the training set n_{train}, the size of the test set n_{test}, the number of PLS components q chosen by 10-fold cross validation, and the number of predictors p^* estimated to have nonzero coefficients by the Lasso. Columns 7–9 give the root mean squared prediction error from the test data for the Lasso, PLS, and OLS using generalized inverses.

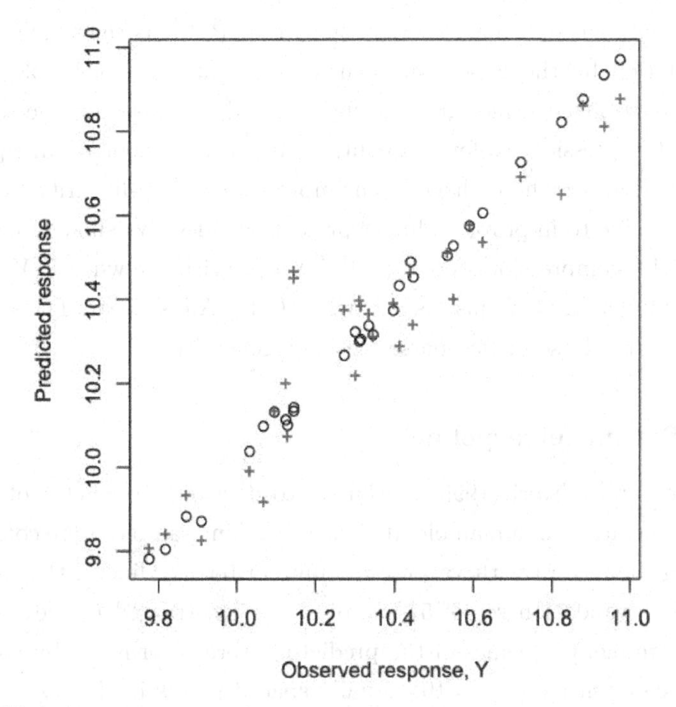

FIGURE 1.1
Corn moisture: Plots of the PLS fitted values as ∘ and lasso fitted values as + versus the observed response for the test data.

1.1.2 Meat protein

This dataset consists of 240 NIR absorbance spectra of meat samples recorded on a Tecator Infratec Food and Feed Analyzer (e.g. Boggaard and Thodberg, 1992; Allegrini and Olivieri, 2016). The response variable for this example is the protein content in each meat sample as determined by analytic chemistry. The spectra ranged from 850 to 1050 nm, discretized into $p = 100$ wavelength values. Training and testing datasets contain $n_{\text{train}} = 170$ and $n_{\text{test}} = 70$ samples. As in Section 1.1.1, the goal is to use the training data, (Y_i, X_i), $i = 1, \ldots, 170$, to estimate a rule, $\widehat{E}(Y \mid X) = \widehat{\alpha} + \widehat{\beta}^T X$, for predicting the protein content of meat. One difference in this example is that $n_{\text{train}} > p$, so a relatively small sample size is not an issue, and the usual OLS estimator was used.

The third row of Table 1.1 shows the results of lasso, PLS, and OLS fitting. PLS compressed the data into $q = 13$ components, while the lasso judged that 19 wavelengths are relevant for predicting protein. The RMSE given in the last three columns of Table 1.1 again show that PLS has the smallest overall error, followed by the lasso and then OLS. Figure 1.2 shows plots of the PLS and lasso fitted values against the observed responses for the test data. Our visual impression conforms qualitatively with the root mean squares in Table 1.1. However, here there is an impression of non-linearity, so that it may be possible to improve on linear prediction rules. We show in Chapter 9 that the PLS compression step $X \mapsto W^T X$ is serviceable when $\text{E}(Y \mid X)$ is a non-linear function of X, just as it is when $\text{E}(Y \mid X)$ is linear. The same may not hold for the lasso or for sparse methods generally.

1.1.3 Serum tetracycline

Goicoechea and Oliveri (1999) used PLS to develop a predictor of tetracycline concentration in human blood. The 50 training samples were constructed by spiking blank sera with various amounts of tetracycline in the range 0–4 $\mu\text{g mL}^{-1}$. A validation set of 57 samples was constructed in the same way. For each sample, the values of the predictors were determined by measuring fluorescence intensity at $p = 101$ equally spaced points in the range 450–550 nm. Goicoechea and Oliveri (1999) determined using leave-one-out cross validation that the best predictions of the training data were obtained with $q = 8$ components, but 10-fold cross validation gave $q = 4$ components.

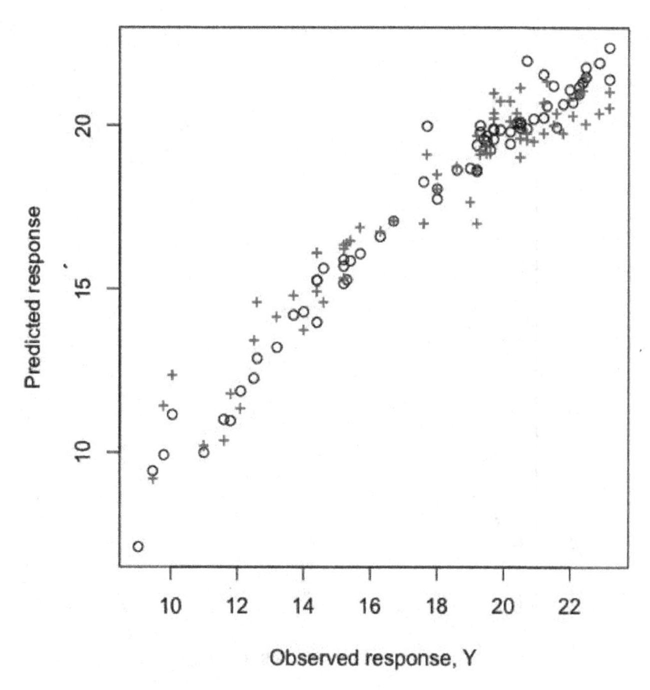

FIGURE 1.2
Meat protein: Plot of the PLS fitted values as ∘ and lasso fitted values as + versus the observed response for the test data.

The fourth row of Table 1.1 shows the results of lasso, PLS, and OLS fitting. PLS compressed the data into $q = 4$ components, while the lasso judged that 10 of the original 101 predictors are relevant. The RMSE in the last three columns of Table 1.1 again show that PLS has the smallest overall error, followed by the lasso and then OLS. Figure 1.3 shows plots of the PLS and lasso fitted values versus the observed response for the test data. Our visual impression conforms qualitatively with the root mean squares in Table 1.1. In this example, the lasso and PLS fitted values seem in better agreement than those of the previous two examples, an observation that is supported by the root mean squared values in Table 1.1.

1.2 The multivariate linear model

Consider the multivariate regression of a response vector $Y \in \mathbb{R}^r$ on a vector of non-stochastic predictors $X \in \mathbb{R}^p$. The standard linear model for describing

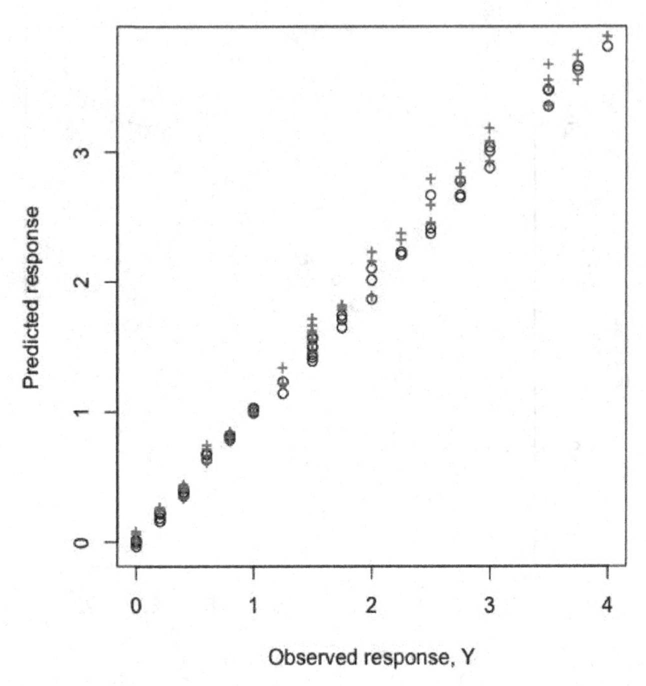

FIGURE 1.3

Serum tetracycline: Plots of the PLS fitted values as ∘ and lasso fitted values as + versus the observed response for the test data.

a sample (Y_i, X_i) can be represented in vector form as

$$Y_i = \alpha + \beta^T(X_i - \bar{X}) + \varepsilon_i, \quad i = 1, \ldots, n, \tag{1.1}$$

where we have centered the predictors $\sum_{i=1}^{n}(X_i - \bar{X}) = 0$ about the sample mean $\bar{X}$, the error vectors $\varepsilon_i \in \mathbb{R}^r$ are independently and identically distributed normal vectors with mean 0 and covariance matrix $\Sigma_{Y|X} > 0$, $\alpha \in \mathbb{R}^r$ is an unknown vector of intercepts and $\beta \in \mathbb{R}^{p \times r}$ is an unknown matrix of regression coefficients. We can think of the responses Y_i as an observation from the conditional distributions of $Y \mid (X = X_i)$, $i = 1, \ldots, n$. Centering the predictors in the model facilitates discussion and presentation of some results, but is technically unnecessary. The normality requirement for ε is used for certain foundations and likelihood-based estimation but is not essential, as discussed in later chapters.

If X is stochastic, so X and Y have a joint distribution, it is standard practice to condition on the observed values of X since the predictors are ancillary under model (1.1). However, when PLS regression is used to compress

X, the predictors are stochastic and not ancillary, as the marginal distribution of X is hypothesized to carry relevant information about the regression. With stochastic predictors, we replace X_i in model 1.1 with $X_i - \mu_X$, centering the predictors about the population mean μ_X, and we add the assumption that the predictor vector X is independent of the error vector ε. Let $\Sigma_X = \text{var}(X)$ and $\Sigma_{X,Y} = \text{cov}(X, Y)$. With this structure, we can represent β in terms of parameters of the joint distribution

$$\beta = \Sigma_X^{-1} \Sigma_{X,Y}. \tag{1.2}$$

Stochastic predictors will come into play starting in Chapter 2. We continue to follow standard practice in this review and regard the predictors as non-stochastic.

1.2.1 Notation and some algebraic background

Let $\mathbb{R}^{r \times c}$ denote the space of all real $r \times c$ matrices and let $\mathbb{S}^{r \times r}$ denote the space of all real symmetric $r \times r$ matrices. For random vectors $A \in \mathbb{R}^a$ and $C \in \mathbb{R}^c$, we use $\Sigma_{A,C} \in \mathbb{R}^{a \times c}$ to denote the matrix of covariances between the elements of A and the elements of C:

$$\Sigma_{A,C} = \text{E}\left\{(A - \text{E}(A))(C - \text{E}(C))^T\right\}.$$

Similarly, Σ_A denotes the variance-covariance matrices for A:

$$\Sigma_A = \text{E}\left\{(A - \text{E}(A))(A - \text{E}(A))^T\right\}.$$

When the vectors are enclosed in parentheses,

$$\Sigma_{(A,C)} = \Sigma_B,$$

where B is the vector constructed by concatenating A and C:

$$B = \begin{pmatrix} A \\ C \end{pmatrix}.$$

This extends to multiple vectors: $\Sigma_{(A_1,\ldots,A_k)} = \Sigma_B$, where B is now constructed by concatenating $A_1, \ldots, A_k$.

Let $A \in \mathbb{R}^{r \times c}$ and let $\mathcal{A} = \text{span}(A)$ denote the subspace of $\mathbb{R}^r$ spanned by the columns of A. Then the projection onto $\mathcal{A}$ in the usual inner product will be indicated by either P_A or $P_{\mathcal{A}}$. The projection onto $\mathcal{A}$ in the $\Delta \in \mathbb{S}^{r \times r}$ inner product, $\Delta > 0$, will be indicated by either $P_{\mathcal{A}(\Delta)}$ or $P_{A(\Delta)}$. This projection can be represented in matrix form as

$$P_{A(\Delta)} = A(A^T \Delta A)^{-1} A^T \Delta, \tag{1.3}$$

provided $A^T \Delta A > 0$. The usual inner product arises when $\Delta = I$ and then $P_A = A(A^T A)^{-1} A^T$. Regardless of the inner product, $Q_{\cdot(\cdot)} = I_r - P_{\cdot(\cdot)}$ denotes the orthogonal projection.

We will occasionally encounter a conditional variate of the form $N \mid C^T N$, where $N \in \mathbb{R}^r$ is a normal vector with mean μ and variance Δ, and $C \in \mathbb{R}^{r \times q}$ is a non-stochastic matrix with $q < r$. Assuming that $C^T \Delta C > 0$, the mean and variance of this conditional form are as follows (Cook, 1998, Section 7.2.3).

$$\begin{aligned} \mathrm{E}(N \mid C^T N) &= \mu + P^T_{C(\Delta)}(N - \mu) & (1.4) \\ \mathrm{var}(N \mid C^T N) &= \Delta - \Delta C (C^T \Delta C)^{-1} C^T \Delta \\ &= \Delta Q_{C(\Delta)} \\ &= Q^T_{C(\Delta)} \Delta Q_{C(\Delta)}. & (1.5) \end{aligned}$$

Let $\mathbb{Y}$ denote the $n \times r$ centered matrix with rows $(Y_i - \bar{Y})^T$, let $\mathbb{Y}_0$ denote the $n \times r$ uncentered matrix with rows Y_i^T, let $\mathbb{X}$ denote the $n \times p$ matrix with rows $(X_i - \bar{X})^T$ and let $\mathbb{X}_0$ denote the $n \times p$ matrix with rows X_i^T, $i = 1, \ldots, n$. With this notation, model (1.1) can be represented in full matrix form as

$$\mathbb{Y}_{0;\, n \times r} = 1_n \alpha^T + \mathbb{X}\beta + E, \tag{1.6}$$

where E is an $n \times r$ matrix with rows ε_i^T.

Let

$$\begin{aligned} S_{Y,X} &= \mathbb{Y}^T \mathbb{X}/n = n^{-1} \sum_{i=1}^n (Y_i - \bar{Y})(X_i - \bar{X})^T \\ S_X &= \mathbb{X}^T \mathbb{X}/n = n^{-1} \sum_{i=1}^n (X_i - \bar{X})(X_i - \bar{X})^T \\ S_Y &= \mathbb{Y}^T \mathbb{Y}/n = n^{-1} \sum_{i=1}^n (Y_i - \bar{Y})(Y_i - \bar{Y})^T, \end{aligned}$$

where S_Y is the usual estimator of Σ_Y. When X is stochastic, S_X and $S_{Y,X}$ are the usual estimators of Σ_X and $\Sigma_{X,Y}$. When X is non-stochastic, S_X is non-stochastic as well, and in this case, we define $\Sigma_X = \lim_{n\to\infty} S_X$ and

$$\begin{aligned}
\Sigma_{Y,X} &= \lim_{n\to\infty} \mathrm{E}(S_{Y,X}|\mathbb{X}_0) \\
&= \lim_{n\to\infty} n^{-1} \sum_{i=1}^{n} \mathrm{E}(Y_i - \bar{Y})(X_i - \bar{X})^T \\
&= \lim_{n\to\infty} n^{-1} \sum_{i=1}^{n} \beta^T (X_i - \bar{X})X_i - \bar{X})^T \\
&= \beta^T \lim_{n\to\infty} S_X = \beta^T \Sigma_X.
\end{aligned}$$

The Kronecker product $\otimes$ between two matrices $A \in \mathbb{R}^{r\times s}$ and $B \in \mathbb{R}^{t\times u}$ is the $rt \times su$ matrix defined in blocks as

$$A \otimes B = (A)_{ij}B, \quad i = 1, \ldots, r, \; j = 1, \ldots, s,$$

where $(A)_{ij}$ denotes the ij-th element of A. Kronecker products are not in general commutative, $A \otimes B \neq B \otimes A$. The vec operator transforms a matrix $A \in \mathbb{R}^{r\times u}$ to a vector $\mathrm{vec}(A) \in \mathbb{R}^{ru}$ by stacking its columns. Representing A in terms of its columns a_j, $A = (a_1, \ldots, a_u)$, then

$$\mathrm{vec}(A) = \begin{pmatrix} a_1 \\ a_2 \\ \vdots \\ a_u \end{pmatrix}.$$

The vech operator transforms a symmetric matrix $A \in \mathbb{S}^{r\times r}$ to a vector $\mathrm{vech}(A) \in \mathbb{R}^{r(r+1)/2}$ by stacking its unique elements on and below the diagonal. If

$$A = \begin{pmatrix} a_{11} & a_{12} & a_{13} \\ a_{12} & a_{22} & a_{23} \\ a_{13} & a_{23} & a_{33} \end{pmatrix},$$

then

$$\mathrm{vech}(A) = \begin{pmatrix} a_{11} & a_{12} & a_{13} & a_{22} & a_{23} & a_{33} \end{pmatrix}^T.$$

The commutation matrix $K_{pm} \in \mathbb{R}^{pm\times pm}$ is the unique matrix that transforms the vec of a matrix into the vec of its transpose: For $A \in \mathbb{R}^{p\times m}$, $\mathrm{vec}(A^T) = K_{pm}\mathrm{vec}(A)$. For background on Kronecker products, vec operators, and commutation matrices, see Henderson and Searle (1979), Magnus and Neudecker (1979), Neudecker and Wansbeek (1983). Properties of these operators were summarized by Cook (2018, Appendix A).

For an $m \times n$ matrix A and a $p \times q$ matrix B, their direct sum is defined as the $(m+p) \times (n+q)$ block diagonal matrix

$$A \oplus B = \operatorname{diag}(A, B) = \begin{pmatrix} A & 0 \\ 0 & B \end{pmatrix}.$$

We will also use the $\oplus$ operator for two subspaces. If $\mathcal{S} \subseteq \mathbb{R}^p$ and $\mathcal{R} \subseteq \mathbb{R}^q$ then $\mathcal{S} \oplus \mathcal{R} = \operatorname{span}(S \oplus R)$ where S and R are basis matrices for $\mathcal{S}$ and $\mathcal{R}$.

1.2.2 Estimation

With normality of the errors in model (1.1) and $n > p$, the maximum likelihood estimator of α is $\bar{Y}$ and the maximum likelihood estimator of β, which is also the OLS estimator, is

$$\widehat{\beta}_{\text{ols}} = (\mathbb{X}^T\mathbb{X})^{-1}\mathbb{X}^T\mathbb{Y} = (\mathbb{X}^T\mathbb{X})^{-1}\mathbb{X}^T\mathbb{Y}_0 = S_X^{-1}S_{X,Y}, \tag{1.7}$$

where the second equality follows because the predictors are centered and, as defined previously, $\mathbb{Y}_0$ denotes the $n \times r$ uncentered matrix with rows Y_i^T. A justification for this result is sketched in Appendix Section A.1.1. We use $\widehat{Y}_i = \bar{Y} + \widehat{\beta}_{\text{ols}}^T(X_i - \bar{X})$ and $\widehat{R}_i = Y_i - \widehat{Y}_i$ to denote the i-th vectors of fitted values and residuals, $i = 1, \ldots, n$. Notice from (1.7) that $\widehat{\beta}_{\text{ols}}$ can be constructed by doing r separate univariate linear regressions, one for each element of Y on X. The coefficients from the j-th regression then form the j-th column of $\widehat{\beta}_{\text{ols}}$, $j = 1, \ldots, r$. This observation will be useful when discussing the differences between PLS1 algorithms, which apply to regressions with a univariate response, and PLS2 algorithms, which apply to regressions with multiple responses.

The sample covariance matrices of $\widehat{Y}$, $\widehat{R}$, and Y – which are denoted $S_{Y \circ X}$, $S_{Y|X}$, and S_Y – can be expressed as

$$\begin{aligned} S_{Y \circ X} &= n^{-1}\mathbb{Y}^T P_{\mathbb{X}}\mathbb{Y} = S_{Y,X}S_X^{-1}S_{X,Y}, & (1.8)\\ S_{Y|X} &= n^{-1}\sum_{i=1}^{n}\widehat{R}_i\widehat{R}_i^T = n^{-1}\mathbb{Y}^T Q_{\mathbb{X}}\mathbb{Y}, & (1.9)\\ &= S_Y - S_{Y,X}S_X^{-1}S_{XY}, \\ &= S_Y - S_{Y \circ X}, \\ S_Y &= n^{-1}\mathbb{Y}^T\mathbb{Y} = S_{Y \circ X} + S_{Y|X}, & (1.10) \end{aligned}$$

where, as defined at (1.3), $P_{\mathbb{X}} = \mathbb{X}(\mathbb{X}^T\mathbb{X})^{-1}\mathbb{X}^T$ denotes the projection onto the column space of $\mathbb{X}$, $Q_{\mathbb{X}} = I_n - P_{\mathbb{X}}$ and $S_{Y|X}$ is the maximum likelihood estimator of $\Sigma_{Y|X}$. We will occasionally encounter a standardized version of $\widehat{\beta}_{\text{ols}}$,

$$\widehat{\beta}_{\text{ols}}^{\text{std}} = S_X^{1/2} \widehat{\beta}_{\text{ols}} S_{Y|X}^{-1/2}, \tag{1.11}$$

which corresponds to the estimated coefficient matrix from the OLS fit of the standardized responses $S_{Y|X}^{-1/2} Y$ on the standardized predictors $S_X^{-1/2} X$.

The joint distribution of the elements of $\widehat{\beta}_{\text{ols}}$ can be found by using the vec operator to stack the columns of $\widehat{\beta}_{\text{ols}}$: $\text{vec}(\widehat{\beta}_{\text{ols}}) = \{I_r \otimes (\mathbb{X}^T\mathbb{X})^{-1}\mathbb{X}^T\}\text{vec}(\mathbb{Y}_0)$. Although X is stochastic in PLS applications, properties of $\widehat{\beta}_{\text{ols}}$ are typically described conditional on the observed values of X. We emphasize that in the following calculations by conditioning on the observed matrix of uncentered predictors $\mathbb{X}_0$. Since $\text{vec}(\mathbb{Y}_0)$ is normally distributed with mean $\alpha \otimes 1_n + (I_r \otimes \mathbb{X})\text{vec}(\beta)$ and variance $\Sigma_{Y|X} \otimes I_n$, it follows that $\text{vec}(\widehat{\beta}_{\text{ols}}) \mid \mathbb{X}_0$ is normally distributed with mean and variance

$$\begin{aligned} \text{E}\{\text{vec}(\widehat{\beta}_{\text{ols}}) \mid \mathbb{X}_0\} &= \text{vec}(\beta) && (1.12) \\ \text{var}\{\text{vec}(\widehat{\beta}_{\text{ols}}) \mid \mathbb{X}_0\} &= \Sigma_{Y|X} \otimes (\mathbb{X}^T\mathbb{X})^{-1} = n^{-1}\Sigma_{Y|X} \otimes S_X^{-1}. && (1.13) \end{aligned}$$

The covariance matrix can be represented also in terms of $\widehat{\beta}_{\text{ols}}^T$ by using the $rp \times rp$ commutation matrix K_{rp} to convert $\text{vec}(\widehat{\beta}_{\text{ols}})$ to $\text{vec}(\widehat{\beta}_{\text{ols}}^T)$: $\text{vec}(\widehat{\beta}_{\text{ols}}^T) = K_{rp}\text{vec}(\widehat{\beta}_{\text{ols}})$ and

$$\text{var}\{\text{vec}(\widehat{\beta}_{\text{ols}}^T) \mid \mathbb{X}_0\} = n^{-1}K_{rp}(\Sigma_{Y|X} \otimes S_X^{-1})K_{rp}^T = n^{-1}S_X^{-1} \otimes \Sigma_{Y|X}.$$

The unconditional mean and variance are then

$$\begin{aligned} \text{E}\{\text{vec}(\widehat{\beta}_{\text{ols}}\} &= \text{E}\left\{\text{E}\{\text{vec}(\widehat{\beta}_{\text{ols}}) \mid \mathbb{X}_0\}\right\} = \text{vec}(\beta) \\ \text{var}\{\text{vec}(\widehat{\beta}_{\text{ols}})\} &= \text{E}\{\text{var}[\text{vec}(\widehat{\beta}_{\text{ols}}) \mid \mathbb{X}_0]\} + \text{var}\left\{\text{E}[\text{vec}(\widehat{\beta}_{\text{ols}}) \mid \mathbb{X}_0]\right\} \\ &= n^{-1}\Sigma_{Y|X} \otimes \text{E}\left\{S_X^{-1}\right\}. \end{aligned}$$

The conditional and unconditional variances $\text{var}\{\text{vec}(\widehat{\beta}_{\text{ols}})\}$ are typically estimated by substituting the residual covariance matrix for $\Sigma_{Y|X}$ and S_X^{-1} for $\text{E}\left\{S_X^{-1}\right\}$.

$$\widehat{\text{var}}\{\text{vec}(\widehat{\beta}_{\text{ols}})\} = n^{-1}S_{Y|X} \otimes S_X^{-1}. \tag{1.14}$$

Let $e_i \in \mathbb{R}^r$ denote the indicator vector with a 1 in the i-th position and 0's elsewhere. Then, the covariance matrix for the i-th column of $\widehat{\beta}_{\text{ols}}$ is

$$\text{var}\{\text{vec}(\widehat{\beta}_{\text{ols}}e_i) \mid \mathbb{X}_0\} = (e_i^T \otimes I_p)\text{var}\{\text{vec}(\widehat{\beta}_{\text{ols}}) \mid \mathbb{X}_0\}(e_i \otimes I_p) = n^{-1}S_X^{-1}(\Sigma_{Y|X})_{ii}.$$

We commented following (1.7) that the j-th column of $\widehat{\beta}_{\text{ols}}$ is the same as doing the linear regression of the j-th response on X, $j = 1, \ldots, r$. Consistent with that observation, we see from this that the covariance matrix for the j-th column of $\widehat{\beta}_{\text{ols}}$ is the same as that from the marginal linear regression of $(Y)_j$ on X. We refer to the estimate $(\widehat{\beta}_{\text{ols}})_{ij}$ divided by its standard error $\{n^{-1}(S_X^{-1})_{jj}(S_{Y|X})_{ii}\}^{1/2}$ as a Z-score:

$$Z = \frac{(\widehat{\beta}_{\text{ols}})_{ij}}{\{n^{-1}(S_X^{-1})_{jj}(S_{Y|X})_{ii}\}^{1/2}}. \tag{1.15}$$

This statistic will be used from time to time for assessing the magnitude of $(\widehat{\beta}_{\text{ols}})_{ij}$, sometimes converting to a p-value using the standard normal distribution.

The usual log-likelihood ratio statistic for testing that $\beta = 0$ is

$$\Lambda = n \log \frac{|S_Y|}{|S_{Y|X}|}, \tag{1.16}$$

which is asymptotically distributed under the null hypothesis as a chi-square random variable with pr degrees of freedom. This statistic is sometimes reported with an adjustment that is useful when n is not large relative to r and p (Muirhead, 2005, Section 10.5.2).

The Fisher information J for $(\text{vec}^T(\beta), \text{vech}^T(\Sigma_{Y|X}))^T$ in the conditional model (1.1) is

$$J = \begin{pmatrix} \Sigma_{Y|X}^{-1} \otimes \Sigma_X & 0 \\ 0 & \frac{1}{2}\mathbb{E}_r^T(\Sigma_{Y|X}^{-1} \otimes \Sigma_{Y|X}^{-1})\mathbb{E}_r \end{pmatrix}, \tag{1.17}$$

where $\mathbb{E}_r$ is the expansion matrix that satisfies $\text{vec}(A) = \mathbb{E}_r \text{vech}(A)$ for $A \in \mathbb{S}^{r \times r}$, and $\Sigma_X = \lim_{n\to\infty} \sum_{i=1}^n S_X > 0$. It follows from standard likelihood theory that $\sqrt{n}(\text{vec}(\widehat{\beta}_{\text{ols}}) - \text{vec}(\beta))$ is asymptotically normal with mean 0 and variance given by the upper left block of J^{-1},

$$\text{avar}(\sqrt{n}\text{vec}(\widehat{\beta}_{\text{ols}}) \mid \mathbb{X}_0) = \Sigma_{Y|X} \otimes \Sigma_X^{-1}. \tag{1.18}$$

Asymptotic normality also holds without normal errors but with some technical conditions: if the errors have finite fourth moments and if the maximum leverage converges to 0, $\max_{1 \le i \le n}(P_{\mathbb{X}})_{ii} \to 0$, then $\sqrt{n}(\text{vec}(\widehat{\beta}_{\text{ols}}) - \text{vec}(\beta))$ converges in distribution to a normal vector with mean 0 (e.g. Su and Cook, 2012, Theorem 2).

1.2.3 Partitioned models and added variable plots

A subset of the predictors may occasionally be of special interest in multivariate regression. Partition X into two sets of predictors $X_1 \in \mathbb{R}^{p_1}$ and $X_2 \in \mathbb{R}^{p_1}$, $p_1 + p_2 = p$, and conformably partition the rows of β into β_1 and β_2. Then, dropping the subscript i, model (1.1) can be rewritten as

$$Y = \mu + \beta_1^T(X_1 - \bar{X}_1) + \beta_2^T(X_2 - \bar{X}_2) + \varepsilon, \tag{1.19}$$

where the elements of β_1 are the coefficients of interest. We next reparameterize this model to force the new predictors to be uncorrelated in the sample and to focus attention on β_1.

Let $\widehat{R}_{1|2} = X_1 - \bar{X}_1 - S_{X_1,X_2}S_{X_2}^{-1}(X_2 - \bar{X}_2)$ denote a typical residual vector from the ordinary least squares fit of X_1 on X_2, and let $\beta_{2*}^T = \beta_1^T S_{X_1,X_2}S_{X_2}^{-1} + \beta_2^T$. Then the partitioned model can be re-expressed as

$$Y = \mu + \beta_1^T \widehat{R}_{1|2} + \beta_{2*}^T(X_2 - \bar{X}_2) + \varepsilon. \tag{1.20}$$

In this version of the partitioned model, the parameter vector β_1 is the same as that in (1.19), while $\beta_2 \neq \beta_{2*}$ unless $S_{X_1,X_2} = 0$. The predictors – $\widehat{R}_{1|2}$ and X_2 – in (1.20) are uncorrelated in the sample $S_{\widehat{R}_{1|2},X_2} = 0$, and consequently the maximum likelihood estimator of β_1 is obtained by regressing Y on $\widehat{R}_{1|2}$. The maximum likelihood estimator of β_1 can also be obtained by regressing $\widehat{R}_{Y|2}$, the residuals from the regression of Y on X_2, on $\widehat{R}_{1|2}$. A plot of $\widehat{R}_{Y|2}$ versus $\widehat{R}_{1|2}$ is called an added variable plot (Cook and Weisberg, 1982). These plots are often used in univariate linear regression ($r = 1$) as general graphical diagnostics for visualizing how hard the data are working to fit individual coefficients.

Model (1.20) will be used in Chapter 6 when developing partial PLS and partial envelope methods.

1.3 Invariant and reducing subspaces

Invariant and reducing subspaces (e.g. Conway, 1994) are essential constituents of envelopes and PLS methodology. In this section we cover aspects of these subspaces that will be useful later. Much of the material in this section is based on the results of Cook et al. (2010). The statistical relevance of the constructions in this section will be addressed in Chapter 2.

If $\mathcal{R}$ is a subspace of $\mathbb{R}^r$ and $A \in \mathbb{R}^{r\times c}$, then we define

$$\begin{aligned} A^T\mathcal{R} &= \{A^T R \mid R \in \mathcal{R}\} \\ \mathcal{R}^T A &= \{R^T A \mid R \in \mathcal{R}\} \\ \mathcal{R}^\perp &= \{S \in \mathbb{R}^r \mid S^T\mathcal{R} = 0\}. \end{aligned}$$

If $\mathcal{R}_1$ and $\mathcal{R}_2$ are subspaces of $\mathbb{R}^r$, then

$$\mathcal{R}_1 + \mathcal{R}_2 = \{R_1 + R_2 \mid R_j \in \mathcal{R}_j, j = 1, 2\}.$$

We begin with the definitions of invariant and reducing subspaces.

Definition 1.1. *A subspace $\mathcal{R}$ of $\mathbb{R}^r$ is an* invariant subspace *of $M \in \mathbb{R}^{r\times r}$ if the linear transformation $M\mathcal{R} \subseteq \mathcal{R}$; so M maps $\mathcal{R}$ to a subset of itself. $\mathcal{R}$ is a* reducing subspace *of M if $\mathcal{R}$ and $\mathcal{R}^\perp$ are both invariant subspaces of M. If $\mathcal{R}$ is a reducing subspace of M, we say that $\mathcal{R}$ reduces M.*

The next lemma describes a matrix equation that characterizes invariant subspaces. A justification is given in Appendix A.1.2.

Lemma 1.1. *Let $\mathcal{R}$ be an s-dimensional subspace of $\mathbb{R}^r$ and let $M \in \mathbb{R}^{r\times r}$. Then $\mathcal{R}$ is an invariant subspace of M if and only if, for any $A \in \mathbb{R}^{r\times s}$ with* span$(A) = \mathcal{R}$, *there exists a $B \in \mathbb{R}^{s\times s}$ such that $MA = AB$.*

Recall from Section 1.2.1 that $\mathbb{S}^{r\times r}$ denotes the space of all real, symmetric $r \times r$ matrices. The next lemma tells us that if $M \in \mathbb{S}^{r\times r}$ and $\mathcal{R}$ is an invariant subspace of M then $\mathcal{R}$ reduces M. This fact is handy in proofs because to show that $\mathcal{R}$ reduces a symmetric M, we need to show only that $\mathcal{R}$ is an invariant subspace of M. In this book we will be concerned almost exclusively with real symmetric linear transformations M and reducing subspaces.

Lemma 1.2. *Let $\mathcal{R}$ be an s-dimensional subspace of $\mathbb{R}^r$ and let $M \in \mathbb{S}^{r\times r}$. Then $\mathcal{R}$ reduces M if and only if $\mathcal{R}$ is an invariant subspace of M.*

Proof. By definition, if $\mathcal{R}$ reduces M then $\mathcal{R}$ is an invariant subspace of M. Next, suppose that $\mathcal{R}$ is an invariant subspace of M. Then, by definition $M\mathcal{R} \subseteq \mathcal{R}$, which implies that $(\mathcal{R}^\perp)^T M\mathcal{R} = \{0\}$. Since M is symmetric, this implies $\mathcal{R}^T M\mathcal{R}^\perp = \{0\}$. Consequently, we must have $M\mathcal{R}^\perp \subseteq \mathcal{R}^\perp$. □

Examples of invariant subspaces of $M \in \mathbb{S}^{r \times r}$ are $\{0\}$, $\mathbb{R}^r$, and span(M). For a less obvious construction that will be relevant later, let $x \in \mathbb{R}^r$ and consider the subspace

$$\mathcal{R} = \text{span}\{x, Mx, M^2x, \ldots, M^{r-1}x, \ldots\}.$$

To avoid trivial cases, we assume $r \geq 2$. The Cayley-Hamilton theorem states that $\sum_{i=0}^{r} a_i M^i = 0$, where the a_i's are the coefficients of the characteristic polynomial $|\lambda I - M|$ of M. In consequence, for $j \geq r$, $M^j x$ is a linear combination of the r vectors $\{x, Mx, \ldots, M^{r-1}x\}$ and so $\mathcal{R}$ can be represented as a subspace spanned by a finite set of r vectors:

$$\mathcal{R} = \text{span}\{x, Mx, M^2x, \ldots, M^{r-1}x\}. \tag{1.21}$$

Then $\mathcal{R}$ reduces M because M is symmetric and $\mathcal{R}$ is an invariant subspace of M,

$$M\mathcal{R} = \text{span}\{Mx, M^2x, M^3x, \ldots, M^r x\} \subseteq \mathcal{R}.$$

Subspaces of this form are often called *cyclic invariant subspaces* in linear algebra.

Subspaces (1.21) arise in connection with PLS regressions having a univariate response, particularly when allowing for the possibility that the full set $\{x, Mx, \ldots, M^{r-1}x\}$ may not be necessary to span $\mathcal{R}$. For $t \leq r$, let

$$\left.\begin{aligned} K_t(M, x) &= \{x, Mx, M^2x, \ldots, M^{t-1}x\} \\ \mathcal{K}_t(M, x) &= \text{span}\{K_t(M, x)\} \end{aligned}\right\}, \tag{1.22}$$

which are called a *Krylov basis* and a *Krylov subspace* of dimension t in numerical analysis, terminology that we adopt for this book. For example, let $r = 3$, $v_1 = (1, 1, 1)/\sqrt{3}$, $v_2 = (-1, 1, 0)/\sqrt{2}$ and $v_3 = (1, 1, -2)/\sqrt{6}$, and construct $M = 3P_{v_1} + P_{v_2} + P_{v_3}$. Then $\mathcal{K}_1(M, v_1)$ and, for any real scalars a and b, $\mathcal{K}_1(M, av_2 + bv_3)$ are one-dimensional reducing subspace of M. For an arbitrary vector $x \in \mathbb{R}^3$, what is the maximum possible dimension of $\mathcal{K}_q(M, x)$?

If $M\mathcal{R} \subseteq \mathcal{R}$, so $\mathcal{R}$ is an invariant subspace of M, and if $x \in \mathcal{R}$ then clearly $Mx \in \mathcal{R}$. Any invariant subspace of M that contains x must then also contain all of the vectors $\{x, Mx, \ldots, M^{r-1}x\}$. Consider a subspace $\mathcal{R}$ that is unknown, but known to be an invariant subspace of M. If we know or can estimate one vector $x \in \mathcal{R}$, then we can iteratively transform x by M to obtain additional vectors in $\mathcal{R}$: for any t, $\mathcal{K}_t(M, x) \subseteq \mathcal{R}$. Here we can think of the vector x as serving as a seed for gaining information on $\mathcal{R}$. This is one of the key ideas behind the NIPALS regression algorithm.

The next proposition from Cook and Li (2002, Prop. 3) provides motivation for considering Krylov subspaces of dimension less than r:

Proposition 1.1. *If $M^j x \in \mathcal{K}_{j-1}(M, x)$ then $M^s x \in \mathcal{K}_{j-1}(M, x)$ for any $s > j$.*

Proof. If $M^j x \in \mathcal{K}_{j-1}(M, x)$ then

$$M^{j+1} x \in M\mathcal{K}_{j-1}(M, x) = \operatorname{span}\{Mx, M^2 x, \dots, M^j x\} \subseteq \mathcal{K}_{j-1}(M, x)$$

and the desired conclusion follows. □

An important implication of Proposition 1.1 is that the Krylov subspaces are strictly monotonically increasing until a certain dimension t is reached and they are constant thereafter,

$$\mathcal{K}_1(M, x) \subset \mathcal{K}_2(M, x) \subset \cdots \subset \mathcal{K}_t(M, x) = \cdots \mathcal{K}_r(M, x), \tag{1.23}$$

where "$\subset$" indicates strict containment. Estimation of the transition point t at which the sequence stops growing is a key consideration in NIPALS regressions. In Section 1.5 we identify the transition point as an envelope dimension. It is also equal to the number of components in PLS regressions when $r = 1$.

The next proposition, whose proof is given in Appendix A.1.3, gives a different characterization of a reducing subspace. It's necessary and sufficient condition could be used as a definition of a reducing subspace in this context, and it is used in envelope literature because it avoids the need to keep track of multiplicities of eigenvalues (Cook, Li, and Chiaromonte, 2010).

Proposition 1.2. *$\mathcal{R}$ reduces $M \in \mathbb{R}^{r \times r}$ if and only if M can be written in the form*

$$M = P_{\mathcal{R}} M P_{\mathcal{R}} + Q_{\mathcal{R}} M Q_{\mathcal{R}}. \tag{1.24}$$

Corollary 1.1 describes consequences of Proposition 1.2 that will be useful in envelope calculations, including derivations of maximum likelihood estimators. Its proof is given in Appendix A.1.4. Let $A^{\dagger}$ denote the Moore-Penrose inverse of the matrix A.

Corollary 1.1. *Let $\mathcal{R}$ reduce $M \in \mathbb{R}^{r \times r}$, let $A \in \mathbb{R}^{r \times u}$ be a semi-orthogonal basis matrix for $\mathcal{R}$, and let A_0 be a semi-orthogonal basis matrix for $\mathcal{R}^{\perp}$. Then*

1. *M and $P_{\mathcal{R}}$, and M and $Q_{\mathcal{R}}$ commute.*

2. *$\mathcal{R} \subseteq \text{span}(M)$ if and only if $A^T M A$ is full rank.*

3. *$|M| = |A^T M A| \times |A_0^T M A_0|$ when $u < r$.*

4. *If M is full rank then*

$$\begin{aligned} M^{-1} &= A(A^T M A)^{-1} A^T + A_0 (A_0^T M A_0)^{-1} A_0^T \\ &= P_{\mathcal{R}} M^{-1} P_{\mathcal{R}} + Q_{\mathcal{R}} M^{-1} Q_{\mathcal{R}}. \end{aligned}$$

5. *If $\mathcal{R} \subseteq \text{span}(M)$ then*

$$M^{\dagger} = A(A^T M A)^{-1} A^T + A_0 (A_0^T M A_0)^{\dagger} A_0^T.$$

The next lemma will be useful when studying likelihood-based algorithms later in this chapter and in Chapter 3. It shows in part how to write $|A_0^T M A_0|$ as a function of a basis A for the orthogonal complement of $\text{span}(A_0)$. A proof is given in Appendix Section A.1.5.

Lemma 1.3. *Suppose that $M \in \mathbb{S}^{r \times r}$ is positive definite and that the column-partitioned matrix $O = (A, A_0) \in \mathbb{R}^{r \times r}$ is orthogonal. Then*

I. $|A_0^T M A_0| = |M| \times |A^T M^{-1} A|$

II. $\log |A^T M A| + \log |A_0^T M A_0| \geq \log |M|$

III. $\log |A^T M A| + \log |A^T M^{-1} A| \geq 0.$

Parts II and III become equalities if and only if $\text{span}(A)$ *reduces M.*

The next proposition gives a sufficient condition for a projection in an M inner product to be the same as the projection in the usual inner product. It will be helpful when studying how PLS algorithms work. Its proof, as given in Appendix Section A.1.6, follows by application of (1.24) in conjunction with (1.3).

Proposition 1.3. *If $\mathcal{R}$ reduces the positive definite matrix $M \in \mathbb{S}^{r \times r}$ and $\mathcal{R} \subseteq \text{span}(M)$ then $P_{\mathcal{R}(M)} = P_{\mathcal{R}}$.*

Certain minimal reducing subspaces are in effect parameters in PLS regressions. The following proposition ensures that the minimal reducing subspace is well defined. Its proof is in Appendix A.1.7.

Proposition 1.4. *The intersection of any two reducing subspaces of $M \in \mathbb{R}^{r \times r}$ is also a reducing subspace of M.*

1.4 Envelope definition

In this section, we give definitions and basic properties of envelopes (from Cook et al., 2010; Cook, 2018) that will facilitate an understanding of their role in PLS regression. The formal statements are devoid of statistical content, but that will become apparent as the book progresses.

We are now ready to define an envelope:

Definition 1.2. *Let $M \in \mathbb{S}^{r\times r}$ and let $\mathcal{S} \subseteq \text{span}(M)$. Then, the M-envelope of $\mathcal{S}$, denoted by $\mathcal{E}_M(\mathcal{S})$, is the intersection of all reducing subspaces of M that contain $\mathcal{S}$.*

This definition is concerned only with reducing subspaces that contain a specified subspace $\mathcal{S}$, which will typically be spanned by a parameter vector or matrix. The requirement that $\mathcal{S} \subseteq \text{span}(M)$ is necessary to ensure that envelopes are well defined and that they capture the quantity of interest $\mathcal{S}$. This definition does not require that M be non-singular, a property that will be useful in understanding how PLS operates in high-dimensional regressions. In that case, the condition $\mathcal{S} \subseteq \text{span}(M)$ constrains the regression structures that can be identified in high dimensions that exceed the sample size.

The next two propositions give certain relationships between envelopes that will be useful later. They were proven originally by Cook et al. (2010). Proofs are available also from (Cook, 2018, Appendix A.3.1).

Proposition 1.5. *Let $K \in \mathbb{S}^{r\times r}$ commute with $M \in \mathbb{S}^{r\times r}$ and let $\mathcal{S} \subseteq \text{span}(M)$. Then, $K\mathcal{S} \subseteq \text{span}(M)$ and*

$$\mathcal{E}_M(K\mathcal{S}) = K\mathcal{E}_M(\mathcal{S}).$$

If, in addition, $\mathcal{S} \subseteq \text{span}(K)$ and $\mathcal{E}_M(\mathcal{S})$ reduces K, then

$$\mathcal{E}_M(K\mathcal{S}) = \mathcal{E}_M(\mathcal{S}).$$

Proposition 1.6. *Let $M \in \mathbb{S}^{r\times r}$ and let $\mathcal{S} \subseteq \text{span}(M)$ Then*

$$\mathcal{E}_M(M^k\mathcal{S}) = \mathcal{E}_M(\mathcal{S}) \text{ for all } k \in \mathbb{R}$$
$$\mathcal{E}_{M^k}(\mathcal{S}) = \mathcal{E}_M(\mathcal{S}) \text{ for all } k \in \mathbb{R} \text{ with } k \neq 0.$$

In reference to model (1.1) with random predictors, we will in later chapters frequently encounter the smallest reducing subspace of $\Sigma_X := \text{var}(X)$ that contains $\mathcal{B} := \text{span}(\beta)$, denoted $\mathcal{E}_{\Sigma_X}(\mathcal{B})$. When $\Sigma_X > 0$, it follows immediately from Proposition 1.6 that $\mathcal{E}_{\Sigma_X}(\mathcal{B}) = \mathcal{E}_{\Sigma_X}(\text{span}(\Sigma_{X,Y}))$, since $\beta = \Sigma_X^{-1}\Sigma_{X,Y}$. For future use, we let $\mathcal{C}_{X,Y} = \text{span}(\Sigma_{X,Y})$ and $\mathcal{C}_{Y,X} = \text{span}(\Sigma_{Y,X})$. With this notation, we can write $\mathcal{E}_{\Sigma_X}(\mathcal{B}) = \mathcal{E}_{\Sigma_X}(\mathcal{C}_{X,Y})$.

Proposition 1.7. *Let* $M \in \mathbb{S}^{r\times r}$, *let* $\mathcal{S}_j \subseteq \text{span}(M)$, $j = 1, 2$. *Then* $\mathcal{E}_M(\mathcal{S}_1 + \mathcal{S}_2) = \mathcal{E}_M(\mathcal{S}_1) + \mathcal{E}_M(\mathcal{S}_2)$.

Proof. $\mathcal{E}_M(\mathcal{S}_1 + \mathcal{S}_2)$ is a reducing subspace of M that contains both $\mathcal{S}_1$ and $\mathcal{S}_2$. Since $\mathcal{E}_M(\mathcal{S}_j)$ is the smallest reducing subspace of M that contains $\mathcal{S}_j$, we have that $\mathcal{E}_M(\mathcal{S}_j) \subseteq \mathcal{E}_M(\mathcal{S}_1 + \mathcal{S}_2)$, $j = 1, 2$, and consequently

$$\mathcal{E}_M(\mathcal{S}_1) + \mathcal{E}_M(\mathcal{S}_2) \subseteq \mathcal{E}_M(\mathcal{S}_1 + \mathcal{S}_2).$$

Next, let B_j be a basis matrix for $\mathcal{E}_M(\mathcal{S}_j)$, $j = 1, 2$ and let $B = (B_1, B_2)$, so $\text{span}(B) = \mathcal{E}_M(\mathcal{S}_1) + \mathcal{E}_M(\mathcal{S}_2)$. Then, from Lemma 1.1 there are matrices A_1 and A_2 so that $MB_j = B_jA_j$ and

$$MB = (B_1, B_2)\begin{pmatrix} A_1 & 0 \\ 0 & A_2 \end{pmatrix} = B\begin{pmatrix} A_1 & 0 \\ 0 & A_2 \end{pmatrix}.$$

In consequence, $\mathcal{E}_M(\mathcal{S}_1) + \mathcal{E}_M(\mathcal{S}_2)$ reduces M and contains $\mathcal{S}_1 + \mathcal{S}_2$, and so must contains the smallest reducing subspace of M that contains $\mathcal{S}_1 + \mathcal{S}_2$,

$$\mathcal{E}_M(\mathcal{S}_1 + \mathcal{S}_2) \subseteq \mathcal{E}_M(\mathcal{S}_1) + \mathcal{E}_M(\mathcal{S}_2).$$

The conclusion follows. □

The direct sum of $A \in \mathbb{R}^{m\times n}$ and $B \in \mathbb{R}^{p\times q}$ is defined as the $(m+p)\times(n+q)$ block diagonal matrix $A \oplus B = \text{diag}(A, B)$. We will also use the $\oplus$ operator for two subspaces. If $\mathcal{S} \subseteq \mathbb{R}^p$ and $\mathcal{R} \subseteq \mathbb{R}^q$ then $\mathcal{S} \oplus \mathcal{R} = \text{span}(S \oplus R)$ where S and R are basis matrices for $\mathcal{S}$ and $\mathcal{R}$. The next lemma (Cook and Zhang, 2015b) shows how to interpret the direct sum of envelopes. In preparation, let $M_1 \in \mathbb{S}^{p_1\times p_1}$, $M_2 \in \mathbb{S}^{p_2\times p_2}$, and let $\mathcal{S}_1$ and $\mathcal{S}_2$ be subspaces of $\text{span}(M_1)$ and $\text{span}(M_2)$. Then

Lemma 1.4. $\mathcal{E}_{M_1}(\mathcal{S}_1) \oplus \mathcal{E}_{M_2}(\mathcal{S}_2) = \mathcal{E}_{M_1 \oplus M_2}(\mathcal{S}_1 \oplus \mathcal{S}_2)$.

The proof of this lemma is in Appendix A.1.8.

We next consider settings in which envelopes are invariant under certain simultaneous changes in M and $\mathcal{S}$. The proof is in Appendix A.1.9.

Proposition 1.8. *Let $\Delta \in \mathbb{S}^{r\times r}$ be a positive definite matrix and let $\mathcal{S}$ be a u-dimensional subspace of $\mathbb{R}^r$. Let $G \in \mathbb{R}^{r\times u}$ be a semi-orthogonal basis matrix for $\mathcal{S}$ and let $V \in \mathbb{S}^{u\times u}$ be positive semi-definite. Define $\Psi = \Delta + GVG^T$. Then $\Delta^{-1}\mathcal{S} = \Psi^{-1}\mathcal{S}$ and*

$$\mathcal{E}_\Delta(\mathcal{S}) = \mathcal{E}_\Psi(\mathcal{S}) = \mathcal{E}_\Delta(\Delta^{-1}\mathcal{S}) = \mathcal{E}_\Psi(\Psi^{-1}\mathcal{S}) = \mathcal{E}_\Psi(\Delta^{-1}\mathcal{S}) = \mathcal{E}_\Delta(\Psi^{-1}\mathcal{S}).$$

For a first application of Proposition 1.8, consider the multivariate linear regression model (1.1) but now assume that the predictors X are random with covariance matrix Σ_X, so that Y and X have a joint distribution. Because the predictors are random we can express the marginal variance of Y as

$$\Sigma_Y = \Sigma_{Y|X} + \beta^T\Sigma_X\beta = \Sigma_{Y|X} + GVG^T, \tag{1.25}$$

where $\Sigma_{Y|X} > 0$, G is a semi-orthogonal basis matrix for $\mathcal{B}' = \text{span}(\beta^T)$, $\beta^T = GA$, $\dim(\mathcal{B}') \le \min(r,p)$, and $V = A\Sigma_X A^T$ is positive definite since A must have full row rank. These forms match the decomposition required for Proposition 1.8 with $\Psi = \Sigma_Y$ and $\Delta = \Sigma_{Y|X}$. Consequently, we have

$$\begin{aligned} \mathcal{E}_{\Sigma_{Y|X}}(\mathcal{B}') &= \mathcal{E}_{\Sigma_Y}(\mathcal{B}') = \mathcal{E}_{\Sigma_{Y|X}}(\Sigma_{Y|X}^{-1}\mathcal{B}') \\ &= \mathcal{E}_{\Sigma_Y}(\Sigma_Y^{-1}\mathcal{B}') = \mathcal{E}_{\Sigma_Y}(\Sigma_{Y|X}^{-1}\mathcal{B}') \\ &= \mathcal{E}_{\Sigma_{Y|X}}(\Sigma_Y^{-1}\mathcal{B}'). \end{aligned} \tag{1.26}$$

1.5 Algebraic methods of envelope construction

In this section we describe various population-level methods for constructing the envelope $\mathcal{E}_M(\mathcal{A})$ with $M \in \mathbb{S}^{r\times r}$ and $\mathcal{A} \subseteq \text{span}(M)$. Some of these algorithms will be seen to be closely related to algorithms for PLS regression.

1.5.1 Algorithm $\mathbb{E}$

The next proposition (Cook et al., 2010) describes a method of constructing an envelope in terms of the eigenspaces of M. We use h generally to denote the number of eigenspaces of a real symmetric matrix. Recalling the subspace computations introduced at the outset of Section 1.3,

Proposition 1.9. (Algorithm $\mathbb{E}$.) *Let $M \in \mathbb{S}^{r\times r}$, let P_i, $i = 1,\ldots,h \leq r$, be the projections onto the eigenspaces of M with distinct non-zero eigenvalues λ_i, $i = 1,\ldots,h$, and let $\mathcal{A} \subseteq \mathrm{span}(M)$. Then $\mathcal{E}_M(\mathcal{A}) = \sum_{i=1}^h P_i\mathcal{A}$.*

Proof. To prove that $\sum_{i=1}^h P_i\mathcal{A}$ is the smallest reducing subspace of M that contains $\mathcal{A}$, it suffices to prove the following statements:

1. $\sum_{i=1}^h P_i\mathcal{A}$ reduces M.
2. $\mathcal{A} \subseteq \sum_{i=1}^h P_i\mathcal{A}$.
3. If $\mathcal{T}$ reduces M and $\mathcal{A} \subseteq \mathcal{T}$, then $\sum_{i=1}^h P_i\mathcal{A} \subseteq \mathcal{T}$.

Let A be a basis matrix for $\mathcal{A}$. Statement 1 follows because $M = \sum_{i=1}^h \lambda_i P_i$ and

$$\begin{aligned} M\sum_{i=1}^h P_i\mathcal{A} &= \mathrm{span}\{MP_1\mathcal{A},\ldots,MP_h\mathcal{A}\} \\ &= \mathrm{span}\{\lambda_1 P_1\mathcal{A},\ldots,\lambda_h P_h\mathcal{A}\} = \sum_{i=1}^h P_i\mathcal{A}. \end{aligned}$$

Statement 2 holds because $\mathcal{A} = \{P_1\mathbf{v} + \cdots + P_h\mathbf{v} : \mathbf{v} \in \mathcal{A}\} \subseteq \sum_{i=1}^h P_i\mathcal{A}$. Turning to statement 3, if $\mathcal{T}$ reduces M, it can be written as $\mathcal{T} = \sum_{i=1}^h P_i\mathcal{T}$. If, in addition, $\mathcal{A} \subseteq \mathcal{T}$ then we have $P_i\mathcal{A} \subseteq P_i\mathcal{T}$ for $i = 1,\ldots,h$. Statement 3 follows since $\sum_{i=1}^h P_i\mathcal{A} \subseteq \sum_{i=1}^h P_i\mathcal{T} = \mathcal{T}$.

□

We refer to the algorithm implied by this proposition as Algorithm $\mathbb{E}$, since it is based directly on the eigenstructure of M.

1.5.2 Algorithm $\mathbb{K}$

When $\dim(\mathcal{A}) = 1$, the dimension of the envelope is bounded above by the number h of eigenspaces of M: $\dim(\mathcal{E}_M(\mathcal{A})) \leq h$. We use Proposition 1.9 in this case to gain further insights into the Krylov subspaces discussed in Section 1.3 and to motivate a new algebraic method of constructing envelopes. As we will see in Chapter 3, Algorithm $\mathbb{K}$ is a foundation for the NIPALS regression algorithm with a univariate response, $Y \in \mathbb{R}^1$.

Let $a \in \mathbb{R}^r$ be a basis vector for the one-dimensional subspace $\mathcal{A}$, let $\lambda_i \neq 0$ be the eigenvalue for the i-th eigenspace of M, let $q = \dim(\mathcal{E}_M(\mathcal{A}))$ so that a

projects non-trivially onto q eigenspaces of M. We take these to be the first q eigenspaces without loss of generality. Let $C_q = (P_1 a, \ldots, P_q a)$. Then

$$\begin{aligned}
\mathcal{E}_M(\mathcal{A}) &= \sum_{i=1}^{q} P_i \mathcal{A} = \operatorname{span}(C_q) \\
M^t a &= \sum_{j=1}^{h} \lambda_j^t P_j a = \sum_{j=1}^{q} \lambda_j^t P_j a \\
a &= \sum_{j=1}^{q} P_j a.
\end{aligned}$$

Using these relationships, we have

$$\begin{aligned}
K_t(M,a) &= (a, Ma, M^2 a, \ldots, M^{t-1} a) \\
&= \left(\sum_{j=1}^{q} P_j a, \sum_{j=1}^{q} \lambda_j P_j a, \sum_{j=1}^{q} \lambda_j^2 P_j a, \ldots, \sum_{j=1}^{q} \lambda_j^{t-1} P_j a \right)
\end{aligned}$$

Writing this as a product of C_q and the Vandermonde matrix

$$V = \begin{pmatrix} 1 & \lambda_1 & \lambda_1^2 & \cdots & \lambda_1^{t-1} \\ 1 & \lambda_2 & \lambda_2^2 & \cdots & \lambda_2^{t-1} \\ \vdots & \vdots & \vdots & & \vdots \\ 1 & \lambda_q & \lambda_q^2 & \cdots & \lambda_q^{t-1} \end{pmatrix}_{q \times t}$$

we have

$$K_t(M,a) = C_q V.$$

The λ_j's are distinct eigenvalues by construction and so, if $t < q$, $\operatorname{rank}(V) = t$. This implies that when $t < q$, $\mathcal{K}_t(M,a) \subset \operatorname{span}(C_q) = \mathcal{E}_M(\mathcal{A})$. Similarly, if $t \geq q$ then $\mathcal{K}_t(M,a) = \operatorname{span}(C_q) = \mathcal{E}_M(\mathcal{A})$. We summarize this result in the following proposition.

Proposition 1.10. (Algorithm $\mathbb{K}$.) *Let $M \in \mathbb{S}^{r \times r}$ be positive semi-definite. Let $a \in \mathbb{R}^r$, $a \neq 0$, $\mathcal{A} = \operatorname{span}(a) \subseteq \operatorname{span}(M)$ and $q = \dim(\mathcal{E}_M(\mathcal{A}))$. Then for the Krylov subspaces are strictly monotonically increasing until dimension q is reached and they are constant thereafter:*

$$\mathcal{K}_1(M,a) \subset \mathcal{K}_2(M,a) \subset \cdots \subset \mathcal{K}_q(M,a) = \mathcal{E}_M(\mathcal{A}) = \mathcal{K}_{q+1}(M,a) \cdots = \mathcal{K}_r(M,a), \tag{1.27}$$

where "$\subset$" indicates strict containment.

We refer to the algorithm implied by this proposition as Algorithm $\mathbb{K}$, since it is based on Krylov subspaces.

Proposition 1.10 describes a method of enveloping a one-dimensional subspace with the reducing subspaces of a positive semi-definite matrix $M \in \mathbb{S}^{r\times r}$. The guidance it offers when the dimension of $\mathcal{A}$ is greater than 1 is not as helpful. Let $h = \dim(\mathcal{A})$, let $A = (a_1, \ldots, a_h)$ be a basis matrix for $\mathcal{A}$ and let $\mathcal{A}_j = \text{span}(a_j)$. The Krylov spaces

$$\left.\begin{aligned} K_j(M, A) &= \{A, MA, M^2A, \ldots, M^{j-1}A\} \\ \mathcal{K}_j(M, A) &= \text{span}\{K_j(M, A)\} \end{aligned}\right\}, \tag{1.28}$$

are monotonic

$$\mathcal{K}_1(M, A) \subset \mathcal{K}_2(M, A) \subset \cdots \subset \mathcal{K}_t(M, A) = \cdots \mathcal{K}_r(M, A), \tag{1.29}$$

as they are when $h = 1$ (1.23). However, the stopping point t is no longer equal to the dimension q of $\mathcal{E}_M(\mathcal{S})$. We can bound $q \leq th$, but there is no way to get a better handle on q without imposing additional structure. The Krylov space $\mathcal{K}_t(M, A)$ at termination is related to the individual Krylov spaces $\mathcal{K}_r(M, A_j)$, $j = 1, \ldots, h$, corresponding to the columns of A as follows,

$$\mathcal{K}_t(M, A) = \sum_{j=1}^{h} \mathcal{K}_t(M, A_j) = \sum_{j=1}^{h} \mathcal{K}_{q_j}(M, A_j) = \sum_{j=1}^{h} \mathcal{E}_M(\mathcal{A}_j) = \mathcal{E}_M(\mathcal{A})$$

where the final two equalities follow from Propositions 1.10 and 1.7. In consequence, the columns of $\mathcal{K}_t(M, A)$ span the desired envelope $\mathcal{E}_M(\mathcal{A})$, but they do not necessarily form a basis for it, leading to inefficiencies in statistical applications. For that reason a different method is needed to deal with multi-dimensional subspaces.

1.5.3 Algorithm $\mathbb{N}$

Let $v_1(\cdot)$ denote the largest eigenvalue of the argument matrix with corresponding eigenvector $\ell_1(\cdot)$ of length 1, and let q denote the dimension of the M-envelope of $\mathcal{A}$, $\mathcal{E}_M(\mathcal{A})$. Then, the following algorithm generates $\mathcal{E}_M(\mathcal{A})$. We refer to it as Algorithm $\mathbb{N}$ and give its proof here since one instance of it corresponds to the NIPALS partial least squares regression algorithm discussed in Section 3.1. Recall from Section 1.2.1 that $Q_{A(\Delta)} = I - P_{A(\Delta)}$, where $P_{A(\Delta)}$ denotes the projection onto span(A) in the Δ inner product.

As described in the next proposition, the population version of Algorithm $\mathbb{N}$ requires positive semi-definite matrices $A, M \in \mathbb{S}^{r\times r}$ as inputs with $\mathcal{A} = \text{span}(A) \subseteq \text{span}(M)$. We describe specific instances of the algorithm as $\mathbb{N}(A, M)$ using A and M as arguments. Sample versions of the algorithm are similarly described as $\mathbb{N}(\widehat{A}, \widehat{M})$, where $\widehat{A}$ and $\widehat{M}$ are consistent estimators of A and M.

Proposition 1.11. *(*Algorithm $\mathbb{N}$.) Let $A, M \in \mathbb{S}^{r\times r}$ be positive semi-definite matrices with $\mathcal{A} = \text{span}(A) \subseteq \text{span}(M)$, let $u_1 = \ell_1(A)$ and let $U_1 = (u_1)$. For $i = 1, 2, \ldots,$ construct

$$\begin{aligned} u_{i+1} &= \ell_1(Q^T_{U_i(M)} A Q_{U_i(M)}) \\ U_{i+1} &= (U_i, u_{i+1}) \\ \text{If } Q^T_{U_{i+1}(M)} A Q_{U_{i+1}(M)} &= 0, \text{ stop and set } k = i+1. \end{aligned}$$

At termination, $k = q$, $\text{span}(U_q) = \mathcal{E}_M(\mathcal{A})$,

$$\begin{aligned} A &= P_{U_q} A P_{U_q} = A P_{U_q} = P_{U_q} A \\ M &= P_{U_q} M P_{U_q} + Q_{U_q} M Q_{U_q}. \end{aligned}$$

Proof. For clarity we divide the proof into a number of distinct claims. Let $\lambda_{i+1} = v_1(Q^T_{U_i(M)} A Q_{U_i(M)})$. Prior to stopping, $\lambda_{i+1} > 0$.

Claim 1: $U_i^T U_i = I$ and $U_i^T u_{i+1} = 0$, $i = 1, \ldots, k-1$.

Since $Q^T_{U_i(M)} A Q_{U_i(M)} u_{i+1} = \lambda_{i+1} u_{i+1}$, we have

$$U_i^T Q^T_{U_i(M)} A Q_{U_i(M)} u_{i+1} = \lambda_{i+1} U_i^T u_{i+1}$$

or equivalently, since $U_i^T Q^T_{U_i(M)} = 0$, $\lambda_{i+1} U_i^T u_{i+1} = 0$, which implies that $U_i^T u_{i+1} = 0$ since $\lambda_{i+1} \neq 0$. Since u_{i+1} is an eigenvector chosen with length 1, $U_i^T U_i = I$.

In the claims that follow, let k be the point at which the algorithm terminates.

Claim 2: $Q^T_{U_k(M)} A = 0$.

Since A is positive semi-definite there is a full rank matrix V so that $A = VV^T$. When the algorithm terminates, $Q^T_{U_k(M)} V V^T Q_{U_k(M)} = 0$, and so $Q^T_{U_k(M)} V = 0$ and $Q^T_{U_k(M)} V V^T = Q^T_{U_k(M)} A = 0$.

Claim 3: span(U_k) reduces M.

By Lemmas 1.1 and 1.2, it is sufficient to show that there is a $k \times k$ matrix Ω_k so that $MU_k = U_k\Omega_k$. From claim 2, $A = P^T_{U_k(M)}A$, meaning

$$A = MU_k(U_k^T MU_k)^{-1}U_k^T A. \tag{1.30}$$

Next,

$$\begin{aligned} \lambda_{i+1}u_{i+1} &= Q^T_{U_i(M)}AQ_{U_i(M)}u_{i+1} \\ &= \left(A - P^T_{U_i(M)}A\right) Q_{U_i(M)}u_{i+1}. \end{aligned} \tag{1.31}$$

Substituting the right-hand side of (1.30) for the first A in (1.31) and using $U_i = U_k(I_i, 0)^T$ gives

$$\begin{aligned} \lambda_{i+1}u_{i+1} &= (MU_k(U_k^T MU_k)^{-1}U_k^T A - MU_k(I_i, 0)^T(U_i^T MU_i)^{-1}U_i^T A) \\ &\quad \times Q_{U_i(M)}u_{i+1} \\ &= MU_k\left((U_k^T MU_k)^{-1}U_k^T A - (I_i, 0)^T(U_i^T MU_i)^{-1}U_i^T A\right) \\ &\quad \times Q_{U_i(M)}u_{i+1} \\ &= MU_kV_{i+1} \text{ for } i = 0, \ldots, k-1 \end{aligned}$$

where $V_{i+1} \in \mathbb{R}^r$ is defined implicitly. Let $S = (V_1, \ldots, V_k)$ and $\Lambda_k = \text{diag}(\lambda_1, \ldots, \lambda_k)$. Therefore $MU_kS = U_k\Lambda_k$. Now, since rank U_k is k and $\Lambda_k > 0$, we have that S is invertible and therefore $MU_k = U_k\Lambda_kS^{-1} = U_k\Omega_k$, with nonsingular $\Omega_k = \Lambda_kS^{-1}$.

Claim 4: span$(A) \subseteq$ span(U_k).

Substituting Claim 3 into (1.30), we have

$$\begin{aligned} A &= MU_k(U_k^T MU_k)^{-1}U_k^T A \\ &= U_k\Omega_k(U_k^T U_k\Omega_k)^{-1}U_k^T A \\ &= U_kU_k^T A = P_{U_k}A. \end{aligned}$$

And the conclusion follows.

Claim 5: span(U_k) is the smallest reducing subspace of M that contains $\mathcal{A}$.

From Claims 3 to 4, we know that span(U_k) is a reducing subspace of M that contains span(A). It remains to show that span(U_k) is the smallest such subspace.

Let $\Gamma : r \times q$ be a semi-orthogonal basis matrix for $\mathcal{E}_M(\mathcal{A})$ and let (Γ, Γ_0) be an orthogonal $r \times r$ matrix. Then $M = \Gamma\Omega\Gamma^T + \Gamma_0\Omega_0\Gamma_0^T$ and $A = \Gamma\Gamma^T A$. We have shown that $\text{span}(\Gamma) = \mathcal{E}_M(\mathcal{A}) \subseteq \text{span}(U_k)$. We now use a contradiction that shows that $\text{span}(U_k) \subseteq \text{span}(\Gamma)$.

Suppose that $\text{span}(U_k) \not\subseteq \text{span}(\Gamma)$, so that $\text{span}(U_k) \cap \text{span}(\Gamma_0)$ contains a nontrivial subspace. From the sequence of vectors $\{u_i\}$, let d be the index of the first vector such that $u_d = \Gamma B_d + \Gamma_0 C_d$ with $C_d \neq 0$, reflecting that U_k intersects Γ_0. First, we show that $B_d \neq 0$. To do so, we first prove that

$$A^T Q_{U_{d-1}(M)} \Gamma_0 = 0. \tag{1.32}$$

Since $U_{d-1} \subset \Gamma$ and $\Gamma^T M \Gamma_0 = 0$, we have

$$U_{d-1}^T M \Gamma_0 = U_{d-1}^T \Gamma\Gamma^T M \Gamma_0 = 0,$$

and

$$\begin{aligned} A^T Q_{U_{d-1}(M)} \Gamma_0 &= A^T \Gamma_0 - A^T U_{d-1}(U_{d-1}^T M U_{d-1})^{-1} U_{d-1}^T M \Gamma_0 \\ &= A^T \Gamma_0 = A^T \Gamma\Gamma^T \Gamma_0 \\ &= 0, \end{aligned}$$

from which (1.32) follows. Now, since $d \leq q$, we have that

$$u_d^T Q_{U_{d-1}(M)}^T A A^T Q_{U_{d-1}(M)} u \neq 0,$$

and since $u_d = \Gamma B_d + \Gamma_0 C_d$, (1.32) implies $B_d \neq 0$, and since $C_d^T C_d + B_d^T B_d = 1$ with $C_d \neq 0$, we have $B_d^T B_d < 1$.

Now, let us call $\tilde{u}_d = \Gamma B_d / (B_d^T B_d)^{1/2}$. Then $\tilde{u}_d^T \tilde{u}_d = 1$ and by (1.32)

$$\begin{aligned} u_d^T Q_{U_{d-1}(M)}^T A &= B_d^T \Gamma^T Q_{U_{d-1}(M)}^T A + C_d^T \Gamma_0^T Q_{U_{d-1}(M)}^T A \\ &= B_d^T \Gamma^T Q_{U_{d-1}(M)}^T A \\ &\neq 0. \end{aligned} \tag{1.33}$$

With this result, we have

$$\begin{aligned} \tilde{u}_d^T Q_{U_{d-1}(M)}^T A A^T Q_{U_{d-1}(M)} \tilde{u}_d &= u_d^T Q_{U_{d-1}(M)}^T A A^T Q_{U_{d-1}(M)} u_d / (B_d^T B_d) \\ &> u_d^T Q_{U_{d-1}(M)}^T A A^T Q_{U_{d-1}(M)} u_d, \end{aligned}$$

where the inequality follows because $B_d^T B_d < 1$ and (1.33). Therefore, the maximum is not at u_d, which is the contradiction we seek.

It follows that $\text{span}(U_k)$ is the smallest reducing subspace of M that contains $\mathcal{A}$ and consequently $\mathcal{E}_M(\mathcal{A}) = \text{span}(U_k)$ and $q = k$. □

The next proposition describes a connection between Algorithms $\mathbb{K}$ and $\mathbb{N}$ when $\mathcal{A}$ has dimension 1. We give the following lemma in preparation for the proposition.

Lemma 1.5. *Let $M \in \mathbb{S}^{r\times r}$ be a positive semi-definite matrix and let $A \in \mathbb{R}^{r\times s}$ with $\text{rank}(A) = s$ and $\mathcal{A} = \text{span}(A) \subseteq \text{span}(M)$. Then*

(i) $A^T MA$ is positive definite.

(ii) for $s = 1$ and $t \leq q = \dim(\mathcal{E}_M(\mathcal{A}))$ we have that $K_t^T(M, A)MK_t(M, A) > 0$.

Proof. Let $m = \text{rank}(M)$ and construct the spectral decomposition of M as $M = GDG^T$, where $G \in \mathbb{R}^{r\times m}$ is semi-orthogonal and $D > 0$ is diagonal. Since $\text{span}(A) \subseteq \text{span}(M)$, there is a $B \in \mathbb{R}^{m\times s}$ with rank s and $A = GB$. Conclusion (i) now follows since

$$A^T MA = B^T G^T GDG^T GB = B^T DB > 0.$$

For conclusion (ii), first it follows from Proposition 1.10 that $K_t(M, A)$ has full column rank if and only if $t \leq q$. Since $\mathcal{A} \subseteq \text{span}(M)$ there is a non-zero vector $b \in \mathbb{R}^r$ so that $A = Mb$ and

$$\begin{aligned} K_t(M, A) &= (A, MA, \ldots, M^{t-1}A) \\ &= (Mb, M^2b, \ldots, M^t b) \\ &= MK_t(M, b). \end{aligned}$$

Consequently, $\mathcal{K}_t(M, A) \subseteq \text{span}(M)$. Conclusion (ii) now follows from conclusion (i). □

Proposition 1.12. *If $\dim(\mathcal{A}) = 1$, $M \geq 0$, $\mathcal{A} \subseteq \text{span}(M)$ and $\mathcal{A}$ is spanned by the vector $a \in \mathbb{R}^r$, then Algorithms $\mathbb{N}$ and $\mathbb{K}$ are equivalent; that is, $\text{span}(U_q) = \mathcal{K}_q(M, a) = \mathcal{E}_M(\mathcal{A})$. Further, the columns of $U_q = (u_1, \ldots, u_q)$ correspond to a sequential orthogonalization of the columns of the Krylov matrix $K_q(M, a) = (a, Ma, M^2a, \ldots, M^{q-1}a)$.*

Proof. Let $A = aa^T$ and assume without loss of generality that $a^T a = 1$. Recall that Proposition 1.11 requires $M \in \mathbb{S}^{r\times r}$ to be positive semi-definite and $a \in \text{span}(M)$. The first u-vector from Algorithm $\mathbb{N}$ in Proposition 1.11 is $u_1 = a$, which is also the first vector in $K_q(M, a)$. The second u-vector is

$$\begin{aligned} u_2 &= Q^T_{u_1(M)}a = Q^T_{a(M)}a \\ &= a - Ma/(a^T Ma), \end{aligned}$$

where $a^T MA > 0$ by Lemma 1.5. Clearly, u_1 and u_2 are orthogonal vectors, $u_1^T u_2 = 0$, that span the same space as the first two columns in the Krylov matrix, $\text{span}(u_1, u_2) = \text{span}(a, Ma) = \mathcal{K}_2(M, a)$. Assume that $\text{span}(U_j) = \mathcal{K}_j(M, a)$, $j = 1, 2, \ldots, t < q$. Then, there is a nonsingular matrix $H_t \in \mathbb{R}^{t \times t}$ so that $U_t = K_t H_t$, where for notational simplicity we are suppressing the arguments to $K_t(M, a)$. Arguing by induction, we need to show that $\text{span}(U_{t+1}) = \text{span}(U_t, u_{t+1}) = \mathcal{K}_{t+1}(M, a) \in \mathbb{R}^{r \times (t+1)}$. Write

$$U_{t+1} = (U_t, u_{t+1}) = (K_t H_t, u_{t+1}) = \left(K_{t+1} \begin{pmatrix} H_t \\ 0_{1 \times t} \end{pmatrix}, u_{t+1} \right).$$

Turning to u_{t+1}, we have

$$\begin{aligned} u_{t+1} &= Q^T_{U_t(M)}a = Q^T_{K_t(M)}a \\ &= a - MK_t(K_t^T M K_t)^{-1} K_t^T a, \end{aligned}$$

where $K_t^T M K_t > 0$ by Lemma 1.5 (ii). Let $h_t = (K_t^T M K_t)^{-1} K_t^T a \in \mathbb{R}^t$. Then

$$\begin{aligned} u_{t+1} &= a - (Ma, M^2 a, \ldots, M^t a) h_t \in \mathbb{R}^r \\ &= K_{t+1} \begin{pmatrix} 1 \\ -h_t \end{pmatrix}. \end{aligned}$$

Substituting this into the above representation for $U_{t+1} \in \mathbb{R}^{r \times (t+1)}$ gives

$$U_{t+1} = \left(K_{t+1} \begin{pmatrix} H_t \\ 0_{1 \times t} \end{pmatrix}, u_{t+1} \right) = K_{t+1} \begin{pmatrix} H_t & 1 \\ 0 & -h_t \end{pmatrix}.$$

Since $t < q$, U_{t+1} and K_{t+1} both have full column rank $t + 1$, so

$$L_t := \begin{pmatrix} H_t & 1 \\ 0 & -h_t \end{pmatrix}$$

must be nonsingular and thus $\text{span}(U_{t+1}) = \mathcal{K}_{t+1}$. The orthogonality of the column of U_t follows from Claim 1 in the proof of Proposition 1.11. Consequently, the columns of U_t can be obtained by orthogonalizing the columns of

K_t. That is, since $U_{t+1} = K_{t+1}L_t$ and the columns of U_{t+1} are orthogonal with length 1, the multiplication of K_{t+1} on the right by L_t serves to orthogonalize the columns of K_{t+1}.

□

1.5.4 Algorithm $\mathbb{S}$

The algorithm indicated in the following proposition is called Algorithm $\mathbb{S}$ because a special case of it yields the SIMPLS algorithm discussed in Section 3.3. Like Algorithm $\mathbb{N}$, its population version requires positive semi-definite matrices $A, M \in \mathbb{S}^{r\times r}$ as inputs with $\mathcal{A} = \text{span}(A) \subseteq \text{span}(M)$. We may describe specific instances of the algorithm as $\mathbb{S}(A, M)$ using A and M as arguments. Sample versions of the algorithm are similarly described as $\mathbb{S}(\widehat{A}, \widehat{M})$, where $\widehat{A}$ and $\widehat{M}$ are consistent estimators of A and M.

Proposition 1.13. (Algorithm $\mathbb{S}$.) *Let $A, M \in \mathbb{S}^{r\times r}$ be positive semi-definite matrices with $\mathcal{A} = \text{span}(A) \subseteq \text{span}(M)$. Let $u_1 = \ell_1(A)$ and $U_1 = (u_1)$. For $i = 1, 2, \ldots$, construct*

$$\begin{aligned} u_{i+1} &= \ell_1(Q_{MU_i} A Q_{MU_i}) \\ U_{i+1} &= (U_i, u_{i+1}) \\ \text{If } Q_{MU_i} A &= 0 \text{ stop and set } k = i. \end{aligned}$$

At termination, $k = q$, $\text{span}(U_q) = \mathcal{E}_M(\mathcal{A})$ and

$$\begin{aligned} A &= P_{U_q} A P_{U_q} = A P_{U_q} = P_{U_q} A \\ M &= P_{U_q} M P_{U_q} + Q_{U_q} M Q_{U_q}. \end{aligned}$$

The proof of this proposition parallels that for Algorithm $\mathbb{N}$ in Proposition 1.11 and is therefore omitted. Details on the SIMPLS version of Algorithm $\mathbb{N}$ are available in Sections 3.3 and 3.4. Those details can be used as a template for the proof of this proposition. Algorithm $\mathbb{S}$ uses the identity inner product while Algorithm $\mathbb{N}$ uses the M inner product. Otherwise, the structure of these algorithms is similar.

None of the algorithms discussed in this section requires M to be nonsingular, but all require that $\mathcal{A}$ be contained in span(M). In statistical applications described in Chapter 2 and in subsequent chapters, M will often be a sample covariance matrix that, under these algorithms, may be singular. In consequence, the algorithms in this section are adaptable to dimension reduction in regressions with $p > n$.

Algorithm $\mathbb{S}$ can be described also as follows. Let $u_0 = 0$ and, as in Proposition 1.13, let $U_{i+1} = (u_0, u_1, \ldots, u_i)$. Then, given U_k, $k < q$, U_{k+1} is constructed by concatenating U_k with

$$\begin{aligned} u_{k+1} &= \max_{u \in \mathbb{R}^p} u^T A u \text{ subject to} \\ & u^T M U_k = 0 \text{ and } u^T u = 0. \end{aligned} \tag{1.34}$$

This can be seen as equivalent to Algorithm $\mathbb{S}$ described in Proposition 1.13 by introducing Q_{MU_k} into (1.34) to incorporate the constraint $u^T MU_k = 0$:

$$\begin{aligned} u_{k+1} &= \max_{u \in \mathbb{R}^p} u^T Q_{MU_k} A Q_{MU_k} u \text{ subject to } u^T u = 0 \\ &= \ell_1(Q_{MU_k} A Q_{MU_k}). \end{aligned}$$

The form of the algorithm given in (1.34) is not directly constructive, but it connects with a version of the SIMPLS algorithm in the literature; see Section 3.3.

The following corollary is provided in summary; its proof straightforward and omitted.

Corollary 1.2. *Under the conditions of Proposition 1.12, algorithms $\mathbb{S}$, $\mathbb{N}$, and $\mathbb{K}$ are equivalent. That is, they result in the same subspace.*

1.5.5 Algorithm $\mathbb{L}$

The Algorithms $\mathbb{K}$, $\mathbb{N}$, and $\mathbb{S}$ all allow for M to be positive semi-definite. In contrast, the algorithm needed to implement Proposition 1.14 requires M to be positive definite. It is referred to as Algorithm $\mathbb{L}$ because it is inspired by likelihood-based estimation discussed in Sections 2.3.1 and 2.5.3.

Proposition 1.14. (Algorithm $\mathbb{L}$.) *Let $M > 0$ and $A \geq 0$ be symmetric $r \times r$ matrices. Then the M-envelope of $\mathcal{A} = \text{span}(A)$ can be constructed as*

$$\begin{aligned} \mathcal{E}_M(\mathcal{A}) &= \text{span}\{\arg\min_G [\log |G^T M G| + \log |G_0^T (M + A) G_0|]\} \\ &= \text{span}\{\arg\min_G [\log |G^T M G| + \log |G^T (M + A)^{-1} G|]\}, \end{aligned}$$

where $\min_G$ is taken over all semi-orthogonal matrices $G \in \mathbb{R}^{r \times q}$ and (G, G_0) is an orthogonal matrix.

Proof. This proof relies on repeated use of Lemma 1.3. First,

$$\log|G_0^T(M+A)G_0| = \log|G^T(M+A)^{-1}G| + \log|M+A|,$$

which establishes the equivalence the two minimization problems. Now,

$$\begin{aligned}\log|G^TMG| + \log|G_0^T(M+A)G_0| &= \log|G^TMG| + \log|G_0^TMG_0 + G_0^TAG_0| \\ &\geq \log|G^TMG| + \log|G_0^TMG_0| \\ &= \log|M| + \log|G^TMG| + \log|G^TM^{-1}G| \\ &\geq \log|M|,\end{aligned}$$

where the final step follows from Lemma 1.3 (III). To achieve the lower bound, equality in the first inequality requires that $\mathcal{A} \subseteq \text{span}(G)$. The second equality follows from Lemma 1.3. Consequently, achieving equality in the second inequality requires that $\text{span}(G)$ reduce M. Overall then, $\text{span}(G)$ must be a reducing subspace of M that contains $\mathcal{A}$. The conclusion follows since q is the dimension of the smallest subspace that satisfies these two properties. □

1.5.6 Other envelope algorithms

We briefly introduce three other algorithms in this section. They are not included in detail because their relationship with PLS and algorithms $\mathbb{N}$, $\mathbb{K}$, and $\mathbb{S}$ are still under study. Nevertheless, adaptations of these algorithms have the potential to add materially to the body of PLS-type methods.

The optimization required in Algorithm $\mathbb{L}$ is essentially over a Grassmannian, the set of all q dimensional subspaces of $\mathbb{R}^r$. Although good algorithms are available, they are time-consuming and prone to getting stuck in a local optimum. Cook, Forzani, and Su (2016) developed an algorithm that does not require optimization over a Grassmannian and is less likely to get caught in a local optimum. Their algorithm is based on a new parameterization that arises from identifying q active rows of G so that optimization can be carried out over the remaining rows.

Cook and Zhang (2016) proposed a sequential likelihood-based algorithm for estimating a general envelope $\mathcal{E}_M(\mathcal{A})$. Called the *1D algorithm*, it requires optimization in r dimensions in the first step and reduces the optimization dimension by 1 in each subsequent step until an estimate of a basis for $\mathcal{E}_M(\mathcal{A})$ is obtained. Compared to direct Grassmannian algorithms for maximizing the log likelihood, it is straightforward to implement and has the potential to reduce the computational burden with little loss of effectiveness. It could be used to produce a stand-alone estimator, or as an alternative method for getting starting values for a Grassmann optimization of the objective function in Proposition 1.14.

Cook and Zhang (2018) proposed a screening algorithm called Envelope Component Screening (ECS) that reduces the original dimension r to a more manageable dimension $d \leq n$, without losing notable structural information on the envelope. The ECS algorithm can be a useful tool for reducing the dimension prior to application of a full optimization algorithm. Adaptations might compete directly with PLS algorithms, although that possibility has apparently not been studied.

2

Envelopes for Regression

Envelopes are dimension reduction constructs that come in several varieties depending on the goals of the analysis. In reference to model (1.1), the two most common are response envelopes for reducing the response vector Y and predictor envelopes for reducing the vector of predictors X. Envelope theory and methodology were developed first by Cook, Li, and Chiaromonte (2010) for response reduction and later extended to predictor reduction by Cook, Helland, and Su (2013), who also established first the connections with PLS regression. Cook and Zhang (2015b) developed envelopes for simultaneous response and predictor reduction. As mentioned at the outset of Chapter 1, the goal of envelope methods, stated informally, is to separate with clarity the information in the data that is material to the study goals from that which is immaterial, which is in the spirit of Fisher's notion of sufficient statistics (Fisher, 1922).

There are now many articles extending and refining various aspects of envelope methodology, including partial envelopes (Su and Cook, 2011) for inference on a subset of the regression coefficients, envelopes for quantile regression (Ding, Su, Zhu, and Wang, 2021), spatial modeling (Rekabdarkolaee, Wang, Naji, and Fluentes, 2020), and matrix-variate and tensor-variate regressions (Ding and Cook, 2018; Li and Zhang, 2017). Forzani and Su (2020) extended the normal likelihood theory underlying envelopes to elliptically contoured distributions. An envelope version of reduced rank regression was developed by Cook, Forzani, and Zhang (2015). Wang and Ding (2018) developed an envelope version of vector autoregression and Samadi and Herath (2021) developed a reduced rank version of envelope vector autoregression.

Several articles have been devoted to using envelopes to aid further understanding and development of partial least squares regression. Cook and Su (2016) showed how to use envelopes to develop scaling methods for partial

DOI: 10.1201/9781003482475-2

least squares regression. Envelopes were used as a basis for studying the high dimensional behavior of partial least squares regression by Cook and Forzani (2018, 2019). Although partial least squares regression is usually associated with the multivariate linear model, Cook and Forzani (2021) showed that PLS dimension reduction may be serviceable in the presence of non-linearity. The role of envelopes in furthering the application of PLS in chemometrics was discussed by Cook and Forzani (2020).

Bayesian versions of envelopes were developed by Khare, Pal, and Su (2016) and Chakraborty and Su (2023). Su and Cook (2011) developed the notion of inner envelopes for capturing part of the regression. Each of these and others demonstrate a potential for envelope methodology to achieve reduction in estimative and predictive variation beyond that attained by standard methods, sometimes by amounts equivalent to increasing the sample size many times over. An introduction to envelopes is available from the monograph by Cook (2018) and a brief overview from Cook (2020).

Our goal in this chapter is to give envelope foundations that allow us to connect with PLS regression methodology.

2.1 Model-free predictor envelopes defined

Predictor envelopes are designed to reduce the dimension of the predictor vector so that the reduced predictors carry unambiguously all of the information that is *material* to the prediction of Y, leaving behind a portion of the predictor vector that is *immaterial* to prediction. To develop this idea, consider the regression of a vector-valued response $Y \in \mathbb{R}^r$ on a stochastic vector-valued predictor $X \in \mathbb{R}^p$ based on a simple random sample $(Y_i, X_i), i = 1, \ldots, n$, of size n, without necessarily invoking model (1.1). Regression methods generally become less effective when n is not large relative to p, perhaps p is large relative to n which is indicated as $p \gg n$, or the predictors are highly collinear. The difficulties caused by a relatively small sample size or high collinearity can often be mitigated effectively by linearly reducing the dimension of X without losing information about Y, either prior to or in conjunction with developing a method of prediction. Denote the reduced predictors as $P_{\mathcal{S}}X$, where $\mathcal{S}$ represents a subspace of $\mathbb{R}^p$. We expressed the reduced predictors as projected vectors in $\mathbb{R}^p$ to avoid the ambiguity involved in selecting a basis at this level of discussion.

2.1.1 Central subspace

The reduced predictors $P_{\mathcal{S}}X$ must hold all of the information that X has about Y. This is interpreted to mean that, holding $P_{\mathcal{S}}X$ fixed, any remaining variation in X should be independent of Y, a condition that is expressed symbolically as

$$Y \perp\!\!\!\perp X \mid P_{\mathcal{S}}X. \tag{2.1}$$

This condition is the driving force behind *sufficient dimension reduction (SDR)*, the goal of which is to reduce the predictor dimension without requiring a pre-specified model for the regression of Y on X. The intersection of all subspaces for which (2.1) holds is called the central subspace, first formulated by Cook (1994) and symbolized $\mathcal{S}_{Y|X}$. It has been the target of much theoretical and methodological inquiry, and there is now a substantial body of effective model-free dimension-reduction methodology based on the central subspace and related constructions. This notion is summarized by the following definition. .

Definition 2.1. *A subspace $\mathcal{S} \subseteq \mathbb{R}^p$ that satisfies (2.1) is called a dimension reduction subspace for the regression of Y on X. If the intersection of all dimension-reduction subspaces is itself a dimension reduction subspace it is called the central subspace and denoted as $\mathcal{S}_{Y|X}$.*

Early work that marks the beginning of the area is available from Cook (1998). Relatively recent advances are described in the monograph by Li (2018). To help fix ideas, consider the following four standard models, each with stochastic predictor $X \in \mathbb{R}^p$ and β, β_1, and β_2 all $p \times 1$ vectors.

1. Single-response linear regression: $Y = \alpha + \beta^T X + \sigma\varepsilon$ with $\varepsilon \perp\!\!\!\perp X$, $\mathrm{E}(\varepsilon) = 0$, and $\mathrm{var}(\varepsilon) = 1$,
2. Logistic regression: $\mathrm{logit}(p(X)) = \beta^T X$.
3. Cox regression with hazard function $\Lambda_0(t)\exp(\beta^T X)$.
4. Single-response heteroscedastic linear regression: $Y = \alpha + \beta_1^T X + \sigma(\beta_2^T X)\varepsilon$.

In models 1–3, the response depends on the predictor only via $\beta^T X$, and so $Y \perp\!\!\!\perp X \mid \beta^T X$ and $\mathcal{S}_{Y|X} = \text{span}(\beta)$. From a model-free SDR perspective, these models look the same because each depends on only one linear combination of the predictors. Model 4 depends on two linear combinations of the predictor, $\mathcal{S}_{Y|X} = \text{span}(\beta_1, \beta_2)$. In the multivariate linear model (1.1) with $\beta \in \mathbb{R}^{p\times r}$, we have $\mathcal{S}_{Y|X} = \text{span}(\beta) \subseteq \mathbb{R}^p$.

2.1.2 Model-free predictor envelopes

Methodology based on condition (2.1) alone may be ineffective when n is not large relative to p or the predictors are highly collinear because then it is hard in application to distinguish the part $P_{\mathcal{S}}X$ of X that is material to the prediction of Y from the complementary part $Q_{\mathcal{S}}X$. Because it may take a large sample size to sort out correlated predictors, requiring that $n \gg p$ is one way to compensate for these potential shortcomings. We may compensate also by requiring, in addition to (2.1), that $P_{\mathcal{S}}X \perp\!\!\!\perp Q_{\mathcal{S}}X$, so the part of X that is selected as material $P_{\mathcal{S}}X$ is independent of the complementary immaterial part $Q_{\mathcal{S}}X$. This condition is then used to induce a measure of clarity in the separation of X into material and immaterial parts. We now have two conditions that characterize a predictor envelope $\mathcal{S}$:

$$(a)\ Y \perp\!\!\!\perp X \mid P_{\mathcal{S}}X \text{ and (b) } P_{\mathcal{S}}X \perp\!\!\!\perp Q_{\mathcal{S}}X \iff \text{(c) } (Y, P_{\mathcal{S}}X) \perp\!\!\!\perp Q_{\mathcal{S}}X. \quad (2.2)$$

As indicated in this equation, conditions (2.2a) and (2.2b) are together equivalent to condition (2.2c), which says that jointly $(Y, P_{\mathcal{S}}X)$ is independent of $Q_{\mathcal{S}}X$. The material components are still represented as $P_{\mathcal{S}}X$ and the immaterial components as $Q_{\mathcal{S}}X$. See Cook (2018) for further discussion of the role of conditional independence in envelope foundations.

In full generality, (2.2) represents a strong ideal goal that is not currently serviceable as a basis for methodological development because of the need to assess condition (2.2b). However, if the predictors are normally distributed, condition (2.2b) is manageable since it then holds if and only if $\text{cov}(P_{\mathcal{S}}X, Q_{\mathcal{S}}X) = P_{\mathcal{S}}\Sigma_X Q_{\mathcal{S}} = 0$, where Σ_X denotes the covariance matrix of the predictor vector X. This connects with the reducing subspaces of Σ_X as described in the next lemma. Normality is not required, and its proof is in Appendix A.2.1.

Lemma 2.1. *Let* $\mathcal{S} \subseteq \mathbb{R}^p$. *Then* $\mathcal{S}$ *reduces* Σ_X *if and only if* $\text{cov}(P_{\mathcal{S}}X, Q_{\mathcal{S}}X) = 0$.

Replacing (2.2b) with $\text{cov}\{P_{\mathcal{S}}X, Q_{\mathcal{S}}X\} = 0$ relaxes independence, requiring only that there be no linear association between $P_{\mathcal{S}}X$ and $Q_{\mathcal{S}}X$. This modification somewhat weakens the strength of (2.2), but experience has shown that it nevertheless leads to methodology that has the potential to outperform standard methods. These arguments lead to the following tempered version of (2.2) that is more amenable to methodological development and is equivalent to (2.2) with normal predictors:

$$(a)\ Y \perp\!\!\!\perp X \mid P_{\mathcal{S}}X \text{ and (b) } P_{\mathcal{S}}\Sigma_X Q_{\mathcal{S}} = 0. \tag{2.3}$$

By Definition 2.1, any subspace that satisfies (2.2) or (2.3) is a dimension reduction subspace. We are now in a position to give a formal definition of an envelope in the context of (2.3). This is the form that connects with PLS regression.

Definition 2.2. *Assume that $\mathcal{S}_{Y|X} \subseteq \text{span}(\Sigma_X)$. The predictor envelope for the regression of Y on X is then the intersection of all dimension reduction subspaces that reduce Σ_X and contain $\mathcal{S}_{Y|X}$. The predictor envelope is denoted as $\mathcal{E}_{\Sigma_X}(\mathcal{S}_{Y|X})$ and represented as just $\mathcal{E}_X$ for use in subscripts.*

If the eigenvalues of Σ_X are unique then $P_{\mathcal{E}_X}\Sigma_X Q_{\mathcal{E}_X} = 0$ holds when $\mathcal{E}_{\Sigma_X}(\mathcal{S}_{Y|X})$ is spanned by *any* subset of q eigenvectors of Σ_X. Further, if $\dim(\mathcal{E}_{\Sigma_X}(\mathcal{S}_{Y|X})) = q$ then $\mathcal{E}_{\Sigma_X}(\mathcal{S}_{Y|X})$ is spanned by q eigenvectors of Σ_X. This implies that the eigenstructure of Σ_X plays a population role in envelope regressions. However, reduction by principal components has no special role: ensuring that $P_{\mathcal{E}_X}\Sigma_X Q_{\mathcal{E}_X} = 0$ by selecting $\mathcal{E}_{\Sigma_X}(\mathcal{S}_{Y|X})$ to be the span of the leading q eigenvectors of Σ_X, which are often preferred, requires a leap of faith that is not supported by these foundations. A useful takeaway from this discussion is that the spectral structure of Σ_X arises as a constituent of envelope foundations because of requirement (2.3b), which was included to enhance the distinction between material and immaterial predictors.

Ensuring that there is clarity between the material and immaterial predictors by imposing condition (2.3b) comes with a potential cost. Since $\mathcal{S}_{Y|X} \subseteq \mathcal{E}_{\Sigma_X}(\mathcal{S}_{Y|X})$, the envelope is an upper bound on the central subspace. In consequence, the number of reduced predictors $P_{\mathcal{E}_X}X$ dictated by the envelope is at least as large as the reduced predictors given by central subspace $P_{\mathcal{S}_{Y|X}}X$. In the extreme, it is possible to have regressions where $\dim(\mathcal{S}_{Y|X}) = 1$ while $\dim(\mathcal{E}_{\Sigma_X}(\mathcal{S}_{Y|X})) = p$, although in our experience such an occurrence is the exception and not the rule.

TABLE 2.1

Conceptual contrasts between three dimension reduction paradigms. 'Construction' means that methodology attempts to select $\mathcal{S}$ to satisfy the condition. 'Assumption' means that methodology requires the condition without any attempt to ensure its suitability. 'No constraint' means that methodology does not address the condition per se.

	Requirement	
Paradigm	(a) $Y \perp\!\!\!\perp X \mid P_{\mathcal{S}}X$	(b) $P_{\mathcal{S}}X \perp\!\!\!\perp Q_{\mathcal{S}}X$
SDR	Construction	No constraints
Variable Selection	Construction	Assumption
Envelopes	Construction	Construction

2.1.3 Conceptual contrasts

Table 2.1 contrasts the conceptual basis for three dimension reduction paradigms. The paradigms all use requirement (a) as stated in the table. Requirement (b) is intended as a representative stand-in for various methods that might be used to control the dependence between the material $P_{\mathcal{S}}X$ and immaterial $Q_{\mathcal{S}}X$ parts of X. For instance, condition (b) might be replaced by $\mathrm{cov}\{P_{\mathcal{S}}X, Q_{\mathcal{S}}X\} = 0$, as discussed in the previous section.

Sufficient dimension reduction methods like sliced inverse regression (SIR; Li, 1991), sliced average variance estimation (SAVE; Cook and Weisberg, 1991; Cook, 2000), likelihood acquired directions (LAD; Cook and Forzani, 2009) and many other SDR methods (Li, 2018) operate under requirement (a) only. Under various technical conditions, they are designed to estimate the central subspace $\mathcal{S}_{Y|X}$. These methods pay no attention to condition (b), $P_{\mathcal{S}}X \perp\!\!\!\perp Q_{\mathcal{S}}X$. As a consequence, when there is high collinearity in X, a relatively large sample may be needed to estimate $\mathcal{S}_{Y|X}$ accurately.

Variable selection is conceptually similar to SDR, except bases for $\mathcal{S}$ are constrained to be subsets of the columns of I_p so $P_{\mathcal{S}}$ projects only onto coordinate axes, which has the effect of selecting variables from X. To ensure consistency, variable selection methods controlled by assumption the dependence between the selected variables $P_{\mathcal{S}}X$ and the variables eliminated $Q_{\mathcal{S}}X$. For instance, Goh and Dey (2019) studied asymptotic properties of marginal least-square estimators in ultrahigh-dimensional linear regression models with correlated errors. They assume in effect that $n^{1/2}$ times the sample correlation between the selected and eliminated predictors is bounded, which exerts firm control over the dependence between $P_{\mathcal{S}}X$ and $Q_{\mathcal{S}}X$ as p and n grow.

Envelopes handle the dependence between $P_{\mathcal{S}}X$ and $Q_{\mathcal{S}}X$ by construction, the selected subspaces $\mathcal{S}$ being required to satisfy both conditions (a) and (b) of Table 2.1. This has the advantage of ensuring a crisp distinction between material and immaterial parts of X, but it has the potential disadvantage of leading to a larger subspace since $\mathcal{S}_{Y|X} \subseteq \mathcal{E}_{\Sigma_X}(\mathcal{S}_{Y|X})$. Nevertheless, experience has indicated that this tradeoff is very often worthwhile, particularly in model-based analyses and in high-dimensional settings where $n \ll p$.

2.2 Predictor envelopes for the multivariate linear model

In the context of model (1.1) with stochastic predictors, $\mathrm{E}(X) = \mu_X$ and $\mathrm{var}(X) = \Sigma_X > 0$. Let $\mathcal{B} = \mathrm{span}(\beta)$, the subspace of $\mathbb{R}^p$ spanned by the columns of β. As stated along with the (1.1), normality is not required unless needed for specific methods like likelihood-based estimation. Since (1.1) depends on X only via $\beta^T X$, it follows immediately that $\mathcal{S}_{Y|X} = \mathcal{B}$ and that the envelope under Definition 2.2 becomes $\mathcal{E}_{\Sigma_X}(\mathcal{S}_{Y|X}) = \mathcal{E}_{\Sigma_X}(\mathcal{B})$, the smallest reducing subspace of Σ_X that contains $\mathcal{B}$. From Proposition 1.6, this envelope can be expressed equivalently as $\mathcal{E}_{\Sigma_X}(\mathcal{B}) = \mathcal{E}_{\Sigma_X}(\mathcal{C}_{X,Y})$, the smallest reducing subspace of Σ_X that contains $\mathcal{C}_{X,Y} = \mathrm{span}(\Sigma_{X,Y})$. The distinction between these representations of the predictor envelope in the context of model (1.1),

$$\mathcal{E}_{\Sigma_X}(\mathcal{S}_{Y|X}) = \mathcal{E}_{\Sigma_X}(\mathcal{B}) = \mathcal{E}_{\Sigma_X}(\mathcal{C}_{X,Y}), \tag{2.4}$$

will be useful when introducing response envelopes in Section 2.4.

We are now in a position to write the envelope version of (1.1). Let $q = \dim\{\mathcal{E}_{\Sigma_X}(\mathcal{B})\}$, let $\Phi \in \mathbb{R}^{p\times q}$ denote a semi-orthogonal basis matrix for $\mathcal{E}_{\Sigma_X}(\mathcal{B})$ and let $(\Phi, \Phi_0) \in \mathbb{R}^{p\times p}$ be an orthogonal matrix. Then we can represent $\beta = \Phi\eta$, where $\eta \in \mathbb{R}^{q\times r}$ gives the coordinates of β in terms of basis Φ. The linear model then becomes

$$\left.\begin{aligned} Y_i &= \alpha + \eta^T\Phi^T X_i + \varepsilon_i, i = 1, \ldots, n \\ \Sigma_X &= \Phi\Delta\Phi^T + \Phi_0\Delta_0\Phi_0^T \end{aligned}\right\}, \tag{2.5}$$

where $\Delta = \mathrm{var}(\Phi^T X) = \Phi^T\Sigma_X\Phi \in \mathbb{R}^{q\times q}$ is the covariance matrix of the material components and $\Delta_0 = \mathrm{var}(\Phi_0^T X) = \Phi_0^T\Sigma_X\Phi_0 \in \mathbb{R}^{(p-q)\times(p-q)}$ is the covariance matrix of the immaterial components. No relationship is assumed between the eigenvalues of Δ and those of Δ_0. Conditions (2.3) are satisfied straightforwardly under this model: $Y \perp\!\!\!\perp X \mid \Phi^T X$ and $\text{cov}(\Phi^T X, \Phi_0^T X) = 0$. The number of free real parameters in this model is the sum of r for α, rq for η, $q(p-q)$ for $\mathcal{E}_{\Sigma_X}(\mathcal{B})$, $q(q+1)/2$ for Δ, $(p-q)(p-q+1)/2$ for Δ_0 and $r(r+1)/2$ for $\Sigma_{Y|X}$, giving a total of

$$N_q = r + rq + p(p+1)/2 + r(r+1)/2.$$

This amounts to a reduction of $N_p - N_q = r(p-q)$ parameters over the standard model.

The constituent parameters η, Φ, Δ and Δ_0 in model (2.5) are not identifiable. For instance, for any conforming orthogonal matrix O, $\Phi\eta = (\Phi O)(O^T\eta) = \Phi^*\eta*$ and, in consequence, Φ and η are not identifiable. However, the key parameters – the envelope $\mathcal{E}_{\Sigma_X}(\mathcal{B}) = \text{span}(\Phi)$, $\beta = \Phi\eta$ and Σ_X – are all identifiable and the model itself is identifiable. These parameters are the ones that are typically important and the same as those in the base multivariate linear model (1.1). The lack of identifiability of the constituent parameters will be of little consequence in application. The eigenvalues of Δ and Δ_0 are also identifiable, which can aid an assessment of a particular fit: Model (2.5) will give the greatest efficiency gains when the eigenvalues of Δ are mostly larger than those of Δ_0 as described in the following discussion of (2.8). This structure is a partial consequence of incorporating a subspace as an unknown parameter. A similar structure holds for all of the envelope and PLS methods for reducing the predictors described in this book.

If the envelope $\mathcal{E}_{\Sigma_X}(\mathcal{B})$ were known then, according to (2.5), we could pick any basis matrix Φ and fit the linear regression of Y on the compressed predictors $\Phi^T X$, giving the estimated coefficient matrix $\widehat{\beta}_{Y|\Phi^T X}$ from the multivariate regression of Y on $\Phi^T X$. The maximum likelihood envelope estimator $\widehat{\beta}_\Phi$ of β with known basis Φ is then

$$\widehat{\beta}_\Phi = \Phi\widehat{\beta}_{Y|\Phi^T X}. \tag{2.6}$$

Asymptotic variances can provide insights into the potential gain offered by the envelope estimator: Repeating (1.18) for ease of comparison,

$$\begin{aligned} \text{avar}(\sqrt{n}\text{vec}(\widehat{\beta}_{\text{ols}})) &= \Sigma_{Y|X} \otimes \Sigma_X^{-1} \\ \text{avar}(\sqrt{n}\text{vec}(\widehat{\beta}_\Phi)) &= \Sigma_{Y|X} \otimes \Phi\Delta^{-1}\Phi^T, \end{aligned} \tag{2.7}$$

where $\text{avar}(\sqrt{n}\text{vec}(\widehat{\beta}_\Phi))$ is based on Cook (2018, Section 4.1.3). This gives a difference of

$$\Sigma_{Y|X} \otimes (\Sigma_X^{-1} - \Phi\Delta^{-1}\Phi^T) = \Sigma_{Y|X} \otimes \Phi_0\Delta_0^{-1}\Phi_0^T,$$

which indicates that the difference in asymptotic variances will be large when the variability of the immaterial predictors $\Delta_0 = \text{var}(\Phi_0^T X)$ is small relative to $\Delta = \text{var}(\Phi^T X)$. This can be seen also from the ratio of variance traces:

$$\frac{\text{tr}\{\Sigma_{Y|X} \otimes \Phi\Delta^{-1}\Phi^T\}}{\text{tr}\{\Sigma_{Y|X} \otimes \Sigma_X^{-1}\}} = \frac{\text{tr}\{\Delta^{-1}\}}{\text{tr}\{\Delta^{-1}\} + \text{tr}\{\Delta_0^{-1}\}}. \tag{2.8}$$

2.3 Likelihood estimation of predictor envelopes

Following Cook, Helland, and Su (2013), we develop in this section maximum likelihood estimation of the parameters in model (2.5). We assume throughout that the dimension q of the envelope $\mathcal{E}_{\Sigma_X}(\mathcal{B})$ is known.

As mentioned at the outset of Section 1.2, it is traditional in regression to base estimation on the conditional likelihood from the distribution of Y given X, treating the predictors as fixed even if they may have been randomly sampled. This practice arose because in many regressions the predictors provide only ancillary information and consequently estimation and inference should be conditioned on their observed values. (See Aldrich (2005) for a review and a historical perspective.) In contrast, PLS and the likelihood-based envelope method developed in this section both postulate a link – represented here by the envelope $\mathcal{E}_{\Sigma_X}(\mathcal{B})$ – between β, the parameter of interest, and Σ_X. As a consequence, X is not ancillary and we pursue estimation through the joint distribution of Y and X.

2.3.1 Likelihood-inspired estimators

Let $C = (X^T, Y^T)^T$ denote the random vector constructed by concatenating X and Y, and let S_C denote the sample version of $\Sigma_C = \text{var}(C)$. Given q, we base estimation on the objective function $F_q(S_C, \Sigma_C) = \log|\Sigma_C| + \text{tr}(S_C\Sigma_C^{-1})$ that stems from the log likelihood of the multivariate normal family after replacing the population mean vector with the vector of sample means, although here we do not require C to have a multivariate normal distribution.

Rather, we are using F_q as a multi-purpose objective function in the same spirit as least squares objective functions are often used.

The structure of the envelope $\mathcal{E}_{\Sigma_X}(\mathcal{B})$ can be introduced into F_q by using the parameterizations $\Sigma_X = \Phi\Delta\Phi^T + \Phi_0\Delta_0\Phi_0^T$ and $\Sigma_{X,Y} = \Phi\gamma$, where $\Phi \in \mathbb{R}^{p\times q}$ is a semi-orthogonal basis matrix for $\mathcal{E}_{\Sigma_X}(\mathcal{B}) = \mathcal{E}_{\Sigma_X}(\mathcal{C}_{X,Y})$, $(\Phi, \Phi_0) \in \mathbb{R}^{p\times p}$ is an orthogonal matrix, and $\Delta \in \mathbb{R}^{q\times q}$ and $\Delta_0 \in \mathbb{R}^{(p-q)\times(p-q)}$ are symmetric positive definite matrices, as defined previously for model (2.5). From Proposition 1.6 or (2.4), $\mathcal{C}_{X,Y} = \text{span}(\Sigma_{X,Y}) \subseteq \mathcal{E}_{\Sigma_X}(\mathcal{B})$, so we wrote $\Sigma_{X,Y}$ as linear combinations of the columns of Φ. The matrix $\gamma \in \mathbb{R}^{q\times r}$ then gives the coordinates of $\Sigma_{X,Y}$ in terms of the basis Φ. With this we have

$$\begin{aligned}
\Sigma_C &= \begin{pmatrix} \Sigma_X & \Sigma_{X,Y} \\ \Sigma_{X,Y}^T & \Sigma_Y \end{pmatrix} = \begin{pmatrix} \Phi\Delta\Phi^T + \Phi_0\Delta_0\Phi_0^T & \Phi\gamma \\ \gamma^T\Phi^T & \Sigma_Y \end{pmatrix} \\
&= \begin{pmatrix} \Phi_0 & \Phi & 0 \\ 0 & 0 & I_r \end{pmatrix} \begin{pmatrix} \Delta_0 & 0 & 0 \\ 0 & \Delta & \gamma \\ 0 & \gamma^T & \Sigma_Y \end{pmatrix} \begin{pmatrix} \Phi_0^T & 0 \\ \Phi^T & 0 \\ 0 & I_r \end{pmatrix} \\
&= O\Sigma_{O^TC}O^T, \qquad (2.9)
\end{aligned}$$

where

$$O = \begin{pmatrix} \Phi_0 & \Phi & 0 \\ 0 & 0 & I_r \end{pmatrix} \in \mathbb{R}^{(p+r)\times(p+r)}$$

is an orthogonal matrix and

$$\Sigma_{O^TC} = \begin{pmatrix} \Delta_0 & 0 & 0 \\ 0 & \Delta & \gamma \\ 0 & \gamma^T & \Sigma_Y \end{pmatrix} \in \mathbb{R}^{(p+r)\times(p+r)}$$

is the covariance matrix of the transformed vector O^TC. The objective function $F_q(S_C, \Sigma_C)$ can now be regarded as a function of the five constituent parameters – Φ, Δ, Δ_0, γ and Σ_Y – that comprise Σ_C. The parameters β and η of model (2.5) can be written as functions of those parameters: $\eta = \Delta^{-1}\gamma$ and $\beta = \Phi\eta = \Phi\Delta^{-1}\gamma$.

To minimize $F_q(S_C, \Sigma_C)$ we first hold Φ fixed and substitute (2.9) giving

$$\begin{aligned}
F_q(S_C, \Sigma_C) &= \log|O\Sigma_{O^TC}O^T| + \text{tr}(O^TS_CO\Sigma_{O^TC}^{-1}) \\
&= \log|\Sigma_{O^TC}| + \text{tr}(S_{O^TC}\Sigma_{O^TC}^{-1}).
\end{aligned}$$

We next write F_q in terms of addends that can be minimized separately. The sample variance of $(\Phi^T X, Y)$ is

$$S_{(\Phi^T X,Y)} = \begin{pmatrix} \Phi^T S_X \Phi & \Phi^T S_{X,Y} \\ S_{Y,X}\Phi & S_Y \end{pmatrix}.$$

Let

$$O_2 = \begin{pmatrix} \Phi & 0 \\ 0 & I_r \end{pmatrix},$$

$$\Sigma_{O_2^T C} = \begin{pmatrix} \Delta & \gamma \\ \gamma^T & \Sigma_Y \end{pmatrix}.$$

Then we have that

$$\Sigma_{O^T C} = \begin{pmatrix} \Delta_0 & 0 \\ 0 & \Sigma_{O_2^T C} \end{pmatrix},$$

and

$$F_q(S_C, \Sigma_C) = \log|\Delta_0| + \operatorname{tr}\left(\Phi_0^T S_X \Phi_0 \Delta_0^{-1}\right) + \log|\Sigma_{O_2^T C}| + \operatorname{tr}\left(S_{\Phi^T X,Y}\Sigma_{O_2^T C}^{-1}\right).$$

The values of the parameters that minimize F_q for fixed Φ are $\Sigma_Y = S_Y$, $\Delta = \Phi^T S_X \Phi$, $\Delta_0 = \Phi_0^T S_X \Phi_0$ and $\gamma = \Phi^T S_{X,Y}$. Substituting these forms into F_q then leads to the following estimator of the envelope when q is assumed known: using G as an optimization argument to avoid confusion with the unknown parameter Φ,

$$\begin{aligned} \widehat{\mathcal{E}}_{\Sigma_X}(\mathcal{B}) &= \operatorname{span}\{\arg\min_G L_q(G)\}, \text{ where} && (2.10)\\ L_q(G) &= \log|G^T S_{X|Y} G| + \log|G^T S_X^{-1} G| \\ &= \log|G^T (S_X - S_{X,Z} S_{X,Z}^T) G| + \log|G^T S_X^{-1} G|, && (2.11) \end{aligned}$$

$Z = S_Y^{-1/2} Y$ is the standardized response vector, the minimization in (2.10) is taken over all semi-orthogonal matrices $G \in \mathbb{R}^{p\times q}$ and we have used Lemma 1.3I,

$$|G_0^T S_X G_0| = |S_X| \times |G^T S_X^{-1} G|,$$

in the derivation of (2.11). Let $\widehat{\Phi}$ be any semi-orthogonal basis of $\widehat{\mathcal{E}}_{\Sigma_X}(\mathcal{B})$. The objective function $L_q(G)$ is an instance of the objective function for the sample version of Algorithm $\mathbb{L}$ introduced in Section 1.5.5. The connection is obtained by setting $M = S_{X|Y}$ and $A = S_{Y\circ X}$. Then from (1.8) to (1.10) we have $S_X = S_{X|Y} + S_{Y\circ X} = M + A$, so

$$\log|G^T M G| + \log|G^T (M+A)^{-1} G| = \log|G^T S_{X|Y} G| + \log|G^T S_X^{-1} G|.$$

The estimators of the constituent parameters, which are maximum likelihood estimators under normality of C, are

$$\begin{aligned}
\widehat{\Sigma}_Y &= S_Y, \\
\widehat{\Delta} &= \widehat{\Phi}^T S_X \widehat{\Phi}, \\
\widehat{\Delta}_0 &= \widehat{\Phi}_0^T S_X \widehat{\Phi}_0, \\
\widehat{\gamma} &= \widehat{\Phi}^T S_{X,Y} \\
\widehat{\eta} &= \widehat{\Delta}^{-1}\widehat{\gamma}.
\end{aligned}$$

From these we construct the estimators of the parameters of interest:

$$\begin{aligned}
\widehat{\Sigma}_{X,Y} &= P_{\widehat{\Phi}} S_{X,Y}, \\
\widehat{\Sigma}_X &= \widehat{\Phi}\widehat{\Delta}\widehat{\Phi}^T + \widehat{\Phi}_0\widehat{\Delta}_0\widehat{\Phi}_0^T = P_{\widehat{\Phi}} S_X P_{\widehat{\Phi}} + Q_{\widehat{\Phi}} S_X Q_{\widehat{\Phi}}, \\
\widehat{\Sigma}_{Y|X} &= S_Y - S_{Y,X} P_{\widehat{\Phi}} S_X^{-1} P_{\widehat{\Phi}} S_{X,Y} \\
\widehat{\beta} &= \widehat{\Phi}\widehat{\Delta}^{-1}\widehat{\gamma} = \widehat{\Phi}(\widehat{\Phi}^T S_X \widehat{\Phi})^{-1}\widehat{\Phi}^T S_{X,Y} = P_{\widehat{\Phi}(S_X)}\widehat{\beta}_{\text{ols}}, \qquad (2.12)
\end{aligned}$$

where $\widehat{\beta}_{\text{ols}} = S_X^{-1}S_{X,Y}$ is the ordinary least squares estimator of β. The estimators $\widehat{\Delta}$, $\widehat{\Delta}_0$ and $\widehat{\gamma}$ depend on the selected basis $\widehat{\Phi}$. The parameters of interest – $\widehat{\Sigma}_{X,Y}$, $\widehat{\Sigma}_X$ and $\widehat{\beta}$ – depend on $\widehat{\mathcal{E}}_{\Sigma_X}(\mathcal{B})$ but do not depend on the particular basis selected.

The following proposition (Cook et al., 2013, Prop. 9) gives the asymptotic distribution of $\widehat{\beta}$ when C is normally distributed. In preparation, let $\widehat{\beta}_\Phi$ and $\widehat{\beta}_\eta$ denote respectively the maximum likelihood estimators of β when Φ and η are known, and let $\dagger$ denote the Moore-Penrose matrix inverse.

Proposition 2.1. *Assume that q is known and that C is normally distributed with mean μ_C and covariance matrix $\Sigma_C > 0$. Then, under model (2.5, $\sqrt{n}\{\text{vec}(\widehat{\beta}) - \text{vec}(\beta)\}$ converges in distribution to a normal random vector with mean 0 and covariance matrix*

$$\begin{aligned}
\text{avar}\left\{\sqrt{n}\text{vec}(\widehat{\beta})\right\} &= \text{avar}\left\{\sqrt{n}\text{vec}(\widehat{\beta}_\Phi)\right\} + \text{avar}\left\{\sqrt{n}\text{vec}(Q_\Phi\widehat{\beta}_\eta)\right\} \\
&= \Sigma_{Y|X} \otimes \Phi\Delta^{-1}\Phi^T + (\eta^T \otimes \Phi_0)M^\dagger(\eta \otimes \Phi_0^T),
\end{aligned}$$

where $M = \eta\Sigma_{Y|X}^{-1}\eta^T \otimes \Delta_0 + \Delta \otimes \Delta_0^{-1} + \Delta^{-1} \otimes \Delta_0 - 2I_q \otimes I_{p-q}$. Additionally, $T_q = n(F(S_C, \widehat{\Sigma}_C) - F(S_C, S_C))$ converges to a chi-squared random variable with $(p-q)r$ degrees of freedom, and

$$\text{avar}\left\{\sqrt{n}\text{vec}(\widehat{\beta})\right\} \leq \text{avar}\left\{\sqrt{n}\text{vec}(\widehat{\beta}_{\text{ols}})\right\}.$$

The first addend $\text{avar}\{\sqrt{n}\text{vec}(\widehat{\beta}_\Phi)\} = \Sigma_{Y|X} \otimes \Phi\Delta^{-1}\Phi^T$ on the right hand side of the asymptotic variance is the asymptotic variance of maximum likelihood estimators of β when a basis Φ for the envelope is known, as discussed previously around (2.7. The second addend $\text{avar}\{\sqrt{n}\text{vec}(Q_\Phi\widehat{\beta}_\eta)\} = (\eta^T \otimes \Phi_0)M^\dagger(\eta \otimes \Phi_0^T)$ can then be interpreted as the cost of estimating the envelope. The final statement of the proposition is that the difference between the asymptotic variance of the ordinary least squares estimator and that for the envelope estimator is always positive semi-definite.

The following corollary, also from Cook et al. (2013), gives sufficient conditions for the envelope and OLS estimators to be asymptotically equivalent.

Corollary 2.1. *Under the conditions of Proposition 2.1, if* $\Sigma_X = \sigma_x^2 I_p$ *and* $\beta \in \mathbb{R}^{p\times r}$ *has rank* r *then*

$$\text{avar}\left\{\sqrt{n}\text{vec}(\widehat{\beta})\right\} = \text{avar}\left\{\sqrt{n}\text{vec}(\widehat{\beta}_{\text{ols}})\right\}.$$

Accordingly, if there is no collinearity among predictors of equal variability then the OLS and envelope estimators are equivalent asymptotically. This finding does not rule out the possibility that envelope estimators may still be superior to OLS in non-asymptotic settings.

Proposition 2.1 can be used to construct an asymptotic standard error for $(\widehat{\beta})_{ij}$, $i = 1, \ldots, p$, $j = 1, \ldots, r$, by first substituting estimates for the unknown quantities on the right side of $\text{avar}\left\{\sqrt{n}\text{vec}(\widehat{\beta})\right\}$ to obtain an estimated asymptotic variance $\widehat{\text{avar}}\{\sqrt{n}\text{vec}(\widehat{\beta})\}$. The estimated asymptotic variance $\widehat{\text{avar}}\{\sqrt{n}(\widehat{\beta})_{ij}\}$ is then the corresponding diagonal element of $\widehat{\text{avar}}\{\sqrt{n}\text{vec}(\widehat{\beta})\}$, and its *asymptotic standard error* is

$$\text{se}\{(\widehat{\beta})_{ij}\} = \frac{[\widehat{\text{avar}}\{\sqrt{n}(\widehat{\beta})_{ij}\}]^{1/2}}{\sqrt{n}}. \tag{2.13}$$

If $C_i, i = 1, \ldots, n$, are independent copies of C, which is not necessarily normal but has finite fourth moments, then $\sqrt{n}\{\text{vec}(\widehat{\beta}) - \text{vec}(\beta)\}$ converges in distribution to a normal random vector with mean 0 and positive definite covariance matrix $\text{avar}\{\sqrt{n}\text{vec}(\widehat{\beta})\}$ that depends on fourth moments of C and is quite complicated (Cook et al., 2013, Prop. 8). The residual bootstrap is a useful option for dealing with variances in this case.

2.3.2 PLS connection

As developed in Chapter 3, the PLS regression algorithms NIPALS and SIMPLS in Tables 3.1 and 3.4 both provide $\sqrt{n}$-consistent estimators of a basis for the predictor envelope $\mathcal{E}_{\Sigma_X}(\mathcal{B})$ when $n \to \infty$ with p and q fixed. And they do so subject to (2.3) (Cook, 2018; Cook et al., 2013). These algorithms proceed sequentially using one-at-a-time estimation of the basis vectors, often called weights in the PLS literature. The dimension q of the envelope corresponds to the number of components in common PLS terminology. The estimated basis for $\mathcal{E}_{\Sigma_X}(\mathcal{B})$ is then used to estimate β. A key point here is that, like other envelope methods, PLS regression stands apart from much of the regression and dimension reduction literature by using both of the conditions (2.3a) and (2.3b) to guide the reduction.

Model (2.5) provides for maximum likelihood estimation of the parameters estimated by PLS algorithms, along with estimation of Σ_X and $\Sigma_{Y|X}$. Since the maximum likelihood estimators inherit optimality properties from general likelihood theory, they must be superior to the PLS estimators as $n \to \infty$ with r and p fixed as long as the model holds. However, the PLS estimators are serviceable when $n < p$ while the maximum likelihood estimators are not, and PLS dimension reduction is serviceable also under certain model deviations (Chapter 9).

The objective function $L_q(G)$ in (2.11) is non-convex and consequently there is always a chance that an optimization algorithm will get caught in a local minimum. On the other hand, computation of the PLS estimator will be seen in Chapter 3 to be relatively straightforward. These contrasts suggest that the maximum likelihood and PLS estimators can play somewhat different roles in application.

Envelope and PLS estimators are compared in Chapter 3.

2.4 Response envelopes, model-free and model-based

So far, we have focused on predictor envelopes guided first by conditions (2.2) and then by conditions (2.3) and finally by model (2.5). We turn to response envelopes in this section, guided by corresponding foundations. Although PLS methods seem concerned mostly with predictor reduction, response reduction is useful in some applications as well, including PLS for discrimination discussed in Chapter 7 and the Nadler-Coifman chemometrics regression discussed in Section 2.6. The inverse regression of X on Y is used in these and related applications. There are strong parallels between response and predictor envelopes, and we make use of them to give a relatively brief exposition on response envelopes in this section. Simultaneous reduction of the predictors and the responses is discussed in Chapter 5. Response envelopes were proposed and developed by Cook et al. (2010).

The overarching idea for response envelopes is to separate with clarity the material part of Y, whose distribution is affected by changing X, from the immaterial part of Y whose distribution is unaffected by changing X. Similar to predictor envelopes, this separation is accomplished by defining (and subsequently estimating) the unique smallest subspace $\mathcal{E}$ of the r-dimensional response space so that the projection of the vector of responses Y onto $\mathcal{E}$ holds the part of Y that is affected by changing X, while the projection onto the orthogonal complement of $\mathcal{E}$ holds the part of Y that is unaffected by X. These aspirations are described formally by requiring $\mathcal{E}$ to be the smallest subspace of the response space that satisfies the following two conditions, which are the counterparts to condition (2.2) for predictor envelopes.

$$(a)\ Q_{\mathcal{E}}Y \perp\!\!\!\perp X \text{ and (b) } P_{\mathcal{E}}Y \perp\!\!\!\perp Q_{\mathcal{E}}Y \mid X \iff \text{(c) } (P_{\mathcal{E}}Y, X) \perp\!\!\!\perp Q_{\mathcal{E}}Y. \quad (2.14)$$

Condition (a) says that $Q_{\mathcal{E}}Y$ is marginally independent of X, so that in the regression context the distribution of $Q_{\mathcal{E}}Y$ is unaffected by changing X. The complementary part of Y, $P_{\mathcal{E}}Y$, then represents the part of Y that is affected by X. Condition (b) formalizes the relationship between $Q_{\mathcal{E}}Y$ and $P_{\mathcal{E}}Y$: $P_{\mathcal{E}}Y$ and $Q_{\mathcal{E}}Y$ are conditionally independent given X, so $Q_{\mathcal{E}}Y$ cannot impact $P_{\mathcal{E}}Y$ through an association. Conditions (a) and (b) together are equivalent to condition (c): $P_{\mathcal{E}}Y$ and X are jointly independent of $Q_{\mathcal{E}}Y$. The distribution of $Q_{\mathcal{E}}Y$ then represents a static component of the regression.

Further, condition (2.14c) is the same as condition (2.2c) with the roles of Y and X interchanged. This means that the discussion of Sections 2.1 and 2.2 hold also for response envelopes by interchanging X and Y. The similarity with (2.2c) can be seen also by expressing (2.14) as

$$(a) X \perp\!\!\!\perp Y \mid P_{\mathcal{E}}Y \text{ and (b) } P_{\mathcal{E}}Y \perp\!\!\!\perp Q_{\mathcal{E}}Y,$$

which is a dual of (2.2) obtained by interchanging X and Y.

Let $\mathcal{S}_{X|Y}$ denote the central subspace for the regression of X on Y. Then, in parallel to Definition 2.2, we have

Definition 2.3. *Assume that $\mathcal{S}_{X|Y} \subseteq \text{span}(\Sigma_Y)$. The response envelope for the regression of Y on X is then the intersection of all reducing subspaces of Σ_Y that contain $\mathcal{S}_{X|Y}$, and is denoted as $\mathcal{E}_{\Sigma_Y}(\mathcal{S}_{X|Y})$, which is represented as $\mathcal{E}_Y$ for use in subscripts.*

Comparing Definitions 2.2 and 2.3 we see that response and predictor envelopes are dual constructions when X and Y are jointly distributed, as summarized in Table 2.2. The predictor envelope $\mathcal{E}_{\Sigma_X}(\mathcal{S}_{Y|X})$ for the regression of Y on X is also the response envelope for the regression of X on Y, and the response envelope $\mathcal{E}_{\Sigma_Y}(\mathcal{S}_{X|Y})$ for the regression of Y on X is also the predictor envelope for the regression of X on Y. These equivalences in model-free envelope construction are shown in Table 2.2(A).

TABLE 2.2
Relationship between predictor and response envelopes when X and Y are jointly distributed: $\mathcal{C}_{X,Y} = \text{span}(\Sigma_{X,Y})$ and $\mathcal{C}_{Y,X} = \text{span}(\Sigma_{Y,X})$. $\mathcal{B} = \text{span}(\beta_{Y|X})$ and $\beta' = \text{span}(\beta_{Y|X}^T)$ are as used previously for model (1.1). $\mathcal{B}_*$ and $\mathcal{B}_*^T$ are defined similarly but in terms of the regression of $X \mid Y$. P-env and R-env stand for predictor and response envelope.

(A) Central subspaces	
$\mathcal{E}_{\Sigma_X}(\mathcal{S}_{Y\|X})$	$\mathcal{E}_{\Sigma_Y}(\mathcal{S}_{X\|Y})$
Predictor envelope for $Y \mid X$ Response envelope for $X \mid Y$	Predictor envelope for $X \mid Y$ Response envelope for $Y \mid X$
(B) Multivariate linear models	
$Y = \alpha + \beta_{Y\|X}^T X + \varepsilon$ $\mathcal{B} = \text{span}(\beta_{Y\|X})$, $\mathcal{B}' = \text{span}(\beta_{Y\|X}^T)$	$X = \mu + \beta_{X\|Y}^T Y + \epsilon$ $\mathcal{B}_* = \text{span}(\beta_{X\|Y})$, $\mathcal{B}_*' = \text{span}(\beta_{X\|Y}^T)$
P-env for $Y\|X$: $\mathcal{E}_{\Sigma_X}(\mathcal{B}) = \mathcal{E}_{\Sigma_X}(\mathcal{C}_{X,Y})$ R-env for $Y\|X$: $\mathcal{E}_{\Sigma_Y}(\mathcal{B}') = \mathcal{E}_{\Sigma_Y}(\mathcal{C}_{Y,X})$	R-env for $X\|Y$: $\mathcal{E}_{\Sigma_X}(\mathcal{B}_*') = \mathcal{E}_{\Sigma_X}(\mathcal{C}_{X,Y})$ P-env for $X\|Y$: $\mathcal{E}_{\Sigma_Y}(\mathcal{B}_*) = \mathcal{E}_{\Sigma_Y}(\mathcal{C}_{Y,X})$
R-env for $Y\|X$: $\mathcal{E}_{\Sigma_Y}(\mathcal{B}') = \mathcal{E}_{\Sigma_{Y\|X}}(\mathcal{B}')$	R-env for $X\|Y$: $\mathcal{E}_{\Sigma_{X\|Y}}(\mathcal{B}_*) = \mathcal{E}_{\Sigma_{X\|Y}}(\mathcal{B}_*')$

Table 2.2(B) shows envelope relationships in the context of multivariate linear models. From there we see that the predictor envelope in the regression of Y on X – $\mathcal{E}_{\Sigma_X}(\mathcal{C}_{X,Y})$– is the same as the response envelope for the linear regression of X on Y– again, $\mathcal{E}_{\Sigma_X}(\mathcal{C}_{X,Y})$: using Proposition 1.8 and the

discussion that follows,

$$\mathcal{E}_{\Sigma_X}(\mathcal{B}) = \mathcal{E}_{\Sigma_X}(\mathrm{span}(\Sigma_X^{-1}\Sigma_{X,Y})) = \mathcal{E}_{\Sigma_X}(\mathcal{C}_{X,Y}).$$

Similarly, the response envelope for the linear regression of Y on X – $\mathcal{E}_{\Sigma_Y}(\mathcal{B}')$, where $\mathcal{B}' = \mathrm{span}(\beta^T)$ as defined near (1.26) – is the same as the predictor envelope for the linear regression of X on Y:

$$\mathcal{E}_{\Sigma_Y}(\mathcal{B}') = \mathcal{E}_{\Sigma_Y}(\mathcal{C}_{Y,X}). \tag{2.15}$$

The envelope forms $\mathcal{E}_{\Sigma_X}(\mathcal{C}_{X,Y})$ and $\mathcal{E}_{\Sigma_Y}(\mathcal{C}_{Y,X})$ reflect the duality of response and predictor envelopes when X and Y are jointly distributed, while forms $\mathcal{E}_{\Sigma_X}(\mathcal{B})$ and $\mathcal{E}_{\Sigma_Y}(\mathcal{B}')$ are in terms of model (1.1) to facilitate reparametrization.

So far, we have required that X and Y be jointly distributed. This implies, for example, that PLS predictor reduction would not normally be useful in designed experiments . For instance, if predictor values are selected to follow a 2^k factorial or if center points are included to allow estimation of quadratic effects, there seems no reason to suspect that the predictor distribution carries useful information about the responses. However, response reduction may still be of value.

The conditions (a) and (b) in (2.14) can be stated alternatively as

$$\left.\begin{array}{l}\text{(a) For all } x_1, x_2,\ F_{Q_{\mathcal{E}}Y|X=x_1}(z) = F_{Q_{\mathcal{E}}Y|X=x_2}(z) \\ \text{(b) For all } x,\ P_{\mathcal{E}}Y \perp\!\!\!\perp Q_{\mathcal{E}}Y \mid X = x,\end{array}\right\}, \tag{2.16}$$

where $F_{Q_{\mathcal{E}}Y|X=x}(z)$ denotes the conditional CDF of $Q_{\mathcal{E}}Y \mid X = x$. These conditions do not require X to be stochastic. In consequence, X is allowed to be ancillary in PLS and envelope response reduction and, using (1.26), the only envelope that is relevant from Table 2.2 is

$$\mathcal{E}_{\Sigma_Y}(\mathcal{B}') = \mathcal{E}_{\Sigma_{Y|X}}(\mathcal{B}').$$

This relation allows us to parameterize the multivariate linear model (1.1) by thinking in terms of either a basis for $\mathcal{E}_{\Sigma_Y}(\mathcal{B}')$ or a basis for $\mathcal{E}_{\Sigma_{Y|X}}(\mathcal{B}')$. In the next section we use $\mathcal{E}_{\Sigma_{Y|X}}(\mathcal{B}')$ as our guide since that seems most convenient in terms of model (1.1).

2.5 Response envelopes for the multivariate linear model

We are now in a position to parameterize model (1.1) in terms of the response envelope $\mathcal{E}_{\Sigma_{Y|X}}(\mathcal{B}')$. We first introduce the model for response envelopes and then discuss background motivation that arises from longitudinal data.

2.5.1 Response envelope model

Let u denote the dimension of the response envelope, let Γ be an $r \times u$ semi-orthogonal matrix whose columns form a basis for it, and let (Γ, Γ_0) be an orthogonal matrix. Then, we have the response envelope model

$$\left.\begin{aligned} Y_i &= \alpha + \Gamma\eta_1 X_i + \varepsilon_i, i = 1, \ldots, n, \\ \Sigma_{Y|X} &= \Gamma\Omega\Gamma^T + \Gamma_0\Omega_0\Gamma_0^T, \end{aligned}\right\}, \tag{2.17}$$

where Ω and Ω_0 are $u \times u$ and $(r-u) \times (r-u)$ positive definite matrices and the errors are independent $N(0, \Sigma_{Y|X})$ random vectors. The coordinates η_1 are subscripted here to distinguish them from the coordinates used in predictor envelopes (2.5). The predictors are ancillary under this model and thus are treated as non-stochastic. The parameterization $\beta^T = \Gamma\eta_1$ in (2.17) arises because we must have $\text{span}(\beta^T) \subseteq \mathcal{E}_{\Sigma_{Y|X}}(\mathcal{B}')$, so the columns of η_1 give the coordinates of the columns of β^T in terms of basis Γ. The likelihood stemming from (2.17) can be maximized to obtain envelope estimators of β and $\Sigma_{Y|X}$ (Cook et al., 2010; Cook, 2018).

That model (2.17) satisfies conditions (2.16) can be seen as follows. Condition (2.14a) stipulates that the conditional distribution of $\Gamma_0^T Y \mid X$ must not depend on the value of X. This follows because the distribution of $\Gamma_0^T Y \mid X$ is characterized as $\Gamma_0^T Y = \Gamma_0^T \alpha + \Gamma_0^T \varepsilon$, where $\Gamma_0^T \varepsilon \sim N_{r-u}(0, \Omega_0)$. Condition (2.16b) requires that $\Gamma^T Y$ be independent of $\Gamma_0^T Y$ given X. This follows from the normality of the errors and $\Gamma^T \Sigma_{Y|X} \Gamma_0 = 0$.

If the envelope $\mathcal{E}_{\Sigma_{Y|X}}(\mathcal{B}')$ were known then, according to (2.17), we could pick any basis matrix Γ and fit the linear regression of the reduced response $\Gamma^T Y$ on predictors X, giving the estimated coefficient matrix $\widehat{\beta}_{\Gamma^T Y|X}$ from the multivariate regression of $\Gamma^T Y$ on X. The maximum likelihood envelope estimator $\widehat{\beta}_\Gamma$ of β with known basis Γ is then

$$\widehat{\beta}_\Gamma = \widehat{\beta}_{\Gamma^T Y|X}\Gamma^T = \widehat{\beta}_{\text{ols}} P_\Gamma. \tag{2.18}$$

Variances can provide insights into the potential gain offered by the envelope estimator:

$$\begin{aligned}
\mathrm{var}(\mathrm{vec}(\widehat{\beta}_\Gamma)) &= \mathrm{var}(\mathrm{vec}(\widehat{\beta}_{\mathrm{ols}} P_\Gamma)) \\
&= (P_\Gamma \otimes I_p)\mathrm{var}(\mathrm{vec}(\widehat{\beta}_{\mathrm{ols}}))(P_\Gamma \otimes I_p) \\
&= n^{-1}(P_\Gamma \otimes I_p)\left(\Sigma_{Y|X} \otimes S_X^{-1}\right)(P_\Gamma \otimes I_p) \\
&= n^{-1}\Gamma\Omega\Gamma^T \otimes S_X^{-1},
\end{aligned}$$

giving a difference with the ordinary least squares estimator of

$$\mathrm{var}(\mathrm{vec}(\widehat{\beta}_{\mathrm{ols}})) - \mathrm{var}(\mathrm{vec}(\widehat{\beta}_\Gamma)) \geq n^{-1}\Gamma_0\Omega_0\Gamma_0^T \otimes S_X^{-1} \geq 0.$$

From this we conclude that if Γ is known and if $\mathrm{var}(\Gamma_0^T Y) = \Omega_0$ is large relative to $\mathrm{var}(\Gamma^T Y) = \Omega$ then the gains from the envelope model can be substantial. While predicated on Γ being known, this results provides useful intuition in to the more common setting where Γ is unknown. Using the spectral norm $\|\cdot\|$ as a measure of overall size, the envelope model will be advantageous when $\|\Omega\| \ll \|\Omega_0\|$. The variance of the maximum likelihood estimator of β under model (2.17) is given in Proposition 2.2.

Recall from the discussion at the end of Section 2.2 that predictor envelopes can be expected to reduce the estimative and predictive variation substantially when the variability of the immaterial predictors $\Delta_0 = \mathrm{var}(\Phi_0^T X)$ is *small* relative to $\Delta = \mathrm{var}(\Phi^T X)$. But here we have found that the gains from using response envelopes can be substantial when variability of the immaterial responses $\Omega_0 = \mathrm{var}(\Gamma_0^T Y)$ is *large* relative to the variation in the material responses $\Omega = \mathrm{var}(\Gamma^T Y)$.

2.5.2 Relationship to models for longitudinal data

Models of the form (2.17) with known Γ, not necessarily semi-orthogonal, occur in regressions where the elements of $Y_i \in \mathbb{R}^r$ are r longitudinal measurements on the i-th subject. For example, suppose that we are studying the effects of a treatment and it is known that a subjects expected response to the treatment is quadratic over time. Perhaps the quadratic is concave and we are ultimately interested in estimating the point at which the response is maximized. In this case, $\Gamma \in \mathbb{R}^{r \times 3}$ and it can be set as

$$\Gamma = \begin{pmatrix} 1 & t_1 & t_1^2 \\ 1 & t_2 & t_2^2 \\ \vdots & \vdots & \vdots \\ 1 & t_r & t_r^2 \end{pmatrix},$$

where t_j is the time at which the j-th measurement is taken. The model in this case is simply $Y_i = \Gamma\alpha + \varepsilon_i$, where $\alpha = (\alpha_0, \alpha_1, \alpha_2)^T \in \mathbb{R}^3$ contains the coefficients of the quadratic response. For instance, the expected responses at times t_1 and t_r are $\alpha_0 + \alpha_1 t_1 + \alpha_2 t_1^2$ and $\alpha_0 + \alpha_1 t_r + \alpha_2 t_r^2$, respectively.

Expanding the context of the illustration, suppose now that we are comparing two treatments indicated by $X = 0, 1$ and that the effect of a treatment is to alter the coefficients of the quadratic. Let the columns of the matrix

$$A = \begin{pmatrix} \alpha_{00} & \alpha_{01} \\ \alpha_{10} & \alpha_{11} \\ \alpha_{20} & \alpha_{21} \end{pmatrix} = (\alpha_{\bullet 0}, \alpha_{\bullet 1})$$

represent the coefficients of the quadratic response for $X = 0$ and $X = 1$, and let $\delta = \alpha_{\bullet 1} - \alpha_{\bullet 0}$. Then the model allowing for the two treatments can be written

$$\begin{aligned} Y_i &= \Gamma A \begin{pmatrix} 1 - X \\ X \end{pmatrix} + \varepsilon \\ &= \Gamma\alpha_{\bullet 0} + \Gamma\delta X + \varepsilon. \end{aligned} \tag{2.19}$$

This model is now in the form of the response envelope model (2.17), except that Γ is known. Depending on the context, variations of this model may be appropriate. For example, allowing for an unconstrained intercept, the model becomes

$$Y_i = \delta_0 + \Gamma\delta X + \varepsilon.$$

If $t_1 = 0$ corresponds to a basal measurement prior to treatment application then we know that the first element of δ is 0.

Kenward (1987) reported on an experiment to compare two treatments for the control of gut worm in cattle. The treatments were each randomly assigned to 30 cows whose weights were measured at 2, 4, 6, ..., 18 and 19 weeks after treatment. The goal of the experiment was to see if a differential treatment effect could be detected and, if so, estimate the time point when the difference was first manifested. Judging from the plot of average weight by treatment and time (Cook, 2018, Fig. 1.5), it is not clear what functional form could be used to describe the weight profiles. In such cases, treating Γ as unknown and adopting a response envelope model (2.17) may be a viable analysis strategy. Details of such an analysis are available from Cook (2018). PLS and maximum likelihood envelope estimators of the parameters in models of the form given in (2.19) are discussed in Section 6.4.4.

We next turn to maximum likelihood estimation of the parameters in response envelope (2.17).

2.5.3 Likelihood-based estimation of response envelopes

Following Cook, Li, and Chiaromonte (2010), we develop in this section maximum likelihood estimation of the parameters in model (2.17). We assume throughout this section that the dimension u of the envelope $\mathcal{E}_{\Sigma_{Y|X}}(\mathcal{B}')$ is known. The connection with PLS algorithms is discussed in Section 2.5.6.

In our derivation of maximum likelihood estimators for predictor reduction via model (2.5), we assumed that X and Y have a joint distribution that was assumed to be normal for the purpose of deriving the asymptotic distribution of $\widehat{\beta}$. However, the predictors are ancillary under model (2.17) for response reduction and, in consequence, we condition on the predictors, treating them as known constants in our development of the maximum likelihood estimators. In asymptotic considerations, recall from Section 1.2.1 that in this setting we define $\Sigma_X = \lim_{n\to\infty} S_X > 0$.

The log likelihood $L(\alpha, \beta, \Sigma_{Y|X})$ for model (1.1) is

$$L = -(nr/2)\log(2\pi) - (n/2)\log|\Sigma_{Y|X}| - (1/2) \times \sum_{i=1}^{n}(Y_i - \alpha - \beta^T X_i)^T \Sigma_{Y|X}^{-1}(Y_i - \alpha - \beta^T X_i).$$

Substituting the parametric forms from envelope model (2.17), $\beta^T = \Gamma\eta_1$ and $\Sigma_{Y|X} = \Gamma\Omega\Gamma^T + \Gamma_0\Omega_0\Gamma_0^T$, we obtain the log likelihood $L_u(\alpha, \eta_1, \mathcal{E}_{\Sigma_{Y|X}}(\mathcal{B}'), \Omega, \Omega_0)$ for the envelope model with dimension u,

$$\begin{aligned} L_u &= -(nr/2)\log(2\pi) - (n/2)\log|\Gamma\Omega\Gamma^T + \Gamma_0\Omega_0\Gamma_0^T| \\ &\quad -(1/2)\sum_{i=1}^{n}(Y_i - \alpha - \Gamma\eta_1 X_i)^T(\Gamma\Omega\Gamma^T + \Gamma_0\Omega_0\Gamma_0^T)^{-1}(Y_i - \alpha - \Gamma\eta_1 X_i). \end{aligned}$$

In model (2.17), Γ is not identifiable but its span $\mathcal{E}_{\Sigma_{Y|X}}(\mathcal{B}')$ is identifiable. For this reason we write the log likelihood L_u in terms of the envelope, but computationally we must work in terms of a basis, and for this reason optimization is carried out using Γ. The original derivation by Cook et al. (2010) uses projections instead of bases. We next sketch how to find the parameter values that maximize L_u.

From conclusions 3 and 4 of Corollary 1.1, we have

$$\begin{aligned} |\Gamma\Omega\Gamma^T + \Gamma_0\Omega_0\Gamma_0^T| &= |\Omega| \times |\Omega_0| \\ (\Gamma\Omega\Gamma^T + \Gamma_0\Omega_0\Gamma_0^T)^{-1} &= \Gamma\Omega^{-1}\Gamma^T + \Gamma_0\Omega_0^{-1}\Gamma_0^T. \end{aligned}$$

Substituting these into the log likelihood we get

$$\begin{aligned} L_u = &-(nr/2)\log(2\pi) - (n/2)\log|\Omega| - (n/2)\log|\Omega_0| \\ &- (1/2)\sum_{i=1}^n (Y_i - \alpha - \Gamma\eta_1 X_i)^T(\Gamma\Omega^{-1}\Gamma^T + \Gamma_0\Omega_0^{-1}\Gamma_0^T)(Y_i - \alpha - \Gamma\eta_1 X_i). \end{aligned}$$

The maximum likelihood estimator of α is $\widehat{\alpha} = \bar{Y}$ because the predictors in model (1.1) are centered. Substituting this into the likelihood function and then decomposing

$$Y_i - \bar{Y} = P_\Gamma(Y_i - \bar{Y}) + Q_\Gamma(Y_i - \bar{Y})$$

and simplifying, we arrive at the first partially maximized log likelihood,

$$\begin{aligned} L_u^{(1)}(\eta_1, \mathcal{E}_{\Sigma_{Y|X}}(\mathcal{B}'), \Omega, \Omega_0) &= -(nr/2)\log(2\pi) + L_u^{(11)}(\eta_1, \mathcal{E}_{\Sigma_{Y|X}}(\mathcal{B}'), \Omega) \\ &\quad + L_u^{(12)}(\mathcal{E}_{\Sigma_{Y|X}}(\mathcal{B}'), \Omega_0), \end{aligned}$$

where

$$\begin{aligned} L_u^{(11)} &= -(n/2)\log|\Omega| \\ &\quad -(1/2)\sum_{i=1}^n \{\Gamma^T(Y_i - \bar{Y}) - \eta_1 X_i\}^T\Omega^{-1}\{\Gamma^T(Y_i - \bar{Y}) - \eta_1 X_i\} \\ L_u^{(12)} &= -(n/2)\log|\Omega_0| - (1/2)\sum_{i=1}^n (Y_i - \bar{Y})^T\Gamma_0\Omega_0^{-1}\Gamma_0^T(Y_i - \bar{Y}). \end{aligned}$$

Holding Γ fixed, $L_u^{(11)}$ can be seen as the log likelihood for the multivariate regression of $\Gamma^T(Y_i - \bar{Y})$ on X_i, and thus $L_u^{(11)}$ is maximized over η_1 at the value $\eta_1 = (\widehat{\beta}_{\text{ols}}\Gamma)^T$. Substituting this into $L_u^{(11)}$ and simplifying we obtain a partially maximized version of $L_u^{(11)}$

$$L_u^{(21)}(\mathcal{E}_{\Sigma_{Y|X}}(\mathcal{B}'), \Omega) = -(n/2)\log|\Omega| - (1/2)\sum_{i=1}^{n}(\Gamma^T \widehat{R}_i)^T \Omega^{-1}\Gamma^T \widehat{R}_i,$$

where, as defined in Section 1.2, $\widehat{R}_i$ is the i-th residual vector from the fit of the standard model (1.1). From this it follows immediately that, still with Γ fixed, $L_u^{(21)}$ is maximized over Ω at $\Omega = \Gamma^T S_{Y|X}\Gamma$. Consequently, we arrive at the third partially maximized log likelihood $L_u^{(31)}(\mathcal{E}_{\Sigma_{Y|X}}(\mathcal{B}')) = -(n/2)\log|\Gamma^T S_{Y|X}\Gamma| - nu/2$. By similar reasoning, the value of Ω_0 that maximizes $L_u^{(2)}(\mathcal{E}_{\Sigma_{Y|X}}(\mathcal{B}'), \Omega_0)$ is $\Omega_0 = \Gamma_0^T S_Y \Gamma_0$. This leads to the maximization of $L_u^{(2)}(\mathcal{E}_{\Sigma_{Y|X}}(\mathcal{B}')) = -(n/2)\log|\Gamma_0^T S_Y \Gamma_0| - n(r-u)/2$.

Combining the above steps, we arrive at the partially maximized form

$$\begin{aligned} L_u^{(2)}(\mathcal{E}_{\Sigma_{Y|X}}(\mathcal{B}')) = &-(nr/2)\log(2\pi) - nr/2 - (n/2)\log|\Gamma^T S_{Y|X}\Gamma| \\ &- (n/2)\log|\Gamma_0^T S_Y \Gamma_0|. \end{aligned}$$

Next, by Lemma 1.3, $\log|\Gamma_0^T S_Y \Gamma_0| = \log|S_Y| + \log|\Gamma^T S_Y^{-1}\Gamma|$. Consequently, we can express $L_u^{(2)}(\mathcal{E}_{\Sigma_{Y|X}}(\mathcal{B}'))$ as a function of Γ alone:

$$\begin{aligned} L_u^{(2)}(\mathcal{E}_{\Sigma_{Y|X}}(\mathcal{B}')) = &-(nr/2)\log(2\pi) - nr/2 - (n/2)\log|S_Y| \\ &- (n/2)\log|\Gamma^T S_{Y|X}\Gamma| - (n/2)\log|\Gamma^T S_Y^{-1}\Gamma|. \end{aligned} \tag{2.20}$$

Summarizing, the maximum likelihood estimators $\widehat{\mathcal{E}}_{\Sigma_{Y|X}}(\mathcal{B}')$ of $\mathcal{E}_{\Sigma_{Y|X}}(\mathcal{B}')$ and of the remaining parameters are determined as

$$\widehat{\mathcal{E}}_{\Sigma_{Y|X}}(\mathcal{B}') = \text{span}\{\arg\min_G (\log|G^T S_{Y|X} G| + \log|G^T S_Y^{-1} G|)\} \tag{2.21}$$

$$\begin{aligned} \widehat{\eta}_1 &= (\widehat{\beta}_{\text{ols}}\widehat{\Gamma})^T, \\ \widehat{\beta} &= (\widehat{\Gamma}\widehat{\eta})^T = \widehat{\beta}_{\text{ols}} P_{\widehat{\Gamma}}, \\ \widehat{\Omega} &= \widehat{\Gamma}^T S_{Y|X}\widehat{\Gamma}, \\ \widehat{\Omega}_0 &= \widehat{\Gamma}_0^T S_Y \widehat{\Gamma}_0, \\ \widehat{\Sigma}_{Y|X} &= \widehat{\Gamma}\widehat{\Omega}\widehat{\Gamma}^T + \widehat{\Gamma}_0\widehat{\Omega}_0\widehat{\Gamma}_0^T, \end{aligned} \tag{2.22}$$

where $\min_G$ is over all semi-orthogonal matrices $G \in \mathbb{R}^{r\times u}$, $\widehat{\Gamma}$ is any semi-orthogonal basis matrix for $\widehat{\mathcal{E}}_{\Sigma_{Y|X}}(\mathcal{B}')$ and $\widehat{\Gamma}_0$ is any semi-orthogonal basis matrix for the orthogonal complement $\widehat{\mathcal{E}}^{\perp}_{\Sigma_{Y|X}}(\mathcal{B}')$ of $\widehat{\mathcal{E}}_{\Sigma_{Y|X}}(\mathcal{B}')$. The fully maximized log-likelihood for fixed u is then

$$\begin{aligned} \widehat{L}_u = &-(nr/2)\log(2\pi) - nr/2 - (n/2)\log|S_Y| \\ &-(n/2)\log|\widehat{\Gamma}^T S_{Y|X}\widehat{\Gamma}| - (n/2)\log|\widehat{\Gamma}^T S_Y^{-1}\widehat{\Gamma}|. \end{aligned} \tag{2.23}$$

Comparing the objective function in (2.21) to the corresponding objective function (2.10) for predictor envelopes, we see that one can be obtained from the other by interchanging the roles of Y and X. This is consistent with our previous comparison of Definitions 2.2 and 2.3 from which we concluded that response and predictor envelopes are dual constructions when X and Y are jointly distributed, as summarized in Table 2.2. The dual nature of likelihood-based envelope constructions is a theme that will occur elsewhere in this book, without necessarily requiring that X and Y be jointly distributed. See, for example, the discussion of partial predictor and response envelopes following (6.16). The partially maximized log likelihood in (2.23) is covered by Algorithm $\mathbb{L}$ discussed in Section 1.5.5.

The optimization required in (2.21) can be sensitive to starting values. Discussions of starting values and optimization algorithms are available from Cook (2018). The estimators $\widehat{\beta}$ and $\widehat{\Sigma}_{Y|X}$ are invariant to the selection of a basis $\widehat{\Gamma}$ and thus are unique, but the remaining estimators $\widehat{\eta}$, $\widehat{\Omega}$, and $\widehat{\Omega}_0$ are basis dependent and thus not unique. However, the eigenvalues of $\widehat{\Omega}$ and $\widehat{\Omega}_0$ are not basis dependent, which enables us to judge their sizes using a standard norm.

Asymptotic distributions of various quantities under model (2.17) were derived by Cook et al. (2010) and reviewed by Cook (2018, Section 1.6). Here we give only the asymptotic distribution of $\widehat{\beta}$ from Cook et al. (2010) since that will typically be of interest in applications.

Proposition 2.2. *Under the envelope model (2.17) with non-stochastic predictors, normal errors and known $u = \dim\{\mathcal{E}_{\Sigma}(\mathcal{B})\}$, $\sqrt{n}(\widehat{\beta} - \beta)$ is asymptotically normal with mean 0 and variance*

$$\text{avar}\{\sqrt{n}\text{vec}(\widehat{\beta})\} = \Gamma\Omega\Gamma^T \otimes \Sigma_X^{-1} + (\Gamma_0 \otimes \eta^T)U^{\dagger}(\Gamma_0^T \otimes \eta), \tag{2.24}$$

where

$$U = \Omega_0^{-1} \otimes \eta\Sigma_X\eta^T + \Omega_0^{-1} \otimes \Omega + \Omega_0 \otimes \Omega^{-1} - 2I_{r-u} \otimes I_u.$$

These results can be used in practice to construct an asymptotic standard error for $(\widehat{\beta})_{ij}$, $i = 1, \ldots, p$, $j = 1, \ldots, r$, by first substituting estimates for the unknown quantities on the right side of (2.24) to obtain an estimated asymptotic variance $\widehat{\text{avar}}\{\sqrt{n}\text{vec}(\widehat{\beta})\}$. The estimated asymptotic variance $\widehat{\text{avar}}\{\sqrt{n}(\widehat{\beta})_{ij}\}$ is then the corresponding diagonal element of

$\widehat{\text{avar}}\{\sqrt{n}\text{vec}(\widehat{\beta})\}$, and its *asymptotic standard error* is

$$\text{se}\{(\widehat{\beta})_{ij}\} = \frac{[\widehat{\text{avar}}\{\sqrt{n}(\widehat{\beta})_{ij}\}]^{1/2}}{\sqrt{n}}. \tag{2.25}$$

Under mild regularity conditions, $\sqrt{n}(\widehat{\beta}-\beta)$ is asymptotically normal with non-normal errors, but $\text{avar}\{\sqrt{n}\text{vec}(\widehat{\beta})\}$ is considerably more complicated than that given in Proposition 2.2. However, the bootstrap can still be used to estimate its variance. See Cook (2018, Section 1.9) for further discussion.

Parallel to our discussion of the asymptotic variance for predictor envelopes following Proposition 2.1, the first addend on the right side of (2.24) is the asymptotic variance of $\widehat{\beta}_\Gamma$, the maximum likelihood estimator of β when Γ is known:

$$\text{avar}\{\sqrt{n}\text{vec}(\widehat{\beta}_\Gamma)\} = \Gamma\Omega\Gamma^T \otimes \Sigma_X^{-1}.$$

The second addend on the right side of (2.24) can be interpreted as the cost of estimating Γ. See Cook (2018) for further discussion.

We present an illustrative analysis in the next section to help fix the ideas behind response reduction. Many more illustrations on the use of response envelopes are available from Cook (2018).

2.5.4 Gasoline data

We use a subset of the well-known gasoline data (Kalivas, 1997) to illustrate the operating characteristics of response envelopes, particularly how they deal with collinearity in the responses. Due to the relationships presented in Table 2.2(B), this will also provide intuition on how predictor envelopes deal with collinearity in the predictors. The full dataset consists of measurements of the octane number and diffuse reflectance measured as log(1/R) from 900 to 1700 nm in 2 nm intervals from $n = 60$ gasoline samples. This data set is too large to allow for straightforward informative graphics, so we selected two reflectance measures to comprise the continuous bivariate $r = 2$ response vector Y, referring to them generically as wavelengths 1 and 2. We constructed a binary predictor $X = 0$ or 1 by dividing the data into 31 cases with relatively low octane numbers and 29 cases with high octane numbers. This simplified setting allows for an informative low-dimensional graphical representation.

The models we consider are as given in (1.1) and (2.17), and the 60 observations are plotted in Figure 2.1 along with various enhancements to be explained. The nominal goal of the analysis is to understand how low and

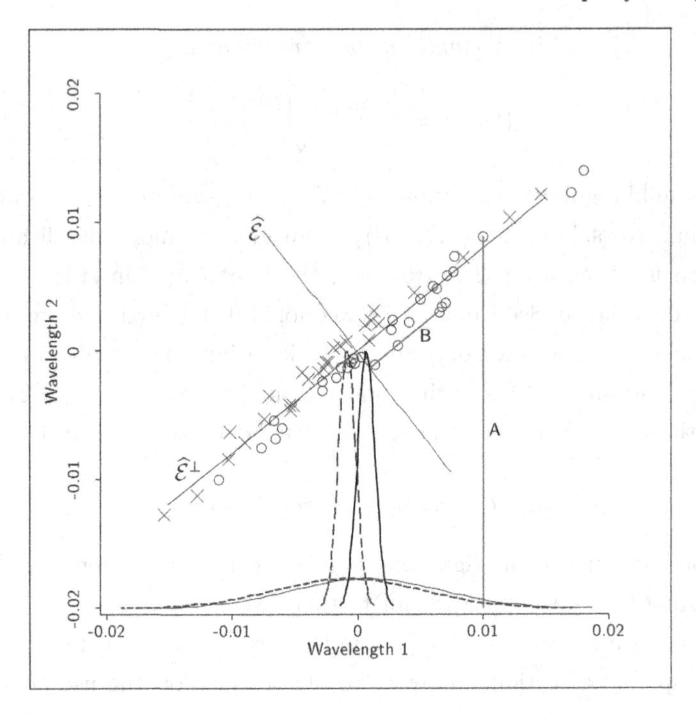

FIGURE 2.1
Illustration of response envelopes using two wavelengths from the gasoline data. High octane data marked by x and low octane numbers by o. $\widehat{\mathcal{E}} = \widehat{\mathcal{E}}_{\Sigma_{Y|X}}(\mathcal{B}')$: estimated envelope. $\widehat{\mathcal{E}}^{\perp} = \widehat{\mathcal{E}}^{\perp}_{\Sigma_{Y|X}}(\mathcal{B}')$: estimated orthogonal complement of the envelope. Marginal distributions of high octane numbers are represented by dished curves along the horizontal axis. Marginal envelope distributions of low octane numbers are represented by solid curves along the horizontal axis. (From the Graphical Abstract for Cook and Forzani (2020) with permission.)

high octane numbers affect the bivariate distribution of the wavelengths by estimating the elements of $\beta = (\beta_1, \beta_2)^T$. Consider estimating β_1, the mean of wavelength 1 for high octane numbers minus the mean of wavelength 1 for low octane numbers. For a likelihood analysis under model (1.1), the data are projected onto the horizontal axis, as represented by path A. This gives rise to the two widely dispersed marginal distributions represented by dashed and solid curves along the horizontal axis in Figure 2.1. These distributions overlap substantially and in consequence it would take a large sample size to infer that $\beta_1 \neq 0$.

An envelope analysis based on model (2.17) results in the inference that the distribution of high and low octane numbers are identical along the orthogonal complement $\mathcal{E}^{\perp}_{\Sigma_{Y|X}}(\mathcal{B}')$ of the true one-dimensional envelope $\mathcal{E}_{\Sigma_Y}(\mathcal{B}')$. The substantial variation along $\widehat{\mathcal{E}}^{\perp}_{\Sigma_{Y|X}}(\mathcal{B}')$ is thus inferred to be immaterial to the analysis and, in consequence, all differences between high and low octane numbers are inferred to lie in the envelope subspace. To estimate β_1, the envelope analysis first projects the data onto the estimated envelope $\widehat{\mathcal{E}}_{\Sigma_{Y|X}}(\mathcal{B}')$ to remove the immaterial variation, as represented by path B in Figure 2.1. The resulting point is then projected onto the horizontal axis for inference on β_1. This produces two narrowly dispersed distributions represented by dashed and solid curves in Figure 2.1. The difference in variation between the narrowly and widely dispersed distributions represents the gain that envelopes can achieve over standard methods of analysis, which in this case is substantial.

In this illustration we have use two wavelengths as the continuous bivariate response Y with a binary predictor X. The response envelope illustrated in Figure 2.1 is $\mathcal{E}_{\Sigma_Y}(\mathcal{B}') = \mathcal{E}_{\Sigma_{Y|X}}(\mathcal{B}')$, which corresponds to the lower left entry in Table 2.2B. This is the predictor envelope in the lower right entry in Table 2.2B. Thus, Figure 2.1 serves to illustrate both response and predictor envelopes. The estimation process illustrated is for response envelopes, however. The estimated envelope aligns closely with the second eigenvector of the sample variance-covariance matrix of the wavelengths. That arises because we are working in only $r = 2$ dimensions. In higher dimensions the envelope does not have to align closely with any single eigenvector but may align with a subset of them.

2.5.5 Prediction

We think of $\Gamma^T Y$ as subject to refined prediction since its distribution depends on X, while $\Gamma_0^T Y$ is only crudely predictable since its distribution does not depends on X. Assuming that the predictors are centered, the actual predictions at a new value X_N of X are

$$\begin{pmatrix} \widehat{\Gamma}^T \widehat{Y} \\ \widehat{\Gamma}_0^T \widehat{Y} \end{pmatrix} = \begin{pmatrix} \widehat{\Gamma}^T \bar{Y} + \widehat{\eta} X_N \\ \widehat{\Gamma}_0^T \bar{Y} \end{pmatrix}.$$

Multiplying by $(\widehat{\Gamma}, \widehat{\Gamma}_0)$ we get simply

$$\widehat{Y}_N = \bar{Y} + \widehat{\Gamma}\widehat{\eta} X_N.$$

2.5.6 PLS connection

The NIPALS and SIMPLS algorithms summarized in Tables 3.1 and 3.4 can be used for response reduction by interchanging the roles of X and Y. They provide $\sqrt{n}$-consistent estimators $\widehat{\Gamma}_{\text{pls}}$ of a basis for $\mathcal{E}_{\Sigma_Y}(\mathcal{B}')$ when $n \to \infty$ with p and q fixed. And they do so subject to (2.14). The estimated basis $\widehat{\Gamma}_{\text{pls}}$ is then used to estimate β by substituting into (2.18) to give

$$\widehat{\beta}_{\text{pls}} = \widehat{\beta}_{\text{ols}} P_{\widehat{\Gamma}_{\text{pls}}}. \tag{2.26}$$

The context for this estimator is regressions with many responses and few predictors so $\widehat{\beta}_{\text{ols}}$ is well defined. Model (2.17) allows maximum likelihood estimation of these same quantities along with estimation of $\Sigma_{Y|X}$. In other words, model (2.17) provides for maximum likelihood estimation of the quantities estimated by the PLS algorithms.

Another envelope-based estimator of β is the maximum likelihood estimator based on model (2.17). That estimator (2.22) is of the same form as (2.26), except the estimated basis derives from maximum likelihood estimation. Adding a subscript 'env' to envelope estimators, we see a parallel between the likelihood and PLS estimators (2.22) and (2.26):

$$\widehat{\beta}_{\text{env}} = \widehat{\beta}_{\text{ols}} P_{\widehat{\Gamma}_{\text{env}}}. \tag{2.27}$$

2.6 Nadler-Coifman model for spectroscopic data

As mentioned in Chapter 1, PLS regression flourishes within the chemometrics community, largely for predicting the concentration of analytes Y from spectral data X (Martens and Næs, 1989). Training samples (X_i, Y_i), $i = 1, \ldots, n$, consist typically of vectors Y_i of analyte concentrations and digitized log spectra X_i of hundreds of perhaps highly collinear predictors. Nadler and Coifman (2005) used Beer's law to guide the formulation of a model for the regression of a spectrum X on concentrations Y. Beer's law states that, in regular settings with a single analyte ($r = 1$) under study, the logarithm of the spectrum is proportional to the analyte concentration Y multiplied by its unique response spectrum. Additivity is typically assumed when several substances are present, enabling a typical measured spectrum to be represented as

$$\begin{aligned} X_{p\times 1} &= \mu + V_{p\times r}Y_{r\times 1} + U_{p\times s}Z_{s\times 1} + \xi \\ &= \mu + A_{p\times(r+s)}W_{(r+s)\times 1} + \xi, \end{aligned} \tag{2.28}$$

where the second line is an abbreviated representation of the model with $A = (V, U)$ and $W^T = (Y^T, Z^T)$. In version (2.28), the constant vector μ may represent a non-stochastic baseline shift or machine drift. The non-stochastic columns V_j, $j = 1, \dots, r$, of V are the unknown characteristic spectra of the r analytes Y of interest. The elements of the stochastic vector Z are centered concentrations of any unknown substances that may be present; their corresponding characteristic spectra are represented by the columns U_j, $j = 1, \dots, s$, of U. The elements of ξ are uncorrelated random errors with means 0 and variances σ^2. They represent noise and other errors introduced by the measurement process. To connect this construction with the response envelope model (2.17), we assume that $p \gg r+s$, as is typical in chemometrics applications. The number of analytes Y of interest, r, is typically small. The number s of unknown substances is also imagined to be small. We assume that $\text{rank}(A) = r + s$.

Let $\Gamma \in \mathbb{R}^{p\times(r+s)}$ be a semi-orthogonal basis matrix for $\text{span}(A)$ and let (Γ, Γ_0) be an orthogonal matrix. Then model (2.28) is an instance of model (2.17) with the roles of the predictors and responses reversed. To see this we assume that all rows of $\Sigma_{W,Y}$ are non-zero to avoid distracting technical diversions and then first consider the regression coefficients from (2.28).

$$\begin{aligned} \Sigma_{X,Y} &= A\Sigma_{W,Y} \\ \beta^T &:= \Sigma_{X,Y}\Sigma_Y^{-1} \\ &= A\Sigma_{W,Y}\Sigma_Y^{-1} = P_\Gamma A\Sigma_{W,Y}\Sigma_Y^{-1} \\ &= \Gamma\eta_1, \end{aligned}$$

where $\eta_1 = \Gamma^T A\Sigma_{W,Y}\Sigma_Y^{-1}$. Turning to the variances,

$$\begin{aligned} \Sigma_X &= A\Sigma_W A^T + \sigma^2 I_p \\ &= P_\Gamma(A\Sigma_W A^T + \sigma^2 I_p)P_\Gamma + Q_\Gamma\sigma^2 \\ \Sigma_{X|Y} &= A\Sigma_{W|Y}A^T + \sigma^2 I_p \\ &= \Gamma\Omega\Gamma^T + \Gamma_0\Omega_0\Gamma_0^T \end{aligned}$$

where $\Omega = \Gamma^T A\Sigma_{W|Y}A^T\Gamma + \sigma^2 I_{r+s}$ and $\Omega_0 = \sigma^2 I_{p-(r+s)}$. It follows immediately from these results that $\mathcal{C}_{X,Y} = \text{span}(A)$ is the smallest reducing subspace of Σ_X that contains $\mathcal{C}_{X,Y}$. Using Table 2.2, we see that the Nadler-Coifman model (2.28), which was developed specifically for spectroscopic application, and its forward regression counterpart both have a natural envelope structure via $\mathcal{E}_{\Sigma_X}(\mathcal{C}_{X,Y})$. Further, both NIPALS and SIMPLS discussed in Chapter 3 are serviceable methods for fitting these models. Methods for the simultaneous compression of the predictor and response vectors, which are applicable in the particular envelope structure for (2.28), are discussed in Chapter 5.

2.7 Variable scaling

The selection of estimators for use in multivariate regression can depend in part on the measurement scales or units involved. The responses and predictors can be represented in different base quantities like length, time, mass, temperature and current, and these quantities could be measured in different rather arbitrary units. Units of length could be inches, feet, meters, miles, kilometers, and perhaps light years. If the regression involves rather arbitrary measurement systems, we may want to insure that our results do not depend materially on the particular systems selected. This can be done by selecting estimators that are invariant or equi-variant under classes of measurement transformations. For example, the OLS estimator $\widehat{\beta}_{\text{ols}}$ for the regression of Y on X is equi-variant under the linear transformation $X \mapsto AX$, where A is full rank; that is, the OLS estimator from the regression of Y on AX is $A^{-1}\widehat{\beta}_{\text{ols}}$, so $\widehat{\beta}_{\text{ols}} \mapsto A^{-1}\widehat{\beta}_{\text{ols}}$. And the fitted values from an OLS fit are invariant under the same class of transformations. The class of rescaling transformations, defined as $X \mapsto \Lambda^{-1}X$ where Λ is a diagonal matrix with positive diagonal elements, could also be relevant in some regressions.

If all responses are measured in the same units or all predictors are measured in the same units and if maintaining the same units is fundamental to the regression, then requiring invariance or equi-variant under a rather arbitrary class of transformations seems like an unnecessary restriction on the regression. Indeed, in such cases we may not wish to restrict attention to invariant or equi-variant estimators and some estimators may not work well in the presence of arbitrarily selected units. For instance, principal components are not equi-variant and Hotelling (1933) cautioned against using arbitrary units:

> The method of principal components can therefore be applied ***only if*** ...a metric – a definition of distance – [is] assumed in the [p]-dimensional space, and not simply a set of axes ... [Emphasis added]

Similarly, the estimators based on the envelope methods introduced in this chapter and the PLS estimators studied in the next chapter are not invariant or equi-variant under rescaling of the response vector or the predictor vector. They too require a purposefully-selected metric. Their applications are less encumbered and they tend to work best when the predictors or responses are measured in the same units, because then the issue of a metric is not pertinent. More specifically, response envelopes and PLS methods for response compression tend to work best when the responses are in the same or similar units. Similarly, predictor envelopes and PLS methods for predictor compression tend to work best when the predictors are measured in the same or similar units. In each of the three examples of Section 1.1, the predictors are all spectral measurements. The responses for the cattle data discussed in Section 2.5.2 are all weights in kilograms taken at various times after treatment. Is it necessary or useful to restrict attention to estimators that are equi-variant under rescaling, say measuring weights at some times in Pounds, Stones, Milligrams and Metric Tons? The responses for the gasoline data discussed in Section 2.5.4 are all spectral measures.

To broaden the applicability of envelopes, Cook and Su (2013, 2016) developed methods to estimate an appropriate scaling of the responses or predictors. Suppose that we rescale Y by multiplication by a non-singular diagonal matrix Λ. Let $Y_\Lambda = \Lambda Y$ denote the new response, let $(\beta_\Lambda, \Sigma_\Lambda)$ and $(\widehat{\beta}_\Lambda, \widehat{\Sigma}_\Lambda)$ denote the corresponding parameters and their envelope estimators based on the envelope model (2.17) for the regression of Y_Λ on X. Then we do not generally have invariance, $\widehat{\beta}_\Lambda = \widehat{\beta}$, $\widehat{\Sigma}_\Lambda = \widehat{\Sigma}$, or equivariance, $\widehat{\beta}_\Lambda = \Lambda\widehat{\beta}$, $\widehat{\Sigma}_\Lambda = \Lambda\widehat{\Sigma}\Lambda$, and the dimension of the envelope subspace may change because of the transformation. This is illustrated schematically in Figure 2.2 for $r = 2$ responses and a binary predictor X. The left panel, which shows a structure similar to that in Figure 2.1, has an envelope that aligns with the second eigenvector of $\Sigma_{Y|X}$. Beginning in the left panel, suppose we rescale Y to Y_Λ. This may change the relationship between the eigenvectors and $\mathcal{B}$, as illustrated by the distributions shown in Figure 2.2b, where $\mathcal{B}_\Lambda = \mathrm{span}(\Lambda\beta)$. Since $\mathcal{B}_\Lambda$ aligns with neither eigenvector of $\Sigma_{Y|X}$, the envelope for the regression of Y_Λ on X

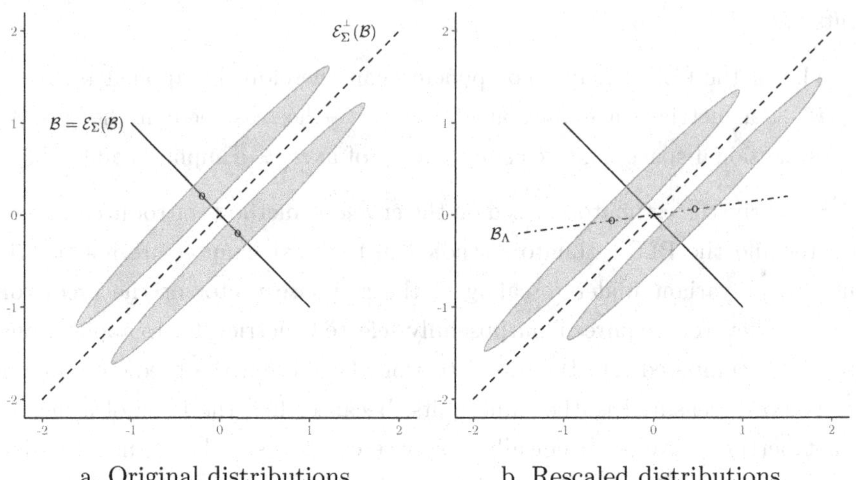

a. Original distributions b. Rescaled distributions

FIGURE 2.2

Illustration of how response rescaling can affect a response envelope from the regression of a bivariate response on a binary predictor, $X = 0, 1$. The two distributions represented in each plot are the distributions of $Y \mid (X = 0)$ and $Y \mid (X = 1)$. The axes represent the coordinates of $Y = (y_1, y_2)^T$.

is two dimensional: $\mathcal{E}_{\Sigma_\Lambda}(\mathcal{B}_\Lambda) = \mathbb{R}^2$. In this case, all linear combinations of Y are material to the regression, the envelope model is the same as the standard model and no efficiency gains are achieved. Cook and Su (2013) developed likelihood-based methodology for estimating the diagonal scaling matrix Λ to take us from Figure 2.2b, where there is no useful envelope structure, to Figure 2.2a where envelopes offer substantial gain. They assumed that the scaled response vector $\Lambda^{-1}Y$ follows the response envelope model (2.17), so the scaled envelope model for the original responses Y becomes

$$Y = \alpha + \Lambda\Gamma\eta X + \varepsilon, \text{ with } \Sigma_{Y|X} = \Lambda\Gamma\Omega\Gamma^T\Lambda + \Lambda\Gamma_0\Omega_0\Gamma_0^T\Lambda. \tag{2.29}$$

The scaled response $\Lambda^{-1}Y$ conforms to an envelope model with u-dimensional envelope $\mathcal{E}_{\Lambda^{-1}\Sigma_{Y|X}\Lambda^{-1}}(\Lambda^{-1}\mathcal{B}')$ and semi-orthogonal basis matrix Γ. From this starting point, they developed likelihood-based methods for estimating Λ and the corresponding envelopes, and showed good results in simulations and data analyses.

Cook and Su (2016) also developed likelihood-based methodology for estimating the diagonal scaling matrix Λ in predictor envelopes. They assumed that the regression of Y on $\Lambda^{-1}X$ follows the q dimensional envelope model

(2.5) and allowed Λ to be structured so groups of predictors can be scaled in the same way:

$$\Lambda = \text{diag}(1, \ldots, 1, \lambda_2, \ldots, \lambda_2, \ldots, \lambda_s \ldots, \lambda_s) \in \mathbb{R}^{p \times p},$$

where there are s distinct positive scaling parameters, $(\lambda_1, \lambda_2, \ldots, \lambda_s)$, and the group structure of Λ is known. The resulting extension of envelope model (2.5) is then

$$\begin{aligned} Y &= \alpha + \eta^T \Phi^T \Lambda^{-1}(X - \mu_X) + \varepsilon \\ \Sigma_X &= \Lambda \Phi \Delta \Phi^T \Lambda + \Lambda \Phi_0 \Delta_0 \Phi_0^T \Lambda, \end{aligned} \quad (2.30)$$

where $\Phi \in \mathbb{R}^{p \times q}$ is a semi-orthogonal basis matrix for the envelope $\mathcal{E}_{\Lambda^{-1}\Sigma_X\Lambda^{-1}}(\Lambda\mathcal{B})$, and the remaining parameter structure is as discussed previously. From this starting point, Cook and Su (2016) developed likelihood-based methods for estimating Λ and the corresponding envelopes, and showed good results in simulations and data analyses. See the original papers and Cook (2018) for additional discussion and details of the methodology.

In Chapter 3, we study PLS algorithms for estimating the response and predictor envelopes in multivariate regressions. As mentioned earlier in this section, these PLS methods are not invariant or equi-variant under rescaling and consequently tend to work best when the predictor or responses are measured in the same units. There is a tradition in some areas of standardizing the variables so they all have the same sample standard deviations. However, there is no theoretical support for this operation and it may make matters worse. For instance, we know that in model (2.29) we can rescale the elements of $Y = (y_j)$ by the diagonal elements of $\Lambda = \text{diag}(\lambda_1, \ldots, \lambda_r)$ – $y_j \mapsto y_i/\lambda_j$, $j = 1, \ldots, r$ – to get an envelope model with response envelope $\mathcal{E}_{\Sigma_Y}(\mathcal{B}') = \text{span}(\Gamma)$. But this scaling is very different from the usual scaling to get unit standard deviations. In the population, the usual scaling operation is equivalent to rescaling by the marginal population standard deviation of y_j:

$$(\text{var}(y_j))^{1/2} = \lambda_j \left[e_j^T \left\{ \Gamma(\eta\eta^T + \Omega)\Gamma^T + \Gamma_0 \Omega_0 \Gamma_0^T \right\} e_j \right]^{1/2},$$

where $e_j \in \mathbb{R}^r$ has a 1 in position j and 0's elsewhere. Clearly, starting with model (2.29) and rescaling by the marginal standard deviations of the responses will not have a result that benefits the analysis. Suppose instead that the original data Y follows envelope model (2.17). Then standardizing by the marginal standard deviations of the responses,

$$(\text{var}(y_j))^{1/2} = \left[e_j^T \left\{ \Gamma(\eta\eta^T + \Omega)\Gamma^T + \Gamma_0 \Omega_0 \Gamma_0^T \right\} e_j \right]^{1/2},$$

may well destroy the envelope structure and result in estimators that are more variable than necessary. In short, the usual scaling to get unit standard deviations is generally questionable.

Aside from the proposal by Cook and Su (2016), there does not seem to be PLS methodology for estimating proper rescaling parameters. Further study along these lines is needed to expand the scope of PLS as envelope methodology has been expanded by Cook and Su (2013, 2016).

3

PLS Algorithms for Predictor Reduction

PLS algorithms were not historically associated with specific models and, in consequence, were not typically viewed as methods for estimating identifiable population parameters, but were seen instead as methods for prediction. We study in this chapter several algorithms for PLS regression, describing corresponding statistical models and their connection with envelopes. For the major algorithms we begin each section with a synopsis that highlights their main points. Each synopsis contains a description of the sample algorithm and the corresponding population version. Subsequent to each synopsis, we study characteristics of the sample algorithms, prove the relationship between the population and sample algorithms and establish the connection with envelopes.

PLS regression algorithms are classified as PLS1 algorithms, which apply to regressions with a univariate response, or PLS2 algorithms, which apply to regressions with multiple responses. The algorithms presented here are all of the PLS2 variety and become PLS1 algorithms when the response is univariate. Their similarities and differences are highlighted later in this chapter. To help maintain a distinction between PLS1 and PLS2 algorithms, we use $\Sigma_{X,Y} = \mathrm{cov}(X, Y)$ when the response Y can be multivariate but we use $\sigma_{X,Y} = \mathrm{cov}(X, Y)$ when there can be only a single real response.

While the PLS algorithms in this chapter are presented and studied for the purpose of reducing the dimension of the predictor vector, with minor modifications they apply also to reducing the dimension of the response vector in the multivariate (multiple response) linear model (1.1). The justification for this arises from the discussion of Section 2.5.1, particularly Table 2.2. PLS for

DOI: 10.1201/9781003482475-3

response reductions is discussed in Section 3.11 after our discussion of PLS for predictor reductions.

3.1 NIPALS algorithm

According to Martens and Næs (1989, p. 118), NIPALS – *nonlinear iterative partial least squares* – arose from Herman Wold's ideas (Wold, 1975b) for iterative fitting of bilinear models in econometrics and social sciences. Such models were often applied to data with several collinear predictors and a number of observations that was insufficient for standard method to work reliably. NIPALS was conceived as a dimension reduction method for reducing the predictors and thereby mitigating the collinearity and sample size issues.

3.1.1 Synopsis

We begin by centering and organizing the data (Y_i, X_i), $i = 1, \ldots, n$, into matrices $\mathbb{X} \in \mathbb{R}^{n\times p}$ and $\mathbb{Y} \in \mathbb{R}^{n\times r}$, as defined in Section 1.2.1. Table 3.1(a) gives steps in the NIPALS algorithm for reducing the predictors. The table serves also to define notation used in this synopsis, like $\mathbb{X}_d$, $\mathbb{Y}_d$ and m_d. It may not be easy to see what the algorithm is doing and the tradition of referring to loadings l_d and m_d, weights w_d and scores s_d does not seem to help with intuition. The weight vectors require calculation of a first eigenvector $\ell_1(\cdot)$ normalized to have length 1. Early versions of the NIPALS algorithm (e.g. Martens and Næs, 1989, Frame 3.6) contained subroutines for calculating ℓ_1 (Stocchero, 2019), which tended to obscure the underlying ideas. We have dropped the eigenvector subroutine in Table 3.1(a). The algorithm is not generally associated with a particular model, but it does involve PLS estimation of $\beta = \Sigma_X^{-1}\Sigma_{X,Y}$.

For simplicity, we use the same notation for weight vectors w_d and weight matrices W_d in a sample and in the population. The context should be clear from the frame of reference under discussion.

As a first step in developing the population version shown in Table 3.1(b), we rewrite selected steps in terms of the sample covariance matrix $S_{X_d,Y_d} = n^{-1}\mathbb{X}_d^T\mathbb{Y}_d \in \mathbb{R}^{p\times r}$ between the deflated predictors and the deflated

TABLE 3.1

NIPALS algorithm: (a) sample version adapted from Martens and Næs (1989) and Stocchero (2019). The $n \times p$ matrix $\mathbb{X}$ contains the centered predictors and the $n \times r$ matrix $\mathbb{Y}$ contains the centered responses; (b) population version derived herein.

(a) *Sample Version*	
Initialize	$\mathbb{X}_1 = \mathbb{X}$, $\mathbb{Y}_1 = \mathbb{Y}$.
	Delete quantities with 0 subscript, e.g. $W_1 = w_1$.
Select	$q \leq \min\{\text{rank}(S_X), n-1\}$
For $d = 1, \ldots q$, compute	
sample covariance matrix	$S_{X_d,Y_d} = n^{-1}\mathbb{X}_d^T\mathbb{Y}_d$
weights	$w_d = \ell_1(S_{X_d,Y_d}S_{X_d,Y_d}^T)$
scores	$s_d = \mathbb{X}_d w_d$
X loadings	$l_d = \mathbb{X}_d^T s_d / s_d^T s_d$
Y loadings	$m_d = \mathbb{Y}_d^T s_d / s_d^T s_d$
X deflation	$\mathbb{X}_{d+1} = \mathbb{X}_d - s_d l_d^T$
Y deflation	$\mathbb{Y}_{d+1} = \mathbb{Y}_d - s_d m_d^T$
Append	w_d, l_d, m_d, s_d,
onto matrices	$W_d = (W_{d-1}, w_d) \in \mathbb{R}^{p\times d}$, $L_d = (L_{d-1}, l_d) \in \mathbb{R}^{p\times d}$, $M_d = (M_{d-1}, m_d) \in \mathbb{R}^{r\times d}$, $S_d = (S_{d-1}, s_d)$
End	
Compute reg. coefficients	$\widehat{\beta}_{\text{npls}} = W_q(L_q^T W_q)^{-1}M_q^T$
(b) *Population Version*	
Initialize	$w_1 = \ell_1(\Sigma_{XY}\Sigma_{XY}^T)$, $W_1 = (w_1)$
For $d = 1, 2 \ldots$	
End if	$Q_{W_d(\Sigma_X)}^T \Sigma_{X,Y} = 0$ and then set $q = d$
Compute weights	$w_{d+1} = \ell_1(Q_{W_d(\Sigma_X)}^T \Sigma_{X,Y}\Sigma_{X,Y}^T Q_{W_d(\Sigma_X)})$
Append	$W_{d+1} = (W_d, w_{d+1})$
Compute reg. coefficients	$\beta_{\text{npls}} = W_q(W_q^T\Sigma_X W_q)^{-1}W_q^T\Sigma_{X,Y}$ $= W_q\beta_{Y\mid W_q^T X}$
(c) *Notes*	
Orthogonal weights	$W_q^T W_q = I_q$.
Projection	$Q_{W_d(\Sigma_X)} = I - W_d(W_d^T\Sigma_X W_d)^{-1}W_d^T\Sigma_X = I - P_{W_d(\Sigma_X)}$.
Envelope connection	$\text{span}(W_q) = \mathcal{E}_{\Sigma_X}(\mathcal{B})$, the Σ_X-envelope of $\mathcal{B} := \text{span}(\beta)$.
Score matrix S_d	These are traditional intermediaries, although they are not needed in the computation of $\widehat{\beta}_{\text{npls}}$.
Algorithm $\mathbb{N}$	This is an instance of Algorithm $\mathbb{N}$ discussed in §§1.5.3 &11.1.
PLS1 v. PLS2	Algorithm is applicable for PLS1 or PLS2 fits; see §3.10.

response and the sample covariance matrix $S_{X_d} = n^{-1}\mathbb{X}_d^T\mathbb{X}_d \in \mathbb{R}^{p\times p}$ of the deflated predictors:

$$\begin{aligned}
w_d &= \ell_1(S_{X_d,Y_d}S_{X_d,Y_d}^T) \in \mathbb{R}^p \\
s_d &= \mathbb{X}_d w_d \in \mathbb{R}^{n\times 1} \\
m_d &= S_{X_d,Y_d}^T w_d (w_d^T S_{X_d} w_d)^{-1} \in \mathbb{R}^{r\times 1} \\
l_d &= S_{X_d} w_d (w_d^T S_{X_d} w_d)^{-1} \in \mathbb{R}^{p\times 1} \\
\mathbb{X}_{d+1} &= \mathbb{X}_d - \mathbb{X}_d w_d (w_d^T S_{X_d} w_d)^{-1} w_d^T S_{X_d} \\
&= \mathbb{X}_d Q_{w_d(S_{X_d})} \qquad (3.1) \\
&= Q_{\mathbb{X}_d w_d}\mathbb{X}_d \in \mathbb{R}^{n\times p} \qquad (3.2) \\
\mathbb{Y}_{d+1} &= \mathbb{Y}_d - \mathbb{X}_d w_d (w_d^T S_{X_d} w_d)^{-1} w_d^T S_{X_d Y_d} \\
&= Q_{\mathbb{X}_d w_d}\mathbb{Y}_d \in \mathbb{R}^{n\times r} \qquad (3.3) \\
S_{X_{d+1}} &= Q_{w_d(S_{X_d})}^T S_{X_d} Q_{w_d(S_{X_d})} \qquad (3.4) \\
S_{X_{d+1},Y_{d+1}} &= Q_{w_d(S_{X_d})}^T S_{X_d,Y_d}, \qquad (3.5)
\end{aligned}$$

where $Q_{w_d(S_{X_d})}$ the operator that projects onto the orthogonal complement of span(w_d) in the S_{X_d} inner product,

$$Q_{w_d(S_{X_d})} = I - w_d(w_d^T S_{X_d} w_d)^{-1} w_d^T S_{X_d} = I - P_{w_d(S_{X_d})},$$

as defined previously at (1.3). From (3.1) we see that the deflation of $\mathbb{X}_d$ consists of the projection of the rows of $\mathbb{X}_d$ onto span$^\perp(w_d)$ in the S_{X_d} inner product, and that $S_{X_{d+1}} = Q_{w_d(S_{X_d})}^T S_{X_d} Q_{w_d(S_{X_d})}$. When S_{X_d} is nonsingular, the deflation $\mathbb{Y}_{d+1}$ can be expressed as a residual matrix

$$\mathbb{Y}_{d+1} = \mathbb{Y}_d - \mathbb{X}_d P_{w_d(S_{X_d})}\widehat{\beta}_d = \mathbb{Y}_d - \mathbb{X}_d w_d \widehat{\beta}_{Y|w_d^T X_d},$$

where $\widehat{\beta}_d = S_{X_d}^{-1} S_{X_d Y_d}$. The deflation steps essentially involve computing residuals sequentially from rank 1 approximations of $\mathbb{X}_d$ and $\mathbb{Y}_d$ to give the residual matrices $\mathbb{X}_{d+1}$ and $\mathbb{Y}_{d+1}$ at the next iteration. They may have computational or data-analytic advantages, but are unnecessary for the computation of $\widehat{\beta}_{\text{npls}}$, as will be shown later.

The population version of the NIPALS algorithm given in Table 3.1(b) is a special case of Algorithm $\mathbb{N}$ described in Proposition 1.11 obtained by setting $A = \Sigma_{X,Y}\Sigma_{X,Y}^T$ and $M = \Sigma_X$. The population stopping criterion $Q_{W_d(\Sigma_X)}^T \Sigma_{X,Y} = 0$ in Table 3.1(b) depends on finding a reducing subspace of Σ_X that contains span$(\Sigma_{X,Y})$, as shown in Section 3.2.1. The population algorithm continues to accumulate weight vectors until it finds a weight matrix W_d

that reduces Σ_X and that spans a subspace containing span$(\Sigma_{X,Y})$. The stopping criterion is then met because, by Proposition 1.3, $Q_{W_d(\Sigma_X)} = Q_{W_d}$ and, since span$(\Sigma_{X,Y}) \subseteq$ span(W_d), $Q_{W_d}\Sigma_{X,Y} = 0$. This represents the essence of the links connecting envelopes with PLS, as discussed in Section 3.2.

The sample NIPALS algorithm of Table 3.1(a) does not contain an explicit mechanism for choosing the stopping point q and so it must be determined by some external criterion. The algorithm is typically run for several values of q and then the stopping point is chosen by predictive cross validation or a holdout sample. Additionally, the sample version does not require S_X to be nonsingular. However, when S_X is nonsingular, the NIPALS estimator can be represented as the projection of the OLS estimator $\widehat{\beta}_{\text{ols}}$ onto span(W_q) in the S_X inner product:

$$\widehat{\beta}_{\text{npls}} = P_{W_q(S_X)}\widehat{\beta}_{\text{ols}}.$$

This requires that S_X be positive definite and so does not have a direct sample counterpart when $n < p$. If in addition, $q = p$ then $P_{W_q(S_X)} = I_p$ and the NIPALS estimator reduces to the OLS estimator, $\widehat{\beta}_{\text{npls}} = \widehat{\beta}_{\text{ols}}$.

The NIPALS estimator $\widehat{\beta}_{\text{npls}}$ shown in Table 3.1(a) depends only on the final weight matrix $W_q = (w_1, \ldots, w_q)$ and the two corresponding loading matrices $L_q = (l_1, \ldots, l_q)$ and $M_q = (m_1, \ldots, m_q)$. The columns of these matrices depend on the data only via S_{X_d} and S_{X_d,Y_d}, $d = 1, \ldots, q$. In consequence, the population version of the algorithm shown in Table 3.1(b) can be deduced by replacing S_{X_d,Y_d} and S_{X_d} with their population counterparts, which we represent as Σ_{X_d,Y_d} and Σ_{X_d}. There is no crisp population counterpart of q associated with the sample algorithm, although later we will see that, as indicated in Table 3.1(c), this is in fact the dimension of $\mathcal{E}_{\Sigma_X}(\mathcal{B})$, the Σ_X-envelope of $\mathcal{B}$. We will show later in Section 3.1.4 that the sample version of the population algorithm in Table 3.1(b) is the NIPALS algorithm.

Starting with the population algorithm in Table 3.1(b), $\widehat{\beta}_{\text{npls}}$ is obtained by replacing Σ_X and $\Sigma_{X,Y}$ by their sample counterparts, S_X and $S_{X,Y}$, and stopping at the selected value of q. Additionally, the population algorithm shows that the intermediate quantities in Table 3.1(a) – $l_d, m_d, s_d, \mathbb{X}_d, \mathbb{Y}_d$, and S_{X_d,Y_d} – are not necessary to get the PLS estimator $\widehat{\beta}_{\text{pls}}$, although they might be useful for computational or diagnostic purposes.

3.1.2 NIPALS example

Before turning to details of the NIPALS algorithm in the next section, we give an illustrative example of the population algorithm to perhaps help fix ideas. With $p = 3$ predictors and $r = 2$ responses, let

$$\Sigma_X = 10\begin{pmatrix} 1 & 4/3 & 0 \\ 4/3 & 4 & 0 \\ 0 & 0 & 5 \end{pmatrix},\ \Sigma_{X,Y} = \begin{pmatrix} 5 & 0 \\ 0 & 4 \\ 0 & 0 \end{pmatrix} \text{ and } \Sigma_Y = 10\begin{pmatrix} 1 & -0.6 \\ -0.6 & 4 \end{pmatrix}.$$

The response covariance matrix Σ_Y is not used here but will be used in subsequent illustrations. The last row of $\Sigma_{X,Y}$ is zero, but this alone does not provide clear information about the contribution of x_3 to the regression. If x_3 is correlated with the other two predictors, it may well be material. However, we see from Σ_X that in this regression x_3 is uncorrelated with x_1 and x_2 and this enables a clear conclusion about the role of x_3. In short, since the last row of $\Sigma_{X,Y}$ is zero and since the third predictor is uncorrelated with the other two, we can conclude immediately that at most two predictors, x_1 and x_2, are needed. One goal of this example is to illustrate how the computations play out to reach that conclusion.

According to the initialization step in Table 3.1(b), the first eigenvector of $\Sigma_{X,Y}\Sigma_{X,Y}^T = \text{diag}(25, 16, 0)$ is $w_1 = (1, 0, 0)^T$. To compute the second weight vector we need

$$Q_{w_1(\Sigma_X)} = \begin{pmatrix} 0 & -4/3 & 0 \\ 0 & 1 & 0 \\ 0 & 0 & 1 \end{pmatrix} \text{ and } Q_{w_1(\Sigma_X)}^T\Sigma_{X,Y} = \begin{pmatrix} 0 & 0 \\ -20/3 & 4 \\ 0 & 0 \end{pmatrix}.$$

Since $Q_{w_1(\Sigma_X)}^T\Sigma_{X,Y}$ has rank 1, the second weight vector is $w_2 = (0, 1, 0)^T$. For the third weight vector we need $Q_{W_2(\Sigma_X)}^T\Sigma_{X,Y}$, where $W_2 = (w_1, w_2)$. It can be seen by direct calculation that $Q_{W_2(\Sigma_X)}^T\Sigma_{X,Y} = 0$ and so the algorithm terminates, giving $q = 2$. And, as described in Table 3.1, $W_2^TW_2 = I_2$.

We can reach the conclusion that $q = 2$ also by reasoning that $\text{span}(W_2)$ is a reducing subspace of Σ_X:

$$\Sigma_X = W_2\Delta W_2^T + w_3\Delta_0 w_3^T,$$

where $w_3 = (0, 0, 1)^T$, $\Delta_0 = 5$ and

$$\Delta = \begin{pmatrix} 1 & 4/3 \\ 4/3 & 4 \end{pmatrix}.$$

It follows from Proposition 1.3 that $Q_{W_2(\Sigma_X)} = Q_{W_2}$, which can be seen also by direct calculation:

$$\begin{aligned} Q_{W_2(\Sigma_X)} &= I_2 - W_2(W_2^T \Sigma_X W_2)^{-1} W_2^T \Sigma_X \\ &= I_2 - W_2 \Delta^{-1} \Delta W_2^T = Q_{W_2}. \end{aligned}$$

The conclusion that we have reached the stopping point now follows since $\text{span}(\Sigma_{X,Y}) = \text{span}(W_2)$.

In short, only two reduced predictors $W_2^T X = (x_1, x_2)^T$ are needed to describe fully the regression of Y on X.

3.1.3 Sample structure of the NIPALS algorithm

We see from (3.2) and (3.3) that $\mathbb{X}_{d+1}$ and $\mathbb{Y}_{d+1}$ are constructed by projecting the columns of $\mathbb{X}_d$ and $\mathbb{Y}_d$ onto the orthogonal complement of $\text{span}(\mathbb{X}_d w_d)$. Repeating this operation, we see that the columns of $\mathbb{X}_{d+1}$ and $\mathbb{Y}_{d+1}$ are constructed from a sequence of projections:

$$\left.\begin{aligned} \mathbb{X}_{d+1} = Q_{\mathbb{X}_d w_d} Q_{\mathbb{X}_{d-1} w_{d-1}} \cdots Q_{\mathbb{X}_2 w_2} Q_{\mathbb{X}_1 w_1} \mathbb{X}_1 \\ \mathbb{Y}_{d+1} = Q_{\mathbb{X}_d w_d} Q_{\mathbb{X}_{d-1} w_{d-1}} \cdots Q_{\mathbb{X}_2 w_2} Q_{\mathbb{X}_1 w_1} \mathbb{Y}_1 \end{aligned}\right\} \tag{3.6}$$

To better understand these projections it is necessary to characterize the relationships between the score vectors $s_j = \mathbb{X}_j w_j$ and between the weights w_j. Justifications for the following two lemmas are given in Appendices A.3.1 and A.3.2

Lemma 3.1. *Following the notation from Table 3.1(a), for the sample version of the NIPALS algorithm*

$$W_d^T W_d = I_d, d = 1, \ldots, q.$$

Lemma 3.2. *For $q > 1$, $d = 1, \ldots, q-1$ and $j = 1, \ldots, d$,*

$$\mathbb{X}_{d+1}^T s_j = \mathbb{X}_{d+1}^T \mathbb{X}_j w_j = 0.$$

Consequently,

$$\begin{aligned} P_{s_{d+1}} s_j &= P_{\mathbb{X}_{d+1} w_{d+1}} \mathbb{X}_j w_j = 0 \\ Q_{s_{d+1}} Q_{s_j} &= I - P_{s_{d+1}} - P_{s_j}, \end{aligned}$$

again for $q > 1$, $d = 1, \ldots, q-1$ and $j = 1, \ldots, d$.

It follows from this lemma that the score vectors s_d are mutually orthogonal, which allows a more informative version of the deflations (3.6). Recall from Table 3.1 that $S_d = (s_1, \ldots, s_d)$ and from (1.3) that

$$Q_{S_d} = I - P_{S_d} = I - S_d(S_d^T S_d)^{-1} S_d^T.$$

Then, as proven in Appendix A.3.3,

Lemma 3.3. *For* $d = 1, \ldots, q-1$,

$$\begin{aligned}
\mathbb{X}_{d+1} &= (I - P_{s_d} - P_{s_{d-1}} - \ldots - P_{s_2} - P_{s_1})\mathbb{X}_1 \\
&= (I - P_{S_d})\mathbb{X}_1 = Q_{S_d}\mathbb{X}_1 \\
\mathbb{Y}_{d+1} &= (I - P_{s_d} - P_{s_{d-1}} - \ldots - P_{s_2} - P_{s_1})\mathbb{Y}_1 \\
&= (I - P_{S_d})\mathbb{Y}_1 = Q_{S_d}\mathbb{Y}_1 \\
S_{X_d, Y_d} &= S_{X_d, Y_1} \\
l_d &= \mathbb{X}_1^T s_d / \|s_d\|^2 \\
m_d &= \mathbb{Y}_1^T s_d / \|s_d\|^2 \\
w_{d+1} &= \ell_1(\mathbb{X}_1^T Q_{S_d} \mathbb{Y}_1 \mathbb{Y}_1^T Q_{S_d} \mathbb{X}_1) \\
S_d &= \mathbb{X}_1 W_d \\
\widehat{\mathbb{Y}}_{\text{npls}} &= \mathbb{X}_1 \widehat{\beta}_{\text{npls}} = P_{S_q}\mathbb{Y}_1,
\end{aligned}$$

where $\widehat{\mathbb{Y}}_{\text{npls}}$ *denotes the* $n \times r$ *matrix of fitted responses from the NIPALS fit and* S_d *is the score matrix as defined in Table 3.1.*

With the NIPALS fitted values $\widehat{\mathbb{Y}}_{\text{npls}}$ given in Lemma 3.3, we define the NIPALS residuals as $\widehat{\mathbb{R}}_{\text{npls}} = \mathbb{Y}_1 - \widehat{\mathbb{Y}}_{\text{npls}}$. These quantities allow the construction of many standard diagnostic and summary quantities, like residual plots and the multiple correlation coefficient. Standard formal inference procedures are problematic, however, because the scores are stochastic.

The form of w_{d+1} given in Lemma 3.3 shows that the deflation of $\mathbb{Y}_1$ is unnecessary for computing the weight matrix:

$$\begin{aligned}
w_{d+1} &= \ell_1(\mathbb{X}_1^T Q_{S_d} \mathbb{Y}_1 \mathbb{Y}_1^T Q_{S_d} \mathbb{X}_1) \\
&= \ell_1(\mathbb{X}_{d+1}^T \mathbb{Y}_1 \mathbb{Y}_1^T \mathbb{X}_{d+1}).
\end{aligned}$$

Consequently, the Y-deflation steps in the NIPALS algorithm of Table 3.1 is unnecessary for computing the PLS estimator of the regressions coefficients, which gives rise to the bare bones version of version of the NIPALS algorithm

TABLE 3.2

Bare bones version of the NIPALS algorithm given in Table 3.1(a). The notation $\mathbb{Y}_1 = \mathbb{Y}$ of Table 3.1(a) is not used since here there is no iteration over Y.

Initialize	$\mathbb{X}_1 = \mathbb{X}$. Quantities subscripted with 0 are to be deleted
Select	$q \leq \min\{\text{rank}(S_X), n-1\}$
For $d = 1, \ldots, q$, compute	
sample covariance matrix	$S_{X_d,Y} = n^{-1}\mathbb{X}_d^T\mathbb{Y}$
X weights	$w_d = \ell_1(S_{X_d,Y}S_{X_d,Y}^T)$
X scores	$s_d = \mathbb{X}_d w_d$
X loadings	$l_d = \mathbb{X}_d^T s_d / s_d^T s_d$
X deflation	$\mathbb{X}_{d+1} = \mathbb{X}_d - s_d l_d^T = Q_{s_d}\mathbb{X}_d = \mathbb{X}_d Q_{w_d(S_{X_d})}$
Append	$W_d = (W_{d-1}, w_d) \in \mathbb{R}^{p \times d}$
End	Set $W = W_q$
Compute regression coefficients	$\widehat{\beta}_{\text{npls}} = W(W^T S_X W)^{-1} W^T S_{X,Y}$.

shown in Table 3.2. The estimator $\widehat{\beta}_{\text{npls}}$ given in Table 3.2 is the sample counterpart of the population version given in Table 3.1(b). It is shown in (3.12), following the justification given in Section 3.1.4 for the population version.

Lemma 3.4 indicates also how the NIPALS algorithms deals operationally with rank-deficient regressions in which $\text{rank}(S_X) < p$. Let the columns of the $p \times p_1$ matrix V be the eigenvectors of S_X with non-zero eigenvalues, $p_1 \leq p$. Then we can express $\mathbb{X}_1^T = V\mathbb{Z}_1^T$, where $\mathbb{Z}_1^T$ is an $p_1 \times n$ matrix that contains the coordinates of $\mathbb{X}_1^T$ in terms of the eigenvectors of S_X. Let w_d^* and s_d^* denote the weights and scores that result from applying NIPALS to data $(\mathbb{Z}_1, \mathbb{Y}_1)$. Then

Lemma 3.4. *For $d = 1, \ldots, q$, $s_d = s_d^*$ and $w_d = Vw_d^*$.*

We see from Table 3.1 and Lemma 3.3 that $\widehat{\beta}_{\text{npls}}$ depends only on $\mathbb{X}_1$, $\mathbb{Y}_1$, weights W_q and the scores S_q. As a consequence of Lemma 3.4 we can apply the NIPALS algorithm by first reducing the data to the principal component scores $\mathbb{Z}_1$, running the algorithm on the reduced data $(\mathbb{Z}_1, \mathbb{Y}_1)$ and then transforming back to the original scale. The eigenvectors V of S_X account for 100% of the

TABLE 3.3

Bare bones version in the principal component scale of the NIPALS algorithm given in Table 3.1(a).

Initialize	$p_1 = \text{rank}(S_X) \leq p$, $V = p \times p_1$ matrix of eigenvectors of S_X with non-zero eigenvalues, $\mathbb{X}_1 = \mathbb{X}$, $\mathbb{Z}_1 = \mathbb{X}_1 V$. Delete quantities with 0 subscript.
Select	$q \leq \min\{p_1, n-1\}$
For $d = 1, \ldots, q$, compute	
sample covariance matrices	$S_{Z_d} = n^{-1}\mathbb{Z}_d^T\mathbb{Z}_d$, $S_{Z_d,Y} = n^{-1}\mathbb{Z}_d^T\mathbb{Y}$
X weights	$w_d^* = \ell_1(S_{Z_d,Y}S_{Z_d,Y}^T)$
X scores	$s_d^* = \mathbb{Z}_d w_d^*$
X loadings	$l_d = \mathbb{Z}_d^T s_d^* / s_d^{*T} s_d^*$
X deflation	$\mathbb{Z}_{d+1} = \mathbb{Z}_d - s_d^* l_d^T = Q_{s_d^*}\mathbb{Z}_d = \mathbb{Z}_d Q_{w_d^*(S_{Z_d})}$
Append	$W_d^* = (W_{d-1}^*, w_d^*) \in \mathbb{R}^{p \times d}$
End	Set $W = W_q^*$
Compute regression coefficients	$\widehat{\beta}_{\text{npls}} = W(W^T S_X W)^{-1} W^T S_{X,Y} = V\{W^*(W^{*T} S_{\mathbb{Z}} W^*)^{-1} W^{*T} S_{\mathbb{Z},Y}\}$
Notes	V might be restricted to eigenvectors that account for a high, say 98 percent, of the variation in X

variation in X. Since the eigenvectors of S_X corresponding to its smallest eigenvalues may be relatively unstable, it may be worthwhile to use only the principal components that account for a high percentage, say 98%, of the variation in X. Cross validation could be used to compare fits with various percentages retained. This discussion is summarized in Table 3.3. A proof of Lemma 3.4 is given in Appendix A.3.4.

3.1.4 Population details for the NIPALS algorithm

In this section we show that the population algorithm in Table 3.1(b) corresponds to the sample version given in Table 3.1(a). Justifications in this and subsequent sections make use of the following lemma, whose proof is in Appendix A.3.5. The lemma is useful in characterizing the NIPALS algorithm because, as represented in (3.4) and (3.5), it uses projections in various inner products, depending on the step in the iteration.

Lemma 3.5. *Let V denote a $p \times c$ matrix of rank $c < p$, let v denote a $p \times 1$ vector that is not contained in* span(V), *let Σ denote a $p \times p$ positive definite matrix and let $\Delta = Q^T_{V(\Sigma)} \Sigma Q_{V(\Sigma)}$. Then*

(a) $\Delta = Q^T_{V(\Sigma)} \Sigma = \Sigma Q_{V(\Sigma)}$

(b) $P_{(V,v)(\Sigma)} = P_{V(\Sigma)} + P_{Q_{V(\Sigma)} v(\Sigma)}$

(c) $Q_{V(\Sigma)} Q_{v(\Delta)} = Q_{(V,v)(\Sigma)}$

(d) $Q^T_{v(\Delta)} \Delta Q_{v(\Delta)} = Q^T_{(V,v)(\Sigma)} \Sigma Q_{(V,v)(\Sigma)}$.

This lemma describes various relationships between projections in the Σ inner product. They may be familiar in the usual inner product $\Sigma = I$, but are generally less familiar in the Σ inner product. For instance, (b) describes the projection onto span(V, v) in the Σ inner product as the sum of the projection onto span(V) in the Σ inner product plus the projection in the Σ inner product of the part of v that is orthogonal to span(V) in the Σ inner product. If $\Sigma = I$, (b) reduces to the relatively well-known result $P_{(V,v)} = P_V + P_{Q_V v}$. Similarly, when $\Sigma = I$ the left hand side of (d) reduces to $I - P_V - P_{Q_V v}$, while the right hand side becomes $Q_{(V,v)}$.

We take it as clear that the following population algorithm follows directly from (3.4) and (3.5):

Direct Population NIPALS Algorithm *Construct* $w_1 = \ell_1(\Sigma_{XY} \Sigma^T_{XY})$. *Then for* $d = 1, 2 \ldots,$

$$\left.\begin{aligned} \Sigma_{X_{d+1}} &= Q^T_{w_d(\Sigma_{X_d})} \Sigma_{X_d} Q_{w_d(\Sigma_{X_d})} \\ \Sigma_{X_{d+1}, Y_{d+1}} &= Q^T_{w_d(\Sigma_{X_d})} \Sigma_{X_d, Y_d} \\ w_{d+1} &= \ell_1(\Sigma_{X_{d+1}, Y_{d+1}} \Sigma^T_{X_{d+1}, Y_{d+1}}), \end{aligned}\right\} \tag{3.7}$$

continuing until $\Sigma_{X_{d+1}, Y_{d+1}} = 0$, *at which point we use* $q = d$ *components.*

To demonstrate that the population algorithm in Table 3.1(b) corresponds to the sample version given in Table 3.1(a), it is sufficient to show that the population algorithm in Table 3.1(b) is equivalent to the direct population algorithm (3.7).

Proposition 3.1. *The population algorithm given in Table 3.1(b) is equivalent to the direct population algorithm. That is, they produce the same semi-orthogonal weight matrix and the same population coefficients.*

Proof. The main part of the proof is by induction. Clearly, the algorithms produce the same w_1. At $d+1=2$, we have

$$\begin{aligned}
\Sigma_{X_2} &= Q^T_{w_1(\Sigma_X)}\Sigma_X Q_{w_1(\Sigma_X)} \\
\Sigma_{X_2,Y_2} &= Q^T_{w_1(\Sigma_X)}\Sigma_{X,Y} \\
w_2 &= \ell_1(\Sigma_{X_2,Y_2}\Sigma^T_{X_2,Y_2}) \\
&= \ell_1(Q^T_{w_1(\Sigma_X)}\Sigma_{X,Y}\Sigma^T_{X,Y}Q_{w_1(\Sigma_X)}).
\end{aligned}$$

This step matches the corresponding step of the population NIPALS algorithm in Table 3.1(b).

At $d+1=3$, we again see that the direct NIPALS algorithm matches that from Table 3.1(b).

$$\begin{aligned}
\Sigma_{X_3} &= Q^T_{w_2(\Sigma_{X_2})}\Sigma_{X_2} Q_{w_2(\Sigma_{X_2})} \\
&= Q^T_{(w_1,w_2)(\Sigma_X)}\Sigma_X Q_{(w_1,w_2)(\Sigma_X)} \\
\Sigma_{X_3,Y_3} &= Q^T_{w_2(\Sigma_{X_2})}\Sigma_{X_2,Y_2} \\
&= Q^T_{(w_1,w_2)(\Sigma_X)}\Sigma_{X,Y} \\
w_3 &= \ell_1(\Sigma_{X_3,Y_3}\Sigma^T_{X_3,Y_3}) \\
&= \ell_1(Q^T_{(w_1,w_2)(\Sigma_X)}\Sigma_{X,Y}\Sigma_{X,Y}Q_{(w_1,w_2)(\Sigma_X)}).
\end{aligned}$$

The second equality for Σ_{X_3} follows by Lemma 3.5(d) with $V=(w_1)$, $v=w_2$ and $\Sigma=\Sigma_X$. The second equality for Σ_{X_3,Y_3} follows by replacing $\Sigma_{X_2,Y_2} = Q^T_{w_1(\Sigma_X)}\Sigma_{X,Y}$ to get

$$\Sigma_{X_3,Y_3} = Q^T_{w_2(\Sigma_{X_2})}Q^T_{w_1(\Sigma_X)}\Sigma_{X,Y}$$

and then using Lemma 3.5(c). The second equality for w_3 follows by direct substitution.

Assume that these relationships hold for $d=k-1\le q-1$ and let $W_{k-1}=(w_1,\ldots,w_{k-1})$:

$$\begin{aligned}
\Sigma_{X_k} &= Q^T_{w_{k-1}(\Sigma_{X_{k-1}})}\Sigma_{X_{k-1}} Q_{w_{k-1}(\Sigma_{X_{k-1}})} \\
&= Q^T_{W_{k-1}(\Sigma_X)}\Sigma_X Q_{W_{k-1}(\Sigma_X)} \qquad (3.8) \\
\Sigma_{X_k,Y_k} &= Q^T_{w_{k-1}(\Sigma_{X_{k-1}})}\Sigma_{X_{k-1},Y_{k-1}} \\
&= Q^T_{W_{k-1}(\Sigma_X)}\Sigma_{X,Y} \qquad (3.9) \\
w_k &= \Sigma_{X_k,Y_k}\Sigma^T_{X_k,Y_k} \\
&= \ell_1(Q^T_{W_{k-1}(\Sigma_X)}\Sigma_{X,Y}\Sigma_{X,Y}Q_{W_{k-1}(\Sigma_X)}).
\end{aligned}$$

To show the general result by induction we need to show that the relationships hold for $d = k$. Let $W_k = (W_{k-1}, w_k)$. Then we need to show

$$\begin{aligned}
\Sigma_{X_{k+1}} &= Q^T_{w_k(\Sigma_{X_k})}\Sigma_{X_k}Q_{w_k(\Sigma_{X_k})} \\
&= Q^T_{W_k(\Sigma_X)}\Sigma_X Q_{W_k(\Sigma_X)} \\
\Sigma_{X_{k+1},Y_{k+1}} &= Q^T_{w_k(\Sigma_{X_k})}\Sigma_{X_k,Y_k} \\
&= Q^T_{W_k(\Sigma_X)}\Sigma_{X,Y} \\
w_{k+1} &= \ell_1(\Sigma_{X_{k+1},Y_{k+1}}\Sigma^T_{X_{k+1},Y_{k+1}}) \\
&= \ell_1(Q^T_{W_k(\Sigma_X)}\Sigma_{X,Y}\Sigma^T_{X,Y}Q_{W_k(\Sigma_X)}).
\end{aligned}$$

The first equality for $\Sigma_{X_{k+1}}$ follows by definition. The second equality for $\Sigma_{X_{k+1}}$ follows by first substituting for

$$\Sigma_{X_k} = Q^T_{W_{k-1}(\Sigma_X)}\Sigma_X Q_{W_{k-1}(\Sigma_X)}$$

from the induction hypothesis to get

$$\Sigma_{X_{k+1}} = Q^T_{w_k(\Sigma_{X_k})}Q^T_{W_{k-1}(\Sigma_X)}\Sigma_X Q_{W_{k-1}(\Sigma_X)}Q_{w_k(\Sigma_{X_k})}.$$

The desired conclusion for $\Sigma_{X_{k+1}}$ then follows from Lemma 3.5(c) with $V = W_{k-1}$, $v = w_k$ and $\Sigma = \Sigma_X$. The first equality for $\Sigma_{X_{k+1},Y_{k+1}}$ follows by definition. The second equality for $\Sigma_{X_{k+1},Y_{k+1}}$ follows by replacing $\Sigma_{X_k,Y_k} = Q^T_{W_{k-1}(\Sigma_X)}\Sigma_{X,Y}$ to get

$$\Sigma_{X_{k+1},Y_{k+1}} = Q^T_{w_k(\Sigma_{X_k})}Q^T_{W_{k-1}(\Sigma_X)}\Sigma_{X,Y},$$

and then using Lemma 3.5(c).

It remains to show that the population version of the estimator $\widehat{\beta}_{\text{pls}} = W_q(L_q^T W_q)^{-1}M_q^T$ given in Table 3.1(a) corresponds to β_{npls} in Table 3.1(b), assuming that the stopping points are the same. This task requires that we deal with the population loading matrices $L_q = (l_1, \ldots, l_q)$ and $M_q = (m_1, \ldots, m_q)$. Let $W_0 = 0$. From Table 3.1,

$$l_k = \frac{\mathbb{X}_k^T s_k}{s_k^T s_k} = \frac{\mathbb{X}_k^T\mathbb{X}_k w_k}{w_k^T\mathbb{X}_k^T\mathbb{X}_k w_k} = \frac{S_{X_k}w_k}{w_k S_{X_k} w_k}.$$

For the population version we replace S_{X_k} with Σ_{X_k}, implicitly substitute the population version of w_k and then use (3.8) to get for $k = 0, \ldots, q-1$,

$$\begin{aligned}
l_{k+1} &= \frac{Q^T_{W_k(\Sigma_X)}\Sigma_X Q_{W_k(\Sigma_X)}w_{k+1}}{w^T_{k+1}Q^T_{W_k(\Sigma_X)}\Sigma_X Q_{W_k(\Sigma_X)}w_{k+1}} = \Sigma_X\frac{Q_{W_k(\Sigma_X)}w_{k+1}}{w^T_{k+1}Q^T_{W_k(\Sigma_X)}\Sigma_X Q_{W_k(\Sigma_X)}w_{k+1}} \\
m_{k+1} &= \Sigma^T_{X,Y}\frac{Q_{W_k(\Sigma_X)}w_{k+1}}{w^T_{k+1}Q^T_{W_k(\Sigma_X)}\Sigma_X Q_{W_k(\Sigma_X)}w_{k+1}},
\end{aligned}$$

where the second equality for l_{k+1} follows from Lemma 3.5(a) and the form of m_{k+1} is found following the same general steps as we did for l_{k+1}.

The role of the numerators $Q_{W_k(\Sigma_X)}w_{k+1}$ is to provide a successive orthogonalization of columns of $W_q = (w_1, \ldots, w_q)$, while the denominators provide a normalization in the Σ_X inner product. Thus,

$$\text{span}\left\{ \frac{Q_{W_k(\Sigma_X)}w_{k+1}}{w_{k+1}^T Q_{W_k(\Sigma_X)}^T \Sigma_X Q_{W_k(\Sigma_X)} w_{k+1}} \mid k = 0, 1, \ldots, q-1 \right\} = \text{span}(W_q), \tag{3.10}$$

and in consequence there is a nonsingular $q \times q$ matrix A so that $L_q = \Sigma_X W_q A$. In the same way we have

$$M_q = \Sigma_{X,Y}^T W_q A.$$

Substituting these forms into the population version of the PLS estimator given in Table 3.1(a), we get

$$\begin{aligned} \beta_{\text{pls}} &= W_q(L_q^T W_q)^{-1} M_q^T \\ &= W_q(A^T W_q^T \Sigma_X W_q)^{-1} A^T W_q^T \Sigma_{X,Y} \\ &= W_q(W_q^T \Sigma_X W_q)^{-1} W_q^T \Sigma_{X,Y} \\ &= W_q \beta_{Y|W_q^T X}, \end{aligned} \tag{3.11}$$

which is β_{npls} given in Table 3.1(b). The semi-orthogonality of the weight matrix W_q can be seen as follows. From Table 3.1(b),

$$w_{d+1} = \ell_1(Q_{W_d(\Sigma_X)}^T \Sigma_{X,Y} \Sigma_{X,Y}^T Q_{W_d(\Sigma_X)}).$$

Thus, there is a vector $v \in \mathbb{R}^r$ so that $w_{d+1} = Q_{W_d(\Sigma_X)}^T \Sigma_{X,Y} v$. But $W_d^T Q_{W_d(\Sigma_X)}^T = 0$, and so $W_d^T w_{d+1} = 0$ for all $d = 1, \ldots, q$. □

Recall that the notation $\beta_{Y|W_q^T X}$ means the coefficients from the population OLS fit on Y on the reduced predictors $W_q^T X$. From this we see that the normalization of l_d and m_d in Table 3.1 plays no essential role as it has no effect on the subspace equality in (3.10). The sample version $\widehat{\beta}_{\text{pls}}$ of β_{pls}, as shown in Table 3.2, follows immediately from (3.11):

$$\widehat{\beta}_{\text{pls}} = W(W^T S_X W)^{-1} W^T S_{X,Y}, \tag{3.12}$$

where W is the sample version of W_q as defined in Table 3.2.

3.2 Envelopes and NIPALS

In this section we concentrate on population characteristics to establish first connections between NIPALS and envelopes. The population stopping condition

$$Q^T_{W_q(\Sigma_X)}\Sigma_{X,Y} = 0 \tag{3.13}$$

will play a key role in our discussion. Some results in this section can be deduced from properties of Algorithm $\mathbb{N}$ in Section 1.5.3. We provide separate demonstrations in this section to aid intuition and establish connections with NIPALS.

3.2.1 The NIPALS subspace reduces Σ_X and contains $\mathcal{B}$

The goal of this section is to show that the NIPALS subspace $\mathrm{span}(W_q)$ reduces Σ_X and contains $\mathcal{B}$ without introducing an envelope construction per se. Assume first that the NIPALS algorithm stops after getting w_1, so then we have $\Sigma_{X,Y} = P^T_{w_1(\Sigma_X)}\Sigma_{X,Y}$, which follows directly from the algorithms stopping condition that $Q^T_{w_1(\Sigma_X)}\Sigma_{X,Y} = 0$. This tells us that $\mathrm{span}(\Sigma_{X,Y}) = \mathrm{span}(\Sigma_X w_1)$, so there is an $r \times 1$ vector c that satisfies $\Sigma_{X,Y} = \Sigma_X w_1 c^T$. In consequence, $\beta = \Sigma_X^{-1}\Sigma_{X,Y} = w_1 c^T$, $\mathrm{rank}(\beta) = 1$ and $\mathcal{B} = \mathrm{span}(w_1)$. Further, since w_1 is an eigenvector of $\Sigma_{X,Y}\Sigma^T_{X,Y}$ by construction, we have for some $r \times 1$ vector b, $\Sigma_{X,Y} = w_1 b^T$ and thus $w_1 b^T = \Sigma_X w_1 c^T$, so w_1 must be also an eigenvector of Σ_X. Summarizing, if the population NIPALS algorithm stops at $q = 1$ then

(a) $\beta = w_1 c^T$,
(b) $\mathcal{B} = \mathrm{span}(w_1)$,
(c) $\dim\{\mathcal{B}\} = 1$,
(d) w_1 is an eigenvector of both $\Sigma_{X,Y}\Sigma^T_{X,Y}$ and Σ_X.

The stopping criterion for a general q implies immediately that

$$\Sigma_{X,Y} = \Sigma_X W_q(W_q^T\Sigma_X W_q)^{-1}W_q^T\Sigma_{X,Y}. \tag{3.14}$$

Therefore,

$$\begin{aligned} \beta &= \Sigma_X^{-1}\Sigma_{X,Y} = P_{W_q(\Sigma_X)}\beta \\ \mathcal{B} &\subseteq \mathrm{span}(W_q) \\ \dim(\mathcal{B}) &\leq q. \end{aligned} \tag{3.15}$$

These are the general q counterparts for conclusions (a)–(c) with $q = 1$. To demonstrate an extension of conclusion (d) consider

$$\begin{aligned} Q^T_{W_d(\Sigma_X)}\Sigma_{X,Y} &= (I - \Sigma_X W_d (W_d^T \Sigma_X W_d)^{-1} W_d^T)\Sigma_{X,Y} \\ &= \Sigma_{X,Y} - \Sigma_X W_d (W_d^T \Sigma_X W_d)^{-1} W_d^T \Sigma_{X,Y}. \end{aligned}$$

Next, we substitute (3.14) for the first $\Sigma_{X,Y}$ on the right hand side and write the first $W_d = W_q(I_d, 0)^T$ to get

$$\begin{aligned} Q^T_{W_d(\Sigma_X)}\Sigma_{X,Y} &= \Sigma_X W_q (W_q^T \Sigma_X W_q)^{-1} W_q^T \Sigma_{X,Y} \\ &\quad - \Sigma_X W_q (I_d, 0)^T (W_d^T \Sigma_X W_d)^{-1} W_d^T \Sigma_{X,Y} \\ &= \Sigma_X W_q \left\{ (W_q^T \Sigma_X W_q)^{-1} W_q^T - (I_d, 0)^T (W_d^T \Sigma W_d)^{-1} W_d^T \right\} \Sigma_{X,Y} \\ &:= \Sigma_X W_q A_d, \end{aligned}$$

where the $q \times r$ matrix A_d is defined implicitly. In consequence, we have for $d = 0, 1, \ldots, q-1$

$$w_{d+1} = \ell_1 \left\{ \Sigma_X W_q A_d A_d^T W_q^T \Sigma_X \right\}.$$

Since $w_{d+1} \in \text{span}(\Sigma_X W_q)$, this implies that w_{d+1} can be represented as $w_{d+1} = \Sigma_X W_q c_d$ for some $q \times 1$ vector c_d and thus that there is a $q \times q$ matrix $C = (c_1, \ldots c_q)$ so that $W_q = \Sigma_X W_q C$. Since W_q has full column rank and Σ_X is nonsingular, C must be nonsingular and so

$$\Sigma_X W_q = W_q C^{-1}. \tag{3.16}$$

This result tells us that $\text{span}(W_q)$ must be a reducing subspace of Σ_X, as described in Definition 1.1.

To see that $\text{span}(W_q)$ contains $\text{span}(\Sigma_{X,Y})$ we substitute the right hand side of (3.16) for the first product $\Sigma_X W_q$ on the right hand side of (3.14) to get

$$\Sigma_{X,Y} = W_q C^{-1} (W_q^T \Sigma_X W_q)^{-1} W_q^T \Sigma_{X,Y}.$$

from which it follows that $\text{span}(\Sigma_{X,Y}) \subseteq \text{span}(W_q)$.

Altogether then,

Proposition 3.2. $\text{span}(W_q)$ *is a reducing subspace of* Σ_X *that contains* $\text{span}(\Sigma_{X,Y})$ *and* $\mathcal{B}$.

This result characterizes $\text{span}(W_q)$ as a reducing subspace of Σ_X that contains $\mathcal{B}$. But to establish that $\text{span}(W_q)$ is an envelope, we need to demonstrate also that it is minimal.

3.2.2 The NIPALS weights span the Σ_X-envelope of $\mathcal{B}$

Taken together, results (3.15) and (3.16) characterize $\text{span}(W_q)$ as a reducing subspace of Σ_X that contains $\mathcal{B}$ or, equivalently, contains $\text{span}(\Sigma_{X,Y})$, as mentioned in the discussion following Proposition 1.6. Those results were obtained without making use of envelopes per se. We now make use of that connection to establish further properties of the NIPALS subspaces $\text{span}(W_d)$, $d = 1, \ldots, q$, including that the NIPALS subspace $\text{span}(W_q)$ is the Σ_X-envelope of $\mathcal{B}$; that is, the *smallest* reducing subspace of Σ_X that contains $\text{span}(\Sigma_{X,Y})$.

Let $\Phi \in \mathbb{R}^{p\times q}$ denote a semi-orthogonal basis matrix for the intersection of all reducing subspace of Σ_X that contain $\text{span}(\Sigma_{X,Y})$ and let $(\Phi, \Phi_0) \in \mathbb{R}^{p\times p}$ be an orthogonal matrix. Then we know from Proposition 1.2 that Σ_X can be expressed as

$$\Sigma_X = \Phi\Delta\Phi^T + \Phi_0\Delta_0\Phi_0^T.$$

We proceed by induction to show that the NIPALS weights w_j, $j = 1, \ldots, q$, are all the envelope and that the algorithm terminates with q weights.

By construction $\Sigma_{X,Y} = P_\Phi\Sigma_{X,Y}$ and consequently $w_1 = \ell_1(P_\Phi\Sigma_{X,Y}\Sigma_{X,Y}^T P_\Phi)$ is in the envelope, $w_1 \in \text{span}(\Phi)$. Represent $w_1 = \Phi b_1$. Then

$$\begin{aligned} Q_{w_1(\Sigma_X)}^T\Sigma_{X,Y} &= \Sigma_{X,Y} - P_{w_1(\Sigma_X)}^T\Sigma_{X,Y} \\ &= P_\Phi\Sigma_{X,Y} - \Phi P_{b_1(\Delta)}^T\Phi^T\Sigma_{X,Y} \\ &= \Phi Q_{b_1(\Delta)}^T\Phi^T\Sigma_{X,Y}. \end{aligned}$$

Consequently,

$$\begin{aligned} w_2 &= \ell_1\left\{Q_{w_1(\Sigma_X)}^T\Sigma_{X,Y}\Sigma_{X,Y}^T Q_{w_1(\Sigma_X)}\right\} \\ &= \ell_1\left\{\Phi Q_{b_1(\Delta)}^T\Phi^T\Sigma_{X,Y}\Sigma_{X,Y}^T\Phi Q_{b_1(\Delta)}\Phi^T\right\} \\ &\in \text{span}(\Phi Q_{b_1(\Delta)}^T) \subset \text{span}(\Phi) \end{aligned}$$

and so $\text{span}(W_2) = \text{span}(w_1, w_2) \subseteq \text{span}(\Phi)$. Additionally, since $w_1 = \Phi b_1$ it follows that $w_1^T\Phi Q_{b_1(\Delta)}^T = 0$ and thus $w_1^T w_2 = 0$, as claimed in Table 3.1(c) and shown in Lemma 3.1.

For the induction hypothesis we assume that $W_{d-1} = \text{span}(w_1, \ldots, w_{d-1}) \subseteq \text{span}(\Phi)$, where the w_j's are orthogonal vectors. Then using the previous argument by replacing w_1 with W_{d-1}, it follows that $w_d \in \text{span}(\Phi)$, and that $\text{span}(W_{d-1})$ is a proper subspace of $\text{span}(W_d)$; that is, $\text{span}(W_{d-1}) \subset \text{span}(W_d) \subseteq \text{span}(\Phi)$. This demonstrates that the dimension of the NIPALS subspaces $\text{span}(W_d)$ continue to increase until $d = q$, at which point

$\text{span}(W_q) = \text{span}(\Phi)$ because $\text{span}(W_q) \subseteq \text{span}(\Phi)$ and $q = \dim(\text{span}(\Phi)) = \dim(\text{span}(W_q))$. The growth of the NIPALS subspaces stops at this point because then the stopping criterion (3.13) is met.

Summarizing,

Proposition 3.3. *The population NIPALS algorithm produces a series of properly nested subspaces that settle on the envelope after q steps,*

$$\text{span}(W_1) \subset \text{span}(W_2) \subset \cdots \subset \text{span}(W_q) = \mathcal{E}_{\Sigma_X}(\mathcal{B}),$$

where $q = \dim[\mathcal{E}_{\Sigma_X}\{\mathcal{B}\}]$.

3.3 SIMPLS algorithm

According to de Jong (1993),

> ...the name SIMPLS for [his] PLS algorithm, since it is a straightforward **impl**ememtation of a **s**tatistically **i**nspired **m**odification of the **PLS** method according to the **simple** concept given in Table 1.

SIMPLS then followed in the footsteps of NIPALS as a relatively "simple" method for reducing the predictors to alleviate collinearity and sample size issues.

3.3.1 Synopsis

Table 3.4(a) gives the SIMPLS data-based algorithm as it appears in de Jong (1993, Table 1). The weight vectors and weight matrices are denoted as v and V to distinguish them from the NIPALS weights. The scores and score matrices are denoted as s and S, and the loadings and loading matrices as l and L. de Jong (1993, Appendix) also discussed a more elaborate version of the algorithm that contains steps to facilitate computation. Like the NIPALS algorithm, the SIMPLS sample algorithm does not contain a mechanism for stopping and so q is again typically determined by predictive cross validation or a holdout sample. Also like the NIPALS algorithm, the sample SIMPLS algorithm does not require S_X to be nonsingular. However, when S_X is nonsingular, the SIMPLS estimator can be represented as the projection of the

TABLE 3.4

SIMPLS algorithm: (a) sample version adapted from de Jong (1993, Table 1). The $n \times p$ matrix $\mathbb{X}$ contains the centered predictors and the $n \times r$ vector $\mathbb{Y}$ contains the centered responses; (b) population version derived herein.

(a) *Sample Version*	
Initialize	$S_{X,Y} = n^{-1}\mathbb{X}^T\mathbb{Y}$, $q \le \min\{\text{rank}(S_X), n-1\}$, v_1 = first left singular vector of $S_{X,Y}$. $V_1 = (v_1)$, $s_1 = \mathbb{X}v_1$, $S_1 = (s_1)$, $l_d = \mathbb{X}^T s_d / s_d^T s_d$, $L_1 = (l_1)$.
For $d = 2, \ldots, q$	
Compute weights	v_d = first left singular vector of $Q_{L_{d-1}} S_{X,Y}$
Compute scores	$s_d = \mathbb{X}v_d$
Compute loadings	$l_d = \mathbb{X}^T s_d / s_d^T s_d$
Append	v_d, s_d, l_d onto V_{d-1}, S_{d-1}, L_{d-1}
End	
Compute reg. coefficients	$\widehat{\beta}_{\text{spls}} = V_q(V_q^T S_X V_q)^{-1} V_q^T S_{X,Y}$
(b) *Population Version*	
Initialize	$v_1 = \ell_1(\Sigma_{XY}\Sigma_{XY}^T)$, $V_1 = (v_1)$
For $d = 1, 2 \ldots$	
End if	$Q_{\Sigma_X V_d}\Sigma_{X,Y} = 0$ and then set $q = d$
Compute weights	$v_{d+1} = \ell_1(Q_{\Sigma_X V_d}\Sigma_{X,Y}\Sigma_{X,Y}^T Q_{\Sigma_X V_d})$
Append	$V_{d+1} = (V_d, v_{d+1})$
Compute reg. coefficients	$\beta_{\text{spls}} = V_q(V_q^T\Sigma_X V_q)^{-1}V_q\Sigma_{X,Y} = V_q\beta_{Y\mid V_q^T X}$
(c) *Notes*	
Weights	$V_q^T\Sigma_X V_q$ is a diagonal matrix.
Envelope connection	$\text{span}(V_q) = \mathcal{E}_{\Sigma_X}(\mathcal{B})$, the Σ_X-envelope of $\mathcal{B} = \text{span}(\beta)$.
Projections	$Q_{\Sigma_X V_d}$ and $Q_{L_{d-1}}$ use the usual inner product. See (1.3).
Scores & Loadings	S_d and L_d are traditional computational intermediaries, but are not needed per se for $\widehat{\beta}_{\text{spls}}$.
Algorithm $\mathbb{S}$	This is an instance of Algorithm $\mathbb{S}$ discussed in §§1.5.4 & 11.1.
PLS1 v. PLS2	Algorithm is applicable for PLS1 or PLS2 fits; See §3.10.

OLS estimator $\widehat{\beta}_{\text{ols}}$ onto span(V_q) in the S_X inner product:

$$\widehat{\beta}_{\text{spls}} = P_{V_q(S_X)}\widehat{\beta}_{\text{ols}},$$

which requires that S_X be positive definite and so does not have a direct sample counterpart when $n < p$. Recall that projections are defined at (1.3), so

$$P_{V_q(S_X)} = V_q(V_q^T S_X V_q)^{-1} V_q^T S_X$$

and $Q_{V_q(S_X)} = I - P_{V_q(S_X)}$.

Maintaining the convention established for NIPALS, we use v_d and V_d to denote weight vectors and weight matrices in both the sample and population.

The population version of the SIMPLS algorithm, which will be justified herein, is shown in Table 3.4(b). Substituting S_X and $S_{X,Y}$ for their population counterparts Σ_X and $\Sigma_{X,Y}$ produces the same weights and the same estimated coefficient vector as the sample version in Table 3.4(a), provided the same value of q is used. The score and loading vectors computed in Table 3.4(a) are not really necessary for the algorithm.

The population version in Table 3.4(b) is a special case of Algorithm $\mathbb{S}$ described previously in Section 1.5.4 with $A = \Sigma_{X,Y}\Sigma_{X,Y}^T$ and $M = \Sigma_X$. Consequently, reasoning from (3.17), the SIMPLS algorithm can be described also as follows, still using the notation of Table 3.4(b). Let $V_i = (v_1, \ldots, v_i)$. Then given V_k, $k < q$, V_{k+1} is constructed by concatenating V_k with

$$\begin{aligned} v_{k+1} &= \max_{v \in \mathbb{R}^p} v^T \Sigma_{X,Y}\Sigma_{X,Y}^T v \text{ subject to} \\ & \quad v^T \Sigma_X V_k = 0 \text{ and } v^T v = 0. \end{aligned} \tag{3.17}$$

This is the description of SIMPLS that (Cook, Helland, and Su, 2013, Section 4.3) used to establish the connection between PLS algorithms and envelopes. The population construction algorithm (3.17) is shown in Section 3.3.3, equations (3.18) and (3.19).

3.3.2 SIMPLS example

In this section we revisit the example of Section 3.1.2 using SIMPLS instead of NIPALS. Recall that $p = 3$, $r = 2$,

$$\Sigma_X = 10\begin{pmatrix} 1 & 4/3 & 0 \\ 4/3 & 4 & 0 \\ 0 & 0 & 5 \end{pmatrix}, \quad \Sigma_{X,Y} = \begin{pmatrix} 5 & 0 \\ 0 & 4 \\ 0 & 0 \end{pmatrix},$$

and that the first weight vector is the same as that for NIPALS, $v_1 = (1,0,0)^T$. As in Section 3.1.2, $v_1 \in \text{span}(\Sigma_{X,Y})$ and $\text{span}(\Sigma_{X,Y})$ reduces Σ_X. Direct computation of the second weight vector

$$v_2 = \ell_1(Q_{\Sigma_X v_1}\Sigma_{X,Y}\Sigma_{X,Y}^T Q_{\Sigma_X v_1})$$

is not as straightforward as it was for NIPALS since we are not now working in the Σ_X inner product. However, computation is facilitated by using a change of basis for $\text{span}(\Sigma_{X,Y})$ that explicitly incorporates $\Sigma_X v_1 = (1, 4/3, 0)^T$. Let

$$A = \begin{pmatrix} 5^{-1} & -5^{-1} \\ 3^{-1} & 3/16 \end{pmatrix}.$$

Then,

$$\Sigma_{X,Y}A = \begin{pmatrix} 1 & -1 \\ 4/3 & 3/4 \\ 0 & 0 \end{pmatrix},$$

$$P_{\Sigma_X v_1}\Sigma_{X,Y} = (P_{\Sigma_X v_1}\Sigma_{X,Y}A)A^{-1} = \begin{pmatrix} 1 & 0 \\ 4/3 & 0 \\ 0 & 0 \end{pmatrix} A^{-1},$$

and

$$Q_{\Sigma_X v_1}\Sigma_{X,Y} = (\Sigma_{X,Y}A - P_{\Sigma_X v_1}\Sigma_{X,Y}A)A^{-1} = \begin{pmatrix} 0 & -1 \\ 0 & 3/4 \\ 0 & 0 \end{pmatrix} A^{-1},$$

which has rank 1. It follows that $v_2 = a/\|a\|$ and $V_2 = (v_1, v_2)$, where $a = (-1, 3/4, 0)^T$.

The weight vectors for NIPALS are orthogonal, $w_1^T w_2 = 0$, but here the SIMPLS weight vectors are not orthogonal $v_1^T v_2 \neq 0$. However, they are orthogonal in the Σ_X inner product,

$$V_2^T \Sigma_X V_2 = \begin{pmatrix} 1 & 0 \\ 0 & 9/4^3 - 1 \end{pmatrix}.$$

For the next iteration we need to find $Q_{\Sigma_X V_2}\Sigma_{X,Y}$. But V_2 is a reducing subspace of Σ_X and so $Q_{\Sigma_X V_2} = Q_{V_2}$. Then the algorithm terminates: $Q_{\Sigma_X V_2}\Sigma_{X,Y} = Q_{V_2}\Sigma_{X,Y} = 0$ since $\text{span}(\Sigma_{X,Y}) = \text{span}(V_2)$. Although the NIPALS weights W_2 are not equal to the SIMPLS weights V_2, they span the same subspace and in consequence $\beta_{\text{npls}} = \beta_{\text{spls}}$.

3.3.3 Details underlying the SIMPLS algorithm

In this section we take a closer look at the sample SIMPLS algorithm by stepping through a few values of d, and in doing so we establish justification for its population version shown in Table 3.4(b).

On the first pass ($d = 1$) through the sample SIMPLS algorithm we compute the first left singular vector of $S_{X,Y}$. This is equivalent to computing

$$v_1 = \ell_1(S_{X,Y}S_{X,Y}^T).$$

The first score vector is then computed as $s_1 = \mathbb{X}v_1$, which gives the first linear combination of the predictors. The first loading vector is

$$l_1 = \frac{\mathbb{X}^T s_1}{s_1^T s_1} = \frac{\mathbb{X}^T \mathbb{X} v_1}{v_1^T \mathbb{X}^T \mathbb{X} v_1} = \frac{S_X v_1}{v_1^T S_X v_1}.$$

We summarize the computations for $d = 1$ as follows.

$$\begin{aligned}
v_1 &= \ell_1(S_{X,Y}S_{X,Y}^T) \\
s_1 &= \mathbb{X}v_1 \\
l_1 &= \frac{S_X v_1}{v_1^T S_X v_1} \\
V_1 &= (v_1);\ S_1 = (s_1);\ L_1 = (l_1).
\end{aligned}$$

In the second pass through the algorithm, $d = 2$, we first compute the first left singular vector of $Q_{L_1}S_{X,Y}$. From the computations from the first step $d = 1$, $\text{span}(L_1) = \text{span}(S_X v_1)$. In other words, the normalization by $v_1^T S_X v_1$ in the computation of l_1 is unnecessary. Thus we can compute the first left singular vector of $Q_{S_X v_1}S_{X,Y}$. Following the logic expressed in step $d = 1$, we then have

$$v_2 = \ell_1(Q_{S_X v_1}S_{X,Y}S_{X,Y}^T Q_{S_X v_1}).$$

The rest of the steps in the second pass through the algorithm are similar to those in the first pass, so we summarize the second pass as

$$\begin{aligned}
v_2 &= \ell_1(Q_{S_X v_1}S_{X,Y}S_{X,Y}^T Q_{S_X v_1}) \\
s_2 &= \mathbb{X}v_2 \\
l_2 &= \frac{S_X v_2}{v_2^T S_X v_2} \\
V_2 &= (v_1, v_2);\ S_2 = (s_1, s_2) = \mathbb{X}V_2; \\
L_2 &= (l_1, l_2) = S_X V_2 \,\text{diag}^{-1}(v_1^T S_X v_1, v_2^T S_X v_2).
\end{aligned}$$

For completeness, we now state the results for the a general d-th pass through the algorithm:

$$\begin{aligned}
v_d &= \ell_1(Q_{S_X V_{d-1}} S_{X,Y} S_{X,Y}^T Q_{S_X V_{d-1}}) \\
&= \ell_1(Q_{L_{d-1}} S_{X,Y} S_{X,Y}^T Q_{L_{d-1}}) \\
s_d &= \mathbb{X} v_d \\
l_d &= \frac{S_X v_d}{v_d^T S_X v_d} \\
V_d &= (v_1, v_2 \ldots v_d);\ S_d = (s_1, s_2 \ldots s_d) = \mathbb{X} V_d; \\
L_d &= (l_1, l_2 \ldots l_d) = S_X V_d \,\mathrm{diag}^{-1}(v_1^T S_X v_1, v_2^T S_X v_2 \ldots v_d^T S_X v_d),
\end{aligned}$$

where $Q_{S_X V_{d-1}} = Q_{L_{d-1}}$. From this general step, it can be seen that the final weight matrix V_q, S_X, and $S_{X,Y}$ are all that is needed to compute the SIMPLS estimator $\widehat{\beta}_{\mathrm{spls}}$ of the coefficient matrix after q steps:

$$\widehat{\beta}_{\mathrm{spls}} = V_q (V_q^T S_X V_q)^{-1} V_q^T S_{X,Y} = V_q \widehat{\beta}_{Y|V_q^T X}.$$

The score matrix $S_q = \mathbb{X} V_q$ gives the $n \times q$ matrix of reduced predictor values. It may be of interest for interpretation and graphical studies. The loading matrix L_q is essentially a matrix of normalized scores.

In view of the previous observations, we can reduce the SIMPLS algorithm to the following compact version. Construct $v_1 = \ell_1(S_{XY} S_{XY}^T)$ and $V_1 = (v_1)$. Then for $d = 1, \ldots, q-1$

$$\begin{aligned}
v_{d+1} &= \ell_1(Q_{L_d} S_{X,Y} S_{X,Y}^T Q_{L_d}) && (3.18) \\
&= \underset{h^T h = 1, h^T S_X V_d = 0}{\arg\max}\ h^T S_{X,Y} S_{X,Y}^T h && (3.19) \\
V_{d+1} &= (V_d, v_{d+1}).
\end{aligned}$$

Replacing $S_{X,Y}$ and S_X with their population counterparts $\Sigma_{X,Y}$ and Σ_X and imposing the stopping criterion $Q_{\Sigma_X V_{d+1}} \Sigma_{X,Y} = 0$ gives the population SIMPLS algorithm shown in Table 3.4(b).

To see the equivalence of the two equalities, (3.18) and (3.19), for v_{d+1}, we first show that (3.18) implies (3.19). Form (3.18) can be written as

$$v_{d+1} = \underset{h \neq 0}{\arg\max} \frac{v^T Q_{L_d} S_{X,Y} S_{X,Y}^T Q_{L_d} h}{h^T h}.$$

Decompose $h = Q_{L_d}h + P_{L_d}h := h_1 + h_2$ to get

$$\begin{aligned} v_{d+1} &= \arg\max_{h_1 \in \text{span}^{\perp}(L_d),\, h_2 \in \text{span}(L_d)} \frac{h_1^T S_{X,Y} S_{X,Y}^T h_1}{h_1^T h_1 + h_2^T h_2} \\ &= \arg\max_{h_1 \in \text{span}^{\perp}(L_d),\, h_2 = 0} \frac{h_1^T S_{X,Y} S_{X,Y}^T h_1}{h_1^T h_1 + h_2^T h_2} \\ &= \arg\max_{h_1 \in \text{span}^{\perp}(L_d)} \frac{h_1^T S_{X,Y} S_{X,Y}^T h_1}{h_1^T h_1}. \end{aligned}$$

Then $v_{d+1} \in \text{span}^{\perp}(L_d)$, which holds if and only if $v_{d+1}^T S_X V_d = 0$ and so (3.19) holds.

To see that (3.19) implies (3.18), we must have $v_{d+1} \in \text{span}^{\perp}(L_d)$ from (3.19). Consequently, we can replace h in (3.19) with $Q_{L_d}h$ and (3.19) becomes equivalently

$$v_{d+1} = \arg\max_{h^T h = 1} h^T Q_{L_d} S_{X,Y} S_{X,Y}^T Q_{L_d} h = \ell_1(Q_{L_d} S_{X,Y} S_{X,Y}^T Q_{L_d}),$$

which gives (3.18).

As mentioned in Table 3.4, the weights V_q are orthogonal in the Σ_X inner product, $V_q^T \Sigma_X V_q$ is a diagonal matrix. This can be seen directly from (3.19). It follows also because $v_{d+1} \in \text{span}(Q_{\Sigma_X V_d} \Sigma_{X,Y})$ which implies $V_d^T \Sigma_X v_{d+1} = 0$.

3.4 Envelopes and SIMPLS

Cook et al. (2013) and Cook (2018) showed that the weights V_q from the population version of the SIMPLS algorithm span a reducing subspace of Σ_X that contains $\mathcal{B}$. In this section we sketch a different justification for the same result. Our justification here relies on the SIMPLS stopping criterion

$$Q_{\Sigma_X V_q} \Sigma_{X,Y} = 0. \tag{3.20}$$

Although key details differ, the argument here follows closely that for NIPALS given in Section 3.2. Also, some results in this section may be deduced from properties of Algorithm $\mathbb{S}$ in Section 1.5.4. We provide separate demonstrations in this section to aid intuition and establish connections with SIMPLS.

3.4.1 The SIMPLS subspace reduces Σ_X and contains $\mathcal{B}$

The goal of this section is to use the SIMPLS algorithm itself to deduce that the SIMPLS subspace $\text{span}(V_q)$ reduces Σ_X and contains $\mathcal{B}$. This then will establish a relationship between SIMPLS and reducing subspaces of Σ_X. To establish a connection with envelopes per se, we need to demonstrate that $\text{span}(V_q)$ is minimal, which is done in the next section.

Assume that $\beta \neq 0$ and that the SIMPLS algorithm stops after getting v_1, so then we have $q = 1$ and stopping condition $Q_{\Sigma_X v_1}\Sigma_{X,Y} = 0$, which is equivalent to $\Sigma_{X,Y} = P_{\Sigma_X v_1}\Sigma_{X,Y}$. This tells us that $\text{span}(\Sigma_{X,Y}) = \text{span}(\Sigma_X v_1)$, so there is an $r \times 1$ vector c that satisfies $\Sigma_{X,Y} = \Sigma_X v_1 c^T$. In consequence, $\beta = \Sigma_X^{-1}\Sigma_{X,Y} = v_1 c^T$, $\text{rank}(\beta) = 1$ and $\mathcal{B} = \text{span}(v_1)$. Further, since v_1 is an eigenvector of $\Sigma_{X,Y}\Sigma_{X,Y}^T$, we have for some $r \times 1$ vector b, $\Sigma_{X,Y} = v_1 b^T$ and thus $v_1 b^T = \Sigma_X v_1 c^T$, so v_1 must be also an eigenvector of Σ_X. Summarizing, if the population SIMPLS algorithm stops at $q = 1$ then

(a) $\beta = v_1 c^T$,

(b) $\mathcal{B} = \text{span}(v_1)$,

(c) $\dim\{\mathcal{B}\} = 1$,

(d) v_1 is an eigenvector of both $\Sigma_{X,Y}\Sigma_{X,Y}^T$ and Σ_X.

To study the behavior of the population SIMPLS algorithm at a general stopping point q, we have for $d = 1, \ldots, q-1$ the population weight vectors constructed as

$$v_{d+1} = \ell_1\left\{Q_{\Sigma_X V_d}\Sigma_{X,Y}\Sigma_{X,Y}^T Q_{\Sigma_X V_d}\right\}. \tag{3.21}$$

The stopping criterion (3.20) implies that

$$\Sigma_{X,Y} = \Sigma_X V_q (V_q^T \Sigma_X^2 V_q)^{-1} V_q^T \Sigma_X \Sigma_{X,Y}, \tag{3.22}$$

and therefore

$$\mathcal{B} \subseteq \text{span}(V_q). \tag{3.23}$$

It remains to demonstrate that $\text{span}(V_q)$ is a reducing subspace of Σ_X:

$$\begin{aligned} Q_{\Sigma_X V_d}\Sigma_{X,Y} &= (I - \Sigma_X V_d (V_d^T \Sigma_X^2 V_d)^{-1} V_d^T \Sigma_X)\Sigma_{X,Y} \\ &= \Sigma_{X,Y} - \Sigma_X V_d (V_d^T \Sigma_X^2 V_d)^{-1} V_d^T \Sigma_X \Sigma_{X,Y}. \end{aligned}$$

Next, we substitute (3.22) for the first $\Sigma_{X,Y}$ on the right hand side and write the $V_d = V_q(I_d, 0)^T$ to get

$$\begin{aligned} Q_{\Sigma_X V_d}\Sigma_{X,Y} &= \Sigma_X V_q (V_q^T \Sigma_X^2 V_q)^{-1} V_q^T \Sigma_X \Sigma_{X,Y} \\ &\quad - \Sigma_X V_q (I_d, 0)^T (V_d^T \Sigma_X^2 V_d)^{-1} V_d^T \Sigma_X \Sigma_{X,Y} \\ &= \Sigma_X V_q \left\{ (V_q^T \Sigma_X^2 V_q)^{-1} V_q - (I_d, 0)^T (V_d^T \Sigma_X^2 V_d)^{-1} V_d^T \right\} \Sigma_X \Sigma_{X,Y} \\ &:= \Sigma_X V_q A_d, \end{aligned}$$

where the $q \times r$ matrix A_d is defined implicitly. In consequence, we have for $d = 1, \ldots, q-1$

$$v_{d+1} = \ell_1 \left\{ \Sigma_X V_q A_d A_d^T V_q^T \Sigma_X \right\}.$$

Since $v_{d+1} \in \mathrm{span}(\Sigma_X V_q)$, this implies that v_{d+1} can be represented as $v_{d+1} = \Sigma_X V_q m_d$ for some $q \times 1$ vector m_d and thus that there is a $q \times q$ matrix $M = (m_1, \ldots, m_q)$ so that $V_q = \Sigma_X V_q M$. Since V_q has full column rank and Σ_X is non-singular, M must be nonsingular and so

$$\Sigma_X V_q = V_q M^{-1}. \tag{3.24}$$

This result tells us that $\mathrm{span}(V_q)$ is a reducing subspace of Σ_X. Thus, in addition to (3.22) we also have (see Proposition 1.6)

$$\mathrm{span}(\Sigma_{X,Y}) \subseteq \mathrm{span}(V_q). \tag{3.25}$$

In summary,

Proposition 3.4. *The span* $\mathrm{span}(V_q)$ *of the weight matrix* V_q *from the population SIMPLS algorithm is a reducing subspace of* Σ_X *that contains* $\mathrm{span}(\Sigma_{X,Y})$ *and* $\mathcal{B}$.

3.4.2 The SIMPLS weights span the Σ_X-envelope of $\mathcal{B}$

Taken together, results (3.23) and (3.24) characterize $\mathrm{span}(V_q)$ as a reducing subspace of Σ_X that contains $\mathcal{B}$ or, equivalently, contains $\mathrm{span}(\Sigma_{X,Y})$. Those results were obtained without making use of envelopes per se. We now make use of that connection to establish further properties of the SIMPLS subspaces $\mathrm{span}(V_d)$, $d = 1, \ldots, q$, including that the SIMPLS subspace $\mathrm{span}(V_q)$ is the Σ_X-envelope of $\mathcal{B}$; that is, the *smallest* reducing subspace of Σ_X that contains $\mathrm{span}(\Sigma_{X,Y})$.

Let $\Phi \in \mathbb{R}^{p \times q}$ denote a semi-orthogonal basis matrix for $\mathcal{E}_{\Sigma_X}(\mathcal{B})$, the intersection of all reducing subspace of Σ_X that contain $\mathrm{span}(\Sigma_{X,Y})$ and let $(\Phi, \Phi_0) \in \mathbb{R}^{p \times p}$ be an orthogonal matrix. This the same notation we used when dealing with NIPALS and envelopes in Section 3.2.2. Then we know from Proposition 1.2 that Σ_X can be expressed as

$$\Sigma_X = \Phi\Delta\Phi^T + \Phi_0\Delta_0\Phi_0^T.$$

Turning to the first step in the SIMPLS algorithm, $\Sigma_{X,Y} = P_\Phi \Sigma_{X,Y}$ by construction and consequently v_1 is in the envelope, $v_1 \in \mathrm{span}(\Phi)$ and $\Sigma_X v_1 = \Phi\Delta\Phi^T v_1 := \Phi a_1$, where $a_1 = \Delta\Phi^T v_1$. And

$$\begin{aligned} Q_{\Sigma_X v_1}\Sigma_{X,Y} &= Q_{\Phi a_1}\Sigma_{X,Y} = \Sigma_{X,Y} - P_{\Phi a_1}\Sigma_{X,Y} \\ &= P_\Phi \Sigma_{X,Y} - P_{\Phi a_1}\Sigma_{X,Y} \\ &= \Phi Q_{a_1}\Phi^T \Sigma_{X,Y}. \end{aligned}$$

Consequently,

$$\begin{aligned} v_2 &= \ell_1\left\{Q_{\Sigma_X v_1}\Sigma_{X,Y}\Sigma_{X,Y}^T Q_{\Sigma_X v_1}\right\} \\ &= \ell_1\left\{\Phi Q_{a_1}\Phi^T \Sigma_{X,Y}\Sigma_{X,Y}^T \Phi Q_{a_1}\Phi^T\right\} \\ &\in \mathrm{span}(\Phi), \end{aligned}$$

and so $V_2 = \mathrm{span}(v_1, v_2) \subseteq \mathrm{span}(\Phi)$. Additionally, since $v_2 \in \mathrm{span}(Q_{\Sigma_X v_1}\Sigma_{X,Y})$ and $v_1^T \Sigma_X Q_{\Sigma_X v_1}\Sigma_{X,Y} = 0$ it follows that $v_2^T \Sigma_X v_1 = 0$, as we have seen.

Proceeding by induction, assume that $V_{d-1} = \mathrm{span}(v_1, \ldots, v_{d-1}) \subseteq \mathrm{span}(\Phi)$. Then using the previous argument by replacing v_1 with V_{d-1}, it follows that $v_d \in \mathrm{span}(\Phi)$, that $v_d^T \Sigma_X V_{d-1} = 0$ and that $\mathrm{span}(V_{d-1})$ is a proper subspace of $\mathrm{span}(V_d)$; that is, $\mathrm{span}(V_{d-1}) \subset \mathrm{span}(V_d)$. This demonstrates that the dimension SIMPLS subspaces $\mathrm{span}(V_d)$ continue to increase until $d = q$, at which point $\mathrm{span}(V_q) = \mathrm{span}(\Phi)$ because $\mathrm{span}(V_q) \subseteq \mathrm{span}(\Phi)$ and $q = \dim(\mathrm{span}(\Phi)) = \dim(\mathrm{span}(V_q))$. We used the same rationale when proving Proposition 3.3.

Summarizing,

Proposition 3.5. *The population SIMPLS algorithm of Table 3.4 produces a series of properly nested subspaces that settle on the envelope after q steps,*

$$\mathrm{span}(V_1) \subset \mathrm{span}(V_2) \subset \cdots \subset \mathrm{span}(V_q) = \mathcal{E}_{\Sigma_X}(\mathcal{B}),$$

where $q = \dim[\mathcal{E}_{\Sigma_X}\{\mathcal{B}\}]$.

3.5 SIMPLS v. NIPALS

In this section we compare the NIPALS and SIMPLS algorithms, starting with their population versions shown in Tables 3.1(b) and 3.4(b).

3.5.1 Estimation

While first weight vectors from the population algorithms are the same, $w_1 = v_1$, the second and subsequent weight vectors differ:

$$\begin{aligned} w_2 &= \ell_1(Q^T_{w_1(\Sigma_X)}\Sigma_{X,Y}\Sigma^T_{X,Y}Q_{w_1(\Sigma_X)}) \\ v_2 &= \ell_1(Q_{\Sigma_X w_1}\Sigma_{X,Y}\Sigma^T_{X,Y}Q_{\Sigma_X w_1}), \end{aligned}$$

where we have used w_1 in place of v_1 for the calculation of v_2. These weight vectors are distinct because of the different projections – $Q_{w_1(\Sigma_X)}$ and $Q_{\Sigma_X w_1}$ – used to remove the first weight vector w_1 from consideration. However, we know from results in Sections 3.2 and 3.4 that at termination

$$\text{span}(W_q) = \text{span}(V_q) = \mathcal{E}_{\Sigma_X}(\mathcal{B}).$$

Since β_{npls} and β_{spls} depend only on the subspaces spanned by the corresponding weight matrices, it follows that $\beta_{\text{npls}} = \beta_{\text{spls}}$, and so NIPALS and SIMPLS produce the same result in the population.

The sample estimators are generally different, $\widehat{\beta}_{\text{npls}} \neq \widehat{\beta}_{\text{spls}}$, with two important exceptions as given in the following proposition (de Jong, 1993).

Proposition 3.6. *Recall that $\widehat{\beta}_{\text{npls}}$ and $\widehat{\beta}_{\text{spls}}$ denote the sample NIPALS and SIMPLS estimators. Assume that these estimators are each constructed with q components.*

1. *If a single component is used, so $q = 1$, then the sample $w_1 = v_1$ and consequently $\widehat{\beta}_{\text{npls}} = \widehat{\beta}_{\text{spls}}$.*

2. *If the response is univariate, so $r = 1$, then again $\widehat{\beta}_{\text{npls}} = \widehat{\beta}_{\text{spls}}$.*

Proof. The first conclusion is straightforward. To demonstrate the second conclusion, consider the sequence of weight vectors for NIPALS starting with

$w_1 \propto S_{X,Y}$ and

$$\begin{aligned} w_2 &\propto Q^T_{w_1(S_X)} S_{X,Y} \\ &= S_{X,Y} - S_X S_{X,Y}(w_1^T S_X w_1)^{-1} w_1^T S_{X,Y} \\ &:= S_{X,Y} - S_X S_{X,Y} c_1 \end{aligned}$$

where $c_1 = (w_1^T S_X w_1)^{-1} w_1^T S_{X,Y}$ is a scalar. From this we have the representation

$$\operatorname{span}(W_2) = \operatorname{span}\left\{S_X^0 S_{X,Y}, S_X^1 S_{X,Y}\right\}.$$

We next show by induction that for $d = 1, \ldots, q$,

$$\operatorname{span}(W_d) = \operatorname{span}\left\{S_X^0 S_{X,Y}, S_X^1 S_{X,Y}, \ldots, S_X^{d-1} S_{X,Y}\right\}. \tag{3.26}$$

Under the induction hypothesis,

$$\operatorname{span}(W_{d-1}) = \operatorname{span}\left\{S_X^0 S_{X,Y}, S_X^1 S_{X,Y}, \ldots, S_X^{d-2} S_{X,Y}\right\} \tag{3.27}$$

and

$$\begin{aligned} w_d &\propto Q^T_{W_{d-1}(S_X)} S_{X,Y} \\ &= S_{X,Y} - S_X W_{d-1}(W_{d-1}^T S_X W_{d-1})^{-1} W_{d-1}^T S_{X,Y}. \end{aligned}$$

We need to show that

$$\operatorname{span}\left\{S_X^0 S_{X,Y}, S_X^1 S_{X,Y}, \ldots, S_X^{d-1} S_{X,Y}\right\} = \operatorname{span}(W_{d-1}, w_d)$$

where $\operatorname{span}(W_{d-1})$ is as given under the induction hypothesis (3.27).

Since w_d depends on W_{d-1} only via $\operatorname{span}(W_{d-1})$, we substitute the induction hypothesis for the W_{d-1}'s and let $c = (W_{d-1}^T S_X W_{d-1})^{-1} W_{d-1}^T S_{X,Y} \in \mathbb{R}^d$. Then

$$w_d \propto S_{X,Y} - (S_X^1 S_{X,Y}, \ldots, S_X^{d-1} S_{X,Y})c.$$

It follows that $\operatorname{span}(W_{d-1}, w_d) = \operatorname{span}(W_d)$ as defined in (3.26). Using the same logic for SIMPLS, it follows that $\operatorname{span}(V_d) = \operatorname{span}(W_d)$ for all $d = 1, 2, \ldots, q$ and thus that $\widehat{\beta}_{\text{npls}} = \widehat{\beta}_{\text{spls}}$ as long as the stopping point is the same. □

From this we see that the orthogonal NIPALS weight vectors $w_1, \ldots, w_q$ result from a sequential residualization of the Krylov sequence (see Section 1.3)

$$\{S_X^0 S_{X,Y}, S_X^1 S_{X,Y}, \ldots, S_X^{q-1} S_{X,Y}\},$$

which led Manne (1987) and Helland (1990) to claim that the NIPALS algorithm is a version of the Gram-Schmidt procedure (see Section 3.6).

Changing to the population to complete the discussion, the SIMPLS weight vectors also result in an orthogonalization of the Krylov sequence, except now the weight vectors $v_1, \ldots, v_q$ are orthogonal in the Σ_X inner product. In particular,

- $v_1 = \sigma_{X,Y}$,
- v_2 is the vector of residuals from the regression of $\sigma_{X,Y}$ on $\Sigma_X v_1$,
- v_3 is the vector of residuals from the regression of $\sigma_{X,Y}$ on $\Sigma_X(v_1, v_2)$,

  $\vdots$
- v_k is the vector of residuals from the regression of $\sigma_{X,Y}$ on $\Sigma_X(v_1, v_2, \ldots, v_{k-1})$.

In consequent, we see that for a univariate response, the NIPALS and SIMPLS sample and population vectors can be viewed as different orthogonalization of the Krylov vectors. This implies that in the population and sample $\text{span}(W_k) = \text{span}(V_k)$, $k = 1, \ldots, q$; that is, we have demonstrated that

Proposition 3.7. *For single-response regressions and* $1 \leq k \leq q$,

Population: $\text{span}(W_k) = \text{span}(V_k) = \mathcal{K}_k(\Sigma_X, \sigma_{X,Y})$,

Sample: $\text{span}(W_k) = \text{span}(V_k) = \mathcal{K}_k(S_X, S_{X,Y})$,

where $\mathcal{K}_k$ *was defined at (1.22).*

3.5.2 Summary of common features of NIPALS and SIMPLS

As we have seen, the SIMPLS and NIPALS algorithms share a number of features:

- Both algorithms give the envelope $\mathcal{E}_{\Sigma_X}(\mathcal{B})$ in the population and in that sense are aiming at the same target. However, the algorithms give different sample weight vectors, as discussed previously.

- In the population, the NIPALS algorithm makes use of the Σ_X inner product to generate its orthogonal weights, $W_q^T W_q = I_q$, while the SIMPLS algorithm uses the identity inner product to generate its weight vectors that are orthogonal in the Σ_X inner product, so $V_q^T \Sigma_X V_q$ is a diagonal matrix. Nevertheless, $\text{span}(W_q) = \text{span}(V_q) = \mathcal{E}_{\Sigma_X}(\mathcal{B})$.

- With $q = \dim\{\mathcal{E}_{\Sigma_X}(\mathcal{B})\}$ known and fixed, p fixed and $n \to \infty$, both algorithms after q steps produce $\sqrt{n}$-consistent estimators of β, because the algorithms are smooth functions of $S_{X,Y}$ and S_X, which are $\sqrt{n}$-consistent estimators of $\Sigma_{X,Y}$ and Σ_X.

- The number of components $q \leq \min\{\text{rank}(S_X), n-1)\}$ is often selected by predictive cross or a hold-out sample, particularly in chemometrics. Underestimation of q produces biased estimators, while overestimation results in estimators that are more variable than need be. Our experience indicates that underestimation is the more serious error and that typically a little overestimation may be tolerable.

- Neither algorithms requires that S_X be positive definite and they are generally serviceable in high-dimensional regressions. Asymptotic properties as $n, p \to \infty$ are discussed in Chapter 4.

- $\widehat{\beta}_{\text{npls}} = \widehat{\beta}_{\text{spls}}$ when $q = 1$ or $r = 1$. If S_X is non-singular and $q = p$ then $\widehat{\beta}_{\text{npls}} = \widehat{\beta}_{\text{spls}} = \widehat{\beta}_{\text{ols}}$. Otherwise, $\widehat{\beta}_{\text{npls}} \neq \widehat{\beta}_{\text{spls}}$.

- Both population algorithms are invariant under full rank linear transformations $Y_A = AY$ of the response vector. The coefficient matrix with the transformed responses is $\beta_A := \beta A^T$. Since $\text{span}(\beta_A) = \mathcal{B}$, the envelope is invariant under this transformation,

  $$\mathcal{E}_{\Sigma_X}(\text{span}(\beta_A)) = \mathcal{E}_{\Sigma_X}(\text{span}(\beta A^T)) = \mathcal{E}_{\Sigma_X}(\mathcal{B}).$$

  The population target is thus invariant under choice of A.

  However, the sample versions of the algorithms are susceptible to change under full rank linear transformations of the response. This can be appreciated by noting that $w_1 = v_1$ changes with the transformation. In the original scale $w_1 = v_1 = \ell_1(S_{X,Y}S_{X,Y}^T)$, while in the transformed scale $w_1 = v_1 = \ell_1(S_{X,Y}A^TAS_{X,Y}^T)$. This begs the question: What is the "best" choice of A? The likelihood-based estimators discussed in Chapters 3 and 11 suggest that when $r \ll n$, $A = S_Y^{-1/2}$ is a good choice.

- Neither algorithm is invariant under full rank linear transformations of X.

3.6 Helland's algorithm

Manne (1987) and Helland (1990) emphasized a connection between univariate PLS regression and the Krylov subspace shown in (3.26) (see also Section 1.3). Frank and Frideman (1993) referred the corresponding algorithm as "Helland's algorithm," as described in Table 3.5. While the Krylov basis vectors $S_X^d S_{X,Y}$, $d = 0, \ldots, q-1$ are linearly independent, they can exhibit a degree of multicollinearity that can cause serious problems for the algorithm. The orthogonalized versions NIPALS and SIMPLS should be computationally more stable in such instances, although the orthogonalization may become problematic if the vectors in the Krylov sequence become highly collinear. This possibility plays a role in the asymptotic collinearity measure $\rho(p)$ introduced in Section 4.6.1.

3.7 Illustrative example

In this section, we use an example from Chun and Keleş (2010) to illustrate selected results from the developments in this chapter.

Consider a regression in which the $n \times p$ data matrix $\mathbb{X}$ of predictor values can be represented as

$$\mathbb{X}_{n \times p} = (H_1 1_1^T + E_1, H_2 1_2^T + E_2, H_3 1_3^T + E_3),$$

where H_j is an $n \times 1$ vector of independent standard normal variates, $j = 1, 2, 3$, and 1_1, 1_2, and 1_3 are vectors of ones having dimensions p_1, p_2 and p_3 with $\sum_j p_j = p$. The E_j's are matrices of conforming dimensions also consisting of independent standard normal variates. The random quantities H_j and E_j, $j = 1, 2, 3$, are mutually independent. A typical $p \times 1$ predictor vector can be represented as

$$X = \begin{pmatrix} h_1 1_1 \\ h_2 1_2 \\ h_3 1_3 \end{pmatrix} + e,$$

where the h_j's are independent standard normal random variables and e is a $p \times 1$ vector of independent standard normal variates.

TABLE 3.5

Helland's algorithm for univariate PLS regression: (a) sample version adapted from Table 2 of Frank and Frideman (1993). The $n \times p$ matrix $\mathbb{X}$ contains the centered predictors and the $n \times r$ vector $\mathbb{Y}$ contains the centered responses; (b) population version derived herein.

(a) *Sample Version*	
Compute sample covariance matrices	$S_{X,Y} = n^{-1}\mathbb{X}^T\mathbb{Y}$, $S_X = n^{-1}\mathbb{X}^T\mathbb{X}$
Initialize	$K_1 = (S_{X,Y})$, the first Krylov basis.
Select	$q \leq \min\{\text{rank}(S_X), n-1)\}$
For $d = 2 \ldots q$	
Compute Krylov vector	$k_d = S_X^{d-1} S_{X,Y}$
Append	$K_d = (K_{d-1}, k_d)$
End	
Compute regression coefficients	$\widehat{\beta}_{\text{Hpls}} = K_q(K_q^T S_X K_q)^{-1} K_q^T S_{X,Y}$
(b) *Population Version*	
Initalize	$K_1 = (\Sigma_{X,Y})$
For $d = 2, \ldots$	
Compute Krylov vector	$k_d = \Sigma_X^{d-1}\Sigma_{X,Y}$
Append	$K_d = (K_{d-1}, k_d)$
End when first	$Q_{K_{d-1}} k_d = 0$ and then set $q = d-1$
Compute regression coefficients	$\beta_{\text{Hpls}} = K_q(K_q^T \Sigma_X K_q)^{-1} K_q^T \Sigma_{X,Y}$ $= K_q \beta_{Y\mid K_q^T X}$
Envelope connection	$\text{span}(K_q) = \mathcal{E}_{\Sigma_X}(\mathcal{B})$, the Σ_X-envelope of $\mathcal{B}$ where $\mathcal{B} = \text{span}(\beta)$
Algorithm $\mathbb{K}$	This is an instance of Algorithm $\mathbb{K}$ discussed in §1.5.2.

The variance-covariance matrix of the predictors X is

$$\Sigma_X = \begin{pmatrix} 1_1 1_1^T & 0 & 0 \\ 0 & 1_2 1_2^T & 0 \\ 0 & 0 & 1_3 1_3^T \end{pmatrix} + I_p.$$

From this structure, we see that the predictors consist of three independent blocks of sizes p_1, p_2 and p_3. Let $u_1^T = (1_1^T, 0, 0)$, $u_2^T = (0, 1_2^T, 0)$ and $u_3^T = (0, 0, 1_3^T)$ denote three eigenvectors of Σ_X with non-zero eigenvalues.

Then Σ_X can be expressed equivalently as

$$\Sigma_X = (1+p_1)P_{u_1} + (1+p_2)P_{u_2} + (1+p_3)P_{u_3} + Q,$$

where $P_{(\cdot)}$ is the projection onto the subspace spanned by the indicated eigenvector and Q is the projection onto the $p-3$ dimensional subspace that is orthogonal to $\text{span}(u_1, u_2, u_3)$.

Next, with a single response $r = 1$, the $n \times 1$ vector of responses was generated as the linear combination $\mathbb{Y} = H_1 - H_2 + \epsilon$, where ϵ is an $n \times 1$ vector of independent normal variates with mean 0 and variance σ^2, which gives

$$\begin{aligned} \Sigma_{X,Y} &= u_1 - u_2 \\ \beta &= \frac{1}{1+p_1}u_1 - \frac{1}{1+p_2}u_2. \end{aligned}$$

Consequently, if $p_1 \neq p_2$ then both β and $\Sigma_{X,Y}$ are linear combinations of two eigenvectors of Σ_X and only $q = 2$ components are needed to characterize the regression, specifically $u_1^T X$ and $u_2^T X$. Equivalently, the two-dimensional envelope $\mathcal{E}_{\Sigma_X}(\mathcal{B}) = \text{span}(u_1, u_2)$. If $p_1 = p_2$ then $\mathcal{E}_{\Sigma_X}(\mathcal{B}) = \text{span}(u_1 - u_2)$. This follows because $\text{span}(\Sigma_{X,Y}) = \mathcal{B} = \text{span}(u_1 - u_2)$ has dimension 1.

To calculate the first two NIPALS weight vectors we need

$$\begin{aligned} \Sigma_X \Sigma_{X,Y} &= (1+p_1)u_1 - (1+p_2)u_2 \\ \Sigma_{X,Y}^T \Sigma_X \Sigma_{X,Y} &= (1+p_1)p_1 + (1+p_2)p_2 \\ \Sigma_{X,Y}^T \Sigma_{X,Y} &= p_1 + p_2. \end{aligned}$$

With this we can now calculate the first two NIPALS weight vectors as

$$\begin{aligned} w_1 &\propto \Sigma_{X,Y} = u_1 - u_2 \\ w_2 &\propto Q_{\Sigma_{X,Y}(\Sigma_X)}^T \Sigma_{X,Y} \\ &= \Sigma_{X,Y} - \Sigma_X \Sigma_{X,Y} (\Sigma_{X,Y}^T \Sigma_X \Sigma_{X,Y})^{-1} \Sigma_{X,Y}^T \Sigma_{X,Y} \\ &= \frac{p_2 - p_1}{(1+p_1)p_1 + (1+p_2)p_2}(p_2 u_1 + p_1 u_2). \end{aligned}$$

Clearly, $\text{span}(W_2) = \text{span}(u_1, u_2)$ provided $p_1 \neq p_2$. If $p_1 = p_2$ then Σ_X reduces to

$$\Sigma_X = (1+p_1)P_{(u_1,u_2)} + (1+p_3)P_{u_3} + Q,$$

u_1 and u_2 belong to the same eigen-space and, in consequence, only one component is needed and $\mathcal{E}_{\Sigma_X}(\mathcal{B}) = \text{span}(u_1 - u_2)$. Also, when $p_1 = p_2$, the stopping criterion is met at $w_2 = 0$, giving $q = 1$.

Turning to the SIMPLS weights, we need in addition to the previous preparatory calculations,

$$\Sigma_{X,Y}^T \Sigma_X^2 \Sigma_{X,Y} = (1+p_1)^2 p_1 + (1+p_2)^2 p_2.$$

Then the first two SIMPLS weights are

$$\begin{aligned} v_1 &\propto \Sigma_{X,Y} = u_1 - u_2 \\ v_2 &\propto Q_{\Sigma_X \Sigma_{X,Y}} \Sigma_{X,Y} \\ &= \Sigma_{X,Y} - \Sigma_X \Sigma_{X,Y} (\Sigma_{X,Y}^T \Sigma_X^2 \Sigma_{X,Y})^{-1} \Sigma_{X,Y}^T \Sigma_X \Sigma_{X,Y} \\ &= \frac{p_2 - p_1}{(1+p_1)^2 p_1 + (1+p_2)^2 p_2}((1+p_2)p_2 u_1 + (1+p_1)p_1 u_2). \end{aligned}$$

The interpretation here is similar to that for the NIPALS weights.

The first three Krylov terms in Helland's algorithm are

$$\begin{aligned} \Sigma_{X,Y} &= u_1 - u_2 \\ \Sigma_X \Sigma_{X,Y} &= (1+p_1)u_1 - (1+p_2)u_2 \\ \Sigma_X^2 \Sigma_{X,Y} &= (1+p_1)^2 u_1 - (1+p_2)^2 u_2. \end{aligned}$$

From this we have the Krylov basis $K_2 = (u_1 - u_2, (1+p_1)u_1 - (1+p_2)u_2)$. If $p_1 \neq p_2$ then $\text{span}(K_2) = \text{span}(u_1, u_2)$ and so $Q_{K_2}\{(1+p_1)^2 u_1 - (1+p_2)^2 u_2\} = 0$ and $q = 2$. If $p_1 = p_2$ then $\text{span}(K_1) = \text{span}(u_1 - u_2)$, $Q_{K_1}\{(1+p_1)u_1 - (1+p_2)u_2\} = 0$ and $q = 1$.

The three algorithms – NIPALS, SIMPLS, Helland – all agree that if $p_1 \neq p_2$ then $\mathcal{E}_{\Sigma_X}(\mathcal{B}) = \text{span}(u_1, u_2)$, while $\mathcal{E}_{\Sigma_X}(\mathcal{B}) = \text{span}(u_1 - u_2)$ if $p_1 = p_2$.

3.8 Likelihood estimation of predictor envelopes

As mentioned at the outset of this chapter, PLS regression algorithms were not historically associated with specific models and, in consequence, were not typically viewed as methods for estimating identifiable population parameters but were seen instead as methods for prediction. Nevertheless, the algorithms of this chapter, and more generally those of Section 1.5, can be viewed as moment-based methods for fitting the predictor envelope model (2.5), an approach that has often been called “soft modeling” in the chemometrics literature. Soft modeling by its nature does not require fully laid foundations, resulting in methodology with a rather elusive quality. In contrast, model-based "hard modeling" has the advantage of laying bare all of the structural and stochastic foundations of a method, allowing the investigator to study performance under deviations from those foundations as necessary for a particular application. There seems to be a feeling within the chemometrics community that the success of PLS algorithms over the past four decades does not fully compensate for the lack of adequate foundations (Stocchero, 2019).

Since we have shown in this chapter that NIPALS and SIMPLS give the envelope in the population, we could use likelihood theory to get maximum likelihood estimators when the sample size is sufficiently large. This approach was described in Section 2.3 and a preview of the connection to PLS was given in Section 2.3.2.

3.9 Comparisons of likelihood and PLS estimators

Turning to comparisons of the likelihood-based method with PLS methods, we see first that $L_q(G)$, given in (2.11), depends on the response only through its standardized version $Z = S_Y^{-1/2}Y$. On the other hand, PLS depends on the scale of the response. When $q = 1$, the PLS estimator of $\mathcal{E}_{\Sigma_X}(\mathcal{B})$ is the span of the first eigenvector w_1 of $S_{X,Y}S_{X,Y}^T$. After performing a full rank transformation of the response $Y \mapsto AY$, the PLS estimator of $\mathcal{E}_{\Sigma_X}(\mathcal{B})$ is the span of the first eigenvector $\widetilde{w}_1$ of $S_{X,Y}A^TAS_{X,Y}^T$. Generally, $\text{span}(w_1) \neq \text{span}(\widetilde{w}_1)$, so the estimates of $\mathcal{E}_{\Sigma_X}(\mathcal{B})$ differ, although $\Sigma_{X,Y}\Sigma_{X,Y}^T$ and $\Sigma_{X,Y}A^TA\Sigma_{X,Y}^T$ span the same subspace. It is customary to standardize the individual responses marginally $y_j \mapsto y_j/\{\widehat{\text{var}}(y_j)\}^{1/2}$, $j = 1, \ldots, r$, prior to application of PLS, but it is evidently not customary to standardize the responses jointly $Y_i \mapsto Z_i = S_Y^{-1/2}Y_i$. Of course, the PLS algorithm could be applied after replacing Y with jointly standardized responses Z.

The methods also differ on how they utilize information from S_X. The likelihood objective function (2.11) can be written as (Cook, Li, and Chiaromonte, 2010, Section 4.3.2)

$$L_q(G) = \log|G^T S_X G| + \log|G^T S_X^{-1} G| + \log|S_{Z|G^T X}|.$$

The sum of the first two addends on the right hand side is non-negative and zero when the columns of G correspond to any subset of q eigenvectors of S_X.

To see this, suppose the columns of G correspond to a subset of q eigenvectors of S_X with eigenvalues λ_i given by the diagonal elements of the diagonal matrix Λ. Then

$$\begin{aligned} \log|G^T S_X G| &= \log|G^T G\Lambda| = \sum_{i=1}^{q} \log \lambda_i \\ \log|G^T S_X^{-1} G| &= \log|G^T G\Lambda^{-1}| = -\sum_{i=1}^{q} \log \lambda_i \end{aligned}$$

and the conclusion follows. Consequently, the role of these addends is to pull the solution toward subsets of q eigenvectors of S_X. This in effect imposes a sample counterpart of the characterization in Proposition 1.9, which states that in the population $\mathcal{E}_{\Sigma_X}(\mathcal{B})$ is spanned by a subset of the eigenvectors of Σ_X. There is no corresponding operation in the PLS methods. The first SIMPLS vector v_1 does not incorporate direct information about S_X. The second PLS vector incorporates S_X by essentially removing the subspace $\mathrm{span}(S_X v_1)$ from consideration, but the choice of $\mathrm{span}(S_X v_1)$ is not guided by the relationship between v_1 and the eigenvectors of S_X. Subsequent SIMPLS vectors operate similarly in successively smaller spaces. In application, PLS methods often require more directions to match the performance of the likelihood-based method (Cook et al., 2013).

Our discussion of likelihood-based estimation has so far been restricted to regressions in which $n \gg \max(p, r)$ or $n \to \infty$ with p and r fixed. Rimal, Trygve, and Sæbø (2019) adapted likelihood-based estimation to accommodate regressions with $n < p$ by selecting the principal components of X that account for 97.5 percent of its sample variation and then using likelihood-based estimation on the principal component predictors. Using large-scale simulations and data analyses, they compared the predictive performance of envelope methods, principal component regression and the kernel PLS method (Lindgren, Geladi, and Wold, 1993) from the `PLS` package in `R` (R Core Team, 2022) in contexts that reflect chemometric applications. They summarized their overall findings as follows ...

> Analysis using both simulated data and real data has shown that the envelope methods are more stable, less influenced by [predictor collinearity] and in general, performed better than PCR and PLS methods. These methods are also found to be less dependent on the number of components.

The envelope methods used by Rimal et al. (2019) are likelihood-based and they differ from the PLS methods only on the method of estimating $\mathcal{E}_{\Sigma_X}(\mathcal{B})$. In contrast, principal component regression is not designed specifically to estimate $\mathcal{E}_{\Sigma_X}(\mathcal{B})$ and this may in part account for its relatively poor performance. Subsequently, Rimal, Trygve, and Sæbø (2020) reported the results of a second study, this time focusing on the estimative performance of the methods.

Comparisons based on asymptotic approximations are presented in Chapter 4.

3.10 PLS1 v. PLS2

As mentioned previously, PLS1 and PLS2 refer to PLS regression with univariate and multivariate responses. Helland's algorithm of Table 3.5 is necessarily of the PLS1 type. The NIPALS and SIMPLS algorithms of Tables 3.1 and 3.4 are of the PLS2 variety, but of course they apply when $r = 1$ and so reduce to PLS1 algorithms. The key distinction between PLS1 and PLS2 algorithms is how they are used to fit a regression with a multivariate response, with PLS2 as described in Tables 3.1 and 3.4.

To motivate the use of PLS1 algorithms to fit multivariate response regressions, let $Y_j, j = 1, \dots, r$, denote the individual responses and consider the OLS estimator of β given in (1.7)

$$\begin{aligned} \widehat{\beta}_{\mathrm{ols}} &= S_X^{-1} S_{X,Y} = (S_X^{-1} S_{X,Y_1}, S_X^{-1} S_{X,Y_2}, \dots, S_X^{-1} S_{X,Y_r}) \\ &:= (\widehat{\beta}_{\mathrm{ols},1}, \widehat{\beta}_{\mathrm{ols},2}, \dots, \widehat{\beta}_{\mathrm{ols},r}), \end{aligned}$$

where $\widehat{\beta}_{\mathrm{ols},j} \in \mathbb{R}^p$ denotes the coefficient vector from the OLS fit of Y_j on X. From this we see that the j-th column of the OLS coefficient matrix $\widehat{\beta}_{\mathrm{ols}}$ consists of the coefficient vector from the univariate-response regression of Y_j on X. Following this lead, we can use NIPALS or SIMPLS to construct a different PLS estimator by performing r univariate PLS regressions:

$$\left.\begin{aligned} \widehat{\beta}_{\mathrm{npls1}} &:= (\widehat{\beta}_{\mathrm{npls},1}, \widehat{\beta}_{\mathrm{npls},2}, \dots, \widehat{\beta}_{\mathrm{npls},r}) \\ \widehat{\beta}_{\mathrm{spls1}} &:= (\widehat{\beta}_{\mathrm{spls},1}, \widehat{\beta}_{\mathrm{spls},2}, \dots, \widehat{\beta}_{\mathrm{spls},r}) \end{aligned}\right\} \tag{3.28}$$

where $\widehat{\beta}_{\mathrm{npls},j}$ is the estimator of the coefficient vector from the NIPALS fit of the j-th response on X, with a similar definition for $\widehat{\beta}_{\mathrm{spls},j}$.

Partition β on its columns, $\beta = (\beta_1, \ldots, \beta_r)$ and let $\mathcal{B}_j = \text{span}(\beta_j)$. The columns of the PLS1 estimators $\widehat{\beta}_{\text{npls1}}$ and $\widehat{\beta}_{\text{spls1}}$ are based on estimators of the corresponding envelopes $\mathcal{E}_{\Sigma_X}(\mathcal{B}_j)$, $j = 1, \ldots, r$, while the PLS2 estimators $\widehat{\beta}_{\text{npls}}$ and $\widehat{\beta}_{\text{spls}}$ are based on estimators of overall envelope $\mathcal{E}_{\Sigma_X}(\mathcal{B}) = \mathcal{E}_{\Sigma_X}(\mathcal{B}_1 + \cdots + \mathcal{B}_r)$, since $\mathcal{B} = \sum_{i=1}^{r} \mathcal{B}_j$. The following lemma gives a relationship between these subspace constructions. Its proof is in Appendix A.3.6

Lemma 3.6. *let $\Sigma \in \mathbb{S}^p$ be positive definite and let $\mathcal{B}_j$ be a subspace of $\mathbb{R}^p$, $j = 1, 2$. Then*

$$\mathcal{E}_\Sigma(\mathcal{B}_1 + \mathcal{B}_2) = \mathcal{E}_\Sigma(\mathcal{B}_1) + \mathcal{E}_\Sigma(\mathcal{B}_2),$$

and

$$\begin{aligned} \max\{\dim(\mathcal{E}_\Sigma(\mathcal{B}_1)), \dim(\mathcal{E}_\Sigma(\mathcal{B}_2))\} &\leq \dim(\mathcal{E}_\Sigma(\mathcal{B}_1 + \mathcal{B}_2)) \\ &\leq \dim(\mathcal{E}_\Sigma(\mathcal{B}_1)) + \dim(\mathcal{E}_\Sigma(\mathcal{B}_2)). \end{aligned}$$

This lemma implies a key fact about the relationship PLS1 and PLS2 envelopes:

$$\mathcal{E}_{\Sigma_X}(\mathcal{B}) = \mathcal{E}_{\Sigma_X}(\mathcal{B}_1 + \cdots + \mathcal{B}_r) = \mathcal{E}_{\Sigma_X}(\mathcal{B}_1) + \cdots + \mathcal{E}_{\Sigma_X}(\mathcal{B}_r).$$

According to this result, if our goal is to estimate $\mathcal{E}_{\Sigma_X}(\mathcal{B})$ in order to facilitate estimation of β or prediction then, in the population, it does not matter if we use PLS1 or PLS2 regressions. However, estimating β via an estimator of $\mathcal{E}_{\Sigma_X}(\mathcal{B})$ is not the only way to use envelopes to improve estimation. An alternative is to pursue column-wise estimation of β, using $\mathcal{E}_{\Sigma_X}(\mathcal{B}_j)$ to improve estimation of the j-th column β_j, $j = 1, \ldots, r$, and then assembling the estimators into an estimator of β, as shown in (3.28).

This distinction reflects a potential for PLS1 fits to do better than PLS2 fits, because PLS1 is better able to adapt to the individual response regressions, at least theoretically. For instance, suppose that $r = 2$, $\dim\{\mathcal{E}_{\Sigma_X}(\mathcal{B}_1)\} = 1$ and $\dim\{\mathcal{E}_{\Sigma_X}(\mathcal{B}_2)\} = p$. Then substantial variance reduction is possible when estimating β_1 from the NIPALS, say, fit of Y_1 on X. But no variance reduction is possible when estimating β_2 from the regression of Y_2 on X. Knowing the dimensions and using NIPALS, the PLS1 estimator of β takes the form $\widehat{\beta}_{\text{npls1}} = (\widehat{\beta}_{\text{npls},1}, \widehat{\beta}_{\text{ols},2})$. On the other hand, because $\dim(\mathcal{E}_{\Sigma_X}(\mathcal{B})) = p$, the overall NIPALS estimator must be simply $\widehat{\beta}_{\text{npls}} = \widehat{\beta}_{\text{ols}}$. In this case a PLS1 fit may be better than PLS2.

If $\dim\{\mathcal{E}_{\Sigma_X}(\mathcal{B}_j)\} = 1$ and $\mathcal{B}_j$ is a unique one-dimensional eigenspace of Σ_X, $j = 1, \ldots, r$, so $\dim(\mathcal{E}_{\Sigma_X}(\mathcal{B})) = r$, then, knowing the dimensions, each

NIPALS estimator $\widehat{\beta}_{\mathrm{npls},j}$ for the univariate regressions will be fitted with $q = 1$. But the overall NIPALS estimator will be fitted with $q = r$, since $\dim(\mathcal{E}_{\Sigma_X}(\mathcal{B})) = r$. Again, a PLS1 fit may be better than PLS2.

If $\dim\{\mathcal{E}_{\Sigma_X}(\mathcal{B})\} = 1$ and $\dim\{\mathcal{E}_{\Sigma_X}(\mathcal{B}_j)\} = 1$, $j = 1, \ldots, r$ then, knowing the dimensions, each NIPALS estimator $\widehat{\beta}_{\mathrm{npls},j}$ for the univariate regressions will be fitted with $q = 1$. But the overall NIPALS estimator will also be fitted with $q = 1$, since $\dim(\mathcal{E}_{\Sigma_X}(\mathcal{B})) = 1$. In this case a PLS2 fit may be better than PLS1.

This discussion indicates that, with known envelope dimensions (number of components), either PLS1 or PLS2 may be preferable, depending in the regression. On the other hand, PLS1 methods require in application that each dimension $\dim\{\mathcal{E}_{\Sigma_X}(\mathcal{B}_j)\}$, $j = 1, \ldots, r$, be estimated separately, while PLS2 methods need only one dimensions $\dim\{\mathcal{E}_{\Sigma_X}(\mathcal{B})\}$ estimated. This means that in PLS1 regressions there are more chances for mis-estimation of the number of components. In practice it may be advantageous to try both methods and use cross validation or a hold-out sample to assess their relative strengths.

While the previous discussion was cast in terms of PLS estimators, the same reasoning and conclusions apply to the more general algorithms introduced in Section 1.5. For instance, in multivariate regressions we can think of applying algorithms $\mathbb{N}$ and $\mathbb{S}$ overall one response at a time, leading to estimators of the form

$$\left.\begin{aligned} \widehat{\beta}_{\mathbb{N}1} &:= (\widehat{\beta}_{\mathbb{N},1}, \widehat{\beta}_{\mathbb{N},2}, \ldots, \widehat{\beta}_{\mathbb{N},r}) \\ \widehat{\beta}_{\mathbb{S}1} &:= (\widehat{\beta}_{\mathbb{S},1}, \widehat{\beta}_{\mathbb{S},2}, \ldots, \widehat{\beta}_{\mathbb{S},r}) \end{aligned}\right\} \tag{3.29}$$

where $\widehat{\beta}_{\mathbb{N},j}$ is the estimator of the coefficient vector from using Algorithm $\mathbb{N}$ to fit the j-th response on X, with a similar definition for $\widehat{\beta}_{\mathbb{S},j}$.

3.11 PLS for response reduction

The various algorithms in this chapter were presented as methods for reducing the dimension of the predictor vector in linear regression. However, they also apply to reducing the dimension of the response vector in multivariate (multiple response) linear regression. To reduce a response vector using PLS simply interchange the roles of the predictor vector X and the response vector

Y in any of the algorithms presented in this chapter. In other words, apply an algorithm to the regression of X on Y. The justification for this stems from the envelope theory presented in Section 2.4: The predictor envelope $\mathcal{E}_{\Sigma_X}(\mathcal{B})$ for the regression of Y on X is also the response envelope for the regression of X on Y. Hence, to use PLS for response reduction, we simply specify Y as the predictor vector and X as the response vector in any of the algorithms in this chapter. The resulting semi-orthogonal weight matrix, say $G_u \in \mathbb{R}^{r \times u}$ to distinguish it from the weight matrix for reducing the predictors, is used to reduce the response vector $Y \mapsto G_u^T Y$.

However, the PLS estimators of β given along with reductive algorithms in this chapter are not appropriate following response reduction. Instead, the PLS estimator is

$$\widehat{\beta} = S_X^{-1} S_{X, G_u^T Y} G_u^T.$$

This estimator can be used in conjunction with the multivariate linear model and cross validation to select an appropriate value for the number u of response components, in the same way that cross validation is used to select the number of predictor components.

Response and predictor reductions can be combined to achieve simultaneous response-predictor reduction in the multivariate linear model. Methodology for this is discussed in Chapter 5.

4

Asymptotic Properties of PLS

In this chapter we consider asymptotic properties of PLS estimators, as $n \to \infty$ and as $n, p \to \infty$, in regressions with a real response, $r = 1$. Sections 4.2 through 4.5 are restricted to regressions with one component, $q = \dim(\mathcal{E}_{\Sigma_X}(\mathcal{B})) = 1$. This is a notable limitation. However, there are many instances of regressions requiring only a single component so these results do have practical value. For instance, in chemometrics, application of Beer's law indicates that samples based on a single analyte will follow a one-component regression (Nadler and Coifman, 2005). For additional examples, see Berntsson and Wold (1986, Section 3.2), Cook (2018, Section 4.4.4), and Kettaneh-Wold (1992, pg. 65). Chiappini et al. (2021) designed experiments based on a single analyte. Data from these experiments are thus analyzed appropriately with a one-component PLS fit.

The results for single-component regressions will serve also to set the stage for our study of multi-component consistency in Section 4.6, still with a real response. We consider one-component and multi-component regressions separately because much more is known about one-component regressions and results for multi-component regressions tend to be more intricate. Illustrations, simulations, and examples are used throughout the chapter. Notably missing from this chapter are results on the asymptotic distribution of PLS estimators in multi-component regressions. Such results were apparently unavailable during the writing of this book, but hopefully will be available in the future.

DOI: 10.1201/9781003482475-4

4.1 Synopsis

Our study of asymptotic properties of PLS led us to believe that classical PLS methods can be effective for studying the class of high dimensional abundant regressions (Cook, Forzani, and Rothman, 2012). Abundance is defined for one-component PLS regressions in Definition 4.1. It is defined for multi-component regressions in Definition 4.3 and a characterization is given in Proposition 4.2. Informally, an abundant regression is one in which many predictors contribute information about the response. In some abundant regressions, estimators of regression coefficients and fitted values can converge at the $\sqrt{n}$ rate as $n, p \to \infty$ without regard to the relationship between n and p. This phenomenon is described in Corollaries 4.3–4.5 for one component regressions and in Theorems 4.7 and 4.8 for multi-component regressions, relying mostly on the results of Cook and Forzani (2018, 2019).

Abundant regressions apparently occur frequently in the applied sciences. For instance, in chemometrics calibration, spectra are often digitized to give hundreds or even thousands of predictor variables (Martens and Næs, 1989) and it is generally expect that many points along the spectrum contribute information about the analyte of interest. Wold, Kettaneh, and Tjessem (1996) argued against the tendency to "... drastically reduce the number of variables ...," in effect arguing for abundance in chemometrics. In contrast to abundance, a sparse regression is one in which few predictors contain response information so the signal coming from the predictors is finite and bounded. Classical PLS methods are generally inconsistent in sparse regressions, but modifications have been developed to handle these cases, as discussed in Section 11.4.

In Section 4.3 we discuss traditional asymptotic approximations based on the one-component model and letting $n \to \infty$ with p fixed. This material mostly comes from Cook, Helland, and Su (2013). Although PLS is often associated with high dimensional regressions in which $n < p$, is it still serviceable in traditional regression contexts and knowledge of this setting will help paint an overall picture of PLS. In particular, there are settings in which PLS asymptotically outperforms standard methods like ordinary least squares and there are also settings in which PLS underperforms.

Using result from Basa, Cook, Forzani, and Marcos (2022), in Section 4.5 we describe the asymptotic normal distribution as $n, p \to \infty$ of a user-selected univariate linear combination of the PLS coefficients estimated from a one-component PLS fit and show how to construct asymptotic confidence intervals for mean and predicted values. We concluded from Theorem 4.2 and Corollary 4.6 that the conditions for asymptotic normality are more stringent than those for consistency and, as illustrated in Figure 4.4, that there is a potential for asymptotic bias.

4.2 One-component regressions

4.2.1 Model

We modify certain notation in model (2.5) to emphasize the restriction to a real response. We use $\sigma^2_{Y|X}$ in place of $\Sigma_{Y|X}$, σ^2_Y for var(Y) and, as mentioned at the outset of Chapter 3, we use $\sigma_{X,Y}$ in place of $\Sigma_{X,Y}$. Since model (2.5) is now restricted to a single component, $\Phi \in \mathbb{R}^p$ is a length-one basis vector for $\mathcal{E}_{\Sigma_X}(\mathcal{B})$ and Δ is a real number. To emphasize this structure, we use δ in place of Δ. By direct calculation from (2.5), we get that $\sigma_{X,Y} = \Phi\eta\delta$, and so without loss of generality, we can define $\Phi = \sigma_{X,Y}/\|\sigma_{X,Y}\|$, where for a vector V, $\|V\| = (V^T V)^{1/2}$, as defined in the notation section. The conclusion that this Φ is a basis vector of $\mathcal{E}_{\Sigma_X}(\mathcal{B})$ follows also from the discussion of the NIPALS and SIMPLS algorithms in Sections 3.2 and 3.4.

Construct $\Phi_0 \in \mathbb{R}^{p\times(p-1)}$ so that $(\Phi, \Phi_0) \in \mathbb{R}^{p\times p}$ is an orthonormal matrix. Then by construction $Y \perp\!\!\!\perp X \mid \Phi^T X$ and $\mathrm{cov}(\Phi^T X, \Phi_0^T X) = 0$. The independence condition means that, given $\Phi^T X$, there is no more information in X about Y, while the covariance conditions insures that $\Phi^T X$ cannot be obscured by high correlations. In consequence we again see $\Phi^T X$ as the material part of X, the part that furnishes information about Y, while $\Phi_0^T X$ is the immaterial part of X. Define the scalar var$(\Phi^T X) = \delta$ and the positive definite matrix var$(\Phi_0^T X) = \Delta_0 = \Phi_0^T \Sigma_X \Phi_0 \in \mathbb{R}^{(p-1)\times(p-1)}$. Then, since $\mathcal{E}_{\Sigma_X}(\mathcal{B})$ reduces Σ_X, we have the representation $\Sigma_X = \delta\Phi\Phi^T + \Phi_0\Delta_0\Phi_0^T$, which shows that δ is an eigenvalue of Σ_X with eigenvector Φ. Model (2.5) with a univariate response and a one-dimensional envelope can now be written as

$$\begin{aligned} Y_i &= \mu + (\delta^{-1}\|\sigma_{X,Y}\|)\, \Phi^T X_i + \epsilon_i, \; i = 1, \ldots, n, & (4.1) \\ \Sigma_X &= \delta\Phi\Phi^T + \Phi_0\Delta_0\Phi_0^T & (4.2) \\ \sigma^2_{Y|X} &= \mathrm{var}(\epsilon) > 0. \end{aligned}$$

In reference to the general form of the linear model for predictor envelopes given at (2.5), $\eta = \delta^{-1}\|\sigma_{X,Y}\|$. While δ is an eigenvalue of Σ_X with corresponding normalized eigenvector Φ,

$$\delta = \Phi^T \Sigma_X \Phi = \frac{\sigma^T_{X,Y}\Sigma_X\sigma_{X,Y}}{\|\sigma_{X,Y}\|^2}, \tag{4.3}$$

it need not be the largest eigenvalue of Σ_X. We see also that the vector of regression coefficients

$$\begin{aligned} \beta &= \Sigma_X^{-1}\sigma_{X,Y} = \Sigma_X^{-1}\Phi\|\sigma_{X,Y}\| = \delta^{-1}\sigma_{X,Y} \\ &= \frac{\|\sigma_{X,Y}\|^2}{\sigma^T_{X,Y}\Sigma_X\sigma_{X,Y}}\sigma_{X,Y} \end{aligned}$$

is an (unnormalized) eigenvector of Σ_X. This type of connection between the conditional mean $\mathrm{E}(Y \mid X)$ and the predictor variance Σ_X sets PLS, and envelope methodology generally, apart from other dimension reduction paradigms. It is a consequence of postulating that the marginal distribution of X contains relevant regression to inform on the regression.

It follows from (4.1), (4.2), and (4.3) that

$$\begin{aligned} \sigma^2_Y &= \mathrm{var}(Y) = \delta^{-1}\|\sigma_{X,Y}\|^2 + \sigma^2_{Y|X} & (4.4) \\ \mathrm{cov}(Y, \Phi^T X) &= \|\sigma_{X,Y}\| & (4.5) \end{aligned}$$

and that the squared population correlation coefficient between Y and $\Phi^T X$ is

$$\rho^2(Y, \Phi^T X) = \frac{\delta^{-1}\|\sigma_{X,Y}\|^2}{\delta^{-1}\|\sigma_{X,Y}\|^2 + \sigma^2_{Y|X}}. \tag{4.6}$$

Since Y and X are required to have a joint distribution, the covariance matrix of the concatenated variable $C = (X^T, Y)^T$ is similar to (2.9):

$$\Sigma_C = \begin{pmatrix} \Sigma_X & \sigma_{X,Y} \\ \sigma^T_{X,Y} & \sigma^2_Y \end{pmatrix} = \begin{pmatrix} \delta\Phi\Phi^T + \Phi_0\Delta_0\Phi_0^T & \delta\eta\Phi \\ \delta\eta\Phi^T & \sigma^2_Y \end{pmatrix}. \tag{4.7}$$

4.2.2 Estimation

The usual estimators of Σ_X and $\sigma_{X,Y}$ are S_X and $S_{X,Y}$, as given in Section 1.2. With a single component, the estimated PLS weight vector is

$$\begin{aligned}\widehat{W} &\propto S_{X,Y}, \\ \widehat{\Phi} &= \frac{S_{X,Y}}{\|S_{X,Y}\|}, \\ \widehat{\delta}_{\text{pls}} &= \frac{S_{X,Y}^T S_X S_{X,Y}}{\|S_{X,Y}\|^2}.\end{aligned}$$

Let $(\widehat{\Phi}, \widehat{\Phi}_0) \in \mathbb{R}^{p\times p}$ be an orthonormal matrix. Then, the PLS estimators of Δ_0 and Σ_X are

$$\begin{aligned}\widehat{\Delta}_{0,\text{pls}} &= \widehat{\Phi}_0^T S_X \widehat{\Phi}_0, \\ \widehat{\Sigma}_{X,\text{pls}} &= \widehat{\delta}_{\text{pls}} \widehat{\Phi}\widehat{\Phi}^T + \widehat{\Phi}_0 \widehat{\Delta}_{0,\text{pls}} \widehat{\Phi}_0^T.\end{aligned}$$

The PLS estimator of β is then

$$\widehat{\beta}_{\text{pls}} = \frac{\|S_{X,Y}\|^2}{S_{X,Y}^T S_X S_{X,Y}} \times S_{X,Y} = \widehat{\delta}_{\text{pls}}^{-1} S_{X,Y}.$$

The estimator of μ is $\widehat{\mu}_{\text{pls}} = \bar{Y} - \widehat{\beta}_{\text{pls}}^T \bar{X}$, so with centered predictors the fitted values can be expressed as $\widehat{Y}_i = \bar{Y} + \widehat{\beta}_{\text{pls}}^T (X_i - \bar{X})$. The predicted value of Y at a new value X_N of X is then $\widehat{Y}_N = \widehat{\mathrm{E}}(Y \mid X_N) = \bar{Y} + \widehat{\beta}_{\text{pls}}^T (X_N - \bar{X})$.

4.3 Asymptotic distribution of $\widehat{\beta}_{\text{pls}}$ as $n \to \infty$ with p fixed

In this section we discuss the asymptotic approximations of the distribution of $\widehat{\beta}_{\text{pls}}$ based on letting $n \to \infty$ with p fixed. The discussion is confine to the regression structure of Section 4.2.1: a one-component ($q = 1$) regression with a univariate response ($r = 1$) based on an independent and identically distributed sample $(Y_i, X_i), i = 1, \ldots, n$, which has finite fourth moments. Under this structure Cook, Helland, and Su (2013, Proposition 10) proved the following.

Proposition 4.1. *Assume the one-component regression structure given in (4.1) and (4.7) with p fixed for an independent and identically distributed sample $(Y_i, X_i), i = 1, \ldots, n$, which has finite fourth moments. Then*

(i) The PLS estimator $\widehat{\beta}_{\rm pls}$ of β has the expansion

$$\sqrt{n}(\widehat{\beta}_{\rm pls} - \beta) = n^{-1/2}\delta^{-1}\sum_{i=1}^{n}\{(X_i - \mu_X)\epsilon_i + Q_\Phi(X_i - \mu_X)(X_i - \mu_X)^T\beta\} + O_p(n^{-1/2}),$$

where ϵ is the error for model (4.1).

(ii) $\sqrt{n}(\widehat{\beta}_{\rm pls} - \beta)$ is asymptotically normal with mean 0 and variance

$$\text{avar}(\sqrt{n}\widehat{\beta}_{\rm pls}) = \delta^{-2}\left\{\Sigma_X\sigma^2_{Y|X} + \text{var}(Q_\Phi(X - \mu_X)(X - \mu_X)^T\beta)\right\}.$$

(iii) If, in addition, $P_\Phi X \perp\!\!\!\perp Q_\Phi X$ then

$$\text{avar}(\sqrt{n}\widehat{\beta}_{\rm pls}) = \delta^{-1}\sigma^2_{Y|X}P_\Phi + \delta^{-2}\sigma^2_Y\Phi_0\Delta_0\Phi_0^T.$$

The proof of Proposition 4.1 is rather long and so Cook, Helland, and Su (2013) provided only a few key steps. We provide more details in the proof given in Appendix A.4.1.

According to the results in parts (i) and (ii) of Proposition 4.1, $\widehat{\beta}_{\rm pls}$ is asymptotically normal and that its asymptotic covariance depends on fourth moments of the marginal distribution of X. However, if $P_\Phi X$ is independent of $Q_\Phi X$, as required in part (iii), then only second moments are needed. The condition of part (iii) is implied when X is normally distributed:

$$\text{cov}(P_\Phi X, Q_\Phi X) = P_\Phi \Sigma_X Q_\Phi = 0$$

since the envelope reduces Σ_X.

Under the conditions of Proposition 4.1 (iii),

$$\text{avar}(\sqrt{n}\widehat{\beta}_{\rm ols}) = \sigma^2_{Y|X}\left\{\delta^{-1}P_\Phi + \Phi_0\Delta_0^{-1}\Phi_0^T\right\}.$$

Using this result, the asymptotic covariance in part (iii) of Proposition 4.1 can be expressed equivalently as

$$\text{avar}(\sqrt{n}\widehat{\beta}_{\rm pls}) = \text{avar}(\sqrt{n}\widehat{\beta}_{\rm ols}) + \Phi_0\Delta_0^{-1/2} \times \left\{(\sigma^2_Y/\sigma^2_{Y|X})(\Delta_0^2/\delta^2) - I_{p-1}\right\}\Delta_0^{-1/2}\Phi_0^T\sigma^2_{Y|X}.$$

From this we see that the performance of PLS relative to OLS depends on the strength of the regression as measured by the ratio $\sigma^2_Y/\sigma^2_{Y|X} \geq 1$

and on the level of collinearity as measured by Δ_0^2/δ^2. For every level of collinearity there is a regression so that PLS does worse asymptotically than OLS, and for every regression strength, there is a level of collinearity so that PLS does better than OLS. For instance, if $\Sigma_X = \sigma_X^2 I_p$ then $\text{avar}(\sqrt{n}\widehat{\beta}_{\text{pls}}) = \text{avar}(\sqrt{n}\widehat{\beta}_{\text{ols}}) + \Phi_0\Phi_0^T(\sigma_Y^2 - \sigma_{Y|X}^2)$ and the asymptotic covariance of the PLS estimator is never less than that of the OLS estimator. In contrast, recall from Proposition 2.1 that the maximum likelihood envelope estimator never does worse than OLS, $\text{avar}(\sqrt{n}\widehat{\beta}) \leq \text{avar}(\sqrt{n}\widehat{\beta}_{\text{ols}})$. Since $\beta \neq 0$ and $r = 1$, it follows from Corollary 2.1 that these asymptotic variances are equal when $\Sigma_X = \sigma_X^2 I_p$.

4.3.1 Envelope vs. PLS estimators

Since the envelope estimator $\widehat{\beta}$ of β under model (4.1) with multivariate normality is the maximum likelihood estimator, it follows from general likelihood theory that $\text{avar}(\sqrt{n}\widehat{\beta}) \leq \text{avar}(\sqrt{n}\widehat{\beta}_{\text{pls}})$. The following corollary, which makes use of Proposition 2.1 and is discussed in Appendix 4.1, addresses the difference between the envelope and PLS estimators in a special case of model (4.1).

Corollary 4.1. *Assume that the regression of $Y \in \mathbb{R}^1$ on $X \in \mathbb{R}^p$ follows model (4.1) with $\Delta_0 = \delta_0 I_{p-1}$ and that $C = (Y, X^T)^T$ correspondingly follows a multivariate normal distribution. Then*

$$\text{avar}(\sqrt{n}\widehat{\beta}) = \sigma_{Y|X}^2\delta^{-1}\Phi\Phi^T + \eta^2\left(\eta^2\delta_0/\sigma_{Y|X}^2 + (\delta_0/\delta)(1-\delta/\delta_0)^2\right)^{-1}\Phi_0\Phi_0^T$$

$$\text{avar}(\sqrt{n}\widehat{\beta}_{\text{pls}}) = \sigma_{Y|X}^2\delta^{-1}\Phi\Phi^T + (\sigma_Y^2\delta_0/\delta^2)\Phi_0\Phi_0^T,$$

where $\text{avar}(\sqrt{n}\widehat{\beta})$ *was obtained as a special case of Proposition 2.1 and* $\eta = \delta^{-1}\|\sigma_{X,Y}\|$*, as defined following (4.2).*

The first addends on the right-hand side of the asymptotic variances are the same and correspond to the asymptotic variances when Φ is known. This is as it must be since the envelope and PLS estimators are identical when Φ is known. Thus, the second addends on the right-hand side of the asymptotic variances correspond to the cost of estimating the envelope. To compare the asymptotic variances we compare the costs of estimating the envelope. Let

$$\begin{aligned}\text{Cost for } \widehat{\beta} &= \eta^2\left(\eta^2\delta_0/\sigma_{Y|X}^2 + (\delta_0/\delta)(1-\delta/\delta_0)^2\right)^{-1} \\ \text{Cost for } \widehat{\beta}_{\text{pls}} &= \sigma_Y^2\delta_0/\delta^2.\end{aligned}$$

It follows from Corollary 4.1 that

$$\frac{\text{Cost for } \widehat{\beta}}{\text{Cost for } \widehat{\beta}_{\text{pls}}} = \frac{\tau(1-\tau)}{(\delta_0/\delta - \tau)^2 + \tau(1-\tau)} \leq 1,$$

where $\tau = \sigma^2_{Y|X}/\sigma^2_Y$, which is essentially one minus the population multiple correlation coefficient for the regression of Y on X. The relative cost of estimating the envelope is always less than or equal to one, indicating that PLS is the more variable method asymptotically. This is expected since the envelope estimator inherits optimal properties from general likelihood theory. Otherwise, the cost ratio depends on the signal strength as measured by τ and the level of collinearity, as measured by δ_0/δ. The envelope estimator tends to do much better than PLS in low signal regressions where τ is close to 1 and in high signal regressions where τ is close to 0. If there is a high degree of collinearity so δ_0/δ is small and $(\delta_0/\delta - \tau)^2 \approx \tau^2$ then the cost ratio reduces to $1 - \tau$ and again the envelope estimator will do better than PLS in low signal regressions. On the other hand, if the level of collinearity is about the same as the signal strength $\delta_0/\delta - \tau \approx 0$ then the PLS estimator will do about the same as the envelope estimator asymptotically.

The comparison of this section is based on approximations achieved by letting $n \to \infty$ with p fixed. They may have little relevance in regression where n is not large relative to p. When $n < p$, PLS regression is still serviceable, while the maximum likelihood estimation based on model (4.1) is not.

4.3.2 Mussels muscles

Cook (2018, Section 4.4.4) used PLS to analyze data from an ecological study of horse mussels sampled from the Marlborough Sounds off the coast of New Zealand. The response variable Y is the logarithm of the mussel's muscle mass, and four predictors are the logarithms of shell height, shell width, shell length (each in millimeters), and shell mass (in grams). The sample size is $n = 79$ and $p = 4$. Cook (2018) obtained a clear inference that the regression required only a single component, $q = 1$, so these data fit the requirements of this section.

As background, Figure 4.1 shows a plot of the response versus fitted values $\widehat{Y}_i = \widehat{\mathrm{E}}(Y \mid X_i) = \widehat{\mu} + \widehat{\beta}^T_{\text{pls}} X_i$ from the one-component PLS fit. The plot shows a clear linear trend with no sign of notable anomalies. Figure 4.2 shows a plot of the fitted values from the one-component PLS fit against the OLS fitted values from the fit of Y on all four predictors.

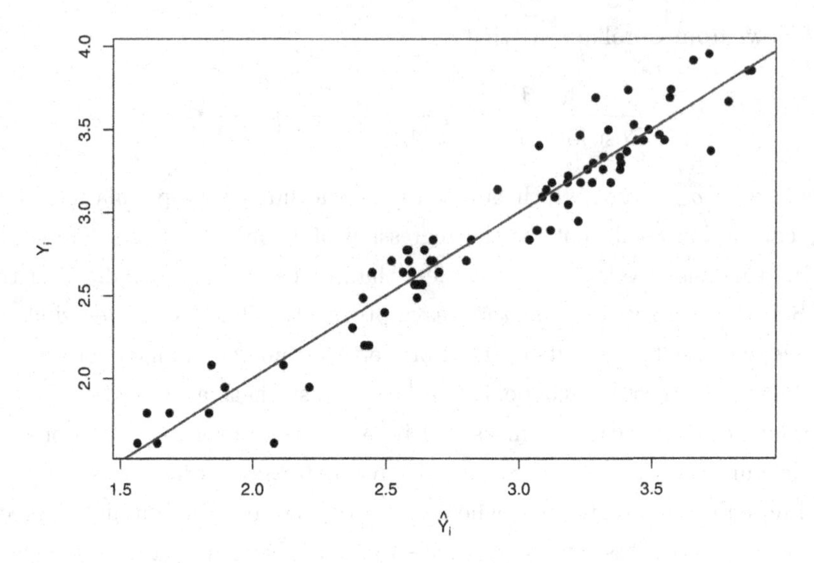

FIGURE 4.1
Mussels' data: Plot of the observed responses Y_i versus the fitted values $\widehat{Y}_i$ from the PLS fit with one component, $q = 1$

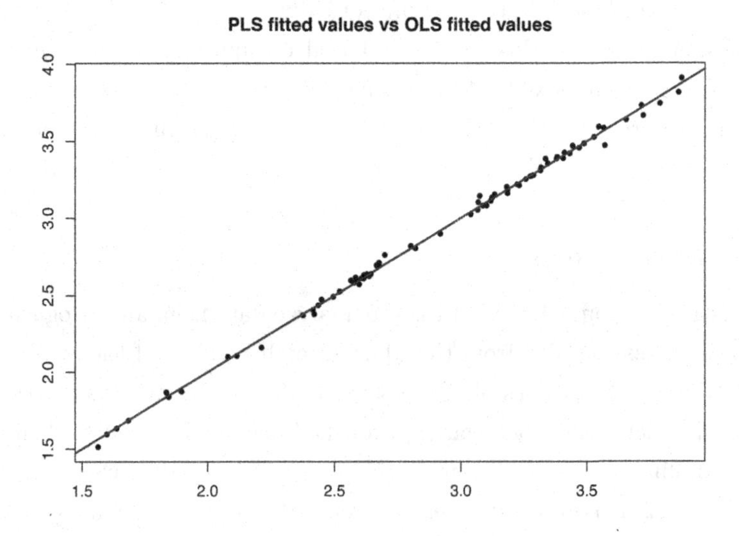

FIGURE 4.2
Mussels' data: Plot of the fitted values $\widehat{Y}_i$ from the PLS fit with one component, $q = 1$, versus the OLS fitted values. The diagonal line $y = x$ was included for clarity.

TABLE 4.1

Coefficient estimates and corresponding asymptotic standard deviations (S.D.) from three fits of the mussels' data: Ordinary least squares, (OLS), partial least squares with one component (PLS), and envelope with one dimension (ENV). Results for OLS and ENV are from Cook (2018).

	OLS		PLS, $q=1$		ENV, $q=1$	
	Estimate	S.D.	Estimate	S.D.	Estimate	S.D.
$\widehat{\beta}_1$	0.741	0.410	0.142	0.0063	0.141	0.0052
$\widehat{\beta}_2$	−0.113	0.399	0.153	0.0067	0.154	0.0056
$\widehat{\beta}_3$	0.567	0.118	0.625	0.0199	0.625	0.0194
$\widehat{\beta}_4$	0.170	0.304	0.206	0.0086	0.206	0.0073

The coefficient estimates from the OLS, PLS, and envelope fits along with their asymptotic standard errors are shown in Table 4.1. The OLS standard errors are the usual ones. The PLS standard errors were obtained by plugging in estimates into the asymptotic variances given in Proposition 4.1. The envelope standard errors were obtained similarly from the asymptotic covariance matrix given in Proposition 2.1. The PLS and envelope standard errors were computed under multivariate normality of X. We see that the OLS standard errors are between about 6 and 60 times those of the PLS estimator. These types of differences are not unusual in the envelope literature, as illustrated in examples from Cook (2018). The PLS standard errors are between about 2 and 20 percent larger than the corresponding envelope standard errors, but are still much smaller than the OLS standard errors.

Shown in Table 4.2 are the estimates of the predictor covariance matrix Σ_X from the envelope, OLS, and PLS fits of the mussels regression. The estimate of Σ_X from the OLS fit is S_X as defined in Section 1.2.1. The estimates from the envelope and PLS fits were obtained by substituting the corresponding estimates into (4.2). Estimation from the PLS fit is discussed in Section 4.2.2. We see that the envelope and PLS estimates are both close to the OLS estimate, which supports the finding that the dimension of the envelope is 1.

TABLE 4.2
Mussels' muscles: Estimates of the covariance matrix Σ_X from the envelope, OLS and PLS ($q = 1$) fits.

	Envelope			
X	$\log H$	$\log L$	$\log S$	$\log W$
$\log H$	0.030	0.031	0.123	0.041
$\log L$		0.035	0.134	0.045
$\log S$			0.550	0.180
$\log W$				0.063
	OLS			
X	$\log H$	$\log L$	$\log S$	$\log W$
$\log H$	0.030	0.031	0.123	0.041
$\log L$		0.035	0.134	0.045
$\log S$			0.550	0.180
$\log W$				0.063
	PLS			
X	$\log H$	$\log L$	$\log S$	$\log W$
$\log H$	0.031	0.032	0.126	0.042
$\log L$		0.036	0.136	0.046
$\log S$			0.557	0.182
$\log W$				0.064

4.4 Consistency of PLS in high-dimensional regressions

In this section we discuss consistency in high-dimensional regressions, essentially reporting on the work of Cook and Forzani (2018). The context begins with that used for Proposition 4.1 – the data $(Y_i, X_i), i = 1, \ldots, n$, arise as independent copies of (Y, X), which follows the single-response, one-component PLS regression described in Section 4.2.1. Here we require, in addition, that (Y, X) follows a non-singular $(p+1)$-dimensional multivariate normal distribution, and we allow n and p to diverge in various alignments. The normality assumption is used to facilitate asymptotic calculations and to connect with related results in the literature; nevertheless, simulations and experience in practice indicate that it is not essential for the results of this section to be useful elucidations. To avoid trivial cases, we assume throughout that $\beta \neq 0$. Estimators are as described in Section 4.2.2.

Recall from Section 1.2 that $\mathbb{Y}_0 = (Y_1, \ldots, Y_n)^T$ denotes the vector of uncentered responses and that $\mathbb{X}$ denotes the $n \times p$ matrix with rows $(X_i - \bar{X})^T$, $i = 1, \ldots, n$. Recall also that $S_{X,Y} = n^{-1}\mathbb{X}^T\mathbb{Y}_0$ and $S_X = n^{-1}\mathbb{X}^T\mathbb{X} \geq 0$ represent the usual moment estimators of $\sigma_{X,Y}$ and Σ_X using n for the divisor. We use $W_q(M)$ to denote the Wishart distribution with q degrees of freedom and scale matrix M. With $W = \mathbb{X}^T\mathbb{X} \sim W_{n-1}(\Sigma_X)$, we can represent $S_X = W/n$, $S_{X,Y} = n^{-1}(W\beta + \mathbb{X}^T\varepsilon)$, where ε is the $n \times 1$ vector with the model errors ϵ_i as elements. (This use of the W notation is different from the weight matrix in PLS.) We use the notation "$a_k \asymp b_k$" to mean that, as $k \to \infty$, $a_k = O(b_k)$ and $b_k = O(a_k)$, and describe a_k and b_k as then being asymptotically equivalent.

4.4.1 Goal

Our general goal is to gain insights into the predictive performance of $\widehat{\beta}_{\text{pls}}$ as n and p grow in various alignments. In this section we describe how that goal will be pursued.

Let $Y_N = \mu + \beta^T(X_N - \mathrm{E}(X)) + \epsilon_N$ denote a new observation on Y at a new independent observation X_N of X. The PLS predicted value of Y_N at X_N is $\widehat{Y}_N = \bar{Y} + \widehat{\beta}^T_{\text{pls}}(X_N - \bar{X})$, giving a difference of

$$\begin{aligned}\widehat{Y}_N - Y_N &= \bar{Y} - \mu + (\widehat{\beta}_{\text{pls}} - \beta)^T(X_N - \mathrm{E}(X)) - (\widehat{\beta}_{\text{pls}} - \beta)^T(\bar{X} - \mathrm{E}(X)) \\ &\quad - \beta^T(\bar{X} - \mathrm{E}(X)) + \epsilon_N.\end{aligned}$$

The first term $\bar{Y} - \mu = O_p(n^{-1/2})$. Since $\sigma^2_Y = \beta^T\Sigma_X\beta + \sigma^2_{Y|X}$ must remain constant as p grows, $\beta \neq 0$ and $\Sigma_X > 0$, we see that $\beta^T\Sigma_X\beta = O(1)$ as $p \to \infty$. Thus the fourth term $\beta^T(\bar{X} - \mathrm{E}(X)) = O_p(n^{-1/2})$ by Chebyschev's inequality:

$$\mathrm{var}(\beta^T(\bar{X} - \mathrm{E}(X))) = \beta^T\Sigma_X\beta/n \to 0 \text{ as } n, p \to \infty.$$

The term $(\widehat{\beta}_{\text{pls}} - \beta)^T(\bar{X} - \mathrm{E}(X))$ must have order smaller than or equal to the order of $(\widehat{\beta}_{\text{pls}} - \beta)^T(X_N - \mathrm{E}(X))$, which will be at least $O_p(n^{-1/2})$.

Consequently, we have the essential asymptotic representation

$$\widehat{Y}_N - Y_N = O_p\left\{(\widehat{\beta}_{\text{pls}} - \beta)^T(X_N - \mathrm{E}(X))\right\} + \epsilon_N \text{ as } n, p \to \infty.$$

Since ϵ_N is the intrinsic error in the new observation, the n, p-asymptotic behavior of the prediction $\widehat{Y}_N$ is governed by the estimative performance of $\widehat{\beta}_{\text{pls}}$ as measured by

$$\begin{aligned}D_N &:= (\widehat{\beta}_{\text{pls}} - \beta)^T\omega_N \\ &= \left(S^T_{X,Y}\widehat{\Phi}(\widehat{\Phi}^TS_X\widehat{\Phi})^{-1}\widehat{\Phi}^T - \sigma^T_{X,Y}\Phi(\Phi^T\Sigma_X\Phi)^{-1}\Phi^T\right)\omega_N, \quad (4.8)\end{aligned}$$

where $\omega_N = X_N - \mathrm{E}(X) \sim N(0, \Sigma_X)$. Consistency of $\widehat{\beta}_{\mathrm{pls}}$ is discussed at the end of Section 4.6.4. Until then, we focus exclusively on predictions via D_N.

We next give characterizations of populations constructions that will be helpful when studying the asymptotic properties of D_N.

4.4.2 Quantities governing asymptotic behavior of D_N

Recall that $\beta = \eta\Phi$, $\eta = \delta^{-1}\|\sigma_{X,Y}\|$ and, from (4.3), $\delta = \Phi^T\Sigma_X\Phi$. Properties of D_N depend in part on the asymptotic behavior of

$$\beta^T\Sigma_X\beta = \eta^2\Phi^T\Sigma_X\Phi = \delta^{-2}\|\sigma_{X,Y}\|^2\Phi^T\Sigma_X\Phi = \delta^{-1}\|\sigma_{X,Y}\|^2 \tag{4.9}$$

as $p \to \infty$. We know that $\sigma_Y^2 = \beta^T\Sigma_X\beta + \sigma_{Y|X}^2$. Since σ_Y^2 remains fixed, $\sigma_Y^2 \geq \beta^T\Sigma_X\beta \geq 0$ as $p \to \infty$. It is technically possible to have $\beta^T\Sigma_X\beta \to \sigma_Y^2$ so in the limit $\sigma_{Y|X}^2 = 0$. We find this implausible as a useful approximation and so we assume that $\sigma_{Y|X}^2$ is bounded away from 0.

On the other extreme, it is possible for $\beta^T\Sigma_X\beta \to 0$ as $p \to \infty$, in which case there is no signal left in the limit and $\rho^2(Y, \Phi^TX) \to 0$, where ρ is as given in (4.6). To gain intuition into scenarios where this might happen, consider

$$\beta^T\Sigma_X\beta = \frac{\|\sigma_{X,Y}\|^2}{\delta} = \frac{\mathrm{cov}^2(\Phi^TX, Y)}{\mathrm{var}(\Phi^TX)},$$

where the second equality follows from (4.5) and $\delta = \mathrm{var}(\Phi^TX) = \Phi^T\Sigma_X\Phi$ from (4.3). The notions of abundant and sparse regression provide useful intuition on the asymptotic behavior of PLS. At the beginnings of Chapters 1 and 4 we characterized an abundant regression informally as one in which many predictors contribute information about the response, while a sparse regression is one in which few predictors contain response information. We now formalize those ideas in the context of one-component regressions. This definition is generalized to multi-component regressions in Section 4.6.1.

Definition 4.1. *A one-component regression (4.1) is said to be abundant if* $\|\sigma_{X,Y}\| \to \infty$ *as* $p \to \infty$. *It is said to be sparse if* $\|\sigma_{X,Y}\|$ *is bounded as* $p \to \infty$.

We assume for illustration and to gain intuition about abundance and sparsity in the context of one-component regressions that X can be represented as $X = V\nu + \zeta$, where ν is a real stochastic latent variable that is related to X via the non-stochastic vector V and the X-errors $\zeta \perp\!\!\!\perp (\nu, Y)$. Without loss of generality we take $\mathrm{var}(\nu) = 1$. We know that, for our one-component model, $\sigma_{X,Y}$ is a basis for the one-dimensional envelope $\mathcal{E}_{\Sigma_X}(\mathcal{B})$. Accordingly, in terms of our illustrative model for X, $\sigma_{X,Y} = V\text{cov}(\nu, Y)$ with $\text{cov}(\nu, Y) \neq 0$ not depending on p, so we can take our normalized basis vector to be $\Phi = V/\|V\|$ and then re-express our illustrative model as $X = \Phi\|V\|\nu + \zeta$. In consequence,

$$\begin{aligned} \beta^T \Sigma_X \beta &= \frac{\text{cov}^2(\Phi^T X, Y)}{\text{var}(\Phi^T X)} \\ &= \frac{\|V\|^2 \text{cov}^2(\nu, Y)}{\|V\|^2 + \Phi^T \Sigma_\zeta \Phi} \\ &= \frac{\|\sigma_{X,Y}\|^2}{\|\sigma_{X,Y}\|^2/\text{cov}^2(\nu, Y) + \Phi^T \Sigma_\zeta \Phi}, \end{aligned} \tag{4.10}$$

where $\Sigma_\zeta = \text{var}(\zeta) > 0$. This result enables us to gain intuition on a few different types of regressions.

Sparse regression: Perhaps the most common instance of a sparse regression is one in which only a finite number of elements of $\sigma_{X,Y}$ are non-zero. Correspondingly, the same elements of V and Φ are non-zero. Result (4.10) shows that in such sparse regressions $\beta^T \Sigma_X \beta$ is bounded away from 0 as $p \to \infty$ and so there will always be a signal present, although that signal may be hard to find in application.

Abundant regression, bounded Σ_ζ: Abundant regressions are those in which $\|V\|^2 \to \infty$ as $p \to \infty$. Since the eigenvalue of Σ_X that goes with eigenvector Φ is

$$\delta = \text{var}(\Phi^T X) = \|V\|^2 + \Phi^T \Sigma_\zeta \Phi,$$

it is generally unreasonable in abundant regressions to require the eigenvalues of Σ_X to be bounded. However, it does seem reasonable to require bounded eigenvalues of Σ_ζ, in which case $\beta^T \Sigma_X \beta$ is again bounded away from 0 as $p \to \infty$.

Abundant regression, unbounded Σ_ζ: It is possible in abundant regressions to have Σ_ζ unbounded and yet $\beta^T \Sigma_X \beta$ is still bounded away from 0. As an instance of this, assume that V is a vector of ones $V = 1_p$ and that Σ_ζ is compound symmetric with off-diagonal element $1 > \rho > 0$ and diagonal element 1. Then, as shown in Appendix A.4.3,

$$\Sigma_\zeta = \rho 1_p 1_p^T + (1 - \rho) I_p = \{(p - 1)\rho + 1\} P_{1_p} + (1 - \rho) Q_{1_p}.$$

Then, as $p \to \infty$,

$$\beta^T \Sigma_X \beta = \frac{p\,\text{cov}^2(\nu, Y)}{p + (p - 1)\rho + 1} \to \frac{\text{cov}^2(\nu, Y)}{\rho + 1}.$$

Since the eigenvalue of Σ_ζ associated with Φ increases linearly with p, $\beta^T \Sigma_X \beta$ is still bounded away from 0.

For $\beta^T \Sigma_X \beta$ to converge to 0 in one-component regressions giving rise to (4.10), we need $\Phi^T \Sigma_\zeta \Phi$ to increase faster than $\|V\|^2$. While that is technically possible, we find it practically implausible. As a consequence of the discussion so far, we assume that $\beta^T \Sigma_X \beta$ is bounded away from zero and infinity as $p \to \infty$, implying that there will always be some regression signal with $\sigma_Y^2 > \sigma_{Y|X}^2 > 0$. PLS has been found to be serviceable in many application areas, particularly chemometrics, so we expect that this is not a severe restriction. These restrictions can be imposed succinctly by requiring that $\beta^T \Sigma_X \beta \asymp 1$ as $p \to \infty$. Since now $\beta^T \Sigma_X \beta \asymp 1$, we have from (4.9) that $\delta^{-1} \|\sigma_{X,Y}\|^2 \asymp 1$ and

$$\delta^{-2} \|\sigma_{X,Y}\|^2 \Phi^T \Sigma_X \Phi = \delta^{-2} \sigma_{X,Y}^T \Sigma_X \sigma_{X,Y} \asymp 1.$$

In consequence,

$$\delta \asymp \|\sigma_{X,Y}\|^2 \text{ and } \sigma_{X,Y}^T \Sigma_X \sigma_{X,Y} \asymp \|\sigma_{X,Y}\|^4.$$

This implies that δ, the eigenvalue of Σ_X associated with Φ, provides a measure of the signal that is asymptotically equivalent to $\|\sigma_{X,Y}\|^2$.

4.4.3 Consistency results

The asymptotic properties of PLS predictions depend also on the relationship between $\mathrm{tr}(\Delta_0)$, which reflects the variation of *the noise in* X, and $\|\sigma_{X,Y}\|^2$, which is a measure of *the signal coming from* X. The following theorem, which stems from Cook and Forzani (2018) and is proven in Appendix A.4.4, provides a characterization of the asymptotic behavior of D_N in terms of the noise to signal ratio $\Delta_\sigma := \Delta_0 / \|\sigma_{X,Y}\|^2$. In preparation, define for $j = 1, 2, 3, 4$,

$$K_j(n, p) = \frac{\mathrm{tr}(\Delta_0^j)}{n \|\sigma_{X,Y}\|^{2j}} = \frac{\mathrm{tr}(\Delta_\sigma^j)}{n}. \tag{4.11}$$

In this section we will use only $K_1(n, p)$ and $K_2(n, p)$. We will make use of $K_3(n, p)$ and $K_4(n, p)$ when considering asymptotic distributions in Section 4.5.

Theorem 4.1. *If model (4.1) holds, $\beta^T \Sigma_X \beta \asymp 1$ and $K_j(n, p)$, $j = 1, 2$, converges to 0 as $n, p \to \infty$ then*

$$D_N = O_p \left\{ n^{-1/2} + K_1(n, p) + K_2^{1/2}(n, p) \right\}.$$

This theorem can be used to gain insights into specific types of regressions. In particular, in many regressions the eigenvalues of Δ_0 may be bounded as $p \to \infty$, reflecting that the noise in X is bounded asymptotically. For instance, this structure can occur when the predictor variation is compound symmetric:

$$\Sigma_X = \pi^2 \left\{(1 - \rho + \rho p)P_{1_p} + (1 - \rho)Q_{1_p}\right\},$$

where $\pi^2 = \text{var}(X_j)$ for all j and $\rho = \text{corr}(X_j, X_k)$ for all $j \neq k$. Since we are restricting consideration to $q = 1$, $\mathcal{E}_{\Sigma_X}(\mathcal{B})$ must fall in one of the two eigenspaces of Σ_X: either $\mathcal{E}_{\Sigma_X}(\mathcal{B}) = \text{span}(1_p)$ or $\mathcal{E}_{\Sigma_X}(\mathcal{B}) \subseteq \text{span}^{\perp}(1_p)$. If $\mathcal{E}_{\Sigma_X}(\mathcal{B}) = \text{span}(1_p)$ then $\Delta_0 = (1 - \rho)I_{p-1}$ which has bounded eigenvalues. We then have $\Phi = 1_p/\sqrt{p}$, $\delta = \pi^2(1 - \rho + \rho p)$, signal $\|\sigma_{X,Y}\|^2 \asymp \delta \asymp p$ and $\text{tr}(\Delta_0) = \text{tr}((1 - \rho)Q_{1_p}) \asymp p$.

If the eigenvalues of Δ_0 are bounded as $p \to \infty$ then so are the eigenvalues of Δ_0^k for any fixed k. Consequently, $\text{tr}(\Delta_\sigma^k) \asymp p/\|\sigma_{X,Y}\|^{2k}$ and we have

$$\begin{aligned} K_1(n,p) &\asymp \frac{p}{n\|\sigma_{X,Y}\|^2} \\ K_2^{1/2}(n,p) &\asymp \frac{\sqrt{p}}{\sqrt{n}\|\sigma_{X,Y}\|^2}. \end{aligned}$$

Comparing the magnitudes of the K_j's as they approach 0, we have

Corollary 4.2. *Assume the conditions of Theorem 4.1 and that the eigenvalues of Δ_0 are bounded as $p \to \infty$. Then*

I.

$$D_N = O_p\left\{\frac{1}{\sqrt{n}} + \frac{\sqrt{p}}{\sqrt{n}\|\sigma_{X,Y}\|^2} + \frac{p}{n\|\sigma_{X,Y}\|^2}\right\}.$$

II. (Abundant) If $\|\sigma_{X,Y}\|^2 \asymp p$ then $D_N = O_p\{1/\sqrt{n}\}$.

III. (Sparse) If $\|\sigma_{X,Y}\|^2 \asymp 1$ then $D_N = O_p\{\sqrt{p}/\sqrt{n}\}$.

The first conclusion tells us that with bounded noise the predictive performance of PLS methods depends on an interplay between the sample size n, the number of predictors p and the signal as measured by $\|\sigma_{X,Y}\|^2$. The second and third conclusions relate to specific instances of that interplay. The second conclusion says informally that if most predictors are correlated with the response then PLS predictions will converge at the usual root-n rate, even if $n < p$. The third conclusion says that if few predictors are correlated with the response or $\|\sigma_{X,Y}\|$ increases very slowly, then for predictive consistency the sample size needs to be large relative to the number of predictors. The third case clearly suggests a sparse solution, while the second case does not. Sparse versions of PLS regression have been proposed by Chun and Keleş (2010) and Liland et al. (2013). In view of the apparent success of PLS regression over the past four decades, it is reasonable to conclude that many regressions are closer to abundant than sparse. The compound symmetry example with $\mathcal{E}_{\Sigma_X}(\mathcal{B}) = \text{span}(1_p)$ is covered by Corollary 4.2 (II) and so the convergence rate is $O_p(n^{-1/2})$.

The next two corollaries give more nuanced results by tying the signal $\|\sigma_{X,Y}\|^2$ and the number of predictors p to the sample size n. Corollary 4.3 is intended to provide insights when the sample size exceeds the number of predictors, while Corollary 4.4 is for regressions where the sample size is less than the number of predictors.

Corollary 4.3. *Assume the conditions of Theorem 4.1 and that the eigenvalues of Δ_0 are bounded as $p \to \infty$. Assume also that $p \asymp n^a$ for $0 < a \leq 1$ and that $\|\sigma_{X,Y}\|^2 \asymp p^s \asymp n^{as}$ for $0 \leq s \leq 1$. Then*

I. $D_N = O_p\left(n^{-1/2} + n^{a(1/2-s)-1/2} + n^{a(1-s)-1}\right) = O_p(n^{-1/2} + n^{a(1/2-s)-1/2})$.

II. (Abundant) $D_N = O_p\left(n^{-1/2}\right)$ *if* $s \geq 1/2$.

III. (In-between) $D_N = O_p\left(n^{-1/2+a(1/2-s)}\right)$ *if* $0 < s \leq 1/2$.

IV. (Sparse) $D_N = O_p\left(n^{-(1-a)/2}\right)$ *if* $s = 0$.

The requirement from Theorem 4.1 that $K_j(n,p)$ converge to 0 forces the terms in the order given in conclusion I to converge to 0 to ensure consistency, which limits the values of a and s jointly. The corollary predicts that $s = 1/2$ is a breakpoint for $\sqrt{n}$-convergence of PLS predictions in high-dimensional regressions. If the signal accumulates at a rate that is greater than $\|\sigma_{X,Y}\|^2 \asymp p^{1/2}$, so $s = 1/2$, then predictions converge at the usual root-n rate. This indicates that there is considerably more leeway in the signal rate to obtain $\sqrt{n}$-convergence than that described by Corollary 4.2(I). Otherwise a price is paid in terms of a slower rate of convergence. For example, if $\|\sigma_{X,Y}\|^2 \asymp p^{1/4}$ and $p \asymp n$, so $s = 1/4$ and $a = 1$, then $D_N = O_p(n^{-1/4})$, which is considerably slower than the root-n rate of case II. This corollary also gives additional characterizations of how PLS predictions will do in sparse regressions. From conclusion IV, we see that if $s = 0$, so $\|\sigma_{X,Y}\|^2 \asymp 1$, and $a = 0.8$, then $D_N = O_p(n^{-0.1})$, which would not normally yield useful results but could likely be improved by using a sparse fit. On the other hand, if a is small, say

$a = .2$ so $p \asymp n^{0.2}$, then $D_N = O_p(n^{-0.4})$, which is a reasonable rate that may yield useful conclusions. The in-between category is meant to signal that those regressions can behave rather like an abundant regression if s is close to 1/2 or like a sparse regression when s is close to 0.

The next corollary allows $p > n$.

Corollary 4.4. *Assume the conditions of Theorem 4.1 and that the eigenvalues of Δ_0 are bounded as $p \to \infty$. Assume also that $p \asymp n^a$ for $a \geq 1$ and that $\|\sigma_{X,Y}\|^2 \asymp p^s \asymp n^{sa}$ for $0 \leq s \leq 1$. Then*

I. $D_N = O_p\left(n^{-1/2} + n^{a(1/2-s)-1/2} + n^{a(1-s)-1}\right) = O_p\left(n^{-1/2} + n^{a(1-s)-1}\right)$.

II. (Abundant) $D_N = O_p(n^{-1/2})$ *if* $a(1-s) \leq 1/2$.

III. (In-between) $D_N = O_p(n^{a(1-s)-1})$ *if* $1/2 \leq a(1-s) < 1$.

IV. (Sparse) No convergence if $s = 0$.

Theorem 4.1 requires that $K_1(n,p) \to 0$. Thus, in the context of this corollary, for consistency we need as $n, p \to \infty$

$$K_1(n,p) \asymp \frac{p}{n\|\sigma_{X,Y}\|^2} \asymp n^{a(1-s)-1} \to 0,$$

which requires $a(1-s) < 1$. This is indicated also in conclusion I of this corollary. The corollary does not indicate an outcome when $a(1-s) \geq 1$, although here PLS might be inconsistent depending on the specific values of a and s. The usual root-n convergence rate is achieved when $a(1-s) \leq 1/2$. For instance, if $a = 2$ so $p = n^2$ then we need $s \geq 3/4$ for root-n convergence. However, in contrast to Corollary 4.3, here there is no convergence with sparsity. If $s = 0$, then we need $a < 1$ for convergence, which violates the corollary's hypothesis.

Figure 4.3 gives a visual representation of the division of the (a, s) plane according to the convergence properties of PLS from parts II and III of Corollaries 4.3 and 4.4. The figure is constructed to convey the main parts of the conclusions; not all aspects of these corollaries are represented. The abundant and in-between categories occupy most of the plane, while sparse fits are indicated only in the upper left corner that represents high-dimensional problems with weak signals, $p > n$ and $\|\sigma_{X,Y}\| \leq \sqrt{p}$.

The previous three corollaries require that the eigenvalues of Δ_0 be bounded. The next corollary relaxes this condition by allowing a finite number of eigenvalues $\delta_{0,j}$ of Δ_0 to be asymptotically similar to p ($\delta_{0,j} \asymp p$ for a finite

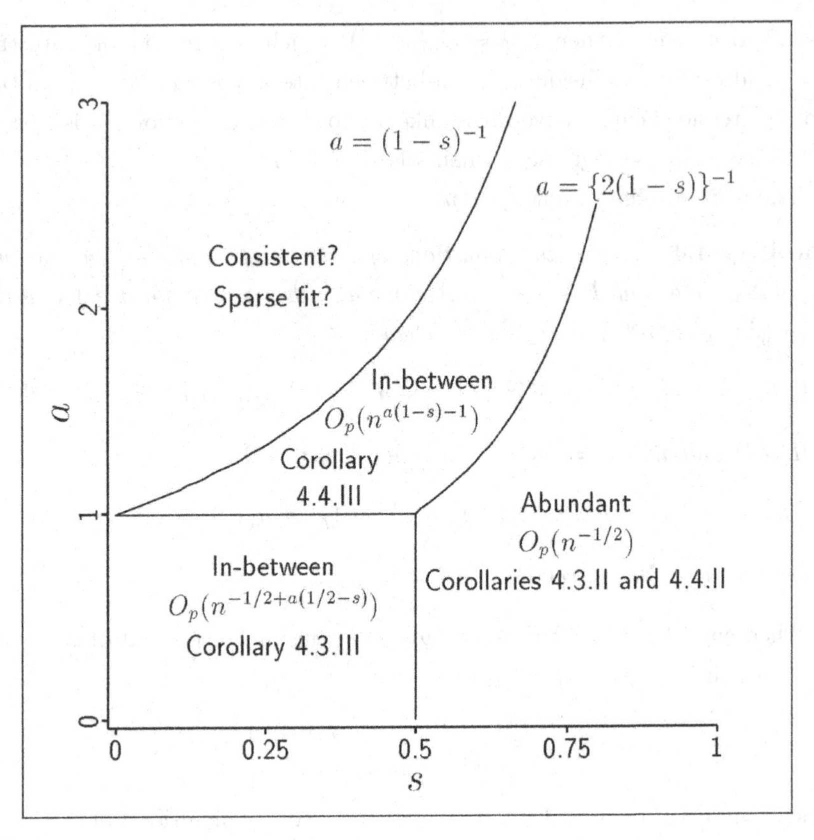

FIGURE 4.3
Division of the (a, s) plane according to the convergence properties given in conclusions II and III of Corollaries 4.3 and 4.4.

collection of indices j), while keeping the remaining eigenvalues bounded. In this case, $\mathrm{tr}(\Delta_0^j) \asymp p^j$, $j = 1, 2, 3$. For instance, returning to the compound symmetry example, we may have $\mathcal{E}_{\Sigma_X}(\mathcal{B}) \subset \mathrm{span}^{\perp}(1_p)$. Then the eigenvalues of Δ_0 are $\pi^2(1 + (p-1)\rho)$ with multiplicity 1, which accounts for the $(1 - \rho + \rho p)P_{1_p}$ term in Σ_X, and $\pi^2(1-\rho)$ with multiplicity $p - 2$. Consequently, $\delta_{0,1} \asymp p$ while the remaining eigenvalues of Σ_X are all bounded.

Corollary 4.5. *Assume the conditions of Theorem 4.1 and that $\delta_{0,j} \asymp p$ for a finite collection of indices j while the other eigenvalues of Δ_0 are bounded as $p \to \infty$. Assume also that $p \asymp n^a$ for $a \geq 1$ and that $\|\sigma_{X,Y}\|^2 \asymp p^s$ for $0 \leq s \leq 1$. Then*

$$D_N = O_p(n^{-1/2+a(1-s)}).$$

The conditions of Theorem 4.1 in the context of Corollary 4.5 imply that for consistency we need $a(1-s) < 1/2$, with the usual root-n convergence rate being essentially achieved when $a(1-s)$ is small. If $s = 1$, then $D_N = O_p(n^{-1/2})$, which agrees with the conclusion of Corollary 4.2. This highlights one important conclusion from Corollary 4.5: PLS predictions can still have root-n convergence when some of the eigenvalues of Δ_0 increase like p, but for this to happen we need an abundant signal, $\|\sigma_{X,Y}\| \asymp p$. Second, Corollary 4.5 shows the interaction between the number of predictors and the signal rate in high-dimensional regression. Write

$$n^{-1/2+a(1-s)} = \frac{1}{\sqrt{n}} \frac{n^a}{n^{as}} \asymp \frac{1}{\sqrt{n}} \frac{p}{p^s}.$$

Thinking of p^s roughly as the number of active predictors, this says that the number of predictors per active predictor must be small relative to the square root of the sample size for a good convergence rate. For instance, with $n = 625$, $p = 1000$ and about 250 active predictors, so $a \sim 1.075$ and $s \sim 0.8$, we get a corresponding convergence rate of about $n^{0.3}$. If we increase the active predictors to 500, the corresponding convergence rate becomes about $n^{0.4}$.

4.5 Distributions of one-component PLS estimators

Let $G \in \mathbb{R}^p$ be a vector of user-selected, non-stochastic constants. We continue our study of PLS by describing in Section 4.5.1 the asymptotic distribution of $(\widehat{\beta}_{\text{pls}} - \beta)^T G$ in one-component regressions as developed in Section 4.2. Our discussion is based largely on the results of Basa et al. (2022). Selecting $G^T = (1, -1, 0, \ldots, 0)$, the asymptotic distribution of $(\widehat{\beta}_{\text{pls}} - \beta)^T G$ could be used to aid inference about the difference between the first two elements of β. In Section 4.5.2 we discuss how the distribution can be used to form asymptotic confidence intervals for $G^T\beta$. We expand the discussion in Section 4.5.3.2 by studying the asymptotic orders of various constituents. A few simulation results are presented in Section 4.5.4.

For a new randomly selected value X_N of X, we report in Section 4.5.5 on the asymptotic distribution for $\hat{\mu} + \widehat{\beta}_{\text{pls}}^T X_N$ as a tool for studying the conditional mean $\mathrm{E}(Y \mid X) = \mu + \beta^T X$. In this section we give only the main results of Basa et al. (2022); full proofs and additional details are available from the supplement to their paper.

4.5.1 Asymptotic distributions and bias

We first introduce quantities that will be used in pursuing the asymptotic distribution of $(\widehat{\beta}_{\text{pls}} - \beta)^T G$. Define

$$\begin{aligned} V(G) &= \frac{G^T\left(\Sigma_X \sigma_Y^2 - \sigma_{X,Y}\sigma_{X,Y}^T\right)G}{n\delta^2} && (4.12)\\ &= \frac{(G^T\Phi)^2\sigma_{Y|X}^2}{n\delta} + \frac{\sigma_Y^2 G^T\Phi_0\Delta_0\Phi_0^T G}{n\delta^2}. && (4.13)\\ b &= -\left(\frac{\|\sigma_{X,Y}\|^2}{\delta} - \sigma_{Y|X}^2\right)K_1 - \frac{\sigma_Y^2\|\sigma_{X,Y}\|^2}{\delta}\left\{K_1^2 + K_2\right\},\\ &= O\left(\max(K_1, K_2)\right). && (4.14)\\ J(n,p) &= n^{-1/2} + K_1(n,p) + K_2(n,p), \end{aligned}$$

where we have suppressed the arguments (n,p) to K_j, $V^{1/2}(G)$ will be used to scale $(\widehat{\beta}_{\text{pls}} - \beta)^T G$ and b is related to a potential for asymptotic bias. We use $\xrightarrow{\mathcal{D}}$ to denote convergence in distribution.

The following theorem describes the asymptotic distribution reported by Basa et al. (2022).

Theorem 4.2. *Assume that the one-component model (4.1) holds with* $\beta^T\Sigma_X\beta \asymp 1$. *Assume also that (a)* $X \sim N_p(\mu_X, \Sigma_X)$, *(b)* $\mathrm{E}(\epsilon^4)$ *from model (4.1) is bounded, (c)* $K_1(n,p)$ *and* $K_2(n,p)$ *converge to 0 as* $n,p \to \infty$, *and (d)* $K_3(n,p)$ *is bounded and* $K_4(n,p)$ *converges to 0 as* $n,p \to \infty$.

(I) If $JV^{-1/2}\beta^T G b \to 0$ *as* $n,p \to \infty$, *then*

$$V^{-1/2}\left\{\widehat{\beta}_{\text{pls}} - \beta(1+b)\right\}^T G \xrightarrow{\mathcal{D}} N(0,1).$$

(II) If, as $n,p \to \infty$, $n^{1/4}K_j(n,p) \to 0$ *for* $j = 1,2$ *then* $JV^{-1/2}\beta^T G b \to 0$.

Condition (a) requires normality for X, and condition (b) is the usual requirement of finite fourth moments. Condition (c) plus the constraint $\beta^T\Sigma_X\beta \asymp 1$ are the same as the conditions for consistency in Theorem 4.1. Condition (d) is new and is needed to insure stable asymptotic distributions. Like conditions (b) and (c), condition (d) is judged to be mild.

Although the non-stochastic sequence represented by b converges to 0 as K_1 and K_2 converge to zero, it nevertheless represents a potential for bias when using Theorem 4.2 as the basis for inference in applications. In the following discussion we assume conditions (a)–(d) of Theorem 4.2 and focus on constraints that control the bias.

Since b is a real scalar, inference about ratios of elements of β will be relatively free of any bias effect. If $G^T\beta = 0$ there is no bias contribution since then the asymptotic mean is 0. If $G^T\beta \neq 0$, bias may play a notable role.

Writing

$$V^{-1/2}G^T\left(\widehat{\beta}_{\text{pls}} - \beta\,(1+b)\right) = V^{-1/2}G^T\left(\widehat{\beta}_{\text{pls}} - \beta\right) - V^{-1/2}G^T\beta b,$$

we see by Slutsky's Theorem that the bias will be unimportant asymptotically when the second addend on the right side $V^{-1/2}G^T\beta b \to 0$ as $p \to \infty$. Since we require $J \to 0$, the condition $V^{-1/2}G^T\beta b \to 0$ implies the sufficient condition of Theorem 4.2(I), $JV^{-1/2}\beta^T Gb \to 0$, but not conversely. We use lower bounds on V to find manageable sufficient conditions under which $V^{-1/2}G^T\beta b \to 0$. Direct calculation gives

$$\begin{aligned} V^{-1}(G^T\beta)^2 &= n\frac{(\|\sigma_{X,Y}\|^2/\delta)(G^T\sigma_{X,Y})^2}{\sigma^2_{Y|X}(G^T\sigma_{X,Y})^2 + \sigma^2_Y(\|\sigma_{X,Y}\|^2/\delta)G^T\Phi_0\Delta_0\Phi_0^TG} \\ &\leq n\,\frac{\|\sigma_{X,Y}\|^2}{\delta\sigma^2_{Y|X}}, \end{aligned} \tag{4.15}$$

with the upper bound being attained when $G \in \text{span}(\beta)$. Since the choice of G does not affect b, this indicates that the bias effects will be the most prominent when $G \in \text{span}(\beta)$.

It follows from (4.15) that

$$V^{-1/2}|G^T\beta b| \leq \sqrt{n}|b|(\|\sigma_{X,Y}\|^2/\delta\sigma^2_{Y|X})^{1/2} \asymp \sqrt{n}|b|. \tag{4.16}$$

In consequence, $\sqrt{n}|b| \to 0$ is a sufficient condition for avoiding the bias effects when $G \notin \text{span}^{\perp}(\beta)$. Inspecting $\sqrt{n}|b|$ we have

$$\sqrt{n}|b| \leq \sqrt{n}\left(\left|\frac{\|\sigma_{X,Y}\|^2}{\delta} - \sigma^2_{Y|X}\right| K_1 + \frac{\sigma^2_Y\|\sigma_{X,Y}\|^2}{\delta}\left\{K_1^2 + K_2\right\}\right).$$

This inequality gives rise to conditions for $\sqrt{n}|b| \to 0$. Since σ^2_Y and $\left|\|\sigma_{X,Y}\|^2/\delta - \sigma^2_{X,Y}\right|$ are bounded, it is sufficient to require $\sqrt{n}K_j(n,p) \to 0$, $j = 1, 2$. We summarize these findings in the following corollary.

Corollary 4.6. *Under the condition (a)–(d) of Theorem 4.2,*

(I) If $\sqrt{n}|b| \to 0$ as $n, p \to \infty$, then

$$V^{-1/2}\left(\widehat{\beta}_{\text{pls}} - \beta\right)^T G \xrightarrow{\mathcal{D}} N(0,1).$$

(II) If, as $n, p \to \infty$, $\sqrt{n}K_j(n,p) \to 0$, $j = 1, 2$, then $\sqrt{n}|b| \to 0$.

Recall that for consistency Theorem 4.1 requires $K_j(n,p) \to 0$ as $n, p \to 0$, $j = 1, 2$. In contrast, Theorem 4.2 and Corollary 4.6 require that $n^{1/4}K_j(n,p)$ and $\sqrt{n}K_j(n,p)$, $j = 1, 2$, converge to 0 for asymptotic normality with and without the bias term. In consequence, more stringent conditions are needed for asymptotic normality than are needed for consistency alone.

We conclude this section with a result in the next theorem that is closely related to Cook and Forzani (2018, Theorem 1).

Theorem 4.3. *Under the conditions of Theorem 4.2,*

$$(\widehat{\beta}_{\text{pls}} - \beta)^T \Sigma_X (\widehat{\beta}_{\text{pls}} - \beta) = O_p\left(n^{-1} + K_1^2(n,p) + K_2(n,p)\right).$$

Therefore for $X_N - \mu_X \sim N(0, \Sigma_X)$ with Σ_X as specified in (4.2),

$$(\widehat{\beta}_{\text{pls}} - \beta)^T (X_N - \mu_X) = O_p\left(n^{-1/2} + K_1(n,p) + K_2^{1/2}(n,p)\right).$$

This establishes consistency of $\widehat{\beta}_{\text{pls}}$ in the Σ_X inner product.

4.5.2 Confidence intervals

Theorem 4.2 allows the straightforward construction of asymptotic confidence intervals for $(1+b)\beta^T G$:

$$\text{CI}_\alpha((1+b)\beta^T G) = \left[\widehat{\beta}_{\text{pls}}^T G \pm z_{\alpha/2} V^{1/2}(G)\right], \tag{4.17}$$

where $z_{\alpha/2}$ denotes a selected percentile of the OLS normal distribution. Under Corollary 4.6 this same interval becomes a confidence interval for $\beta^T G$, in which case we refer to the interval as $\text{CI}_\alpha(\beta^T G)$.

It is not possible to use (4.17) in applications because V is unknown. However, a sample version of interval (4.17) can be constructed by using plug-in estimators for V. To construct an estimator of V we simply plug in the estimators of the constituents of representation (4.12),

$$\widehat{V}(G) = \frac{G^T(S_X S_Y - S_{X,Y} S_{X,Y}^T)G}{n\widehat{\delta}_{\text{pls}}^2}. \tag{4.18}$$

where

$$\widehat{\delta}_{\text{pls}} = \widehat{\Phi}^T S_X \widehat{\Phi} = S_{X,Y}^T S_X S_{X,Y} / S_{X,Y}^T S_{X,Y},$$

as described in Section 4.2.2. The following theorem implies by Slutsky's Theorem that asymptotically it is reasonable to replace V with $\widehat{V}$ in the construction of confidence intervals:

Theorem 4.4. *Under the conditions of Theorem 4.2, $\widehat{V}/V \to 1$ in probability as $n, p \to \infty$.*

Additionally, we have conditions that allow us to replace b with $\hat{b}$, and therefore obtain a confidence interval $\text{CI}_\alpha(\beta^T G)$ for $\beta^T G$ under only Theorem 4.2(I). Let $\hat{b}$ denote the plug-in estimator of b:

$$\hat{b} = -\left(\frac{\|S_{X,Y}\|^2}{\widehat{\delta}_{\text{pls}}} - S_{Y|X}^2\right)\widehat{K}_1 - \frac{S_Y^2 \|S_{X,Y}\|^2}{\widehat{\delta}_{\text{pls}}}\left\{\widehat{K}_1^2 + \widehat{K}_2\right\},$$

where

$$\widehat{K}_j = \frac{\text{tr}(\widehat{\Delta}_{0,\text{pls}}^j)}{n\|S_{X,Y}\|^{2j}}$$

and $\widehat{\Delta}_{0,\text{pls}}$ is as defined in Section 4.2.2.

Theorem 4.5. *Under all conditions of Theorem 4.2(II), $\sqrt{n}(\hat{b} - b) \to 0$ in probability as $n, p \to \infty$. In consequence, an asymptotic confidence interval for $\beta^T G$ is*

$$\text{CI}_\alpha(\beta^T G) = \frac{1}{1+\hat{b}}\left[\widehat{\beta}_{\text{pls}}^T G \pm z_{\alpha/2}\widehat{V}^{1/2}(G)\right].$$

4.5.3 Theorem 4.2, Corollary 4.6 and their constituents

In this section we assume that the one-component model (4.1) holds with $\beta^T \Sigma_X \beta \asymp 1$, and that conditions (a) and (b) of Theorem 4.2 hold. Our objective is to study the remaining constituents of the theorem, including conditions (c) and (d), and its Corollary 4.6 in plausible scenarios that may help with intuition in practice. We now consider settings in which,

1. The one-component model (4.1) holds with $\beta^T \Sigma_X \beta \asymp 1$, and (a) and (b) of Theorem 4.2.

2. $\|\sigma_{X,Y}\|^2 \asymp p^t$ for $0 \le t \le 1$.

3. Eigenvalue constraints,

(i). The eigenvalues of Δ_0 are bounded, so $\lambda_{\max}(\Delta_0) \asymp 1$ and $\text{tr}(\Delta_0^k) \asymp p$ for all k. In this case we redefine $K_j(n,p) = p/np^{jt}$, which is equivalent to Definition 4.11 for the purpose of assessing condition (c) and (d) of Theorem 4.2.

(ii). A finite number of eigenvalues of Δ_0 increase as p, so $\lambda_{\max}(\Delta_0) \asymp p$, and $\text{tr}(\Delta_0^k) \asymp p^k$ for all k. In this case we redefine $K_j(n,p) = p^j/np^{jt}$.

Conditions 1 and 2, and either 3(i) or 3(ii) are assumed for the rest of this section.

4.5.3.1 $\mathbf{K_j(n,p)}$

It should be understood in the following that we deal with $K_j(n,p)$ for only $j = 1, \ldots, 4$. Under condition 3(i), we have for $j \geq 1$

$$\left.\begin{array}{l} K_j(n,p) = \dfrac{p}{np^{jt}} = \dfrac{1}{np^{jt-1}} \\ K_1(n,p) \geq K_j(n,p) \text{ for all } n,p. \end{array}\right\} \tag{4.19}$$

From here we seen that $K_j(n,p) \to 0$ if $p^{1-t} = o(n)$.

Under condition 3(ii),

$$\left.\begin{array}{l} K_j(n,p) = \dfrac{p^j}{np^{jt}} = \dfrac{1}{np^{j(t-1)}}, \\ K_i(n,p) \leq K_j(n,p),\ i \leq j. \end{array}\right\} \tag{4.20}$$

From here we seen that $K_j(n,p) \to 0$ if $K_4(n,p) \to 0$ or equivalently if $p^{4(1-t)} = o(n)$. Sparse regressions with $t = 0$ place the most severe requirements on n: under condition 3(i) we must have $p/n \to 0$ and under condition 3(ii) we must have $p^4/n \to 0$.

To summarize for later reference, conditions (c) and (d) of Theorem 4.2 will hold when

$$\frac{p^{1-t}}{n} \to 0 \quad \text{under 3(i)}, \tag{4.21}$$

$$\frac{p^{4(1-t)}}{n} \to 0 \quad \text{under 3(ii)}. \tag{4.22}$$

4.5.3.2 Scaling quantity $V(G)$

Two equal forms for $V(G)$ were presented in (4.12) and (4.13). The form given in (4.12) was used in previous developments. The form in (4.13) derives immediately from the asymptotic variance for the fixed p case presented in Proposition 4.1(iii):

$$V(G) = \frac{G^T \text{avar}\left(\sqrt{n}\widehat{\beta}_{\text{pls}}\right) G}{n} = \frac{\text{avar}\left(\sqrt{n} G^T \widehat{\beta}_{\text{pls}}\right)}{n}.$$

That is, the $V(G)$ derived from n-asymptotics is the same as that in n, p-asymptotics. When letting $n \to \infty$ with p fixed, nV is static; it does not change since p is static. In consequence, with p static, the result from Proposition 4.1 can be stated alternatively as $\sqrt{n}G^T(\widehat{\beta}_{\text{pls}} - \beta) \xrightarrow{\mathcal{D}} N(0, nV)$, leading to the conclusion that convergence is at the usual $\sqrt{n}$ rate.

The same scaling quantity V is appropriate for our n, p-asymptotic results:

$$V^{-1/2}(\widehat{\beta}_{\text{pls}} - \beta(1+b))^T G \xrightarrow{\mathcal{D}} N(0,1) \text{ and } V^{-1/2}(\widehat{\beta}_{\text{pls}} - \beta)^T G \xrightarrow{\mathcal{D}} N(0,1).$$

However, we do not necessarily have $\sqrt{n}$ convergence because, in our context, nV is dynamic, changing as $p \to \infty$. Let $\lambda_{\max}(\cdot)$ denote the maximum eigenvalue of the argument matrix. Then

$$\begin{aligned} V(G) &= \frac{1}{n\delta}\left\{(G^T\Phi)^2\sigma^2_{Y|X} + \delta^{-1}\sigma^2_Y G^T \Phi_0 \Delta_0 \Phi_0^T G\right\} \\ &\le \frac{1}{n\delta}\left\{\|G\|^2\sigma^2_{Y|X} + \delta^{-1}\sigma^2_Y \|G\|^2 \lambda_{\max}(\Delta_0)\right\} \\ &= \|G\|^2 \frac{\|\sigma_{X,Y}\|^2}{\delta}\left\{\frac{\sigma^2_{Y|X}}{n\|\sigma_{X,Y}\|^2} + \frac{\|\sigma_{X,Y}\|^2}{\delta}\sigma^2_Y \frac{\lambda_{\max}(\Delta_0)}{n\|\sigma_{X,Y}\|^4}\right\}. \end{aligned}$$

Under conditions 1, 2 and 3(i), as $n, p \to \infty$,

$$\begin{aligned} V(G) &= O\left(\|G\|^2 n^{-1} \max\left\{\|\sigma_{X,Y}\|^{-2}, \|\sigma_{X,Y}\|^{-4}\right\}\right) \\ &= O\left(\|G\|^2 n^{-1} \max\left\{1/p^t, 1/p^{2t}\right\}\right) = O(\|G\|^2/np^t). \end{aligned}$$

In consequence, the variance of the limiting distribution is approached at rate at least $\sqrt{np^{t/2}}/\|G\|$. If the regression is sparse, $t = 0$, and $\|G\|$ is bounded, we get the usual $\sqrt{n}$ convergence rate. But if $t > 0$, then there is a synergy between the n and p, with the sample size being magnified by p^t. If $t = 1$ then the convergence rate is $\sqrt{np}/\|G\|$, for instance. In many applications we may have $\|G\| \le \sqrt{p}$ and then the convergence rate is at least $\sqrt{n}$. This discussion is summarized in the first column of the second row of Table 4.3.

Under conditions 1, 2 and 3(ii), as $n, p \to \infty$,

$$\begin{aligned} V(G) &= O\left(\|G\|^2 n^{-1} \max\{\|\sigma_{X,Y}\|^{-2}, p\|\sigma_{X,Y}\|^{-4}\}\right) \\ &= O\left(\|G\|^2 n^{-1} \max\{1/p^t, 1/p^{2t-1}\}\right) = O\left(\|G\|^2/np^{2t-1}\right), \end{aligned}$$

and now the variance of the limiting distribution is approached at least at rate $\sqrt{n}p^{t-1/2}/\|G\|$. This is summarized in the first column of the third row of Table 4.3.

This discussion is summarized in the first column of Table 4.3. The necessary conditions (4.21) and (4.22) to guarantee conditions (c) and (d) of Theorem 4.2 also insure that $V(G) \to 0$ under conditions 3(i) and 3(ii).

4.5.3.3 Bias b

It follows from (4.14) and (4.19) that under condition 3(i) $b = O(1/np^{t-1})$, which is reported in the second column, second row of Table 4.3. That $1/np^{t-1} \to 0$ is implied by (4.21). For $t = 0$, $b = O(p/n)$, for $t = 1/2$, $b = O(\sqrt{p}/n)$ and for $t = 1$, $b = O(1/n)$. When hypotheses (a) - (d) of Theorem 4.2 are met in this example, $n^{1/4}K_j \to 0$, $j = 1, 2$ is sufficient to have the limiting distribution stated in the theorem. Under condition 3(i), $K_1 \geq K_2$ and so

$$n^{1/4}K_1 = \frac{1}{n^{3/4}p^{t-1}} \to 0$$

is sufficient for Theorem 4.2. This condition is listed in the third column, second row of Table 4.3.

It follows from (4.14) and (4.20) that under condition 3(ii), $b = O\left(1/np^{2(t-1)}\right)$ as given in the second column, third row of Table 4.3. That $1/np^{2(t-1)} \to 0$ is implied by (4.22). For $t = 0$, $b = O(p^2/n)$, for $t = 1/2$, $b = O(p/n)$ and for $t = 1$, $b = O(1/n)$. Still under condition 3(ii), $n^{1/4}K_j \to 0$, $j = 1, 2$ is sufficient to have the limiting distribution stated in the theorem. But in our example, $K_1 \leq K_2$ and so

$$n^{1/4}K_2 = \frac{1}{n^{3/4}p^{2(t-1)}} \to 0$$

is sufficient for Theorem 4.2 given that the remaining conditions are satisfied. This condition is listed in the third column, third row of Table 4.3.

The conditions from Corollary 4.6 that are needed to eliminate the bias term are shown in the column headed $\not{b}$. The entries in that column are "between" those in columns headed b and $\not{b}$. The factors listed in the first and second columns of Table 4.3 all converge to 0 provided $\|G\|$ does not grow too fast: since (4.21) and (4.22) are required for Theorem 4.2 to hold in the first place, and (4.22) is the same as the last factor listed in the last column of Table 4.3.

TABLE 4.3

The first and second columns give the orders $O(\cdot)$ of $V^{1/2}$ and b under conditions 1–3 as $n, p \to \infty$. The fourth column headed $\not{b}$ gives from Corollary 4.6 the order of quantity $\sqrt{n}|b|$ that must converge to 0 for the bias term to be eliminated.

$V^{1/2}$	b	$n^{1/4}K_j(n,p), j=1,2$	$\not{b}$
Condition 3(i)			
$\lVert G\rVert/\sqrt{n}p^{t/2}$	$1/np^{t-1}$	$1/n^{3/4}p^{t-1}$	$1/\sqrt{n}p^{t-1}$
Condition 3(ii)			
$\lVert G\rVert/\sqrt{n}p^{t-1/2}$	$1/np^{2(t-1)}$	$1/n^{3/4}p^{2(t-1)}$	$1/\sqrt{n}p^{2(t-1)}$

4.5.3.4 Simulation to illustrate bias

It seems clear from Table 4.3 that the bias plays a notable role in the convergence. Nevertheless, in abundant regressions with $\lVert\sigma_{X,Y}\rVert^2 \asymp p$, the quantities in Table 4.3 all converge to 0 at the $\sqrt{n}$ rate or better when $\lVert G\rVert/\sqrt{p} \asymp 1$.

The results in Table 4.3 hint that, as $n, p \to \infty$, the proper scaling of the asymptotic normal distribution may be reached relatively quickly, while achieving close to the proper location may take a larger sample size. Basa et al. (2022) conducted a series of simulations to illustrate the relative impact that bias and scaling can have on the asymptotic distribution of $G^T\widehat{\beta}_{\text{pls}}$. In reference to the one-component model (4.1), they set $n = p/2$, $\mu = 0$, $\sigma_{Y|X} = 1/2$, $\delta = \lVert\sigma_{X,Y}\rVert^2$ and $\Delta_0 = I_{p-1}$. The covariance vector $\sigma_{X,Y}$ was generated with $\lfloor p^{1/2}\rfloor$ standard normal elements and the remaining elements equal to 0, so $\lVert\sigma_{X,Y}\rVert^2 \asymp \sqrt{p}$. Following the discussion of (4.15), they selected $G = \sigma_{X,Y}$ to emphasize bias. Then, X was generated as $N(0, \Sigma_X)$ and $Y \mid X$ was then generated according to (4.1) with $N(0, \sigma^2_{Y|X})$ error. In reference to Table 4.3, this simulation is an instance of condition 3(i) with $t = 1/2$.

For each selection of $n = p/2$ this setup was replicated 500 times and side-by-side histograms drawn of $D_1 = V^{-1/2}G^T\left(\widehat{\beta}_{\text{pls}} - \beta(1+b)\right)$ and $D_2 = V^{-1/2}G^T\left(\widehat{\beta}_{\text{pls}} - \beta\right)$. Their results are shown graphically in Figure 4.4. Since $K_j(n,p) \asymp p^{-j/2}$ conditions (a) – (d) of Theorem 4.2 hold. Further, since $n^{1/4}K_j(n,p) \asymp p^{(-2j+1)/4} \to 0$ for $j = 1, 2, 3, 4$ it follows from Theorem 4.2(II) that D_1 converges in distribution to a standard normal. The convergence rate for the largest of these $n^{1/4}K_1(n,p) \asymp p^{-1/4}$ is quite slow so it may take a

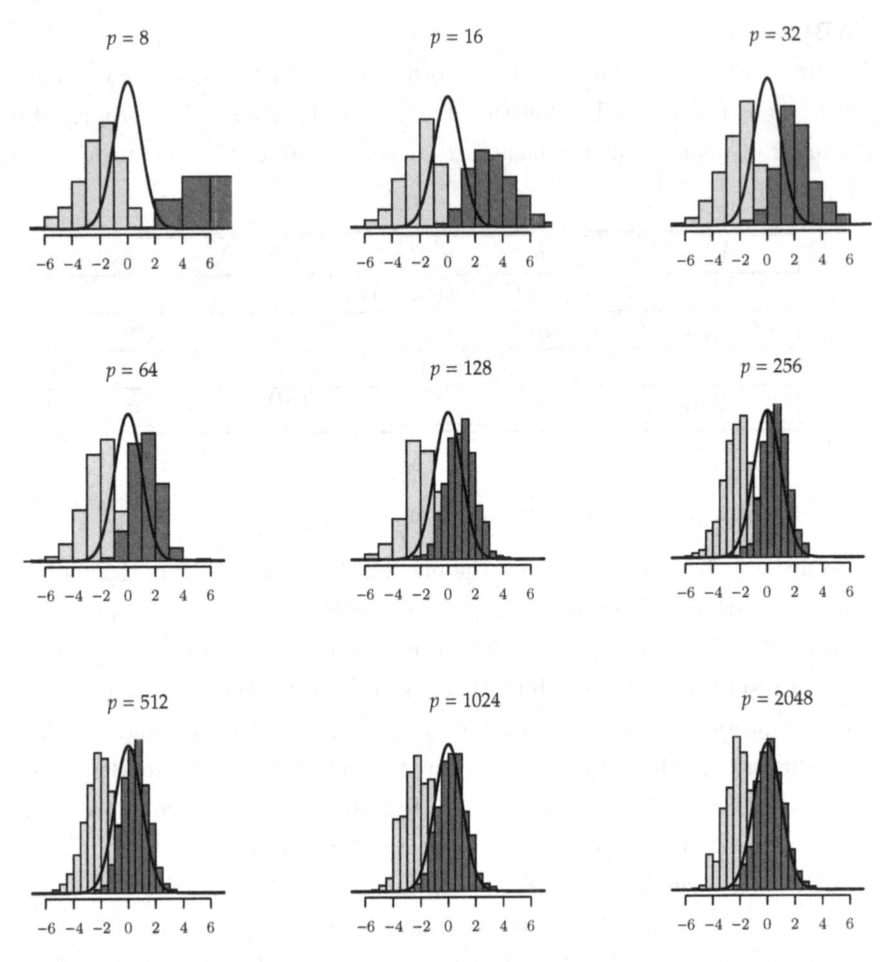

FIGURE 4.4
Simulation results on bias: The right histogram in each plot is of $V^{-1/2}G^T\left(\widehat{\beta}_{\text{pls}} - \beta(1+b)\right)$ and the left histogram is of $V^{-1/2}G^T\left(\widehat{\beta}_{\text{pls}} - \beta\right)$. The standard normal reference density is also shown and in all cases $n = p/2$. (Plot was constructed with permission using the data that Basa et al. (2022) used for their Fig. 1.)

large value of p before the asymptotic distribution is a useful approximation. Because $\sqrt{n}K_1(n,p) \asymp 1$, we cannot use Corollary 4.6 to conclude that D_2 converges to a standard normal.

We see from Figure 4.4 that the histogram of D_1 moves to coincide with the standard normal density as n and p increase, although it takes around 2,000

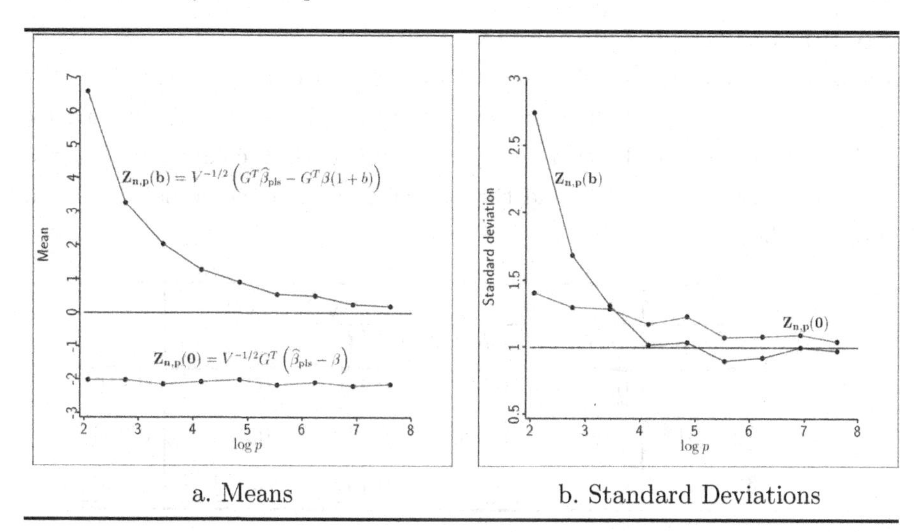

a. Means b. Standard Deviations

FIGURE 4.5
Plots of the means and standard deviations versus $\log p$ corresponding to the simulations of Figure 4.4.

predictors before the approximation seems quite close to the standard normal. This is in qualitative agreement with the slow rate of convergence mentioned previously for this example. The histograms for D_2 do not converge to a standard normal density. Visually, it seems that the histogram of D_1 gets to the right scaling faster than it achieves the right location, which is in agreement with the discussion of Table 4.3. This is supported by the plots in Figure 4.5.

4.5.4 Simulation results

Basa et al. (2022) constructed a series of simulations to illustrate the behavior of the confidence intervals (4.17) as $n = \sqrt{p}$ grows with p. For each p they constructed $\sigma_{X,Y}$ to have $\lfloor p^{3/4} \rfloor$ elements generated independently from a standard normal distribution with the remaining $p - \lfloor p^{3/4} \rfloor$ elements set equal to 0. For $p = 1024$, $\sigma_{X,Y}$ had 181 non-zero elements. They set $\delta = \|\sigma_{X,Y}\|^2$, so the ratio $\delta/\|\sigma_{X,Y}\|^2$ is bounded. They also set $\Delta_0 = I_{p-1}$, constructed Σ_X according to (4.2), $\beta = \delta^{-1}\sigma_{X,Y}$ and generated the data (Y_i, X_i), $i = 1, \ldots, n$, independently as $X_i \sim N(0, \Sigma)$ and $Y \mid X_i \sim N(\beta^T X_i, 1)$. The contrast vector G was generated as a standard normal vector and nominal 95% confidence intervals were constructed according to (4.17). This scenario has relatively small bias because the choices $\delta = \|\sigma_{X,Y}\|^2$ and $\sigma^2_{Y|X} = 1$ have the effect

TABLE 4.4

Estimated coverage rates of nominal 95% confidence intervals (4.17) for the parameters indicated in the first row. The third and fourth columns are for the the setting in which $\sqrt{n}|b| \to 0$, the fifth and sixth columns are for the setting in which $\sqrt{n}|b| \not\to 0$, and the last column is for the adjusted interval given in Theorem 4.5.

		$\sqrt{n}\lvert b\rvert \to 0$		$\sqrt{n}\lvert b\rvert \not\to 0$		
n	p	$\beta^T G$	$(1+b)\beta^T G$	$\beta^T G$	$(1+b)\beta^T G$	$\beta^T G$
16	256	0.924	0.897	0.766	0.887	0.856
22	512	0.949	0.920	0.734	0.927	0.903
32	1024	0.947	0.946	0.742	0.952	0.920

of eliminating the first addend in the bias b (4.14). The part of G that falls in $\mathrm{span}^{\perp}(\beta)$ will also mitigate the bias. The coverage rates, estimated by repeating the procedure 1,000 times and counting the number of intervals that covered $\beta^T G$ and $(1+b)\beta^T G$, are shown in the third and fourth columns of Table 4.4. Theorem 4.2 holds in this simulation scenario. For instance, $K_j(n,p) = 1/p^{3j/4-1/2}$. Corollary 4.6(I) also holds since $\sqrt{n}|b| \asymp n^{-1/2}$. However, the sufficient condition given in Corollary 4.6(II) does not hold since $\sqrt{n}K_1(n,p) = 1$ for all p.

To emphasize the bias, Basa et al. (2022) conducted another simulation with the same settings, except we set $\sigma^2_{Y|X} = 1/2$, $G = \sigma_{X,Y}$ and, to enhance the contrast, the true value of V was used in the interval construction. According to the discussion following (4.15), this choice of G will maximize the bias effect and the first addend of (4.14) will now contribute to the bias. The results are shown in the fifth and sixth columns of Table 4.4. We see now that $\mathrm{CI}_{0.05}(\beta^T G)$ suffers in comparison to $\mathrm{CI}_{0.05}((1+b)\beta^T G)$ since $\sqrt{n}|b|$ does not converge to 0, $\sqrt{n}|b| \asymp 1$. However, by (4.16),

$$JV^{-1/2}|G^T\beta b| \le \sqrt{n}J|b|(\|\sigma_{X,Y}\|^2/\delta\sigma^2_{Y|X})^{1/2} = \sqrt{2n}J|b| \to 0,$$

and thus Theorem 4.2(I) holds. Additionally, the conditions for Theorem 4.2(II) hold and so the adjusted confidence interval of Theorem 4.5 is applicable. The rates for that adjusted interval are shown in the last column of Table 4.4.

4.5.5 Confidence intervals for conditional means

The confidence intervals discussed in Section 4.5.2 require that G be a non-stochastic vector. The estimated conditional mean at a new value X_N of X is $\widehat{\mathrm{E}}(Y \mid X_N) = \bar{Y} + \widehat{\beta}^T_{\mathrm{pls}}(X_N - \bar{X})$. Although X_N is fixed, $\bar{X}$ is stochastic and so the corresponding $G = (X_N - \bar{X})$ is stochastic. In consequence, the confidence intervals of Section 4.5.2 may not be appropriate for a conditional mean $\mathrm{E}(Y \mid X_N)$ or a prediction.

Let $M(Y|X_N) = \mu_Y + (1+b)\beta^T(X_N - \mu_X)$, $F_N = \widehat{\mathrm{E}}(Y \mid X_N) - M(Y \mid X_N)$ and let $\omega(X_N) = V(X_N - \mu_X) + \sigma^2_{Y|X}/n$, where $V(\cdot)$ is as defined at (4.12). The asymptotic distribution of F_N is given in the following theorem from Basa et al. (2022).

Theorem 4.6. *Assume that the data (Y_i, X_i), $i = 1, \ldots, n$, are independent observations on a multivariate normal random vector of dimension $p+1$ that follows the one-component model (4.1) and has the asymptotic properties needed for Theorem 4.2. Then*

$$\omega^{-1/2}(X_N)F_N \xrightarrow{\mathcal{D}} N(0,1). \tag{4.23}$$

Theorem 4.6 allows us to construct confidence statements for $M(Y \mid X_N)$ based on $\omega^{-1/2}(X_N)F_N$. However, our main interest is in confidence statements for $\mathrm{E}(Y \mid X)$ based on $\widehat{\mathrm{E}}(Y \mid X) - \mathrm{E}(Y \mid X) = F_N + b\beta^T(X_N - \mu_X)$. The following corollary addresses these confidence intervals.

Corollary 4.7. *Assume the conditions of Theorem 4.6 and that $\sqrt{n}|b| \to 0$ as $n, p \to \infty$. Then $\omega^{-1/2}(X_N)\left\{\widehat{\mathrm{E}}(Y \mid X) - \mathrm{E}(Y \mid X)\right\} \xrightarrow{\mathcal{D}} N(0,1)$.*

As a consequence of Corollary 4.7, approximate confidence intervals for $\mathrm{E}(Y \mid X_N)$ and $Y_N := \mathrm{E}(Y \mid X_N) + \epsilon_N$ can then be constructed by using the plug-in method to get an estimator $\widehat{\omega}$ of ω:

$$\mathrm{CI}_\alpha(\mathrm{E}(Y \mid X_N)) = \left[\widehat{\mathrm{E}}(Y \mid X_N) \pm z_{\alpha/2}\widehat{\omega}^{1/2}(X_N)\right] \tag{4.24}$$

$$\mathrm{CI}_\alpha(Y_N) = \left[\widehat{\mathrm{E}}(Y \mid X_N) \pm z_{\alpha/2}\left(\widehat{\omega}(X_N) + \widehat{\tau}^2\right)^{1/2}\right]. \tag{4.25}$$

Similar to the interpretation of the confidence intervals in Section 4.5.2, if $\sqrt{n}|b|$ does not converge to zero, then (4.24) and (4.25) can always be interpreted under Theorem 4.6 as confidence intervals for $M(Y \mid X_N)$ and $M(Y \mid X_N) + \epsilon_N$.

4.6 Consistency in multi-component regressions

The asymptotic behavior of PLS predictions in regressions with multiple $q > 1$ components is qualitatively similar to the behavior of single-component regressions described in Section 4.4. However, there are also important differences that serve to distinguish the two cases. In this section we give an expanded discussion of the consistency results of Cook and Forzani (2019) for multiple component regressions. Full technical details are available in the supplement to their article.

Our context for studying the asymptotic behavior of PLS predictions in multi-component regressions is the same as that described for single-component regressions: the single-response model is as given in (2.5) with $r = 1$ and the notational changes indicated at the outset of Section 4.2.1. The goal is still to characterize the asymptotic behavior of D_N as given in (4.8). The main difference is that now we allow the fixed envelope dimension $q = \dim(\mathcal{E}_{\Sigma_X}(\mathcal{B})) > 1$ so that the column dimension of its semi-orthogonal basis matrix $\Phi \in \mathbb{R}^{p \times q}$ is allowed to exceed 1. Subsequent differences arise because the orders for multiple-component regressions are not as tight as those given for one-component regressions. Nevertheless, as will be discussed in this section, there are multiple-component regressions in which PLS predictions can converge at or near the $\sqrt{n}$ rate.

In this section we revert back to usual notation and use $\Sigma_{X,Y}$ to denote the covariance between X and Y, while still using $\sigma_{X,Y}$ for the covariance when $r = 1$. This notational convention will facilitate distinguishing regressions with $r = 1$ from those with $r > 1$.

4.6.1 Definitions of the signal $\eta(p)$, noise $\kappa(p)$ and collinearity $\rho(p)$

Cook and Forzani (2019) found by using its envelope structure that the performance of a multi-component PLS regression in high dimensions depends primarily on the signal coming from the material predictors, the noise arising from the immaterial predictors and collinearity among the Krylov vectors, as discussed in Section 3.6. Of these, the two most important are perhaps the signal as measure by the variation in the material predictors $\Phi^T X$ and the noise arising from the immaterial predictors as measured by the variation in $\Phi_0^T X$:

Definition 4.2. *The signal η and noise κ in a multi-component regression are*

$$\eta(p) = \mathrm{tr}(\mathrm{var}(\Phi^T X)) = \mathrm{tr}(\Delta) = \mathrm{tr}(P_{\mathcal{E}}\Sigma_X),$$

and

$$\kappa(p) = \mathrm{tr}(\mathrm{var}(\Phi_0^T X)) = \mathrm{tr}(\Delta_0) = \mathrm{tr}(Q_{\mathcal{E}}\Sigma_X).$$

In Definition 4.1 and at the outset of Section 4.4.3 we used $\|\sigma_{X,Y}\|^2 \asymp \delta = \mathrm{tr}(\Delta)$ as a measure of the signal in one-component regressions. Our definition of signal in multi-component regressions then reduces to the previous one when $q = 1$. In our treatment of one-component regressions, we used the noise-to-signal ratio $\Delta_\sigma := \Delta_0/\|\sigma_{X,Y}\|^2$ to characterize in Theorem 4.1 the asymptotic behavior of D_N in terms of $K_1(n,p) = \mathrm{tr}(\Delta_\sigma)/n$ and $K_2(n,p) = \mathrm{tr}(\Delta_\sigma^2)/n$. A corresponding fine level of analysis was not maintained in the multi-component case. Instead, Cook and Forzani (2019) consistently used additional bounding of the form $\mathrm{tr}(A^k) \le \mathrm{tr}^k(A)$ when characterizing the asymptotic behavior of PLS predictions. For instance, this means that $K_2(n,p) \le nK_1^2(n,p)$. In this way, we obtain a coarser version of Theorem 4.1 depending only on the sample size and noise-to-signal ratio via an extended definition of K_1:

$$K_1(n,p) = \frac{\kappa(p)}{n\eta(p)}.$$

This is similar to the signal rate found by Cook, Forzani, and Rothman (2013) in their study of abundant high-dimensional linear regression .

It follows from our discussion of Section 3.5.1 that the population NIPALS weight vectors arise from a sequential orthogonalization of the q vectors $\Sigma_X^j \sigma_{X,Y}$, $j = 0, \dots, q-1$, in the Krylov sequence $K_q = (\sigma_{X,Y}, \Sigma_X\sigma_{X,Y}, \dots, \Sigma_X^{q-1}\sigma_{X,Y})$. While these basis vectors are linearly independent by construction, for this orthogonalization to be stable as $p \to \infty$, the Krylov basis vectors cannot be too collinear. Let R_j^2 denote the squared multiple correlation coefficient from the linear regression of the j-th coordinate $\sigma_{X,Y}^T \Sigma_X^{j-1} X$ of $K_q^T X$ onto the rest of its coordinates,

$$(\sigma_{X,Y}^T X, \dots, \sigma_{X,Y}^T \Sigma_X^{j-2} X, \sigma_{X,Y}^T \Sigma_X^j X, \dots, \sigma_{X,Y}^T \Sigma_X^{q-1} X).$$

Then the collinearity among the Krylov basis vectors arises asymptotically as the rate of increase in the sum of variance inflation factors

$$\rho(p) = \sum_{j=1}^{q} (1 - R_j^2)^{-1}.$$

PLS regressions have the best performance asymptotically when $\rho(p)$ is bounded, but can also perform reasonably when $\rho(p)$ increases with p but at a slower rate than $\sqrt{n}$. A sufficient condition for $\rho(p)$ to be bounded comes from the regression of Y on the material predictors scaled by the signal rate $Z_p = \Phi^T X/\sqrt{\eta(p)}$. If, as $p \to \infty$, $\text{var}(Z_p) = \eta^{-1}(p)\Delta > 0$ and the rank of the matrix of Krylov vectors $K_{Z_p} = (\sigma_{Z_p,Y}, \Sigma_{Z_p}\sigma_{Z_p,Y}, \ldots, \Sigma_{Z_p}^{q-1}\sigma_{Z_p,Y})$ for the regression of Y on Z_p is stable at q, then $\rho(p)$ is bounded (Cook and Forzani, 2019, Prop. 1). These conditions ensure that $q = \dim(\mathcal{E}_{\Sigma_X}(\mathcal{B}))$ is stable as $p \to \infty$. Since $\text{tr}(\text{var}(Z_p)) = 1$, the first condition implies that the eigenvalues δ_j of Δ are similarly sized so that $\delta_j \asymp \eta(p)$, $j = 1, \ldots, q$. In contrast, if $\delta_q/\eta(p) \to 0$ as $p \to \infty$ then asymptotically we drop at least one dimension of $\mathcal{E}_{\Sigma_X}(\mathcal{B})$. In application, if the eigenvalues of Δ have quite dissimilar sizes then PLS methods may be inclined to miss the material predictors associates with the smaller eigenvalues. This may not be surprising since all standard regression methods have relative difficulty finding the predictors with quite small effects.

These results imply that collinearity within the material part of X matters, while collinearity within the immaterial part does not. In consequence, observing collinearity among the columns of $\mathbb{X}$ does not necessarily signal problems. The results hint that if the computations are stable then the *basis-vector collinearity* described here will not be an issue.

4.6.2 Abundance v. sparsity

At the beginning of this chapter, we informally defined an abundant regression to be one in which many predictors contribute information about the response, while a sparse regression is one in which few predictors contain response information. Then in the context of one-component regressions we defined an abundant regression as one in which $\|\sigma_{X,Y}\|^2 \to \infty$, while a sparse regression is one in which $\|\sigma_{X,Y}\|^2$ is bounded as $n \to \infty$. We now adapt those ideas to multi-component regressions.

Definition 4.3. *If $\eta(p) \to \infty$ as $p \to \infty$ then the regression is said to be abundant. The regression is said to be sparse if $\eta(p)$ is bounded as $p \to \infty$.*

There is also a more easily interpretable sufficient condition for abundance given in the following proposition. Its proof is in Appendix A.4.5.

Proposition 4.2. *Assume that the eigenvalues of* Δ_0 *are bounded and* $\beta^T \Sigma_X \beta \asymp 1$. *Then* $\kappa(p) \asymp p$ *and,* $\|\sigma_{X,Y}\|^2 \to \infty$ *if and only if* $\eta(p) \to \infty$, *where* $\|\cdot\|^2$ *denotes the Euclidean norm.*

Practically, if there are many predictors that are correlated with the response and if basis-vector collinearity as described above is not an issue then we have an abundant regression.

To perhaps provide a little intuition we next describe how $\|\sigma_{X,Y}\|^2$ connects with $\eta(p)$ and the eigenvalues of Σ_X. Let $\lambda_1(A) \geq \cdots \geq \lambda_m(A)$ denote the ordered eigenvalues of the symmetric $m \times m$ matrix A. Since $\eta(p) = \sum_{j=1}^{q} \lambda_j(\Phi^T \Sigma_X \Phi)$, it follows that (Cook and Forzani, 2019, eq. (4.4))

$$\sigma_Y^2 > \frac{\|\sigma_{X,Y}\|^2}{\eta(p)}. \tag{4.26}$$

The response variance σ_Y^2 is a finite constant and does not change with p. Consequently, regardless of the value of p, the ratio in this relationship is bounded above by σ_Y^2. If $\|\sigma_{X,Y}\|^2$ increases as we add more predictors then $\eta(p)$ must correspondingly increase to maintain the bound (4.26) for all p. If $\|\sigma_{X,Y}\|^2 \to \infty$ as $p \to \infty$ then $\eta(p)$ must also diverge in a way that maintains the bound. Equivalently, if $\eta(p)$ is bounded, so the regression is sparse, then $\|\sigma_{X,Y}\|^2$ must also be bounded. In sum, $\|\sigma_{X,Y}\|^2$ may serve as a useful surrogate for the signal $\eta(p)$.

There is also a relationship between $\eta(p)$ and the eigenvalues of Σ_X (Rao, 1979, Thm. 2.1):

$$\eta(p) \leq \sum_{j=1}^{q} \lambda_j(\Sigma_X). \tag{4.27}$$

Accordingly, if the regression is abundant, so $\eta(p) \to \infty$, then at least the largest eigenvalue $\lambda_1(\Sigma_X) \to \infty$ as $p \to \infty$. And if the eigenvalues of Σ_X are bounded then the signal must be bounded, giving a sparse regression. This result clarifies a potentially mis-interpreted result by Chun and Keleş (2010), as discussed following Theorem 4.8.

4.6.3 Universal conditions

Before discussing asymptotic results in the next section, we summarize the overarching conditions used by Cook and Forzani (2019). Several of these conditions are carryovers from the conditions required for the one-component case discussed in Section 4.4.3.

C1. Model (2.5) holds. The response Y is real ($r = 1$), (Y, X) follows a non-singular multivariate normal distribution and the data (Y_i, X_i), $i = 1, \ldots, n$, arise as independent copies of (Y, X). To avoid the trivial case, we assume that the coefficient vector $\beta \neq 0$, which implies that the dimension of the envelope $q \geq 1$. This is the same as the model adopted for the one-component case, except here the number of components is allowed to exceed one.

C2. The error standard deviation $\sigma_{Y|X}$ is bounded away from 0 as $p \to \infty$, and $\beta^T \Sigma_X \beta \asymp 1$. These conditions were used in the one-component case.

C3. The number of components q, which is the same as the dimension of the envelope $\mathcal{E}_{\Sigma_X}(\mathcal{B})$, is known and fixed for all p. This is the same structure adopted for the one-component case, except here the number of components is allowed to exceed one.

C4. K_1 and $\rho/\sqrt{n} \to 0$ as $n, p \to \infty$, where K_1 and ρ are defined in Section 4.6.1.

C5. $\eta = O(\kappa)$ as $p \to \infty$, where $\eta \geq 1$, and η and κ are defined in Section 4.6.1. Although it is technically possible for κ to be a smaller order than η, we find such situations either implausible or bordering on deterministic. This condition was used implicitly when discussing one-component consistency in Section 4.4.3.

C6. $\Sigma_X > 0$ for all p. This restriction allows S_X to be singular, which is a scenario PLS was designed to handle. We do not require as a universal condition that the eigenvalues of Σ_X are bounded as $p \to \infty$.

Additional conditions will be needed for various results.

4.6.4 Multi-component consistency results

The definition of D_N given at (4.8) holds for multi-component as well as one-component regressions. Depending on properties of the regression, the asymptotic behavior D_N in multi-component regressions can depend crucially on all of the quantities described in Section 4.6.3: n, q, η, κ and K_1. In this section we summarize the main results of Cook and Forzani (2019) along with a few special scenarios that may provide useful intuition in practice. All of the asymptotic results in this section should be understood to hold as $n, p \to \infty$. Proofs are available from the online supplement to Cook and Forzani (2019).

The results of Theorem 4.7 are the most general, requiring for potentially good results in practice only that C1–C6 hold and that the terms characterizing the orders go to zero as $n, p \to \infty$. In particular, the eigenvalues of Σ_X need not be bounded.

Theorem 4.7. *As $n, p \to \infty$,*

$$D_N = O_p(\rho/\sqrt{n}) + O_p\left\{\rho^{1/2} n^{-1/2} (\kappa/\eta)^q\right\}.$$

In particular,

I. If $\rho \asymp 1$ then $D_N = O_p\left\{n^{-1/2}(\kappa/\eta)^q\right\}$.

II. If $\kappa \asymp \eta$ then $D_N = O_p(\rho/\sqrt{n})$.

III. If $q = 1$ then $D_N = O_p(\sqrt{n} K_1)$.

We see from this that the asymptotic behavior of PLS depends crucially on the relative sizes of signal η and noise κ in X. It follows from the general result that if $\kappa \asymp p$, as likely occurs in many applications, particularly spectral applications in chemometrics, and if $\eta \asymp p$, so the regression is abundant, then $D_N = O_p(\rho/\sqrt{n})$. If, in addition, $\rho \asymp 1$ then PLS fitted values converge at the usual $\sqrt{n}$-rate, regardless of the relationship between n and p.

On the other hand, if the signal in X is small relative to the noise in X, so $\eta = o(\kappa)$, then it may take a very large sample size for PLS prediction to be consistent. For instance, suppose that the regression is sparse and only q predictors matter and thus $\eta \asymp 1$. Then it follows reasonably that $\rho \asymp 1$ and, from part I, $D_N = O_p\{n^{-1/2}\kappa^q\}$. If, in addition, $\kappa \asymp p$ then $D_N = O_p\{p^q n^{-1/2}\}$. Clearly, if q is not small, then it could take a huge sample size for PLS prediction to be usefully accurate.

Theorem 4.7 places no constraints on the rate of increase in $\kappa(p) = \mathrm{tr}(\Delta_0)$. In many regressions it may be reasonable to assume that the eigenvalues of Δ_0 are bounded so that $\kappa(p) \asymp p$ as $p \to \infty$. In the next theorem we describe the asymptotic behavior of PLS predictions when the eigenvalues of Δ_0 are bounded. It is a special case of Cook and Forzani (2019, Theorem 2).

Theorem 4.8. *If the eigenvalues of Δ_0 are bounded as $p \to \infty$ then $\kappa \asymp p$ and*

$$D_N = O_p\left(\rho/\sqrt{n}\right) + O_p\left((\rho p/n\eta)^{1/2}\right).$$

In particular,

I. If $\rho \asymp 1$ *or if* $q = 1$ *then*

$$D_N = O_p\left\{\left(\frac{p}{n\eta}\right)^{1/2}\right\}.$$

II. If $\eta \asymp p$ *then* $D_N = O_p\left(\rho/\sqrt{n}\right).$

The order of D_N now depends on a balance between the sample size n, the variance inflation factors as measured through ρ and the noise to signal ratio in K_1, but it no longer depends on the dimension q. We see in the conclusion to case I that there is synergy between the sample size and the signal, the signal serving to multiply the sample size. For instance, if we assume a modest signal of $\eta(p) \asymp \sqrt{p}$ then in case I we must have n large relative to $\sqrt{p}$ for the best results. If $\eta \asymp p$ and $\rho \asymp 1$ then from case II we again get convergence at the $\sqrt{n}$ rate.

Chun and Keleş (2010) concluded that PLS can be consistent in high-dimensional regressions only if $p/n \to 0$. However, they required the eigenvalues of Σ_X to be bounded as $p \to \infty$ and that $\rho(p) \asymp 1$. If the eigenvalues of Σ_X are bounded then the eigenvalues of Δ_0 are bounded so $\kappa \asymp p$ and from (4.27) the signal must be bounded as well $\eta \asymp 1$. It then follows from Theorem 4.8 that $D_N = O_p\{(p/n)^{1/2}\}$, which is the rate obtained by Chun and Keleş (2010). By required the eigenvalues of Σ_X to be bounded, Chun and Keleş (2010) in effect restricted their conclusion to sparse regressions.

So far our focus has been on the rate of convergence of predictions as measured by D_N. There is a close connection between the rate for D_N and the rate of convergence of $\widehat{\beta}_{\text{pls}}$ in the Σ_X inner product. Let

$$V_{n,p} = \{(\widehat{\beta}_{\text{pls}} - \beta)^T \Sigma_X (\widehat{\beta}_{\text{pls}} - \beta)\}^{1/2}.$$

Then, as shown by Cook and Forzani (2019, Supplement Section S8), $V_{n,p}$ and D_N have the same order as $n, p \to \infty$, so Theorems 4.7 and 4.8 hold with D_N replaced by $V_{n,p}$. It follows from that the special cases of Theorems 4.7 and 4.8 and the subsequent discussions apply to $V_{n,p}$ as well. In particular, estimative convergence for $\widehat{\beta}_{\text{pls}}$ as measured in the Σ_X inner product will be at or near the root-n rate under the same conditions as predictive convergence.

4.7 Bounded or unbounded signal?

The distinction between a bounded or unbounded signal $\eta(p)$ has been a common theme in this chapter. If the signal is bounded $\eta(p) \asymp 1$ we would not normally expect PLS to do well in high dimensional $n < p$ regressions, and then methods tailored to sparse regressions would be necessary. But if the signal is abundant $\eta(p) \to \infty$, then we would normally expect PLS to give serviceable results, particularly when $\eta(p) \asymp p^{1/2}$ or greater. How this dichotomy plays out in practice depends on the area of application. For instance, it seems that many Chemometrics regressions are of the abundant variety.

Goicoechea and Oliveri (1999) used PLS to develop a predictor of tetracycline concentration in human blood. Fifty training samples were constructed by spiking blank sera with various amounts of tetracycline in the range 0–4 μg mL^{-1}. A validation set of 57 samples was constructed in the same way. For each sample, the values of the predictors were determined by measuring fluorescence intensity at $p = 101$ equally spaced points in the range 450–550 nm. The authors determined using leave-one-out cross validation that the best predictions of the training data were obtained with $q = 4$ linear combinations of the original 101 predictors.

Cook and Forzani (2019) use these data to illustrate the behavior of PLS predictions in Chemometrics as the number of predictors increases. They used PLS with $q = 4$ to predict the validation data based on p equally spaced spectra, with p ranging between 10 and 101. The root mean squared error (MSE) is shown in Figure 4.6 for five values of p. PLS fits were determined by using *library{pls}* in *R* (R Core Team, 2022). We see a relatively steep drop in MSE for small p, say less than 30, and a slow but steady decrease in MSE thereafter. Since we are dealing with actual out-of-sample prediction, there is no artifactual reason why the root-MSE should continue to decrease with p.

4.8 Illustration

We use in this section variations on the example presented in Section 3.7 to illustrate selected asymptotic results, including the distinction between

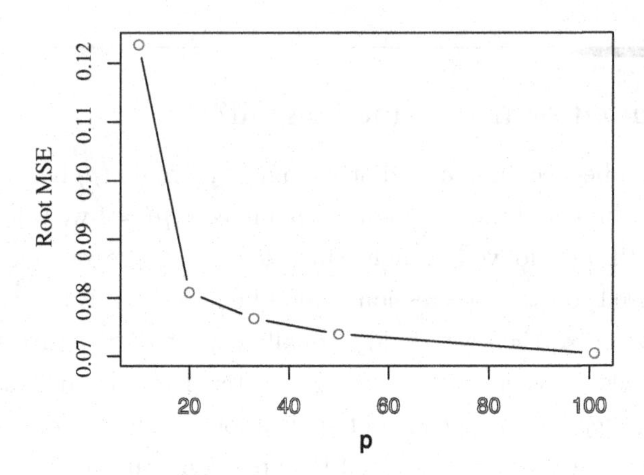

FIGURE 4.6
Tetracycline data: The open circles give the validation root MSE from $10, 20, 33, 50$, and 101 equally spaced spectra. (From Fig. 4 of Cook and Forzani (2019) with permission.)

abundant and sparse regressions (Cook and Forzani, 2020). A detailed analysis of a one-component regression is described in Section 4.8.1, and a cursory analysis of a two-component regression is given in Section 4.8.2.

4.8.1 One-component regression

Beginning with the example of Section 3.7, we now set $p_1 = p_2$ and increase the variance of the predictors from 1 to 25. The $n \times p$ data matrix $\mathbb{X}$ of predictor values can be represented as

$$\mathbb{X}_{n\times p} = (H_1 1_1^T + E_1, H_2 1_2^T + E_2, H_3 1_3^T + E_3),$$

where now H_j is an $n \times 1$ vector of independent normal variates with mean 0 and variance 25, $j = 1, 2, 3$, and 1_1, 1_2, and 1_3 are vectors of ones having dimensions $p_1 = p_2 = (p - d)/2$ and $p_3 = d$, and we restrict $p - d$ to be even. The E_j's are matrices of conforming dimensions also consisting of independent standard normal variates. A typical $p \times 1$ predictor vector can still be represented as

$$X = \begin{pmatrix} h_1 1_1 \\ h_2 1_2 \\ h_3 1_3 \end{pmatrix} + e,$$

but now the h_j's are independent normal random variables with mean 0 and variance 25. The variance-covariance matrix of the predictors X is now

$$\Sigma_X = 25 \begin{pmatrix} 1_1 1_1^T & 0 & \\ 0 & 1_2 1_2^T & 0 \\ 0 & 0 & 1_3 1_3^T \end{pmatrix} + I_p.$$

Again, the predictors consist of three independent blocks of sizes $(p-d)/2$, $(p-d)/2$ and d. The correlation between predictors in the same block is about 0.96. Recall that $u_1^T = (1_1^T, 0, 0)$, $u_2^T = (0, 1_2^T, 0)$, and $u_3^T = (0, 0, 1_3^T)$. Then Σ_X can be expressed equivalently as

$$\begin{aligned} \Sigma_X &= (1+25p_1)P_{u_1} + (1+25p_2)P_{u_2} + (1+25p_3)P_{u_3} + Q, \\ &= (1+25(p-d)/2)P_{(u_1,u_2)} + (1+25d)P_{u_3} + Q, \end{aligned}$$

where $P_{(\cdot)}$ is the projection onto the subspace spanned by the indicated vectors and Q is the projection onto the $p-3$ dimensional subspace that is orthogonal to $\text{span}(u_1, u_2, u_3)$. We see that Σ_X has three eigenspaces $\text{span}(u_1, u_2)$, $\text{span}(u_3)$ and $\text{span}^{\perp}(u_1, u_2, u_3)$. With $r = 1$, the $n \times 1$ vector of responses $\mathbb{Y}$ was generated as the linear combination $\mathbb{Y} = 3H_1 - 4H_2 + \epsilon$, where H_1 and H_2 where ϵ is an $n \times 1$ vector of independent normal variates with mean 0 and finite variance. This gives

$$\begin{aligned} \sigma_{X,Y} &= 75u_1 - 100u_2 \\ \|\sigma_{X,Y}\|^2 &= (75^2 + 100^2)(p-d)/2 \\ \beta &= \frac{75}{1+25p_1}u_1 - \frac{100}{1+25p_2}u_2 \\ &= \frac{1}{1+25(p-d)/2}\sigma_{X,Y} \\ \beta^T \Sigma_X \beta &= (75^2 + 100^2)\frac{(p-d)/2}{1+25(p-d)/2}. \end{aligned}$$

Consequently, both β and $\sigma_{X,Y}$ are eigenvectors of Σ_X since these vectors fall in the one-dimensional envelope $\mathcal{E}_{\Sigma_X}(\mathcal{B}) = \text{span}(3u_1 - 4u_2)$. This implies that X can be replaced with the single predictor $3u_1^T X - 4u_2^T X$ without loss of information on the regression. Normalized bases for the envelope and its two-dimensional orthogonal complement are then

$$\begin{aligned} \Phi &= \frac{3u_1 - 4u_2}{\|3u_1 - 4u_2\|} \\ \Phi_0 &= \left(\frac{4u_1 + 3u_2}{\|4u_1 + 3u_2\|}, \frac{u_3}{\|u_3\|}, U\right), \end{aligned}$$

where the columns of $U \in \mathbb{R}^{p\times(p-3)}$ form an orthonormal basis for $\text{span}^{\perp}(u_1, u_2, u_3)$. All of the results of this chapter are applicable here since this is a one-component regression.

According to Definition 4.1, this regression is abundant if $\|\sigma_{X,Y}\| \to \infty$ and sparse otherwise. The first two blocks of $(p-d)/2$ predictors shown in $\mathbb{X}$ are material to the regression, while the last block of d predictors is immaterial. If the total of $(p-d)$ material predictors is large relative to the total d of immaterial predictors then $\|\sigma_{X,Y}\|$ will be large, so PLS and envelopes should do well, depending on the sample size. Otherwise, there will be some degradation in the performance of the methods, with perhaps extreme failure when d is large relative to $(p-d)$, unless $n \gg p$.

From Definition 4.2, the predictor signal and noise measures are

$$\begin{aligned}\eta(p) &= \Phi^T\Sigma_X\Phi \\ &= 1+25(p-d)/2 \\ \kappa(p) &= \text{tr}(\Phi_0^T\Sigma_X\Phi_0) \\ &= -1+25(p+d)/2+p.\end{aligned}$$

The eigenvalues of Δ_0 are $1+25(p-d)/2$, $1+25d$, and 1 with multiplicity $p-3$. The effectiveness of envelopes and PLS is now governed by n, p, and d. Corollaries 4.2–4.4 are not applicable in this case since one and perhaps two of the eigenvalues of Δ_0 are unbounded. However, Corollary 4.5 is applicable since only a finitely many of the eigenvalues of Δ_0 are unbounded. The following three cases illustrate the application of Corollary 4.5. Recall that a and s are defined as constants such that $p = n^a$ with $a \geq 1$ and $\|\sigma_{X,Y}\|^2 = p^s$, which is informally interpreted as the number of material predictors, and that $D_N = O_p(n^{-1/2+a(1-s)})$, so the convergence rate is $n^{1/2-a(1-s)}$.

Case 1: When d is fixed, $\|\sigma_{X,Y}\|^2 \asymp p$ and so $s=1$ and D_N converges at the rate $\sqrt{n}$, regardless of the relationship between n and p. In this case, all but finitely many predictors are material, rather like an antithesis of sparsity.

Case 2: If $d = p - p^{2/3}$, so there are $p - d = p^{2/3}$ material predictors then $\|\sigma_{X,Y}\|^2 \asymp p^{2/3}$, which gives $s = 2/3$. If, in addition, $p \asymp n$ then $a = 1$ and D_N converges at the rate $n^{1/2-a(1-s)} = n^{1/6}$, which is quite slow.

Case 3: If $d = p-2$ then $s = 0$, $a \geq 1$, there are only 2 material predictors and no convergence is indicated by the corollary.

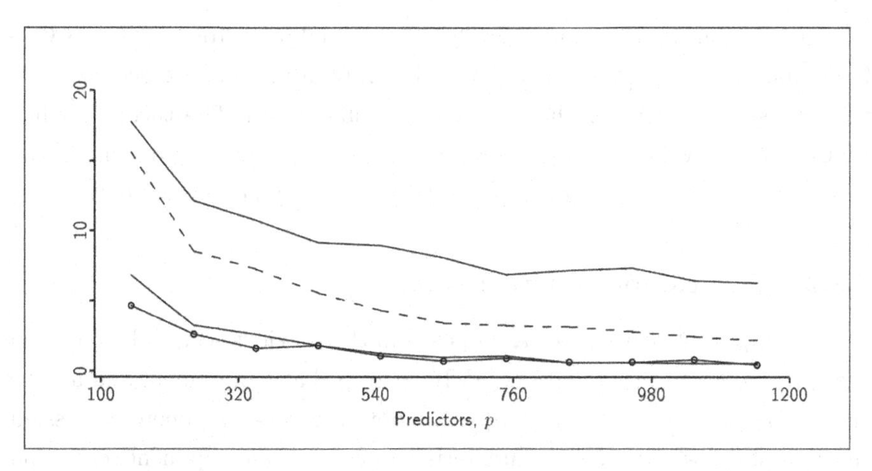

FIGURE 4.7
Illustration of the behavior of PLS predictions in abundant and sparse regressions. Lines correspond to different numbers of material predictors. Reading from top to bottom as the lines approach the vertical axis, the first bold line is for 2 material predictors. The second dashed line is for $p^{2/3}$ material predictors. The third solid lines is for $p - 40$ material predictors and the last line with circles at the predictor numbers used in the simulations is for p material predictors. The vertical axis is the squared norm $\|\beta - \widehat{\beta}\|^2_{S_X}$ and always $n = p/3$. (This figure was constructed with permission using the same data as Cook and Forzani (2020) used for their Fig. 1.)

Figure 4.7 shows the results of a small simulation to reinforce these comments. The vertical axis is the squared difference between centered mean value $\beta^T X$ and its PLS estimator $\widehat{\beta}^T X$ averaged over the sample: $\|\beta - \widehat{\beta}\|^2_{S_X} = n^{-1}\sum_{i=1}^n(\beta^T X_i - \widehat{\beta}^T X_i)^2$. The horizontal axis is the number of predictors p and the sample size is $n = p/3$. The lines on the plot are for different numbers of material predictors. The curve for 2 material predictors corresponds to a sparse regression and, as expected, the results support Case 3, as there is no visual evidence that it is converging to 0. The other three curves are for abundant regressions with varying rates of information accumulation.

For the curve for "p" material predictors is the best that can be achieved in the context of the simulated regression. For the "$p-40$" curve all predictors are material except for 40. As the total number of predictors increases, the 40 immaterial predictors cease to play a role and the "$p - 40$" curve coincides with the "p" curve. These results support Case 1.

The remaining dashed curve for "$p^{2/3}$" material predictors represents Case 2, an abundant regression with a very slow rate of information accumulation that is less than that for the "p" and "$p - 40$" curves. The theory predicts that this curve will eventually coincide with the "p" curve, but demonstrating that result by simulation will surely take a very, very large value of p.

4.8.2 Two-component regression

A one-component regression was imposed in the previous section by requiring that $p_1 = p_2 = (p-d)/2$ and $p_3 = d$. If we drop that requirement and assume that p_1, p_2, and p_3 are distinct, then we obtain a two-component regression. In this section, we give a few characteristics of that two-component regression.

The quantities $\mathbb{X}$, X, and $\mathbb{Y}$ are as given in the previous section, except we no longer require $p_1 = p_2$. Thus we still have

$$\begin{aligned} \Sigma_X &= (1+25p_1)P_{u_1} + (1+25p_2)P_{u_2} + (1+25p_3)P_{u_3} + Q, \\ \sigma_{X,Y} &= 75u_1 - 100u_2 \\ \beta &= \frac{75}{1+25p_1}u_1 - \frac{100}{1+25p_2}u_2, \end{aligned}$$

which are the same as corresponding quantities given in the previous section, but are repeated here for ease of reference. Now, Σ_X has four eigenspaces $\text{span}(u_j)$, $j = 1,2,3$ and $\text{span}^{\perp}(u_1,u_2,u_3)$. The envelope is two-dimensional $\mathcal{E}_{\Sigma_X}(\mathcal{B}) = \text{span}(u_1,u_2)$ and its orthogonal complement $\mathcal{E}^{\perp}_{\Sigma_X}(\mathcal{B}) = \text{span}(u_3,U)$ has dimension $p-2$, where the columns of $U \in \mathbb{R}^{p\times(p-3)}$ form an orthonormal basis for $\text{span}^{\perp}(u_1,u_2,u_3)$, as defined in the previous section. Now, X can be replaced by two univariate predictors $(u_1^T X, u_2^T X)$ without loss of information about the regression. Since this is a two-component regression, we now rely on the results of Section 4.6 to characterize D_N asymptotically. For this we need

$$\begin{aligned} \Phi &= (u_1/\sqrt{p_1}, u_2/\sqrt{p_2}) \\ \eta(p) &= \text{tr}(\Phi^T\Sigma_X\Phi) \\ &= 2 + 25(p_1+p_2) \\ \kappa(p) &= -2 + p + 25p_3, \end{aligned}$$

where $\kappa(p)$ comes from adding the eigenvalues of Δ_0, $(1+25p_3)$ and 1 with multiplicity $p-3$. From here we can appeal to Theorems 4.7 and 4.8 to characterize the asymptotic behavior of PLS predictions.

If $\rho \asymp 1$ and p_3 is bounded, then the eigenvalues of Δ_0 are bounded, $\kappa \asymp p$, and we can appeal to part I of Theorem 4.8 to get D_N. For this we need

$$\frac{p}{n\eta(p)} = \frac{p_1 + p_2 + p_3}{n(2 + 25(p_1 + p_2))} \asymp \frac{1}{n}.$$

In this case then we have root-n convergence, $D_N = O_p(n^{-1/2})$. Further, since p_3 is bounded, $\eta \asymp p$ and part II of Theorem 4.8 holds. That is, $D_N = O_p(\rho/\sqrt{n})$.

If $\rho \asymp 1$ and p_3 is unbounded then we can use part I of Theorem 4.7. For this theorem we need

$$\begin{aligned} \frac{\kappa}{\eta} &= \frac{-2 + p + 25p_3}{2 + 25(p_1 + p_2)} \\ &\asymp 1 + \frac{p_3}{p_1 + p_2}. \end{aligned}$$

In consequence,

$$D_N = O_p\left(\frac{1}{\sqrt{n}}\left(1 + \frac{p_3}{p_1 + p_2}\right)^2\right).$$

This rate depends on the total number of material predictors p_1+p_2 relative to the number of immaterial predictors. If $p_3/(p_1+p_2)$ is bounded as $p \to \infty$ we again achieve root-n convergence. In this case $\kappa \asymp \eta$ and we recover the result of part II of Theorem 4.7. However, if the ratio $p_3/(p_1 + p_2)$ is unbounded, then the sample size needs to be large enough to compensate and again force $n^{-1/2}(\kappa/\eta)^2 \to 0$. This discussion once more reflects the idea of abundance: for the best results, the number of material predictors $p_1 + p_2$ should be large relative to the number of immaterial predictors p_3.

5

Simultaneous Reduction

Response and predictor envelope methods have the potential to increase efficiency in estimation and prediction. It might be anticipated then that combining response and predictor reductions may have advantages over either method applied individually. We discuss simultaneous predictor-response reduction in this chapter. Our discussion is based mostly on two relatively recent papers: Cook and Zhang (2015b) developed maximum likelihood estimation under a multivariate model for simultaneous reduction of the response and predictor vectors, studied asymptotic properties under different scenarios and proposed two algorithms for getting estimators. Cook, Forzani, and Liu (2023b) developed simultaneous PLS estimation based on the same type of multivariate linear regression model used by Cook and Zhang (2015b).

As in previous developments, we first discuss in Section 5.1 conditional independence foundations for simultaneous reduction. We then incorporate the multivariate linear model, turning to likelihood-based estimation in Section 5.2. Simultaneous PLS estimation is discussed in Section 5.3 and other related methods are discussed briefly in Section 5.4. Since PLS is a focal point of this book, the main thrust of our discussion follows Cook, Forzani, and Liu (2023b), with results from Cook and Zhang (2015b) integrated as relevant.

5.1 Foundations for simultaneous reduction

Recall that $X \in \mathbb{R}^p$, $Y \in \mathbb{R}^r$ and that, from (2.2), the condition $(Y, P_{\mathcal{S}}X) \perp\!\!\!\perp Q_{\mathcal{S}}X$ is a foundation for predictor envelopes. It insures that $Q_{\mathcal{S}}X$ is immaterial to the regression since it is independent of Y and $P_{\mathcal{S}}X$ jointly. Similarly, the foundational condition for response envelopes is obtained by

DOI: 10.1201/9781003482475-5

interchanging the roles of X and Y: $(P_{\mathcal{R}}Y, X) \perp\!\!\!\perp Q_{\mathcal{R}}Y$ from (2.14). We combine these two conditions in the following proposition for simultaneous reduction of X and Y considered jointly, without designating one as the response and one as the predictor. Its proof is given in Appendix A.5.1.

Proposition 5.1. *Define the subspaces $\mathcal{S} \subseteq \mathbb{R}^p$ and $\mathcal{R} \subseteq \mathbb{R}^r$. Then the two condition*

$$\text{(a) } Q_{\mathcal{S}}X \perp\!\!\!\perp (Y, P_{\mathcal{S}}X) \text{ and (b) } Q_{\mathcal{R}}Y \perp\!\!\!\perp (X, P_{\mathcal{R}}Y)$$

hold if and only if the following two conditions hold:

$$\text{(I) } Q_{\mathcal{R}}Y \perp\!\!\!\perp Q_{\mathcal{S}}X \text{ and (II) } (P_{\mathcal{R}}Y, P_{\mathcal{S}}X) \perp\!\!\!\perp (Q_{\mathcal{R}}Y, Q_{\mathcal{S}}X).$$

To implement Proposition 5.1 in practice, we replace conditions (I) and (II) with corresponding zero covariance conditions:

$$\text{(I}^*\text{) } \mathrm{cov}(Q_{\mathcal{R}}Y, Q_{\mathcal{S}}X) = 0 \text{ and (II}^*\text{) } \mathrm{cov}\{(P_{\mathcal{R}}Y, P_{\mathcal{S}}X), (Q_{\mathcal{R}}Y, Q_{\mathcal{S}}X)\} = 0. \tag{5.1}$$

These covariance conditions imply straightforwardly that

$$\mathrm{var}\begin{pmatrix} P_{\mathcal{R}}Y \\ P_{\mathcal{S}}X \\ Q_{\mathcal{R}}Y \\ Q_{\mathcal{S}}X \end{pmatrix} = \begin{pmatrix} P_{\mathcal{R}}\Sigma_Y P_{\mathcal{R}} & P_{\mathcal{R}}\Sigma_{Y,X}P_{\mathcal{S}} & 0 & 0 \\ P_{\mathcal{S}}\Sigma_{X,Y}P_{\mathcal{R}} & P_{\mathcal{S}}\Sigma_X P_{\mathcal{S}} & 0 & 0 \\ 0 & 0 & Q_{\mathcal{R}}\Sigma_Y Q_{\mathcal{R}} & 0 \\ 0 & 0 & 0 & Q_{\mathcal{S}}\Sigma_X Q_{\mathcal{S}} \end{pmatrix} \tag{5.2}$$

Recall that $C = (X^T, Y^T)^T \in \mathbb{R}^{p+r}$ denotes the random vector constructed by concatenating X and Y, and that S_C denotes the sample version of $\Sigma_C = \mathrm{var}(C)$. Equation (5.2) is expressed in terms of subspaces $\mathcal{S}$ and $\mathcal{R}$, but for the development of methodology we use it to express the covariance matrix Σ_C of the observable vector C in terms of $\mathcal{S}$ and $\mathcal{R}$:

$$\begin{aligned} \Sigma_X &= \mathrm{var}(X) = \mathrm{var}(P_{\mathcal{S}}X + Q_{\mathcal{S}}X) \\ &= \mathrm{var}(P_{\mathcal{S}}X) + \mathrm{var}(Q_{\mathcal{S}}X) \\ &= P_{\mathcal{S}}\Sigma_X P_{\mathcal{S}} + Q_{\mathcal{S}}\Sigma_X Q_{\mathcal{S}}, \end{aligned}$$

where the second equality follows because from (5.2), $\mathrm{cov}(P_{\mathcal{S}}X, Q_{\mathcal{S}}X) = 0$. Similarly, $\Sigma_Y = P_{\mathcal{R}}\Sigma_Y P_{\mathcal{R}} + Q_{\mathcal{R}}\Sigma_Y Q_{\mathcal{R}}$. This then implies that

$$\begin{aligned} \Sigma_C &= \begin{pmatrix} \Sigma_X & \Sigma_{X,Y} \\ \Sigma_{Y,X} & \Sigma_Y \end{pmatrix} \\ &= \begin{pmatrix} P_{\mathcal{S}}\Sigma_X P_{\mathcal{S}} + Q_{\mathcal{S}}\Sigma_X Q_{\mathcal{S}} & P_{\mathcal{S}}\Sigma_{X,Y}P_{\mathcal{R}} \\ P_{\mathcal{R}}\Sigma_{Y,X}P_{\mathcal{S}} & P_{\mathcal{R}}\Sigma_Y P_{\mathcal{R}} + Q_{\mathcal{R}}\Sigma_Y Q_{\mathcal{R}} \end{pmatrix}. \end{aligned} \tag{5.3}$$

The structure of (5.3) does not specify X or Y as the response. Proceeding with these variables in their traditional roles of regressing Y on X, (5.3) is satisfied by setting $\mathcal{S}$ to be the predictor envelope $\mathcal{S} = \mathcal{E}_{\Sigma_X}(\mathcal{C}_{X,Y}) = \mathcal{E}_{\Sigma_X}(\mathcal{B})$ and $\mathcal{R}$ to be the response envelope $\mathcal{R} = \mathcal{E}_{\Sigma_Y}(\mathcal{C}_{Y,X}) = \mathcal{E}_{\Sigma_{Y|X}}(\mathcal{B}')$, as described in Table 2.2. (Recall from the discussion following Proposition 1.8 that $\mathcal{E}_{\Sigma_Y}(\mathcal{C}_{Y,X}) = \mathcal{E}_{\Sigma_Y}(\mathcal{B}') = \mathcal{E}_{\Sigma_{Y|X}}(\mathcal{B}')$.)

A coordinate representation of Σ_C may be of help when introducing the linear model. Recall from (2.5) that $\Phi \in \mathbb{R}^{p\times q}$ is a semi-orthogonal basis matrix for $\mathcal{E}_{\Sigma_X}(\mathcal{B})$, that (Φ, Φ_0) is orthogonal, that $\Delta = \mathrm{var}(\Phi^T X) = \Phi^T \Sigma_X \Phi \in \mathbb{R}^{q\times q}$ is the covariance matrix of the material predictors and that $\Delta_0 = \mathrm{var}(\Phi_0^T X) = \Phi_0^T \Sigma_X \Phi_0 \in \mathbb{R}^{(p-q)\times(p-q)}$ is the covariance matrix of the immaterial predictors. Similarly, from (2.17), $\Gamma \in \mathbb{R}^{r\times u}$ is a semi-orthogonal basis matrix for $\mathcal{E}_{\Sigma_{Y|X}}(\mathcal{B}')$ and for $\mathcal{E}_{\Sigma_Y}(\mathcal{C}_{Y,X})$, and (Γ, Γ_0) is orthogonal. Define $\Theta = \mathrm{var}(\Gamma^T Y) = \Gamma^T \Sigma_Y \Gamma \in \mathbb{R}^{u\times u}$, which is the variance of the material part of the response vector and $\Theta_0 = \mathrm{var}(\Gamma_0^T Y) = \Gamma_0^T \Sigma_Y \Gamma_0 \in \mathbb{R}^{(r-u)\times(r-u)}$, which is the variance of the immaterial part of the response vector. Define also $K = \Phi^T \Sigma_{X,Y} \Gamma \in \mathbb{R}^{q\times u}$, which contains the coordinates of $\Sigma_{X,Y}$ relative to Φ and Γ. Then the coordinate form of Σ_C can be represented as

$$\begin{aligned}\Sigma_C &= \begin{pmatrix} \Sigma_X & \Sigma_{X,Y} \\ \Sigma_{Y,X} & \Sigma_Y \end{pmatrix} \\ &= \begin{pmatrix} \Phi\Delta\Phi^T + \Phi_0\Delta_0\Phi_0^T & \Phi K \Gamma^T \\ \Gamma K^T \Phi^T & \Gamma\Theta\Gamma^T + \Gamma_0\Theta_0\Gamma_0^T \end{pmatrix}. \end{aligned} \tag{5.4}$$

As in our discussion of predictor envelopes in Section 2.2, no relationship is assumed between the eigenvalues of Δ and Δ_0: The eigenvalues of Δ could be any subset of the eigenvalues of Σ_X. Similar flexibility holds for the eigenvalues of Σ_Y.

It follows straightforwardly from (5.4) that the covariance matrices for the material and immaterial parts of C agree with (5.2),

$$\Sigma_{(\Phi\oplus\Gamma)^T C} = \begin{pmatrix} \Delta & K \\ K^T & \Theta \end{pmatrix}$$

and

$$\Sigma_{(\Phi_0\oplus\Gamma_0)^T C} = \begin{pmatrix} \Delta_0 & 0 \\ 0 & \Theta_0 \end{pmatrix},$$

where the direct sum operator $\oplus$ is defined in Section 1.2.1.

To see the roles of dimension and rank, let

$$d = \text{rank}(\Sigma_{X,Y}) = \text{rank}(K) \leq \min(q, u), \tag{5.5}$$

where $q \leq p$ and $u \leq r$. The equality $\text{rank}(\Sigma_{X,Y}) = \text{rank}(K)$ follows since, from (5.4), $\Sigma_{X,Y} = \Phi K \Gamma^T$ and Φ and Γ both have full column rank. From the definitions of the X- and Y-envelopes, if $d = r < p$ then $\mathcal{E}_{\Sigma_{Y|X}}(\mathcal{B}') = \mathbb{R}^r$, $\Gamma = I_r$ and we may reduce X only. Similarly, if $d = p < r$ then $\mathcal{E}_{\Sigma_X}(\mathcal{B}) = \mathbb{R}^p$, $\Phi = I_p$ and reduction is possible only in the response space. Hence, we will assume $d < \min(r, p)$ from now on and discuss the general situation where simultaneous reduction is possible.

5.1.1 Definition of the simultaneous envelope

Model (5.4) is formulated in terms of two envelopes, the response envelope $\mathcal{E}_{\Sigma_{Y|X}}(\mathcal{B}')$ with basis matrix Γ and the predictor envelope $\mathcal{E}_{\Sigma_X}(\mathcal{B})$ with basis matrix Φ. Model estimation requires that we estimate these envelopes, either individually using the methodology described in Chapter 2 or jointly based on the likelihood stemming from model (5.4). However, using Lemma 1.4, it is possible to describe model (5.4) in terms of a single simultaneous envelope:

$$\mathcal{E}_{\Sigma_{Y|X}}(\mathcal{B}') \oplus \mathcal{E}_{\Sigma_X}(\mathcal{B}) = \mathcal{E}_{\Sigma_Y}(\mathcal{B}') \oplus \mathcal{E}_{\Sigma_X}(\mathcal{B}) = \mathcal{E}_{\Sigma_Y \oplus \Sigma_X}(\mathcal{B}' \oplus \mathcal{B}). \tag{5.6}$$

In consequence, we can view our goal as the estimation of a single envelope $\mathcal{E}_{\Sigma_Y \oplus \Sigma_X}(\mathcal{B}' \oplus \mathcal{B})$, although the methodology described in this chapter still deals with separate envelopes.

5.1.2 Bringing in the linear model

So far we have not made explicit use of a regression model. Assuming now that linear model (1.1) holds, it follows from (5.4) that

$$\begin{aligned}
\beta &= \Sigma_X^{-1} \Sigma_{X,Y} \\
&= \Phi \Delta^{-1} \Phi^T \Phi K \Gamma^T \\
&= \Phi \eta \Gamma^T, \text{ where } \eta = \Delta^{-1} K; \\
\Sigma_{Y|X} &= \Sigma_Y - \text{var}(\beta^T X) \\
&= \Gamma \Theta \Gamma^T + \Gamma_0 \Theta_0 \Gamma_0^T - \Gamma K^T \Delta^{-1} K \Gamma^T \\
&= \Gamma(\Theta - K^T \Delta^{-1} K)\Gamma^T + \Gamma_0 \Theta_0 \Gamma_0^T \\
&= \Gamma \Omega \Gamma^T + \Gamma_0 \Omega_0 \Gamma_0^T,
\end{aligned}$$

where $\Omega = \Theta - K^T \Delta^{-1} K = \Theta - \eta^T \Delta \eta$ and $\Omega_0 = \Theta_0$.

This structure leads to the following linear model for the simultaneous reduction of predictors and responses,

$$\left.\begin{aligned} Y &= \alpha + \Gamma\eta^T\Phi^T X + \varepsilon \\ \Sigma_{Y|X} &= \Gamma\Omega\Gamma^T + \Gamma_0\Omega_0\Gamma_0^T \\ \Sigma_X &= \Phi\Delta\Phi^T + \Phi_0\Delta_0\Phi_0^T, \end{aligned}\right\}, \tag{5.7}$$

where η contains the coordinates of β relative to bases Γ and Φ. In the following sections we discuss estimation under model (5.7) by maximum likelihood, PLS and a related two-block method.

Model (5.7) is the same as that used by Cook and Zhang (2015b) in their development of a likelihood-based simultaneous reduction method. However, instead of starting with general reductive conditions given in Proposition 5.1, Cook and Zhang (2015b) took a different route, which we now describe since it may furnish additional insights.

5.1.3 Cook-Zhang development of (5.7)

The approach used by Cook and Zhang (2015b) essentially combines the envelopes for predictor reduction, model (2.5), and response reduction, model (2.17), to achieve simultaneous reduction in the multivariate linear model (1.1) in which Y and X are still jointly stochastic. To see how they accomplished this union of envelopes, recall that $d = \text{rank}(\beta) = \text{rank}(K)$ and consider the singular value decomposition $\beta = UDV^T$, where $U \in \mathbb{R}^{p\times d}$ and $V \in \mathbb{R}^{r\times d}$ are orthogonal matrices, $U^TU = I_d = V^TV$, and $D = \text{diag}(\lambda_1, \ldots, \lambda_d)$ is a diagonal matrix with elements $\lambda_1 \geq \cdots \geq \lambda_d > 0$ being the d singular values of β. This decomposition of β provides the essential constituents, U and V, for construction the envelopes. We have denoted the column space (the left eigenspace) and row space (the right eigenspace) of β as $\mathcal{B} = \text{span}(U)$ and $\mathcal{B}' = \text{span}(V)$. Parallel to our discussion of K, if $d = r < p$ then $\mathcal{E}_{\Sigma_{Y|X}}(\mathcal{B}') = \mathbb{R}^r$ and we may reduce X only. Similarly, if $d = p < r$ then $\mathcal{E}_{\Sigma_X}(\mathcal{B}) = \mathbb{R}^p$ and reduction is possible only in the response space, which leads to the reminder that we assume $d < \min(r, p)$. Since $\beta = UDV^T$, where $\text{span}(U) = \mathcal{B} \subseteq \text{span}(\Phi)$ and $\text{span}(V) = \mathcal{B}' \subseteq \text{span}(\Gamma)$ by the definition of the X- and Y-envelopes, we can represent

$$\beta = UDV^T = (\Phi A)D(B^T\Gamma^T) = \Phi\eta\Gamma^T$$

for some semi-orthogonal matrices $A \in \mathbb{R}^{q\times d}$ and $B \in \mathbb{R}^{u\times d}$, where $\eta = ADB^T$. Bringing in the structured covariance matrices from the response and predictor envelopes, we again arrive at the simultaneous reductive model (5.7).

5.1.4 Links to canonical correlations

Result (5.2) shows a similarity between simultaneous envelope reduction and canonical correlation analysis: the components that capture the linear relationships between X and Y are uncorrelated with the rest of the components.

Canonical correlation analysis is widely used for the purpose of simultaneously reducing the predictors and the responses. In the population, it finds canonical pairs of directions $\{a_i, b_i\}$, $i = 1, \ldots, d$, so that the correlations between $a_i^T X$ and $b_i^T Y$ are maximized subject to the constraints $a_j^T \Sigma_X a_k = 0$, $a_j^T \Sigma_{X,Y} b_k = 0$ and $b_j^T \Sigma_Y b_k = 0$ for all $j \neq k$, and $a_j^T \Sigma_X a_j = 1$ and $b_j^T \Sigma_Y b_j = 1$ for all j. The solution is then $\{a_i, b_i\} = \{\Sigma_X^{-1/2} e_i, \Sigma_Y^{-1/2} f_i\}$, where $\{e_i, f_i\}$ is the i-th left-right eigenvector pair of the matrix $\rho = \Sigma_X^{-1/2} \Sigma_{X,Y} \Sigma_Y^{-1/2}$.

The following lemma from Cook and Zhang (2015b, Lemma 2) connects envelopes and canonical correlations. A proof is sketched in Appendix A.5.2.

Lemma 5.1. *Under the simultaneous envelope model (5.7), canonical correlation analysis can find at most d directions in the population, where $d = \text{rank}(\Sigma_{X,Y})$ as defined in (5.5). Moreover, the directions are contained in the simultaneous envelope as*

$$\text{span}(a_1, \ldots, a_d) \subseteq \mathcal{E}_{\Sigma_X}(\mathcal{B}), \ \text{span}(b_1, \ldots, b_d) \subseteq \mathcal{E}_{\Sigma_{Y|X}}(\mathcal{B}'). \tag{5.8}$$

Canonical correlation analysis may thus miss some information about the regression by ignoring some material parts of X and/or Y. For example, when r is small, it can find at most r linear combinations of X, which can be insufficient for regression. Cook and Zhang (2015b) found in the simulation studies that the performance of predictions based on estimated canonical correlation reductions $a_1^T X, \ldots, a_d^T X$ and $b_1^T Y, \ldots, b_d^T$ varied widely for different covariance structures and was generally poor.

5.2 Likelihood-based estimation

Assuming that C is normally distributed, Cook and Zhang (2015b) developed maximum likelihood estimation under model (5.7). They also studied asymptotic properties under different scenarios and proposed two algorithms for getting estimators G and W of the semi-orthogonal bases Γ and Φ in model

(5.7). One is a likelihood-based algorithm that alternates between iterations on two non-convex objective functions for predictor and response reductions. That algorithm is generally reliable, but can be slow and can get caught in a local optimum. Their second algorithm optimizes over one basis vector at a time. It is generally faster and nearly as efficient as the likelihood-based algorithm.

They also prove that if the data $C_i, i = 1, \dots, n$, are independent observations on C which has finite fourth moments then jointly the vectorized maximum likelihood estimators of β, Σ_X and $\Sigma_{Y|X}$ are asymptotically distributed as a normal random vector. If C is normally distributed that asymptotic distribution simplifies considerably.

5.2.1 Estimators

As in other envelope model encountered in this book, the most difficult part of parameter estimation is determining estimators for the basis matrices Φ and Γ. Once this is accomplished, the estimators of the remaining parameters in model (5.7) are relatively straightforward to construct.

Let G and W denote semi-orthogonal matrices that estimate versions of Γ and Φ. These may come from maximum likelihood estimation, PLS estimation or some other methods. Estimators of the remaining parameters in (5.7) are given in the following lemma. The derivation is omitted since it is quite similar to the derivations of the estimators for predictor and response envelopes.

Lemma 5.2. *After compressing the centered predictor $X \mapsto W^T X$ and the response $Y \mapsto G^T Y$ vectors and assuming multivariate normality for C, the maximum likelihood estimators of the parameters in model (1.1) via model (5.7) are*

$$
\begin{aligned}
\widehat{\alpha} &= \bar{Y} \\
\widehat{\eta} &= S_{W^T X}^{-1} S_{W^T X, G^T Y} \\
\widehat{\beta} &= W S_{W^T X}^{-1} S_{W^T X, G^T Y} G^T \\
\widehat{\Delta} &= S_{W^T X} \\
\widehat{\Delta}_0 &= S_{W_0^T X} \\
\widehat{\Omega} &= S_{G^T Y | W^T X} \\
\widehat{\Omega}_0 &= S_{G_0^T Y} \\
\widehat{\Sigma}_{Y|X} &= G \widehat{\Omega} G^T + G_0 \widehat{\Omega}_0 G_0^T \\
\widehat{\Sigma}_X &= W \widehat{\Delta} W^T + W_0 \widehat{\Delta}_0 W_0^T,
\end{aligned}
$$

where (W, W_0) and (G, G_0) are orthogonal matrices.

The next lemma gives an instructive form of the residuals from the fit of model (5.7. It proof is in Appendix A.5.3.

Lemma 5.3. *Using the sample estimators from Lemma 5.2, the sample covariance matrix of the residuals from model (5.7),*

$$S_{\text{res}} = n^{-1}\sum_{i=1}^{n}\{(Y_i - \bar{Y}) - G\widehat{\eta}^T W^T X_i\}\{(Y_i - \bar{Y}) - G\widehat{\eta}^T W^T X_i\}^T,$$

can be represented as

$$S_{\text{res}} = \widehat{\Sigma}_{Y|X} + P_G S_{Y|W^T X} Q_G + Q_G S_{Y|W^T X} P_G,$$

where $\widehat{\Sigma}_{Y|X}$ is as given in Lemma 5.2.

From this lemma we see that $S_{\text{res}} = \widehat{\Sigma}_{Y|X}$ if and only if G is a reducing subspace of $S_{Y|W^T X}$. As this holds in the population, it fits well with the basic theory leading to model (5.7). It also suggests that lack-of-fit diagnostics might be developed from the term $P_G S_{Y|W^T X} Q_G$.

An objective function for estimation under model (5.7) can constructed under the same general setup as we used for predictor envelopes in Section 2.3. We again base estimation on the objective function $F_q(S_C, \Sigma_C) = \log|\Sigma_C| + \text{tr}(S_C \Sigma_C^{-1})$ that stems from the log likelihood of the multivariate normal family after replacing the population mean vector with the vector of sample means, although here we do not require C to have a multivariate normal distribution. Rather, as discussed in Section 2.3.1, we are again using F_q as a multi-purpose objective function in the same spirit as least squares objective functions are often used.

We use (5.4), (5.7), $\eta = \Delta^{-1}K$ and $\Theta = \Omega + K^T\Delta^{-1}K$ to structure Σ_C as

$$\begin{aligned}
\Sigma_C &= \begin{pmatrix} \Phi & 0 \\ 0 & \Gamma \end{pmatrix}\begin{pmatrix} \Delta & \Delta\eta \\ \eta^T\Delta & \Omega + \eta^T\Delta\eta \end{pmatrix}\begin{pmatrix} \Phi & 0 \\ 0 & \Gamma \end{pmatrix}^T \\
&\quad + \begin{pmatrix} \Phi_0 & 0 \\ 0 & \Gamma_0 \end{pmatrix}\begin{pmatrix} \Delta_0 & 0 \\ 0 & \Omega_0 \end{pmatrix}\begin{pmatrix} \Phi_0 & 0 \\ 0 & \Gamma_0 \end{pmatrix}^T \\
&= (\Phi \oplus \Gamma)\begin{pmatrix} \Delta & \Delta\eta \\ \eta^T\Delta & \Omega + \eta^T\Delta\eta \end{pmatrix}(\Phi \oplus \Gamma)^T \\
&\quad + (\Phi_0 \oplus \Gamma_0)(\Delta_0 \oplus \Omega_0)(\Phi_0 \oplus \Gamma_0)^T. \qquad (5.9)
\end{aligned}$$

From this we see that the simultaneous envelope is a reducing subspace of Σ_C with special structure, just as the response and predictor envelopes are reducing subspaces of their respective covariance matrices. Then following the general ideas used in Section 2.3 for maximizing the likelihood for predictor envelopes, the objective function $F_q(S_C, \Sigma_C)$ can be minimized explicitly over Δ, Δ_0, Ω, Ω_0 and η with Γ and Φ held fixed. The resulting partially minimized objective function can be written as

$$F(S_C, \Phi \oplus \Gamma) = \log \left|(\Phi \oplus \Gamma)^T S_C(\Phi \oplus \Gamma)\right| + \log \left|(\Phi \oplus \Gamma)^T (S_X \oplus S_Y)^{-1}(\Phi \oplus \Gamma)\right|.$$

Let

$$(\widehat{\Phi}, \widehat{\Gamma}) = \text{argmin}_{W,G} F(S_C, W \oplus G),$$

where the minimization is over $p \times q$ semi-orthogonal matrices W and $r \times u$ semi-orthogonal matrices G. Once these estimators are found, the estimators of the constituent parameters are as given in Lemma 5.2 replacing W with $\widehat{\Phi}$ and G with $\widehat{\Gamma}$.

If the observations on C_i, $1 = 1, \ldots, n$, are independent and identically distributed with finite fourth moments and if the dimensions are known then the joint asymptotic distribution of the maximum likelihood estimators is multivariate normal. The asymptotic covariance matrix is rather complicated and not amenable to use in applications, although it simplifies considerably if C is normally distributed. Nevertheless, the residual bootstrap works well (Cook and Zhang, 2015b) and that is what we recommend for constructing inference statements.

5.2.2 Computing

Cook and Zhang (2015b) discussed three algorithms for minimizing the simultaneous objective function, one of which uses an algorithm that alternates between predictor and response reduction. If we fix Φ as an arbitrary orthogonal basis, then the objective function $F(S_C, \Phi \oplus \Gamma)$ can be re-expressed as an objective function in Γ for response reduction as given in (2.21):

$$F(\Gamma|\Phi) = \log |\Gamma^T S_{Y|\Phi^T X} \Gamma| + \log |\Gamma^T S_Y^{-1} \Gamma|. \tag{5.10}$$

Similarly, if we fix Γ, the objective function $F(S_C, \Phi \oplus \Gamma)$ reduces to the conditional objective function for predictor reduction as given in (2.11),

$$F(\Phi|\Gamma) = \log |\Phi^T S_{X|\Gamma^T Y} \Phi| + \log |\Phi^T S_X^{-1} \Phi|. \tag{5.11}$$

The following alternating algorithm based on (5.10) and (5.11) can be used to obtain a minimizer of the objective function $F(S_C, \Phi \oplus \Gamma)$.

1. Initialization: Set the starting value $\Phi^{(0)}$ and get $\Gamma^{(0)} = \arg\min_\Gamma F(\Gamma|\Phi^{(0)})$.
2. Alternating: For the k-th stage, obtain $\Phi^{(k)} = \arg\min_\Phi F(\Phi|\Gamma = \Gamma^{(k-1)})$ and $\Gamma^{(k)} = \arg\min_\Gamma F(\Gamma|\Phi = \Phi^{(k)})$.
3. Convergence criterion: Evaluate $\{F(\Phi^{(k-1)} \oplus \Gamma^{(k-1)}) - F(\Phi^{(k)} \oplus \Gamma^{(k)})\}$ and return to the alternating step if it is bigger than a tolerance; otherwise, stop the iteration and use $\Phi^{(k)} \oplus \Gamma^{(k)}$ as the final estimator.

According to Cook and Zhang (2015b), as long as good initial values are used, the alternating algorithm, which monotonically decreases $F(\Phi \oplus \Gamma)$, will converge after only a few cycles. Root-n consistent starting values are particularly important to mitigate potential problems caused by multiple local minima and to ensure efficient estimation. For instance, under joint normality of X and Y, one Newton-Raphson iteration from any $\sqrt{n}$-consistent estimator will be asymptotically as efficient as the maximum likelihood estimator, even if local minima are present (Small et al., 2000).

5.3 PLS estimation

The Cook-Zhang likelihood-based analysis under the joint reduction model (5.7) is not serviceable unless the sample size is sufficiently large. This is where PLS comes in. We have seen in Proposition 5.1 that at the population level, the requirements for simultaneous reduction of X and Y are obtained by combining the requirements for separate reductions. This suggests that we proceed in an analogous fashion using PLS: Obtain the weight matrix W for predictor compression by running a PLS algorithm, NIPALS or SIMPLS, as usual. Then interchange the roles of X and Y and run the algorithm again to obtain the weight matrix G for response compression. Since we are running two separate applications of the NIPALS or SIMPLS algorithms, basic asymptotic behavior of the reductions is as described in Propositions 2.1, 2.2 and the discussions that follow them.

The weight matrices W and G can be used in Lemma 5.2 to estimate the remaining parameters. For clarity, NIPALS-type algorithms for constructing W and G are shown in Table 5.1. Table 5.1(a) is the same as the bare-bones version in Table 3.2.

TABLE 5.1

NIPALS algorithm for computing the weight matrices W and G for compressing X and Y. The $n \times p$ matrix $\mathbb{X}$ contains the centered predictors and the $n \times r$ matrix $\mathbb{Y}$ contains the centered responses. $\ell_1(\cdot)$ denotes the eigenvector corresponding to the largest eigenvalue of the matrix argument.

(a) *Predictor reduction, W*	
Initialize	$\mathbb{X}_1 = \mathbb{X}$. Quantities subscripted with 0 are to be deleted.
Select	$q \leq \min\{\text{rank}(S_X), n-1)\}$
For $d = 1, \ldots, q$, compute	
sample covariance matrix	$S_{X_d,Y} = n^{-1}\mathbb{X}_d^T\mathbb{Y}$
X weights	$w_d = \ell_1(S_{X_d,Y}S_{X_d,Y}^T)$
X scores	$s_d = \mathbb{X}_d w_d$
X loadings	$l_d = \mathbb{X}_d^T s_d / s_d^T s_d$
X deflation	$\mathbb{X}_{d+1} = \mathbb{X}_d - s_d l_d^T = Q_{s_d}\mathbb{X}_d = \mathbb{X}_d Q_{w_d(S_{X_d})}$
Append	$W_d = (W_{d-1}, w_d) \in \mathbb{R}^{p \times d}$
End	Set $W = W_q$
(b) *Response reduction, G*	
Initialize	$\mathbb{Y}_1 = \mathbb{Y}$,
Select	$u \leq \min\{\text{rank}(S_Y), n-1)\}$
For $d = 1, \ldots, u$, compute	
sample covariance matrix	$S_{Y_d,X} = n^{-1}\mathbb{Y}_d^T\mathbb{X}$
Y weights	$g_d = \ell_1(S_{Y_d,X}S_{Y_d,X}^T)$
Y scores	$c_d = \mathbb{Y}_d g_d$
Y loadings	$h_d = \mathbb{Y}_d^T c_d / c_d^T c_d$
Y deflation	$\mathbb{Y}_{d+1} = \mathbb{Y}_d - c_d h_d^T = Q_{c_d}\mathbb{Y}_d = \mathbb{Y}_d Q_{g_d(S_{Y_d})}$
Append	$G_d = (G_{d-1}, g_d) \in \mathbb{R}^{p \times d}$
End	Set $G = G_u$
(c) Compute	$\widehat{\beta} = W S_{W^TX}^{-1} S_{W^TX,G^TY} G^T$
(d) Notes	Switching the roles of X and Y, parts (a) and (b) of this algorithm are identical. This is in line with the theory of Section 2.4 and summarized in Table 2.2. A difference arises in how W^TX and G^TY are used in computing $\widehat{\beta}$.

To help fix ideas we revisit the example introduced in Section 3.1.2 in which $p = 3$, $r = 2$,

$$\Sigma_X = 10 \begin{pmatrix} 1 & 4/3 & 0 \\ 4/3 & 4 & 0 \\ 0 & 0 & 5 \end{pmatrix}, \ \Sigma_{X,Y} = \begin{pmatrix} 5 & 0 \\ 0 & 4 \\ 0 & 0 \end{pmatrix}, \ \text{and } \Sigma_Y = 10 \begin{pmatrix} 1 & -0.6 \\ -0.6 & 4 \end{pmatrix}.$$

We demonstrated in Section 3.1.2 using NIPALS and in Section 3.3.2 using SIMPLS that only two linear combinations of the predictors are material to the regression with $\mathrm{span}(W_2) = \mathrm{span}(V_2) = \mathrm{span}((1,0,0)^T, (0,1,0)^T)$. To reduce the 2×1 response vector we interchange the roles of the predictors and responses, treating X as the response and Y as the predictor. The first response weight vector is then the first eigenvector of

$$\Sigma_{Y,X}\Sigma_{X,Y} = \begin{pmatrix} 25 & 0 \\ 0 & 16 \end{pmatrix},$$

which is $w_1 = (1, 0)^T$. The second weight vector$(0, 1)^T$ is constructed using

$$Q_{w_1(\Sigma_Y)} = \begin{pmatrix} 0 & -0.6 \\ 0 & 1 \end{pmatrix}, \ Q^T_{w_1(\Sigma_Y)}\Sigma_{Y,X} = \begin{pmatrix} 0 & 0 & 0 \\ -3 & 4 & 0 \end{pmatrix}.$$

and

$$Q^T_{w_1(\Sigma_Y)}\Sigma_{Y,X}\Sigma_{X,Y}Q_{w_1(\Sigma_Y)} = \begin{pmatrix} 0 & 0 \\ 0 & 1 \end{pmatrix}.$$

In consequence, no response reduction is possible since $\mathrm{span}(w_1, w_2) = \mathbb{R}^2$.

5.4 Other methods

5.4.1 Bilinear models for simultaneous reduction

Bilinear models have also been used as guides to simultaneous reduction of Y and X. A common formulation of a bilinear model for simultaneous reduction is (e.g. Rosipal and Krämer, 2006; Rosipal, 2011; Weglin, 2000; Wold, 1975a)

$$\left.\begin{aligned} \mathbb{X}_{0,n\times p} &= 1_n\mu_X^T + V_{n\times d_v}C^T_{p\times d_v} + E \\ \mathbb{Y}_{0,n\times r} &= 1_n\mu_Y^T + T_{n\times d_t}D^T_{r\times d_t} + F \end{aligned}\right\}, \tag{5.12}$$

where $\mathbb{X}_0$ and $\mathbb{Y}_0$ are the matrices of uncentered predictors and responses, V and T are matrices of random scores with rows v_i^T and t_i^T, $i = 1, \ldots, n$. We assume that the v_i's and t_i's are independent copies random vectors v and t, both with mean 0. The models are the same form after nonsingular linear transformations of t and v and, in consequence, we assume without loss of generality that $\text{var}(t) = I_{d_t}$ and $\text{var}(v) = I_{d_v}$. However, t and u are required to be correlated, $\text{cov}(t, u) = \Sigma_{t,u}$, for the predictors and responses to be correlated. C and D are non-stochastic loading matrices with full column ranks, and E and F are matrices of random errors.

Model (5.12) is an instance of a structural equation model. Investigators using structural equation models have historically taken a rather casual attitude toward the random errors, focusing instead on the remaining structural part of the model (Kruskal, 1983). Often, properties of the error matrices E and F are not mentioned, presumably with the implicit understanding that the errors are small relative to the structural components. That view may be adequate for some purposes, but not here. We assume that the rows e_i^T and f_i^T, $i = 1, \ldots, n$, of E and F are independently distributed with means 0 and constant variance-covariance matrices Σ_e and Σ_f, and $E \perp\!\!\!\perp F$.

Written in terms of the rows X_i^T of $\mathbb{X}_0$ and the rows Y_i^T of $\mathbb{Y}_0$, both uncentered, these models become for $i = 1, \ldots, n$

$$\left.\begin{array}{l} X_i = \mu_X + Cv_i + e_i \\ Y_i = \mu_Y + Dt_i + f_i \end{array}\right\}. \tag{5.13}$$

The vectors v_i and t_i are interpreted as latent vectors that control the extrinsic variation in X and Y. These latent vectors may be seen as imaginable constructs as in some applications like path analyses (e.g. Haenlein and Kaplan, 2004; Tenenhaus and Vinzi, 2005) or as convenient devices to achieve dimension reduction. The X and Y models are connected because t and v are correlated with covariance matrix $\Sigma_{t,v}$. If t and v are uncorrelated then so are X and Y.

To develop a connection with envelopes, let $\mathcal{C} = \text{span}(C)$ and $\mathcal{D} = \text{span}(D)$ and let $\Phi \in \mathbb{R}^{p \times q}$ and $\Gamma \in \mathbb{R}^{r \times u}$ be semi-orthogonal basis matrices for $\mathcal{E}_{\Sigma_e}(\mathcal{C})$ and $\mathcal{E}_{\Sigma_f}(\mathcal{D})$, respectively, $q \geq d_v$, $u \geq d_t$, and let $(\Phi, \Phi_0) \in \mathbb{R}^{p \times p}$ and $(\Gamma, \Gamma_0) \in \mathbb{R}^{r \times r}$ be orthogonal matrices. With this we have

$$\begin{aligned} \Sigma_e &= \Phi W_e \Phi^T + \Phi_0 W_{e,0} \Phi_0^T \\ \Sigma_f &= \Gamma W_f \Gamma^T + \Gamma_0 W_{f,0} \Gamma_0^T \\ \Sigma_{Y,X} &= P_\Gamma \Sigma_{t,v} P_\Phi, \end{aligned} \tag{5.14}$$

where the $W_{(\cdot)}$'s are positive definite matrices what can be expressed in terms of quantities in model (5.13). Such expressions of the $W_{(\cdot)}$'s will play no role in what follows and so are not given. We see from (5.14) that $\Gamma_0^T \Sigma_{Y,X} = \Sigma_{\Gamma_0^T Y,X} = 0$ and that $\Sigma_{Y,X}\Phi_0 = \Sigma_{Y,\Phi_0^T X} = 0$, and thus that the covariance between X and Y is captured entirely by the covariance between $\Phi^T X$ and $\Gamma^T Y$. In effect, $\Gamma^T Y$ represents the predictable part of Y and $\Phi^T X$ represents the material part of X.

In short, the bilinear model (5.12) leads to the envelope structure of (5.3) that stems from Proposition 5.1. See Section 11.2 for additional discussion of bilinear models as a basis for PLS.

5.4.2 Another class of bilinear models

The phrase 'bilinear model' generally means that somehow there are two linear structures represented. The precise meaning of 'bilinear model' can change depending on context. The book *Bilinear Regression Analysis* by vonRosen (2018) contains a through treatment of bilinear models represented, in vonRosen's notation, as $X_{p\times n} = A_{p\times q}B_{q\times k}C_{k\times n} + E$, where X is observable, A and C are known non-stochastic matrices, B is unknown to be estimated and E is a matrix of errors. While this class of models is outside the scope of this book, we show how it can arise in the context of model (1.1) to emphasize how it differs from model (5.12).

Starting with the full matrix representation (1.6) of model (1.1), suppose that we model the columns of β^T as being contained in a known subspace $\mathcal{S}$ with known $p \times k$ basis matrix $\mathbb{Z}$: $\beta^T = \mathbb{Z}B^T$ where the $p \times k$ matrix B contains the coordinates of β^T relative to basis $\mathbb{Z}$. Substituting this into (1.6), we get the model

$$\mathbb{Y}_0 = 1_r \alpha^T + \mathbb{X}B\mathbb{Z}^T + E.$$

This represents linear models in the rows and columns of $\mathbb{Y}$ and is the same model form as that studied by vonRosen (2018). Models of this form arise in analyses of longitudinal data, as discussed in Section 6.4.4.

5.4.3 Two-block algorithm for simultaneous reduction

Staying with the theme of connecting PLS algorithms with envelopes, we next consider the population version of the two-block algorithm that has been used frequently for simultaneous X-Y reduction in conjunction with (5.12) (e.g.

TABLE 5.2

Population version of the common two-block algorithm for simultaneous X-Y reduction based on the bilinear model (5.12). (*Not recommended.*)

Initialize	$(c_1, d_1) = \arg\max_{\lVert c\rVert=\lVert d\rVert=1} c^T \Sigma_{X,Y} d$, $C_1 = (c_1)$, $D_1 = (d_1)$
For $k = 1, 2 \ldots$	
Compute weights	$(c_{k+1}, d_{k+1}) = \arg\max_{\lVert c\rVert=\lVert d\rVert=1} c^T Q^T_{C_k(\Sigma_X)} \Sigma_{X,Y} Q_{D_k(\Sigma_Y)} d$
Append	$C_{k+1} = (C_k, c_{k+1})$, $D_{k+1} = (D_k, d_{k+1})$
End when first	$\max_{\lVert c\rVert=\lVert d\rVert=1} c^T Q^T_{C_k(\Sigma_X)} \Sigma_{X,Y} Q_{D_k(\Sigma_Y)} d = 0$ and then define $\bar{k} = k$
Compute regression coefficients	$\beta_{\mathrm{bipls}} = C_{\bar{k}}(C^T_{\bar{k}} \Sigma_X C_{\bar{k}})^{-1} C^T_{\bar{k}} \Sigma_{X,Y} P_{D_{\bar{k}}}$
Envelope connection	$\mathrm{span}(C_{\bar{k}}) \subseteq \mathcal{E}_{\Sigma_X}(\mathcal{B})$, $\mathrm{span}(D_{\bar{k}}) \subseteq \mathcal{E}_{\Sigma_Y}(\mathcal{B}')$ where $\mathcal{B} = \mathrm{span}(\beta)$, $\mathcal{B}' = \mathrm{span}(\beta^T)$

Rosipal and Krämer, 2006; Rosipal, 2011; Weglin, 2000; Wold, 1975a). The derivation of the population two-block algorithm from the data-based algorithm is available in Appendix A.5.4, since it follows the same general steps as the derivations of the population algorithms for NIPALS and SIMPLS discussed previously in Sections 3.1 and 3.3.

The population version of the two-block algorithm developed in Appendix A.5.4 is shown in Table 5.2. The sample version, which is obtained by replacing Σ_X, Σ_Y, and $\Sigma_{X,Y}$ with their sample versions S_X, S_Y, and $S_{X,Y}$, does not require S_X and S_Y to be nonsingular. Perhaps the main drawback of this algorithm is that the column dimensions of the weight matrices $C_{\bar{k}}$ and $D_{\bar{k}}$ must both equal $\bar{k}$, the number of components, so the number of material response linear combinations will be the same as the number of material predictor linear combinations. In consequence, we are only guaranteed in the population to generate subsets of $\mathcal{E}_{\Sigma_X}(\mathcal{B})$ and $\mathcal{E}_{\Sigma_Y}(\mathcal{B}')$, and a material part of the response or predictors must necessarily be missed, except perhaps in special cases.

We use the example of Sections 3.1.2 and 3.3.2 to illustrate an important limitation of the algorithm in Table 5.2. That example had $p = 3$ predictors and $r = 2$ responses. For ease of reference,

$$\Sigma_X = 10 \begin{pmatrix} 1 & 4/3 & 0 \\ 4/3 & 4 & 0 \\ 0 & 0. & 5 \end{pmatrix}, \ \Sigma_{X,Y} = \begin{pmatrix} 5 & 0 \\ 0 & 4 \\ 0 & 0 \end{pmatrix}, \ \text{and} \ \Sigma_Y = 10 \begin{pmatrix} 1 & -0.6 \\ -0.6 & 4 \end{pmatrix}.$$

According to the initialization step in Table 5.2, we first need to calculate

$$(c_1, d_1) = \underset{\|c\|=\|d\|=1}{\arg\max} \ c^T \Sigma_{X,Y} d$$

to get $c_1 = (1, 0, 0)^T$ and $d_1 = (1, 0)^T$. The next step is to check the stopping criterion, and for that we need to calculate $Q^T_{c_1(\Sigma_X)} \Sigma_{X,Y} Q_{d_1(\Sigma_Y)}$. The projections in this quantity are

$$Q_{c_1(\Sigma_X)} = \begin{pmatrix} 0 & -4/3 & 0 \\ 0 & 1 & 0 \\ 0 & 0 & 1 \end{pmatrix} \ \text{and} \ Q_{d_1(\Sigma_Y)} = \begin{pmatrix} 0 & 0.6 \\ 0 & 1.0 \end{pmatrix}.$$

Direct calculation then gives

$$Q^T_{c_1(\Sigma_X)} \Sigma_{XY} Q_{d_1(\Sigma_Y)} = 0, \tag{5.15}$$

so we stop with (c_1, d_1), which implies that we need only $c_1^T X = x_1$ and $d_1^T Y = y_1$ to characterize the regression of Y on X. However, we know from Sections 3.1.2 and 3.3.2 that two linear combinations of the predictors are required for the regression, and from Section 5.3 that two linear combinations of the response are required. In consequence, the two-block algorithm missed key response and predictor information, and for this reason it cannot be recommended for simultaneous reduction of the predictors and responses. The simultaneous envelope methods discussed in Sections 5.2 and 5.3 are definitely preferred.

5.5 Empirical results

In this section we present simulation results and brief analyses of three datasets from Cook et al. (2023b) and its supplement. Five methods are compared for predicting Y and estimating β in model (1.1): (1) ordinary least squares (OLS), (2) the two block method (Section 5.4.3), (3) sample size permitting, the envelope-based likelihood reduction method for simultaneously reducing X and Y (XY-ENV, Section 5.2), (4) PLS for predictor reduction only (X-PLS, Table 3.1), and (5) the newly proposed PLS method for reducing both X and Y (XY-PLS, Section 5.3).

Our overall conclusions are that any of these five methods can give competitive results depending on characteristics of the regression. However the XY-ENV and XY-PLS methods were judged best overall, with XY-ENV being preferred when n is sufficiently large. We judged XY-PLS to be the best overall. Lacking detailed knowledge of the regression or if n is sufficiently 'large', we would likely choose XY-PLS for use in predictive applications.

5.5.1 Simulations

The performance of a method can depend on the response and predictor dimensions r and p, the dimensions of the response and predictor envelopes u and q, the components Δ, Δ_0, Ω, and Ω_0 of the covariance matrices $\Sigma_{Y|X}$ and Σ_X given at (5.7), $\beta = \Phi\eta\Gamma^T$ and the distribution of C. Following Cook et al. (2023b) we focused on the dimensions and the covariance matrices. Given all dimensions, the elements of the parameters Φ, η, and Γ were generated using independent uniform $(0, 1)$ random variables. The distribution of C was taken to be multivariate normal. This gives an edge to the XY-ENV method so we felt comfortable thinking of it as the gold standard in large samples.

Following Cook and Zhang (2015b) we set

$$\begin{aligned} \Delta &= aI_q, \ \Delta_0 = I_{p-q} \\ \Omega &= bI_u, \ \Omega_0 = 10I_{r-u}, \end{aligned}$$

where a and b were selected constants. We know from our discussion at the end of Section 2.5.1 and from previous studies (Cook and Zhang, 2015b; Cook, 2018) that the effectiveness of predictor envelopes tends to increase with a, while the effectiveness of response envelopes increases as b decreases. Here we are interested mainly in the effectiveness of the joint reduction methods as they compare to each other and to the marginal reduction methods. We compared methods based on the average root mean squared prediction error per response and the total mean squared error

$$\begin{aligned} \text{predRMSE} &= \frac{1}{m}\sum_{i=1}^{m}\sqrt{\sum_{j=1}^{r}(Y_{ij}-\widehat{Y}_{ij})^2}, \\ \text{totalMSE} &= \frac{1}{rn}\sum_{i=1}^{n}\sum_{j=1}^{r}(Y_{ij}-\widehat{Y}_{ij})^2, \end{aligned}$$

where $\widehat{Y}_{ij}$ denotes a fitted value and Y_{ij} is an observation from an independent testing sample of size $m = 1000$ generated with the same parameters as the data for the fitted model. To focus on estimation, we also computed the average root mean squared estimation error per coefficient

$$\text{betaRMSE} = \frac{1}{p}\sum_{i=1}^{p}\sqrt{\sum_{j=1}^{r}(\beta_{ij}-\widehat{\beta}_{ij})^2}.$$

The true dimensions were used for all the methods. Results are given in Tables 5.3–5.4. Some cells in these tables are missing data because a method ran into rank issues and could not be implement cleanly.

5.5.2 Varying sample size

The results in Table 5.3 are for $p = 50$ predictors with $q = 10$ components and $r = 4$ responses with $u = 3$ components. At $n = 1000$ the XY-ENV had the best performance, as expected, with XY-PLS a close second. The performance of all methods deteriorated as the sample size decreased. After $n = 57$, the rank requirements for OLS and XY-ENV were no longer met and so computing ceased for these methods. At $n = 50$ and $n = 25$, the two-block method deteriorated badly, while XY-PLS continued to be serviceable. The difference between the performances of X-PLS and XY-PLS supports response reduction in addition to predictor reduction.

The settings for Table 5.4 are the same as those for Table 5.3 except q was increased to 40. The observed errors generally increased. The two-block method and X-PLS does deteriorate, but the two simultaneous methods again dominate. The performance is similar to the $q = 10$ case.

The results in Table 5.5 are for $p = 30$ predictors with $q = 20$ components and $r = 50$ responses with $u = 2$ components. For $n \geq 85$, the XY-ENV again had the best performance, as expected, with XY-PLS a close second. The

TABLE 5.3

Error measures with $a = 50$, $b = 0.01$, $p = 50, r = 4, q = 10$, and $u = 3$.

$n = 1000$	OLS	Two Block	XY-ENV	X-PLS	XY-PLS
predRMSE	2.64	3.52	2.51	2.63	2.53
totalMSE	2.61	3.30	2.55	2.61	2.55
betaRMSE	0.07	0.04	0.00	0.07	0.01
$n = 200$					
predRMSE	3.38	7.18	2.51	3.07	2.58
totalMSE	2.94	4.78	2.54	2.82	2.59
betaRMSE	0.18	0.08	0.01	0.14	0.03
$n = 100$					
predRMSE	5.17	11.74	2.55	3.39	2.67
totalMSE	3.63	6.06	2.57	2.98	2.66
betaRMSE	0.33	0.11	0.02	0.16	0.04
$n = 70$					
predRMSE	9.48	15.69	2.57	3.74	2.76
totalMSE	4.89	6.95	2.59	3.14	2.73
betaRMSE	0.53	0.13	0.03	0.16	0.05
$n = 57$					
predRMSE	26.99	20.20	2.68	3.93	2.87
totalMSE	7.92	7.84	2.70	3.25	2.81
betaRMSE	0.93	0.15	0.07	0.16	0.06
$n = 50$					
predRMSE		21.12		4.16	2.94
totalMSE		8.02		3.36	2.86
betaRMSE		0.15		0.16	0.06
$n = 25$					
predRMSE		36.93		6.74	4.37
totalMSE		10.54		4.42	3.66
betaRMSE		0.21		0.17	0.10

two-block method does relatively better in this case because there is substantial response reduction possible, while X-PLS does relatively worse because it neglects response reduction. We again conclude that with sufficiently large sample size, XY-ENV performs the best, while XY-PLS is serviceable overall.

TABLE 5.4

Error measures with $a = 50$, $b = 0.01$, $p = 50, r = 4, q = 40$, and $u = 3$.

$n = 1000$	OLS	Two Block	XY-ENV	X-PLS	XY-PLS
predRMSE	2.63	20.86	2.50	2.63	2.51
totalMSE	2.60	8.15	2.54	2.60	2.54
betaRMSE	0.04	0.15	0.00	0.04	0.00
$n = 200$					
predRMSE	3.40	89.17	2.53	3.40	2.57
totalMSE	2.95	16.49	2.55	2.95	2.57
betaRMSE	0.10	0.32	0.01	0.10	0.01
$n = 100$					
predRMSE	5.08	148.64	2.54	5.03	2.65
totalMSE	3.61	21.26	2.56	3.60	2.64
betaRMSE	0.16	0.42	0.01	0.16	0.03
$n = 70$					
predRMSE	9.26	185.52	2.58	8.11	2.83
totalMSE	4.83	23.72	2.60	4.59	2.79
betaRMSE	0.27	0.47	0.02	0.23	0.05
$n = 57$					
predRMSE	27.70	212.60	2.75	13.04	3.35
totalMSE	8.10	25.42	2.73	5.86	3.16
betaRMSE	0.49	0.50	0.04	0.26	0.08
$n = 50$					
predRMSE		229.25		22.42	4.87
totalMSE		26.41		7.75	3.87
betaRMSE		0.52		0.29	0.10
$n = 25$					
predRMSE		317.80			
totalMSE		31.08			
betaRMSE		0.61			

5.5.3 Varying the variance of the material components via a

In Table 5.6, all simulation parameters were held fixed, except for a which was varied. The XY-PLS method is judged to be the best over the table, except at $a = 0.5$. The performance of the two-block method improves as a decreases and is the best at $a = 0.5$.

TABLE 5.5

Error measures with $a = 50, b = 0.01, p = 30, r = 50, q = 20$, and $u = 2$.

	OLS	Two Block	XY-ENV	X-PLS	XY-PLS
$n = 1000$					
predRMSE	9.91	9.89	9.62	9.81	9.64
totalMSE	22.15	22.12	21.82	22.04	21.84
betaRMSE	0.41	0.08	0.00	0.08	0.03
$n = 200$					
predRMSE	11.39	10.99	9.66	10.76	9.76
totalMSE	23.73	23.29	21.86	23.07	21.98
betaRMSE	0.98	0.19	0.01	0.19	0.06
$n = 100$					
predRMSE	13.97	12.19	9.72	12.18	9.95
totalMSE	26.26	24.45	21.93	24.53	22.18
betaRMSE	1.54	0.25	0.03	0.29	0.09
$n = 85$					
predRMSE	15.25	12.63	9.84	12.83	10.01
totalMSE	27.42	24.85	22.06	25.17	22.25
betaRMSE	1.74	0.27	0.06	0.32	0.10
$n = 70$					
predRMSE	17.46	13.06		13.79	10.09
totalMSE	29.31	25.24		26.08	22.34
betaRMSE	2.05	0.29		0.36	0.11
$n = 50$					
predRMSE	26.08	14.27		16.66	10.33
totalMSE	35.61	26.29		28.61	22.60
betaRMSE	2.95	0.34		0.47	0.13
$n = 25$					
predRMSE		16.99		44.08	11.68
totalMSE		28.46		45.61	24.00
betaRMSE		0.42		1.20	0.25

The settings for Table 5.7 are the same as those for Table 5.6 except $b = 0.1$. Now, XY-PLS does the best for $a \geq 50$ and XY-ENV is the best for $a = 5, 0.5$. As we have seen in other tables, the relative sizes of the material and immaterial variation matters.

TABLE 5.6
Error measures with $n = 57, b = 5, p = 50, r = 4, q = 40$, and $u = 3$.

$a = 1000$	OLS	Two Block	XY-ENV	X-PLS	XY-PLS
predRMSE	69.72	4203.06	44.55	23.18	17.06
totalMSE	14.98	112.51	11.99	8.91	7.67
betaRMSE	0.88	0.50	0.67	0.04	0.03
$a = 500$					
predRMSE	69.58	2104.98	44.53	23.59	17.17
totalMSE	14.97	79.66	11.99	8.99	7.70
betaRMSE	0.88	0.50	0.68	0.07	0.05
$a = 50$					
predRMSE	68.54	216.87	45.75	28.69	19.81
totalMSE	14.88	25.75	12.18	9.87	8.24
betaRMSE	0.91	0.50	0.70	0.35	0.25
$a = 5$					
predRMSE	68.64	28.26	48.34	41.53	26.98
totalMSE	14.90	9.62	12.52	11.73	9.52
betaRMSE	1.18	0.52	0.96	0.87	0.66
$a = 0.5$					
predRMSE	69.56	10.69	53.46	46.60	37.95
totalMSE	14.96	6.11	12.99	12.43	11.12
betaRMSE	2.66	0.70	2.25	2.15	1.87

5.5.4 Varying sample size with non-diagonal covariance matrices

In Table 5.3 the matrices Ω, Ω_0, Δ, and Δ_0 were all generated as constants times an identity matrix. For contrast, Table 5.8 was as Table 5.3 but the matrices were generated as MM^T where M had the same size as the corresponding covariance matrix and was filled with independent realizations from a uniform $(0, 1)$ random variable. The conclusions are qualitative similar to those from Table 5.3, with the important exception that the relatively good performance of XY-ENV at small sample sizes is not apparent. At $n = 57$ in Table 5.3, $n = 57$, the predRMSE is 2.68 for XY-ENV and 2.87 for XY-PLS, while in Table 5.8 it is 15.46 for XY-ENV and 2.57 for XY-PLS. The table sustains our general conclusion that XY-PLS is overall the best from among those considered.

TABLE 5.7
Error measures with $n = 57, b = 0.1, p = 50, r = 4, q = 40, u = 3$.

$a = 1000$	OLS	Two Block	XY-ENV	X-PLS	XY-PLS
predRMSE	28.47	4199.55	4.04	9.18	3.04
totalMSE	8.40	112.45	3.50	4.94	2.97
betaRMSE	0.49	0.50	0.10	0.03	0.02
$a = 500$					
predRMSE	28.46	2101.22	4.19	9.51	3.08
totalMSE	8.40	79.56	3.55	5.03	3.01
betaRMSE	0.49	0.50	0.10	0.05	0.03
$a = 50$					
predRMSE	28.38	212.71	3.99	13.29	3.60
totalMSE	8.39	25.43	3.51	5.99	3.34
betaRMSE	0.51	0.50	0.11	0.26	0.08
$a = 5$					
predRMSE	25.27	24.00	3.68	19.19	4.36
totalMSE	7.99	8.75	3.40	7.08	3.70
betaRMSE	0.64	0.51	0.15	0.54	0.18
$a = 0.5$					
predRMSE	25.96	6.33	3.65	22.57	22.45
totalMSE	8.06	4.58	3.39	7.54	7.60
betaRMSE	1.42	0.66	0.33	1.32	1.33

5.5.5 Conclusions from the simulations

Setting $a = 50$ and $b = 0.01$ in Tables 5.3–5.5 was intended to represent regressions in which predictor and response compressions have a reasonable chance to reducing prediction variation. Changing these values enough may certainly change the results materially. For instance, we know from Corollary 2.1 that if $\Sigma_X = \sigma_x^2 I_p$ and $\beta \in \mathbb{R}^{p\times r}$ has rank r then the estimator of β based on predictor envelopes is asymptotically equivalent to the OLS estimator. Nevertheless, we conclude from Tables 5.3–5.5 that, in regressions where response and predictor envelopes are appropriate, joint envelope reduction via XY-ENV and joint PLS reduction via XY-PLS are the clear winners, even doing better than OLS when n is large.

In Tables 5.6 and 5.7 we held the sample size fixed at $n = 57$, varied b between the tables and varied the variance of the material predictors by varying

TABLE 5.8

Settings $p = 50, r = 4, q = 10, u = 3$ with Ω, Ω_0, Δ and Δ_0 generated as MM^T where M had the same size as the corresponding covariances and was filled with uniform $(0, 1)$ observations.

$n = 1000$	OLS	Two Block	XY-ENV	X-PLS	XY-PLS
predRMSE	1.24	1.71	1.28	1.26	1.27
totalMSE	1.98	2.37	2.01	2.00	2.01
betaRMSE	0.33	0.19	0.10	0.10	0.10
$n = 200$					
predRMSE	1.59	3.28	1.57	1.55	1.56
totalMSE	2.24	3.06	2.22	2.23	2.24
betaRMSE	0.77	0.23	0.19	0.14	0.14
$n = 100$					
predRMSE	2.42	4.46	2.11	1.92	1.92
totalMSE	2.76	3.48	2.57	2.49	2.49
betaRMSE	1.34	0.24	0.38	0.17	0.17
$n = 70$					
predRMSE	4.55	8.46	3.66	2.27	2.24
totalMSE	3.76	4.45	3.31	2.71	2.69
betaRMSE	2.21	0.27	1.30	0.19	0.20
$n = 57$					
predRMSE	13.96	9.48	15.46	2.61	2.57
totalMSE	6.38	4.75	6.55	2.89	2.87
betaRMSE	4.34	0.28	4.59	0.21	0.21
$n = 50$					
predRMSE		10.62		2.77	2.74
totalMSE		5.02		2.98	2.96
betaRMSE		0.28		0.22	0.22
$n = 25$					
predRMSE		17.77		4.27	4.20
totalMSE		6.44		3.66	3.62
betaRMSE		0.30		0.25	0.25

a within the tables. At $b = 5$ in Table 5.6, the XY-PLS method performed the best overall. The only exception was that the two-block method did better at $a = 0.5$. At $b = 0.1$ in Table 5.7, the XY-PLS method also performed well, except at $a = 0.5$ when the XY-ENV method performed the best, with the two-block method not far behind.

In Tables 5.3–5.7 the various covariance matrices were set to constants times an identity matrix so their relative magnitudes could be controlled easily. In Table 5.8 the covariance matrices were not diagonal but had the form MM^T, as described previously. Still, the XY-PLS method did best overall.

From this and other simulations not reported, we concluded that the XY-PLS method is preferable at small sample sizes. When n is 'large' XY-PLS and XY-ENV are both serviceable, with XY-PLS being computationally easier to implement.

Rimal et al. (2019) adapted X-ENV to small n regressions by replacing X with its principal components sufficient to absorb 97.5% of the variation. While they reported good empirical results, to our knowledge this method has not been studied for XY compressions.

5.5.6 Three illustrative data analyses

The slump flow of concrete depends on its ingredients: cement, fly ash, slag, water, super plasticizer, coarse aggregate, and fine aggregate. These ingredients, which were measured in kilograms per cubic meter of concrete, comprise the $p = 7$ predictors for this regression. The $r = 3$ response variables were slump, flow, and 28-day compressive strength. The data set (Yeh, 2007) contains 103 multivariate observations, which are comprised of 78 original records and 25 new records that were added later. Cook et al. (2023b) used the first 78 records as a training set and the new 25 records as the testing set. Predictions were evaluated against the testing set using predRMSE. The dimensions were determined via 10-fold cross validation for the two-block, XY-PLS and X-PLS. BIC was used to determine the dimensions for XY-ENV. For completeness, we also included the PLS1 method (see Section 3.10) with dimension determined by 10-fold cross validation. The results shown in Table 5.9 conform to the general conclusions given at the outset of Section 5.5, although here is no advantage to compressing the response in addition to the predictor.

As a second illustration we report the results from a study by Skagerberg et al. (1992) of a low density tubular polyethylene reactor. There are $n = 56$ multivariate observations, each with $p = 22$ predictors and $r = 6$ responses. The predictor variables consist of 20 temperatures measured at equal distances along the reactor together with the wall temperature of the reactor and the feed rate. The responses are the output characteristics of the polymers produced. Because the distributions of the values of all the response variables are

TABLE 5.9

Prediction errors predMSE for three datasets

Dataset	OLS	Two Block	XY-ENV	X-PLS	XY-PLS	X-PLS1
Concrete	25.02	19.58	13.29	13.17	13.17	13.22
Reactor	2.22	1.14	1.88	0.94	0.93	0.95
Biscuit	NA	2.13	NA	2.07	1.29	2.26

highly skewed to the right, the responses were transformed to the log scale and then standardized to have unit variance. Dimensions were determined as described in the concrete example. The prediction errors predRMSE were determined using the leave-one-out method to compare with their results. The results shown in Table 5.9 suggest that for these data there is little advantage in compressing the responses although $q = 3$.

Our third illustration comes from a near-infrared spectroscopy study on the composition of biscuit dough (Osborne et al., 1984). The original data has $n = 39$ samples in a training dataset and 31 samples in a testing set. The two sets were created on separate occasions and are not the result of a random split of a larger dataset. Each sample consists of $r = 4$ responses, the percentages of fat, sucrose, flour and water, and spectral readings at 700 wavelengths. Cook and Zhang (2015b) used a subset of these data to illustrate application of XY-ENV. They constructed the data subset by reducing the spectral range to restrict the number of predictors to 20 from a potential of 700. This allowed them to avoid '$n < p$' issues and again dimension was chosen by cross-validation. In a separate study, Li, Cook, and Tsai (2007) reasoned that the leading and trailing wavelengths contain little information, which motivated them to use middle wavelengths, ending with $p = 64$ predictors. Using the subset of wavelengths constructed by Li et al. (2007), Cook et al. (2023b) applied the four methods that do not require $n > p$, which gave the results in the last row of Table 5.9. We see that XY-PLS again performed the best. Since its performance was better than that of X-PLS, we see an advantage to compressing the responses.

6

Partial PLS and Partial Envelopes

It can happen when striving to reduce the predictors in the regression of $Y \in \mathbb{R}^r$ on $X \in \mathbb{R}^p$ that it is desirable and proper to shield part of X from the reduction. Partition X into two sub-vectors $X_1 \in \mathbb{R}^{p_1}$ and $X_2 \in \mathbb{R}^{p_2}$, $p = p_1 + p_2$, where we wish to compress X_1 while leaving X_2 untouched by the dimension reduction. How do we compress X_1 while correctly adjusting for X_2? This general class of problems is referred to as *partial dimension reduction* (Chiaromonte, Cook, and Li, 2002).

Skagerberg, MacGregor, and Kiparissides (1992) collected $n = 56$ observations to study the polymerization reaction along a reactor. The $r = 6$ response variables were polymer properties: number-average molecular weight, weight-average molecular weight, frequency of long chain branching, frequency of short chain branching, the content of vinyl groups and vinylidene groups in the polymer chain. The predictors were twenty temperatures measured at equal distances along the reactor plus the wall temperature of the reactor and the solvent feed rate. The last two predictors measure different characteristics than the first 20 predictors and it may be desirable to exclude them from the dimension reduction, aiming to find a low dimensional index that characterizes temperature along the reactor. Instead or in addition, it may be relevant to compress the 20 temperature measurements along the reactor in the hope of improving estimation of the effects of wall temperature and the solvent feed rate. In either case, the $p = 22$ predictor would then be partitioned in to the $p_1 = 20$ temperatures along the reactor X_1 and the $p_2 = 2$ remaining predictors X_2.

In this chapter we discuss foundations of partial envelopes, leading to PLS and maximum likelihood methodology for compressing the response or the

DOI: 10.1201/9781003482475-6

predictors X_1 in the partitioned linear model

$$\begin{aligned} Y &= \alpha + \beta^T X + \varepsilon \\ &= \alpha + \beta_1^T X_1 + \beta_2^T X_2 + \varepsilon, \end{aligned} \tag{6.1}$$

where $Y \in \mathbb{R}^r$,

$$X = \begin{pmatrix} X_1 \\ X_2 \end{pmatrix}, \ \beta = \begin{pmatrix} \beta_1 \\ \beta_2 \end{pmatrix},$$

and the errors have mean 0 variance-covariance matrix $\Sigma_{Y|X}$ and are independent of the predictors. We assume throughout this chapter that $p_2 \ll \min(n, p_1)$. There are several contexts in which partial reduction may be useful, in addition to the kinds of application emphasized in the reaction example. For instance, there may be a linear combination $G^T\beta$ of the components of β that is of special interest. In which case reducing the dimension while shielding $G^T\beta$ from the reduction may improve its estimation (see Section 6.2.4).

6.1 Partial predictor reduction

Chiaromonte, Cook, and Li (2002) were the first to consider partial sufficient dimension reduction, defining the partial central subspace $\mathcal{S}_{Y|X_1}^{(X_2)}$ as the intersection of all subspaces $\mathcal{S} \subseteq \mathbb{R}^{p_1}$ that satisfy the conditional independence statement

$$Y \perp\!\!\!\perp X_1 \mid P_{\mathcal{S}} X_1, X_2, \tag{6.2}$$

which is equivalent to requiring that $Y \perp\!\!\!\perp X \mid P_{\mathcal{S}} X_1, X_2$. The partial central subspace does not always exist, as the intersection of all subspaces that satisfy (6.2) does not necessarily satisfy (6.2). But the intersection does satisfy (6.2) under quite mild conditions and so we always assume the existence of $\mathcal{S}_{Y|X_1}^{(X_2)}$.

Chiaromonte et al. (2002) developed partial sliced inverse regression (SIR) for estimating $\mathcal{S}_{Y|X_1}^{(X_2)}$ when X_2 is categorical serving to identify a number of subpopulations. They motivated their approach in part by the regression of lean body mass on logarithms of height, weight and three hematological variables X_1 plus an indicator variable X_2 for gender based on a sample of 202 athletes from the Australian Institute of Sport. They concluded that for this regression $\mathcal{S}_{Y|X_1}^{(X_2)}$ has dimension 1 and thus that a single linear combination of the hematological variables, called the first partial SIR predictor, is sufficient to characterize the regression of lean body mass on (X_1, X_2). Shown in Figure 6.1 is a summary plot, marked by gender, of lean body mass versus the first partial SIR predictor. It can be inferred from their analysis that lean body mass for males and females responds to the same linear combination of the hematological variables, but also that the gender-specific regressions differ. Dimension reduction in this example serves to improve prediction and the estimation of the gender effect. Bura, Duarte, and Forzani (2016) developed a method for estimating the sufficient reductions in regressions with exponential family predictors. Applying their method to the data from the Australian Institute of Sport, they inferred that $\mathcal{S}_{Y|X}$ has dimension 1. They gave a summary plot that is essentially identical to that shown in Figure 6.1.

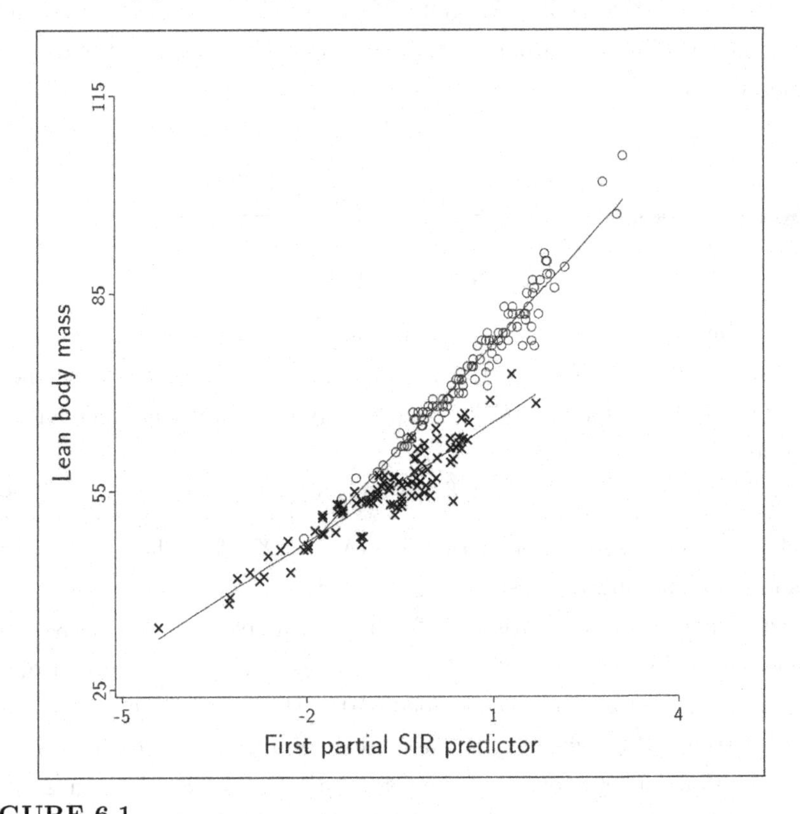

FIGURE 6.1
Plot of lean body mass versus the first partial SIR predictor based on data from the Australian Institute of Sport. circles: males; exes: females.

Shao, Cook, and Weisberg (2009) adapted sliced average variance estimation for estimating the partial central subspace. Wen and Cook (2007) extended the minimum discrepancy approach to SDR developed by Cook and Ni (2005) to allow the partial central subspace to be estimated. In this chapter we describe PLS and maximum likelihood methods for compressing X_1 while leaving X_2 unchanged in the context of linear models. Park, Su, and Chung (2022) recently applied partial PLS methods to the study of cytokine-based biomarker analysis for COVID-19. We will link to their work as this chapter progresses.

6.2 Partial predictor envelopes

6.2.1 Synopsis of PLS for compressing X_1

As developed in this section, to use PLS to compress the predictors X_1 in linear model (6.1), follow these steps. We will use numbers 1 and 2 as substitutes for X_1 and X_2 in subscripts, unless a fuller notation seems necessary for clarity.

Get residuals: Construct the residuals $\widehat{R}_{Y|2i}$ and $\widehat{R}_{1|2i}$, $i = 1, \ldots, n$, from the OLS fit of the multivariate linear regressions of Y on X_2 and X_1 on X_2. Because we are confining attention to regressions in which $p_2 \ll \min(p_1, n)$ there should be no dimension or sample size issues in these regressions.

Set components: Select the number of components $q_1 \leq p_1$.

Run PLS: Run the dimension reduction arm of a PLS algorithm with dimension q_1, response $\widehat{R}_{Y|2i}$ and predictor vector $\widehat{R}_{1|2i}$, and get the output matrix $\widehat{W} \in \mathbb{R}^{p_1 \times q_1}$ of weights.

Compress: $X_1 \mapsto \widehat{W}^T X_1$. This gives the compressed version of X_1.

Predict: Fit the linear regression of Y on $\widehat{W}^T X_1$ and X_2 to get a prediction rule.

This algorithm can be used in conjunction with cross validation and a holdout sample to select the number of components and estimate the prediction error of the final estimated model. It is also possible to use maximum likelihood to estimate a partial reduction, as described in Section 6.2.3.

6.2.2 Derivation

Our pursuit of methods for partial predictor reduction proceeds via partial predictor envelopes which are introduced by adding a second conditional independence condition to (6.2) and then following the reasoning that led to (2.2):

$$(a)\ Y \perp\!\!\!\perp X_1 \mid P_{\mathcal{S}} X_1, X_2 \text{ and (b) } P_{\mathcal{S}} X_1 \perp\!\!\!\perp Q_{\mathcal{S}} X_1 \mid X_2. \tag{6.3}$$

The rationale here is the same as that leading to (2.2): we use condition (b) to induce a measure of clarity in the separation of X_1 into its material and immaterial parts. Conditions (a) and (b) hold if and only if (Cook, 1998, Proposition 4.6)

$$\text{(c) } (Y, P_{\mathcal{S}} X_1) \perp\!\!\!\perp Q_{\mathcal{S}} X_1 \mid X_2. \tag{6.4}$$

Consequently $(Y, P_{\mathcal{S}} X_1)$ is required to be independent of $Q_{\mathcal{S}} X_1$ at every value of X_2. The conditional independence condition (b) is hard to meet in practice so we follow the logic leading to (2.3) and replace it with a conditional covariance condition,

$$\text{cov}(P_{\mathcal{S}} X_1, Q_{\mathcal{S}} X_1 \mid X_2) = P_{\mathcal{S}} \text{var}(X_1 \mid X_2) Q_{\mathcal{S}} = 0.$$

This relaxes independence, requiring only that there be no linear association between $P_{\mathcal{S}} X$ and $Q_{\mathcal{S}} X$ at any value of X_2, and leads to the new pair of conditions

$$(a)\ Y \perp\!\!\!\perp X_1 \mid P_{\mathcal{S}} X_1, X_2 \text{ and (b) } P_{\mathcal{S}} \text{var}(X_1 \mid X_2) Q_{\mathcal{S}} = 0. \tag{6.5}$$

Replacing condition (a) with an operational version, we get the conditions that (Park et al., 2022, eq. (8)) used to motivate their study:

$$\text{cov}(Y, Q_{\mathcal{S}} X_1 \mid P_{\mathcal{S}} X_1, X_2) \text{ and (ii) } P_{\mathcal{S}} \text{var}(X_1 \mid X_2) Q_{\mathcal{S}} = 0.$$

We can now define a partial envelope based on (6.5):

Definition 6.1. *The partial predictor envelope* $\mathcal{E}_{\text{var}(X_1 \mid X_2)}^{(X_2)}\left(\mathcal{S}_{Y \mid X_1}^{(X_2)}\right)$ *for compressing* X_1 *in the regression of* Y *on* (X_1, X_2) *is the intersection of all subspaces that contain the partial central subspace* $\mathcal{S}_{Y \mid X_1}^{(X_2)}$ *and reduce* $\text{var}(X_1 \mid X_2)$ *for all values of* X_2*. Let* q_1 *denote the dimension of the partial envelope.*

This serves as a conceptual starting point for our discussion. We now describe additional structure that facilitates our pursuit of methodology. To facilitate the development, we restrict the rest of our discussion of partial predictor envelopes to regressions with a real response $Y \in \mathbb{R}^1$, but much of the development can be extended rather straightforwardly to multivariate responses. We return to multivariate responses when considering partial response envelopes in Section 6.4.

Let $\beta_{1|2} \in \mathbb{R}^{p_2 \times p_1}$ denote the population OLS regression coefficient from the multivariate linear regression of X_1 on X_2 and let $R_{1|2} = X_1 - \mathrm{E}(X_1) - \beta_{1|2}^T(X_2 - \mathrm{E}(X_2))$ denote a population residual vector from that same regression. In addition, let $\beta_{Y|2} \in \mathbb{R}^{p_1}$ denote the population OLS regression coefficient from the linear regression of Y on X_2 and let $R_{Y|2} = Y - \mathrm{E}(Y) - \beta_{Y|2}^T(X_2 - \mathrm{E}(X_2))$ denote a population residual from that same regression. Then conditions (6.5) are equivalent to the same conditions expressed in terms of the residuals $R_{1|2}$ and $R_{Y|2}$,

$$(a)\ R_{Y|2} \perp\!\!\!\perp R_{1|2} \mid P_{\mathcal{S}} R_{1|2}, X_2 \text{ and (b) } P_{\mathcal{S}} \mathrm{var}(R_{1|2} \mid X_2) Q_{\mathcal{S}} = 0. \qquad (6.6)$$

and, as a consequence, the partial envelopes are identical

$$\mathcal{E}^{(X_2)}_{\mathrm{var}(R_{1|2}|X_2)}\left(\mathcal{S}^{(X_2)}_{R_{Y|2}|R_{1|2}}\right) = \mathcal{E}^{(X_2)}_{\mathrm{var}(X_1|X_2)}\left(\mathcal{S}^{(X_2)}_{Y|X_1}\right).$$

A brief justification for (6.6) is given in Appendix A.6.1.

The replacement of the conditional independence condition (6.3b) with the conditional covariance condition (6.5b) is a minor structural adaptation. We now make an assumption that may be suitable for linear models but may not be so in more general contexts. By construction $\mathrm{cov}((R_{Y|2}, R_{1|2}), X_2) = 0$. We take this a step further and assume that

A1 $\qquad (R_{Y|2}, R_{1|2}) \perp\!\!\!\perp X_2.$

This condition plays a central role in dimension reduction generally. For instance, (Cook, 1998, Section 13.3) used it in the development of Global Net Effects Plots. With assumption A1, conditions (6.6) reduce immediately to

$$(a)\ R_{Y|2} \perp\!\!\!\perp R_{1|2} \mid P_{\mathcal{S}} R_{1|2} \text{ and (b) } P_{\mathcal{S}} \mathrm{var}(R_{1|2}) Q_{\mathcal{S}} = 0, \qquad (6.7)$$

and the partial envelope reduces to an ordinary envelope,

$$\mathcal{E}^{(X_2)}_{\mathrm{var}(R_{1|2}|X_2)}\left(\mathcal{S}^{(X_2)}_{R_{Y|2}|R_{1|2}}\right) = \mathcal{E}_{\mathrm{var}(R_{1|2})}\left(\mathcal{S}_{R_{Y|2}|R_{1|2}}\right).$$

Similar constructions involving just the central subspace were given by (Cook, 1998, Chapter 7).

These conditions are identical to those given in (2.3) with Y and X replaced by $R_{Y|2}$ and $R_{1|2}$. Thus we can learn how to compress X_1 by applying PLS to the regression of $R_{Y|2}$ on $R_{1|2}$, except that these variables are not observed and must be estimated. Recall we are requiring p_2 to be small relative to n and p_1:

A2 $\qquad p_2 \ll \min(n, p_1).$

With this condition, the estimators $\widehat{R}_{Y|2}$ and $\widehat{R}_{1|2}$ of $R_{Y|2}$ on $R_{1|2}$ can be constructed straightforwardly by using OLS fits. Let $B_{Y|2}$ and $B_{1|2}$ denote the coefficient vector and matrix from the OLS fits of Y and X_1 on X_2. Then, for $i = 1, \ldots, n$,

$$\begin{aligned}\widehat{R}_{Y|2i} &= Y_i - \bar{Y} - B_{Y|2}^T(X_{2i} - \bar{X}_2)\\ \widehat{R}_{1|2i} &= X_{1i} - \bar{X}_1 - B_{1|2}^T(X_{2i} - \bar{X}_2).\end{aligned}$$

We now perform PLS dimension reduction with input data $(\widehat{R}_{Y|2i}, \widehat{R}_{1|2i})$, $i = 1, \ldots, n$, or equivalently with the sample covariance matrices $S_{1|2}$ and $S_{R_{1|2}, R_{Y|2}}$ constructed from $\widehat{R}_{1|2i}$ and $\widehat{R}_{Y|2i}$

$$\begin{aligned}S_{1|2} &= S_{11} - S_{12}S_{22}^{-1}S_{21}\\ S_{R_{1|2}, R_{Y|2}} &= S_{R_{1|2}, Y}.\end{aligned}$$

Writing

$$\begin{aligned}S_{R_{1|2}, Y} &= n^{-1}\sum_{i=1}^{n}[X_{1i} - \bar{X}_1 - B_{1|2}^T(X_{2i} - \bar{X}_2)]Y_i^T\\ &= n^{-1}\sum_{i=1}^{n}[X_{1i} - \bar{X}_1 - S_{1|2}S_{22}^{-1}(X_{2i} - \bar{X}_2)]Y_i^T\\ &= S_{1,Y} - S_{1,2}S_{22}^{-1}S_{2,Y},\end{aligned}$$

we have

$$S_{R_{1|2}, R_{Y|2}} = S_{1,Y} - S_{1,2}S_{22}^{-1}S_{2,Y}.$$

This then leads directly to the algorithm given by the synopsis of Section 6.2.1. Since the sample covariance matrix between $\widehat{R}_{1|2}$ and $\widehat{R}_{Y|2}$ is the same as that between $\widehat{R}_{1|2}$ and Y, the algorithm could be run with Y in place of $\widehat{R}_{Y|2}$. In this case the dimension reduction arms of the SIMPLS and NIPALS algorithms are instances of Algorithms $\mathbb{S}$ and $\mathbb{N}$ with $A = S_{R_{1|2}, R_{Y|2}}S_{R_{1|2}, R_{Y|2}}^T$ and $M = S_{1|2}$. This version of Algorithm $\mathbb{S}$ was used by Park et al. (2022) as a partial PLS algorithm.

The algorithm of Section 6.2.1 can be deduced also by proceeding informally based on the partitioned linear model (6.1). Let $\widehat{X}_{1|2i} = \bar{X}_1 + B_{1|2}^T(X_{2i} - \bar{X}_2)$ denote the i-th fitted value from the OLS fit of the linear regression of X_1 on X_2. Then write the partitioned linear model as

$$\begin{aligned} Y &= \alpha + \beta_1^T(X_1 - \widehat{X}_{1|2}) + \beta_1^T \widehat{X}_{1|2} + \beta_2^T X_2 + \varepsilon \\ &= \alpha^* + \beta_1^T \widehat{R}_{1|2} + \beta_2^{*T} X_2 + \varepsilon, \end{aligned}$$

where $\beta_2^* = B_{1|2}\beta_1$ and constants have been absorbed by α^*. Since $\widehat{R}_{1|2}$ and X_2 are uncorrelated in the sample, $\widehat{\beta}_1$ can be obtained from the regression of Y on $\widehat{R}_{1|2}$ and $\widehat{\beta}_2^*$ from the regression of Y on X_2, which leads back to the algorithm of Section 6.2.1. The OLS estimator B_1 of β_1 can then be represented as the coefficients from the OLS fit of $R_{1|2i}$ on Y_i:

$$B_1 = S_{R_{1|2}}^{-1} S_{R_{1|2},Y} = S_{1|2}^{-1} S_{R_{1|2},Y}. \tag{6.8}$$

6.2.3 Likelihood-based partial predictor envelopes

In this section, continuing to restrict discussion to regressions with a real response, we consider likelihood-based partial reduction under model (6.1) as a large-sample counterpart to partial PLS summarized in Section 6.2.1. Under that model, since $Y \perp\!\!\!\perp X_1 \mid \beta_1^T X_1, X_2$, the partial central subspace is

$$\mathcal{S}_{Y|X_1}^{(X_2)} = \text{span}(\beta_1) := \mathcal{B}_1$$

and consequently the partial envelope is

$$\mathcal{E}_{\text{var}(X_1|X_2)}^{(X_2)}\left(\mathcal{S}_{Y|X_1}^{(X_2)}\right) = \mathcal{E}_{\text{var}(X_1|X_2)}^{(X_2)}(\mathcal{B}_1).$$

Since the predictors are not ancillary in this treatment, the likelihood should be based on the joint distribution of (Y, X_1, X_2). Without loss of generality we assume that all variables are centered and so have marginal means of zero. We assume also that in model 6.1

$$\begin{aligned} X_2 &\sim N(0, \Sigma_2) \\ X_1 \mid X_2 &\sim N(\beta_{1|2}^T X_2, \Sigma_{1|2}) \\ Y \mid (X_1, X_2) &\sim N(\beta_1^T X_1 + \beta_2^T X_2, \sigma_{Y|X}^2). \end{aligned}$$

Cast in terms of the parameters of this model, the partial envelope is

$$\mathcal{E}_{\text{var}(X_1|X_2)}^{(X_2)}(\mathcal{B}_1) = \mathcal{E}_{\Sigma_{1|2}}^{(X_2)}(\mathcal{B}_1),$$

with q_1 now denoting the dimension of $\mathcal{E}_{\Sigma_{1|2}}^{(X_2)}(\mathcal{B}_1)$. Let $\Phi \in \mathbb{R}^{p_1 \times q_1}$ be a semi-orthogonal basis matrix for it and let (Φ, Φ_0) be an orthogonal matrix. Parallel to the development in Section 2.2, we then have representations

$$\begin{aligned} \beta_1 &= \Phi\eta \\ \Sigma_{1|2} &= \Phi\Delta\Phi^T + \Phi_0\Delta_0\Phi_0^T. \end{aligned}$$

With this structure we can write our normal model in terms of the envelope basis,

$$\begin{aligned} X_2 &\sim N(0, \Sigma_2) \\ X_1 \mid X_2 &\sim N(\beta_{1|2}^T X_2, \Phi\Delta\Phi^T + \Phi_0\Delta_0\Phi_0^T) \\ Y \mid (X_1, X_2) &\sim N(\eta^T\Phi^T X_1 + \beta_2^T X_2, \sigma_{Y|X}^2). \end{aligned}$$

From here the log likelihood is straightforwardly

$$\begin{aligned} \log L &= -\frac{n}{2}\log|\Sigma_2| - \frac{1}{2}\mathrm{tr}\sum_{i=1}^{n} X_{2i}^T\Sigma_2^{-1}X_{2i} - \frac{n}{2}\log|\Delta| - \frac{n}{2}\log|\Delta_0| \\ &\quad -\frac{1}{2}\sum_{i=1}^{n}(X_{1i} - \beta_{1|2}^T X_{2i})^T(\Phi\Delta^{-1}\Phi^T + \Phi_0\Delta_0^{-1}\Phi_0^T)(X_{1i} - \beta_{1|2}^T X_{2i}) \\ &\quad -\frac{n}{2}\log\sigma_{Y|X}^2 - \frac{1}{2\sigma_{Y|X}^2}\sum_{i=1}^{n}(y_i - \eta^T\Phi^T X_{1i} - \beta_2^T X_{2i})^2. \end{aligned}$$

Details of how to maximize this log likelihood function are available in Appendix A.6.2. Here we report only the final estimators.

After maximizing the log likelihood over all parameters except Φ we find that the MLE of the partial envelope is

$$\widehat{\mathcal{E}}_{\Sigma_{1|2}}^{(X_2)}(\mathcal{B}_1) = \mathrm{span}\left\{\arg\min_G \log|G^T S_{1|Y,2}G| + \log\left|G^T S_{1|2}^{-1}G\right|\right\}, \tag{6.9}$$

where the minimization is over all $p_1 \times q_1$ semi-orthogonal matrices G, $S_{1|Y,2}$ is the sample covariance matrix of the residuals from the multivariate linear regression of X_1 on Y and X_2, and $S_{1|2}$ is the sample covariance matrix of the residuals from the multivariate linear regression of X_1 on X_2. Let $\widehat{\Phi}$ be any semi-orthogonal basis this estimator and let $(\widehat{\Phi}, \widehat{\Phi}_0)$ be an orthogonal matrix. Then the compressed version of X_1 is $\widehat{\Phi}^T X_1$. The MLEs of the location parameters are

$$\begin{aligned}
\widehat{\eta} &= (\widehat{\Phi}^T S_{1|2}\widehat{\Phi})^{-1}\widehat{\Phi}^T S_{R_{1|2},Y} \\
\widehat{\beta}_1 &= \widehat{\Phi}\widehat{\eta} \\
&= P_{\widehat{\Phi}(S_{1|2})} B_1 \\
\widehat{\beta}_2 &= S_2^{-1}[S_{2,y} - S_{2,1}\widehat{\Phi}\widehat{\eta})] \\
&= S_2^{-1}[S_{2,Y} - S_{2,1}\widehat{\beta}_1],
\end{aligned}$$

where B_1 was defined at (6.8). In particular, we see that the MLE of β_1 is the projection of B_1 onto $\text{span}(\widehat{\Phi})$ in the $S_{1|2}$ inner product. The MLE's of the scale parameters are

$$\begin{aligned}
\widehat{\Sigma}_2 &= S_2, \\
\widehat{\Delta} &= \widehat{\Phi}^T S_{1|2}\widehat{\Phi}, \\
\widehat{\Delta}_0 &= \widehat{\Phi}_0^T S_{1|2}\widehat{\Phi}_0, \\
\widehat{\Sigma}_{1|2} &= P_{\widehat{\Phi}} S_{1|2} P_{\widehat{\Phi}} + Q_{\widehat{\Phi}} S_{1|2} Q_{\widehat{\Phi}} \\
\widehat{\sigma}^2_{Y|X} &= S_{Y|2} - S^T_{R_{1|2},Y}\widehat{\Phi}[\widehat{\Phi}^T S_{1|2}\widehat{\Phi}]^{-1}\widehat{\Phi}^T S_{R_{1|2},Y}.
\end{aligned}$$

Our envelope estimator (6.9) is the same as that found by Park, Su, and Chung (2022, eq. (11)), who also gave asymptotic distributions of $\widehat{\beta}_1$ and $\widehat{\beta}_1$ in their Proposition 2. Park et al. (2022) named the subsequent estimators envelope-based partial least squares estimators (EPPLS). We see this as a misnomer. Historically and throughout this book, partial least squares has been associated with specific algorithms, which we have generalized to Algorithms $\mathbb{N}$ and $\mathbb{S}$. Only relatively recently was it found that the dimension reduction arm of PLS estimates an envelope (Cook, Forzani, and Rothman, 2013). We think communication will be clearer by continuing the historical practice of linking PLS with a specific class of algorithms as described in Chapter 3, particularly since the algorithms are serviceable in $n < p$ regressions, while EPPLS is not.

When $n < p_1$ the covariance matrices $S_{1|Y,2}$ and $S_{1|2}$ in (6.9) are singular. Park et al. (2022) suggested to then replace these matrices with their sparse permutation invariant covariance estimators (SPICE, Rothman, Bickel, Levina, and Zhu, 2008), but this option does not seem to have been studied in detail.

6.2.4 Adaptations of partial predictor envelopes

It often happens when studying the regression of a real response Y on X via the linear model $Y = \alpha + \beta^T X + \varepsilon$ that there is a linear combination $G^T\beta$ of the components of β that is of special interest, where the $p \times 1$ vector G is known. The usual estimators of $G^T\beta$ are the OLS estimator $G^T B$, the MLE envelope estimator $G^T\widehat{\beta}$ and the corresponding PLS envelope estimator $G^T\widehat{\beta}_{\text{pls}}$. However, it is possible that a partial envelope or a partial PLS estimator will outperform these possibilities.

To develop a partial estimator of $G^T\beta$, let $G_0 \in \mathbb{R}^{p \times p-1}$ be a semi-orthogonal basis matrix for $\text{span}^{\perp}(G)$. Then

$$\begin{aligned} \beta^T X &= \beta^T Q_G X + \beta^T P_G X \\ &= \beta^T G_0 \left(G_0^T X\right) + \beta^T G \left(G^T X/\|G\|^2\right) \\ &= \phi_1^T Z_1 + \phi_2 Z_2, \end{aligned}$$

where $\phi_2 = \beta^T G$ is the new parameter of interest, $\phi_1 = G_0^T\beta$ are nuisance parameters for the purpose of inferring about ϕ_2, $Z_2 = \left(G^T X/\|G\|^2\right)$ and $Z_1 = G_0^T X$ are the observable transformed predictors that go with the nuisance parameters. The model in terms of the new parameters and predictors is

$$Y = \alpha + \phi_1^T Z_1 + \phi_2 Z_2 + \varepsilon. \tag{6.10}$$

In this model we wish to reduce the dimension of Z_1 without impacting Z_2 to improve the estimation of the parameter of interest ϕ_2. This falls neatly within the context of Sections 6.2 and 6.2.3 from which we learned that

$$\widehat{\phi}_2 = S_{Z_2}^{-1}\left(S_{Z_2,Y} - S_{Z_2,Z_1}\widehat{\phi}_1\right), \tag{6.11}$$

where $\widehat{\phi}_1$ can come from either the partial PLS fit or the MLE of the partial envelope, $\mathcal{E}_{\text{var}(Z_1|Z_2)}^{(Z_2)}(\text{span}(\phi_1))$, where

$$\begin{aligned} Z_1 &= G_0^T X \\ Z_2 &= G^T X/\|G\|^2 \\ S_{Z_2} &= G^T S_X G/\|G\|^4 \\ S_{Z_2,Y} &= G^T S_{X,Y}/\|G\|^2 \\ S_{Z_2,Z_1} &= G^T S_X G_0/\|G\|^2 \\ \widehat{\phi}_2 &= \|G\|^4 (G^T S_X G)^{-1}\{G^T S_{X,Y}/\|G\|^2 - G^T S_X G_0/\|G\|^2 \widehat{\phi}_1\} \\ &= \|G\|^2 (G^T S_X G)^{-1} G^T \{S_{X,Y} - S_X G_0 \widehat{\phi}_1\}. \end{aligned}$$

Informally, if either estimator reduces the variation in the estimator of ϕ_1 then we would expect improved estimation of ϕ_2.

Partial least squares fits are often used for constructing rules to predict Y at a new value X_{new} of X. Partial envelopes fitted by either PLS or maximum likelihood offer a route to an alternative estimator that has the potential to outperform the usual predictions. Centering the predictors to facilitate exposition, the population and sample forms of a prediction based on the linear model $Y = \alpha + \beta^T X + \varepsilon$ are

$$\begin{aligned} \mathrm{E}(Y \mid X_{\text{new}}) &= \mathrm{E}(Y) + \beta^T(X_{\text{new}} - \mathrm{E}(X)) \\ \widehat{Y} &= \widehat{\mathrm{E}}(Y \mid X_{\text{new}}) \\ &= \bar{Y} + \widetilde{\beta}^T(X_{\text{new}} - \bar{X}), \end{aligned}$$

where $\widetilde{\beta}$ stands for a generic estimator of β. In consequence, we can adapt ϕ_2 for prediction by setting $G = X_{\text{new}} - \bar{X}$ and, with this value of G, following the development that lead to (6.10) and (6.11). The prediction is then $\widehat{Y} = \bar{Y} + \widehat{\phi}_2$.

6.2.5 Partial predictor envelopes in the Australian Institute of Sport

We introduced partial predictor reduction in Section 6.1 by using data from the Australian Institute of Sport. Using SIR as an implementation of sufficient dimension reduction, Chiaromonte et al. (2002) concluded that only one linear combinations of the X_1 variates is necessary to explain the regression of lean body mass (LBM) on X_1 and gender X_2, with the plot in Figure 6.1 serving as a summary of the regression of LBM on X_1 and X_2.

We applied partial PLS and partial predictor envelope to the same data and concluded via cross validation that each method gave $q_1 = 4$, indicating little reduction of X_1. How could partial sufficient reduction via SIR lead to $q_1 = 1$ component, while partial PLS and partial envelopes lead to $q_1 = 4$ components? Recall that partial sufficient reduction is grounded in condition (6.3a), while partial PLS and partial envelopes require both conditions (6.3a) and (6.3b). If the variables in X_1 are correlated to such an extent that condition (6.3b) cannot be satisfied with one component then naturally we will end with $q_1 > 1$ components. We expect that is the case here.

In the next section we give a more detailed application of partial predictor envelopes.

6.3 Partial predictor envelopes in economic growth prediction

6.3.1 Background

Governance and institutional quality can play a significant role in a country's economic growth as measured by the growth of its per capita gross domestic product (GDP) Arvin, Pradhan, and Nair, 2021; Emara and Chiu, 2015; Fawaz et al., 2021; Gani, 2011; Han et al., 2014; Radulović, 2020. The Worldwide Governance Indicators (WGI) from the World Bank are usually used to measure aspects of governance. Such indicators have been used as covariates in regression models, either jointly or in separate models due to the high correlations between them (Gani, 2011; Radulović, 2020; Fawaz et al., 2021). Other authors (e.g. AlShiab, Al-Malkawi, and Lahrech, 2020; Arvin, Pradhan, and Nair, 2021; Asongu and Nnanna, 2019; Emara and Chiu, 2015; Han, Khan, and Zhuang, 2014) use the first principal component analysis of the WGI variates as a composite governance index then, along with other socioeconomic control variables, estimate a regression model to analyze the impact of this governance index on economic growth. Here the dimension of the reduction is set to $q_1 = 1$ to facilitate interpretability and then each loading of the principal component is interpreted as the weight of each governance indicator in the index.

In this application we follow the economist's approach in the sense that we apply dimension reduction on the governance variables, but incorporating two variants: First, extensions that use supervised reduction via partial least squares and envelopes; and second, in order to improve predictions, we allow for the possibility that $q_1 > 1$ in both the supervised reduction and non-supervised reduction via PCA. In addition, we contemplate this in a context of partial reductions, i.e. we only want to reduce the set of variables that measure governance, leaving the other socioeconomic variables as a control without reducing their dimension.

6.3.2 Objectives

Given the empirical evidence of governance as a driver of per capita GDP, here we analyze the power of governance to predict the economic growth. Therefore, the objective is the prediction of the per capita GDP using governance indicators. Specifically, we study the impact of governance on economic growth in the twelve South American countries as measured by the percentage change in yearly per capita GDP (Y) using WGI's from the World Bank. This application is novel in several ways:

1. The objective is to obtain a good prediction of economic growth.

2. In particular, this can be extended to problems where you want to predict a variable that is difficult or expensive to calculate from an indicator that uses variables that are easier or less expensive to capture.

3. We compare supervised methodologies for construction of composite governance indicators in search of better predictive results, including methodologies that are serviceable in high-dimensional "$n < p$" settings.

4. The application of this phenomenon focused on South American countries has been little explored.

6.3.3 Model and data

For prediction we consider the linear model

$$Y = \beta_0 + \beta_1^T X_1 + \beta_2^T X_2 + \varepsilon, \tag{6.12}$$

where X_1 are p_1 governance measures and X_2 are p_2 economics variables taken as a control. We have a total of $p_1 = 24$ governance variables and, in view of the limited sample sizes, expect that reducing these to Composite Governance Indicators will improve prediction of the response Y. Additionally, we have included $p_2 = 19$ economic control variables, including dummy variables for specific country effects. Therefore, in (6.12) we have a total of $p = 43$ predictors variables, but hope to reduce the dimension of X_1.

The response and predictors were measured for 12 South American countries over a period of up to 16 years, giving a maximum sample size of $12 \times 16 = 192$. However, data for several year-country combinations are missing leading to the sample sizes and periods listed in the first column of Table 6.1.

6.3.3.1 Governance variables, X_1

The World Bank considers six aggregate WGI's that combine the views of a large number of enterprise, citizen and expert survey respondents: control of

TABLE 6.1
Leave-one-out cross validation MSE for GPD growth prediction.

n	Period	Countries	Method							
			PPLS-q_1	PPCA-q_1	PPLS-1	PPCA-1	LASSO	FULL	PPENV-q_1	PPENV-1
161	2003–2018	12	9.37 ($q_1 = 7$)	**9.04** ($q_1 = 11$)	10.05	10.14	11.63	10.29	9.60 ($q_1 = 5$)	10.36
109	2008–2018	12	6.75 ($q_1 = 8$)	6.80 ($q_1 = 20$)	7.07	7.34	7.54	7.06	**6.58** ($q_1 = 12$)	7.30
86	2010–2018	12	**4.93** ($q_1 = 3$)	5.12 ($q_1 = 12$)	4.99	5.24	6.49	6.61	4.95 ($q_1 = 5$)	5.25
53	2010–2014	12	5.68 ($q_1 = 2$)	5.85 ($q_1 = 8$)	7.72	6.66	6.68	22.40	**5.39** ($q_1 = 5$)	6.66
51	2013–2018	10	7.10 ($q_1 = 3$)	6.97 ($q_1 = 3$)	8.32	8.25	**6.70**	37.8	–	–
41 ($n < p$)	2014–2018	9	**4.98** ($q_1 = 1$)	5.29 ($q_1 = 2$)	4.98	5.30	5.49	–	–	–
32 ($n < p$)	2015–2018	9	**1.80** ($q_1 = 1$)	1.94 ($q_1 = 1$)	1.80	1.94	7.54	–	–	–

corruption, rule of law, regulatory quality, government effectiveness, political stability and voice and accountability. The basic measure of each indicator is a weighted average standardized to have mean zero and standard deviation one, with values from -2.5 to 2.5, approximately, where higher values correspond to better governance. All six are highly positively correlated, and are all positively correlated with the per capita GDP; i.e., economic growth is positively associated with better governance indicators. In addition to these six variables, we have included several other governance measures such as the percentile rank term among all countries. In total we have $p_1 = 24$ governance variables.

6.3.3.2 Economic control variables, X_2

Following the related literature, we include foreign domestic investment (net inflows as % of GPD), domestic credit to private sector (as % of GPD), Age dependency ratio (% of working-age population), population growth (annual %), secondary school enrollment (gross % of population), an inflation factor (GDP deflator) and the logged lag of per capita GPD. Including a trend variable and country specific effects (through dummies variables) we have a total of $p_2 = 19$ control variables that do not enter in the reduction.

6.3.4 Methods and results

To evaluate the predictive performance of partial PLS and envelope methods, we ran a leave-one-out cross validation experiment, comparing the following methodological alternatives:

- `PPLS-`q_1: Partial PLS with optimal q_1 obtained via cross validation, as described in Section 6.2.
- `PPLS-1`: Partial PLS with $q_1 = 1$ as is usual used for composite indicator, as described in Section 6.2.
- `PPCA-`q_1: Partial principal component regression with optimal q_1 obtained via cross validation. In involved simply selecting the q_1 best principal components from the $p_1 = 24$ governance variables.
- `PPCA-1`: This is the usual implementation for governance indicators in regression context. That is, we used the first principal component as a governance index and as a replacement for the 24 governance variables.

- `LASSO`: Regression with variable selection using LASSO with the penalization parameter selected by cross validation. This is intended to represent penalization methods.

- `FULL`: OLS Regression on all $p = 43$ governance and control variables if allowed by sample size.

- `PPENV-d`: Partial predictor envelopes, as described in Section 6.2.3, with q_1 selected by leave-one-out cross validation.

- `PPENV-1`: Partial predictor envelopes, as described in Section 6.2.3, with $q_1 = 1$, as allowed by sample size.

Shown in Table 6.1 are the leave-one-out predictive mean squared errors,

$$\text{predMSE} = \frac{1}{n}\sum_{i=1}^{n} ||Y_i - \hat{Y}_{-i}||_2^2 = \frac{1}{n}\sum_{i=1}^{n}\sum_{j=1}^{r}(Y_{ij} - \hat{Y}_{-ij})^2,$$

where $\hat{Y}_{-i}$ is the prediction of the i-th response vector based on the data without that vector. The second form of predMSE expresses the same quantity in terms of the individual responses Y_{ij}.

Several observations are apparent.

- The full and PPENV methods cannot handle relatively small sample sizes, as expected, while the PLS and PCA methods are apparently serviceable in such cases.

- As expected, there is appreciable gain over the FULL method when the sample size is relatively small.

- The PLS and envelope methods are close competitors, although the PLS methods are serviceable when $n < p$, while the envelope methods are not.

- The methods with $q_1 = 1$ component are dominated by other methods and cannot be recommended based on these results.

- There is little reason to recommend the PCA and LASSO methods, although these methods may be in the running on occasion.

To round out the discussion, Figure 6.2 shows plots of the response Y versus the leave-one-out predicted values from the partial PLS fit with $q_1 = 7$ and the lasso for the data in the first row of Table 6.1 comprising 12 South American countries, 2003–2018, $n = 161$. The visual impression seems to confirm the MSE's shown in Table 6.1.

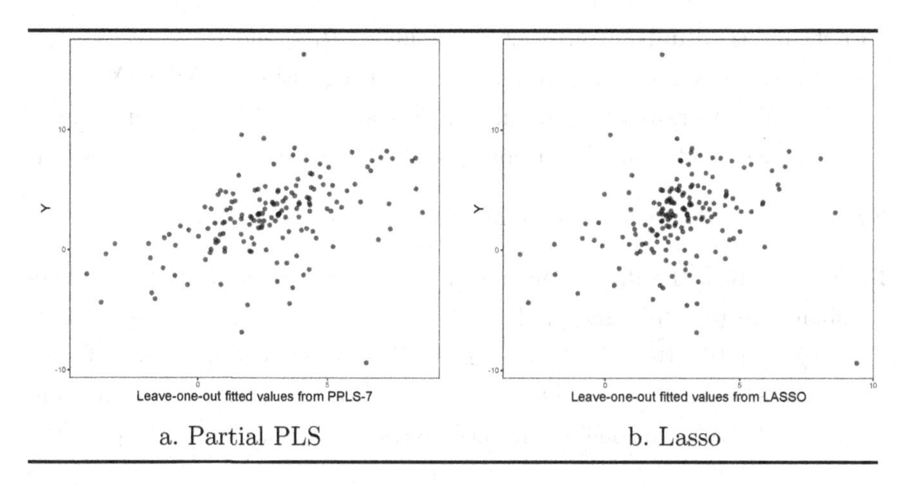

a. Partial PLS b. Lasso

FIGURE 6.2
Economic growth prediction for 12 South American countries, 2003–2018, $n = 161$: Plot of the response Y versus the leave-one-out fitted values from (a) the partial PLS fit with $q_1 = 7$ and (b) the lasso.

6.4 Partial response reduction

In Section 6.2 we considered PLS and maximum likelihood envelope methods for compressing X_1 in linear model (6.1). Starting in Section 6.2.2, that discussion was restricted to regressions with a real response. In this section we return to model (6.1) with multivariate responses $Y \in \mathbb{R}^r$, $r > 1$, and consider methods for compressing Y. Our goal here is the same as it was in Section 6.2, improved estimation of β_1, but now we pursue that goal via compression of Y rather than compression of X_1. The predictors in this section need not be stochastic, as they are not used to inform the reduction of Y.

6.4.1 Synopsis of PLS for Y compression to estimate β_1

To use PLS to compress the response Y for the purpose of estimating β_1 in linear model (6.1), follow these steps.

Center the predictors: This step is technically unnecessary, but it makes subsequent descriptions somewhat easier. Center the predictors so $\bar{X}_1 = 0$ and $\bar{X}_2 = 0$.

Construct residuals: Construct the residuals $\widehat{R}_{1|2i}$ and $\widehat{R}_{Y|2i}$, $i = 1, \ldots, n$ from the OLS fits of the multivariate linear regressions of X_1 on X_2 and of Y on X_2. We restrict attention to regressions in which $p_2 \ll \min(p_1, r, n)$ so there should be no dimension or sample size issues in these regressions.

Set components: Select the number of components $u_1 \leq r$.

Run PLS: Run the dimension reduction arm of a PLS algorithm with response vector $\widehat{R}_{1|2i}$ and predictor vector $\widehat{R}_{Y|2i}$, $i = 1, \ldots, n$, and get the output matrix $\widehat{W} \in \mathbb{R}^{r \times u_1}$ of weights. The projection onto $\text{span}(\widehat{W})$ is a $\sqrt{n}$-consistent estimator of $\mathcal{E}_{\Sigma_{Y|X}}(\text{span}(\beta_1^T))$, the smallest reducing subspace of the conditional covariance matrix $\Sigma_{Y|X}$ that contains $\text{span}(\beta^T)$.

Compress and estimate: Compress $Y \mapsto \widehat{W}^T Y$ and then fit the multivariate regression of $\widehat{W}^T Y$ on (X_1, X_2):

$$\widehat{W}^T Y = a + \eta X_1 + \gamma X_2 + e.$$

If the sample size permits, use OLS; otherwise use PLS to reduce the predictors. Let $\widehat{\eta}$ denote the estimated coefficients of X_1. Then the PLS estimator of β_1^T is $\widehat{W}\widehat{\eta}$.

The estimator of β_2 is the estimated coefficient vector $\widehat{\beta}_2$ in the regression of the residuals $Y - \bar{Y} - \widehat{W}\widehat{\eta}X_1$ on X_2.

Predict: The rule for predicting Y is then $\widehat{Y} = \bar{Y} + \widehat{W}\widehat{\eta}X_1 + \widehat{\beta}_2^T X_2$.

This algorithm can be used in conjunction with cross validation and a holdout sample to select the number of components and estimate the prediction error of the final estimated model. It is also possible to use maximum likelihood to estimate a partial reduction, as described in Section 6.4.3.

6.4.2 Derivation

To compress the response vector using PLS for the purpose of estimating β_1 in model (6.1), we strive to project Y onto the smallest subspace $\mathcal{E}$ or $\mathbb{R}^r$ that satisfies

$$\text{(a) } Q_{\mathcal{E}} Y \perp\!\!\!\perp X_1 \mid X_2 \text{ and } \text{(b) } P_{\mathcal{E}} Y \perp\!\!\!\perp Q_{\mathcal{E}} Y \mid (X_1, X_2). \tag{6.13}$$

Let $\Gamma \in \mathbb{R}^{r \times u_1}$ be a semi-orthogonal basis matrix for $\mathcal{E}$ and let (Γ, Γ_0) be an orthogonal matrix. Condition (a) requires that $\text{span}(\beta_1^T) \subseteq \mathcal{E}$. In consequence,

β_1 can be represented in terms of basis Γ, $\beta_1^T = \Gamma\eta$, Γ_0 is orthogonal to the row space of β_1, and the distribution of $\Gamma_0^T Y \mid (X_1, X_2)$ does not depend on X_1:

$$\begin{aligned} \Gamma^T Y &= \Gamma\alpha + \eta X_1 + \Gamma^T \beta_2^T X_2 + \Gamma^T \varepsilon. \\ \Gamma_0^T Y &= \Gamma_0 \alpha + \Gamma_0^T \beta_2^T X_2 + \Gamma_0^T \varepsilon. \end{aligned}$$

Condition (b) requires that $P_{\mathcal{E}} \Sigma_{Y|X} Q_{\mathcal{E}} = 0$ so $\mathcal{E}$ is a reducing subspace of $\Sigma_{Y|X}$. It follows that $\mathcal{E}$ can be characterized as an envelope. Specifically, $\mathcal{E}$ is the smallest reducing subspace of $\Sigma_{Y|X}$ that contains $\mathcal{B}_1' = \text{span}(\beta_1^T)$, which is denoted $\mathcal{E}_{\Sigma_{Y|X}}(\mathcal{B}_1')$.

Conditions (a) and (b) of (6.13) are equivalent to the combined condition

$$\text{(c) } (X_1, P_{\mathcal{E}} Y) \perp\!\!\!\perp Q_{\mathcal{E}} Y \mid X_2. \tag{6.14}$$

Comparing this to condition (6.4) we see that the two conditions are the same, except the roles of X_1 and Y are interchanged. In consequence, we can achieve partial dimension reduction of the response vector by following the PLS algorithm of Section 6.2.1, but interchanging X_1 and Y and using a different fitting scheme. This then is how we arrived at the algorithm of Section 6.4.1.

6.4.3 Likelihood-based partial response envelopes

It follows from our discussion in Section 6.4.2 that the partial response envelope version of model (6.1) is

$$\left.\begin{aligned} Y &= \alpha + \Gamma\eta X_1 + \beta_2^T X_2 + \varepsilon \\ \Sigma_{Y|X} &= \Gamma\Omega\Gamma^T + \Gamma_0 \Omega_0 \Gamma_0^T \end{aligned}\right\}, \tag{6.15}$$

where $\text{var}(\Gamma^T Y \mid X) = \Omega$ and $\text{var}(\Gamma_0^T Y \mid X) = \Omega_0$. Assuming that $Y \mid X$ is normally distributed, Su and Cook (2011) derived the maximum likelihood estimators under this model. Their results were reviewed by Cook (2018, Chapter 3). Here we give only the estimators. Let $\mathcal{B}' = \text{span}(\beta_1^T)$.

The maximum likelihood estimator of the envelope $\mathcal{E}_{\Sigma_{Y|X}}(\mathcal{B}_1')$ is

$$\begin{aligned} \widehat{\mathcal{E}}_{\Sigma_{Y|X}}(\mathcal{B}_1') &= \text{span}\left\{\arg\min_G \log|G^T S_{R_{Y|2}|R_{1|2}} G| + \log|G_0^T S_{R_{Y|2}} G_0|\right\} \\ &= \text{span}\left\{\arg\min_G \log|G^T S_{Y|X} G| + \log|G^T S_{Y|2}^{-1} G|\right\}, \end{aligned} \tag{6.16}$$

where $\min_G$ is over $r \times u_1$ semi-orthogonal matrices, $(G, G_0) \in \mathbb{R}^{r \times r}$ is an orthogonal matrix. The function to be optimized is of the general form Algorithm $\mathbb{L}$ discussed in Section 1.5.5. Comparing (6.16) with that for partial predictor reduction (6.9) we see that one can be obtained from the other by interchanging the roles of Y and X_1, although X_1 need not be stochastic for (6.16). This is in line with our justification around (6.14) for the PLS compression algorithm described in Section 6.4.

Following determination of $\widehat{\mathcal{E}}_{\Sigma_{Y|X}}(\mathcal{B}_1')$, the maximum likelihood estimators of the remaining parameters are as follows. The maximum likelihood estimator $\widehat{\beta}_1$ of β_1 is obtained from the projection onto $\widehat{\mathcal{E}}_{\Sigma_{Y|X}}(\mathcal{B}_1')$ of the maximum likelihood estimator B_1^T of β_1^T from model (6.1),

$$\widehat{\beta}_1^T = P_{\widehat{\mathcal{E}}_1} B_1^T,$$

where $P_{\widehat{\mathcal{E}}_1}$ denotes the projection operator for $\widehat{\mathcal{E}}_{\Sigma_{Y|X}}(\mathcal{B}_1')$. The maximum likelihood estimator $\widehat{\beta}_2$ of β_2 is the coefficient matrix from the ordinary least squares fit of the residuals $Y - \bar{Y} - \widehat{\beta}_1^T X_1$ on X_2. If X_1 and X_2 are orthogonal then $\widehat{\beta}_2$ reduces to the maximum likelihood estimator of β_2 from the standard model. Let $\widehat{\Gamma}$ be a semi-orthogonal basis matrix for $\widehat{\mathcal{E}}_{\Sigma_{Y|X}}(\mathcal{B}_1')$ and let $(\widehat{\Gamma}, \widehat{\Gamma}_0)$ be an orthogonal matrix. The maximum likelihood estimator $\widehat{\Sigma}_{Y|X}$ of $\Sigma_{Y|X}$ is then

$$\begin{aligned}
\widehat{\Sigma}_{Y|X} &= P_{\widehat{\mathcal{E}}_1} S_{R_{Y|2}|R_{1|2}} P_{\widehat{\mathcal{E}}_1} + Q_{\widehat{\mathcal{E}}_1} S_{R_{Y|2}} Q_{\widehat{\mathcal{E}}_1} = \widehat{\Gamma}\widehat{\Omega}\widehat{\Gamma}^T + \widehat{\Gamma}_0\widehat{\Omega}_0\widehat{\Gamma}_0^T \\
\widehat{\Omega} &= \widehat{\Gamma}^T S_{R_{Y|2}|R_{1|2}} \widehat{\Gamma} \\
\widehat{\Omega}_0 &= \widehat{\Gamma}_0^T S_{R_{Y|2}} \widehat{\Gamma}_0.
\end{aligned}$$

The asymptotic distributions of $\widehat{\beta}_1$ and $\widehat{\beta}_2$ were derived and reported by Su and Cook (2011). Here we give only their result for $\widehat{\beta}_1$ since that will typically be of primary interest in applications. See also Cook (2018, Chapter 3).

As we discussed for response reduction in Section 2.5.3, it is appropriate to treat the predictors as non-stochastic and ancillary. Accordingly, Su and Cook (2011) treated the predictors as non-stochastic when deriving the asymptotic distribution of $\widehat{\beta}_1$. Recall from Section 2.5.3 that when X is non-stochastic we define Σ_X as the limit of S_X as $n \to \infty$. Partition $\Sigma_X = (\Sigma_{i,j})$ according to the partitioning of X $(i, j = 1, 2)$ and let

$$\Sigma_{1|2} = \Sigma_{1,1} - \Sigma_{1,2}\Sigma_{2,2}^{-1}\Sigma_{2,1}.$$

The matrix $\Sigma_{1|2}$ is constructed in the same way as the covariance matrix for the conditional distribution of $X_1 \mid X_2$ when X is normally distributed, although here X is fixed. Define

$$U_{1|2} = \Omega_0^{-1} \otimes \eta\Sigma_{1|2}\eta^T + \Omega_0^{-1} \otimes \Omega + \Omega_0 \otimes \Omega^{-1} - 2I_{u_1(r-u_1)}.$$

The limiting distribution of $\widehat{\beta}_1$ is stated in the following proposition.

Proposition 6.1. *Under the partial envelope model (6.1) with non-stochastic predictors, $\sqrt{n}\{\mathrm{vec}(\widehat{\beta}_1) - \mathrm{vec}(\beta_1)\}$ converges in distribution to a normal random vector with mean 0 and covariance matrix*

$$\mathrm{avar}\left\{\sqrt{n}\mathrm{vec}(\widehat{\beta}_1)\right\} = \Gamma\Omega\Gamma^T \otimes \Sigma_{1|2}^{-1} + (\Gamma_0 \otimes \eta^T)U_{1|2}^{\dagger}(\Gamma_0^T \otimes \eta),$$

where $\dagger$ denotes the Moore-Penrose inverse.

The form of the asymptotic variance of $\widehat{\beta}_1$ given in this proposition is identical to the form of the asymptotic variance of $\widehat{\beta}$ from the full envelope model given in Proposition 2.2, although the definitions of components are different. For instance, $\mathrm{avar}\{\sqrt{n}\mathrm{vec}(\widehat{\beta})\}$ requires Σ_X, while $\mathrm{avar}\{\sqrt{n}\mathrm{vec}(\widehat{\beta}_1)\}$ uses $\Sigma_{1|2}$ in its place. The asymptotic standard error for an element of $\widehat{\beta}_1$ is then

$$\mathrm{se}\left\{(\widehat{\beta}_1)_{ij}\right\} = \frac{\left\{\widehat{\mathrm{avar}}[\sqrt{n}(\widehat{\beta}_1)_{ij}]\right\}^{1/2}}{\sqrt{n}}.$$

6.4.4 Application to longitudinal data

In Section 2.5.2 we motivated response envelopes by pointing out their potential utility when the model matrix connecting the treatments with a longitudinal response vector is unknown. In this section we address a perhaps more typical setting in which the model matrix is presumed known and we use envelopes to improve estimation of the effects vector. The results here are from Cook, Forzani, and Liu (2023a), which contains additional developments along the same lines.

Let Y denote a vector of r longitudinal responses and let $U \in \mathbb{R}^{r\times k}$ denote a known model matrix connecting the predictors $X \in \mathbb{R}^p$ to the response vector via the generalization of model (2.19)

$$Y_i = U\alpha_0 + U\alpha X_i + \varepsilon_i,\ i = 1, \ldots, n, \tag{6.17}$$

where $\alpha_0 \in \mathbb{R}^k$, $\alpha \in \mathbb{R}^{k\times p}$, and the error vectors are independent copies of a $N(0, \Sigma_{Y|X})$ random vector. In longitudinal data analyses we normally imagine X as non-stochastic reflecting the treatment assigned to the i-th experimental unit. We adopt that view here, unless stated otherwise. The first step in developing envelope/PLS estimators for α is to transform model (6.17) to a form that is more amenable to analysis. Let $\mathcal{U} = \mathrm{span}(U)$, let U_0 be a semi-orthogonal basis matrix for $\mathcal{U}^\perp$, and let $W = (U(U^TU)^{-1}, U_0) := (W_1, W_2)$.

Then the transformed model becomes

$$W^TY_i = \begin{pmatrix} W_1^TY \\ W_2^TY \end{pmatrix} := \begin{pmatrix} Y_{Di} \\ Y_{Si} \end{pmatrix} = \begin{pmatrix} \alpha_0 + \alpha X_i \\ 0 \end{pmatrix} + W^T\varepsilon_i,\ i = 1, \ldots, n, \tag{6.18}$$

where $Y_{Di} \in \mathbb{R}^k$ and $Y_{Si} \in \mathbb{R}^{r-k}$. The transformed variance can be represented block-wise as $\Sigma_W := \text{var}(W^T\varepsilon) = (W_i^T\Sigma_{Y|X}W_j)$, $i, j = 1, 2$. The mean $\text{E}(Y_D \mid X)$ depends non-trivially on X and thus, as indicated by the subscript D, we think of Y_D as providing direct information about the regression. On the other hand, $\text{E}(Y_S \mid X) = 0$ and thus Y_S provides no direct information but may provide useful subordinate information by virtue of its association with Y_D.

To find the maximum likelihood estimators from model (6.18), we write the full log likelihood as the sum of the log likelihoods for the marginal model from $Y_S \mid X$ and the conditional model from $Y_D \mid (X, Y_S)$:

$$Y_{Si} \mid X_i = e_{Si} \tag{6.19}$$

$$Y_{Di} \mid (X_i, Y_{Si}) = \alpha_0 + \alpha X_i + \phi_{D|S}Y_{Si} + e_{D|Si}, \tag{6.20}$$

where

$$\begin{aligned} \phi_{D|S} &= (U^TU)^{-1}U^T\Sigma_{Y|X}U_0(U_0^T\Sigma_{Y|X}U_0)^{-1} \in \mathbb{R}^{k\times(r-k)} \\ e_{D|S} &= W_1^T\varepsilon \\ e_S &= W_2^T\varepsilon. \end{aligned}$$

The variances of the errors are $\Sigma_S := \text{var}(e_S) = U_0^T\Sigma_{Y|X}U_0$ and $\Sigma_{D|S} := \text{var}(e_{D|S}) = (U^T\Sigma_{Y|X}^{-1}U)^{-1}$. The number of free real parameters in this conditional model is $N_{\text{cm}}(k) = k(p+1) + r(r+1)/2$. The subscript "cm" is used also to indicate estimators arising from the conditional model (6.20). The maximum likelihood estimator and its asymptotic variance are

$$\begin{aligned} \widehat{\beta}_{\text{cm}} &= U\widehat{\alpha}_{\text{cm}} = US_{D,R_{X|(1,S)}}S_{X|S}^{-1} \\ &= U(S_{D,X} - S_{D,S}S_S^{-1}S_{S,X})S_{X|S}^{-1} \\ \text{avar}(\sqrt{n}\text{vec}(\widehat{\beta}_{\text{cm}})) &= \Sigma_X^{-1} \otimes U\Sigma_{D|S}U^T, \end{aligned}$$

where $\widehat{\alpha}_{\text{cm}}$ and $\widehat{\beta}_{\text{cm}}$ are the MLEs of α and β in model (6.18).

Model (6.20) is in the form of a partitioned multivariate linear model (6.1) and so it is amenable to compression of the response vector for the purpose of estimating α. Let $\mathcal{A} = \text{span}(\alpha)$ and parameterize (6.20) in terms of a semi-orthogonal basis matrix $\Gamma \in \mathbb{R}^{k\times u}$ for $\mathcal{E}_{\boldsymbol{\Sigma}_{D|S}}(\mathcal{A})$, the $\Sigma_{D|S}$-envelope of $\mathcal{A}$. Let

$\eta \in \mathbb{R}^{u \times p}$ be an unconstrained matrix giving the coordinates of α in terms of semi-orthogonal basis matrix Γ, so $\alpha = \Gamma\eta$, and let $(\Gamma, \Gamma_0) \in \mathbb{R}^{k \times k}$ be an orthogonal matrix. Then the envelope version of model (6.20) conforms to our current context:

$$\left.\begin{aligned} Y_{Di} \mid (X_i, Y_{Si}) &= \alpha_0 + \Gamma\eta X_i + \phi_{D|S} Y_{Si} + e_{D|Si} \\ \Sigma_{D|S} &= \Gamma\Omega\Gamma^T + \Gamma_0\Omega_0\Gamma_0^T \end{aligned}\right\}, \tag{6.21}$$

where $\Omega \in \mathbb{R}^{u \times u}$ and $\Omega_0 \in \mathbb{R}^{(k-u)\times(k-u)}$ are positive definite matrices. This model can now be fitted with the PLS methods of Section 6.4.1 or the likelihood methods of Section 6.4.3. Many additional details and results on this envelope-based analyses of longitudinal data are available from Cook et al. (2023a).

6.4.5 Reactor data

The data for this illustration arose from an experiment involving a low density tubular polyethylene reactor (Skagerberg et al., 1992). These data were used by Cook and Su (2016) in their study of scaled predictor envelopes and partial least squares. The predictor variables consist of 20 temperatures measured at equal distances along the reactor together with the wall temperature of the reactor and the feed rate. The 6 responses are the output characteristics of the polymers produced: y_1, the number-average molecular weight; y_2, the weight-average molecular weight; y_3, the frequency of long chain branching; y_4, the frequency of short chain branching; y_5, the content of vinyl groups; y_6, the content of vinylidene groups. Because the distributions of the values of all the response variables are highly skewed to the right, the analysis was performed using the logarithms of their corresponding values. For interpretational convenience all responses were standardized to unit variance. Responses y_1 and y_2 are seen to be strongly correlated, and y_4, y_5 and y_6 form another strongly correlated group. The third response y_3 is more weakly correlated with the others. The predictive accuracy of each method was estimated through leave-one-out cross-validation. Overall, there are a total of $n = 56$ observations on $p = 22$ predictors and $r = 6$ responses.

Since the first $p_1 = 20$ predictors X_1 are temperatures along the reactor, they differ from the remaining two predictors X_2, and so we consider dimension reduction treating X_1 and X_2 differently. Specifically, we use model (6.1) as the basis for partial response reduction using the PLS algorithm of Section 6.4.1

TABLE 6.2
Leave-one-out cross validation of the mean prediction error predRMSE for the reactor data.

		Partial Predictor		Partial Response	
Method	OLS	PLS	Envelope	PLS	Envelope
No. components	$p_1 = 20$	16	16	2	2
predRMSE	0.44	0.37	0.37	0.34	0.33

and the envelope estimator described in Section 6.4.3. For comparison, we also studied partial predictor reduction using the model of Section 6.2.3 as the basis for the PLS algorithm of Section 6.2.1 and the likelihood-based partial envelope estimation of Section 6.2.3. The comparison criterion is prediction of the response vector using leave-one-out cross validation to form the mean prediction error predRMSE as defined in Section 5.5.1.

The results shown in Table 6.2 indicate that partial response reduction is somewhat better than partial predictor reduction and that both reduction types do noticeably better than OLS. The two components indicated in the partial response applications indicate that the basis matrix Γ in model (6.1) is 6×2. Consequently, $\Gamma^T Y$ is the part of the response vector that is influenced by the changes in the predictors, while $\Gamma_0^T Y$ is unaffected by changes in the predictors.

7

Linear Discriminant Analysis

Partial least squares has gained some recognition as a method of discriminating between $r + 1$ populations or classifications based on a vector $X \in \mathbb{R}^p$ of features, although it was not originally designed for that task. Define $Y = (0, \ldots, 0, 1_j, 0, \ldots 0)^T$ as an $r \times 1$ indicator vector with a single one indicating class $j = 1, \ldots, r$ and the zero vector $Y = (0, \ldots, 0)^T$ indicating class $j = 0$, which is occasionally called one-hot encoding. In this way the $r \times 1$ vector Y serves to indicate $r+1$ classes. Using one of the algorithms discussed in Chapter 3, PLS regression is then used to reduce the feature vector X in the data (X_i, Y_i), $i = 1, \ldots, n$, followed by use of Bayes' rule for the actual classification of a new observation. Barker and Ravens (2003) pointed out certain basic connections between this partial least squares discriminant analysis (PLS-DA) and Fisher's linear discriminant analysis (LDA). For instance, if $S_X > 0$ and the number of components $q = p$ then PLS-DA is the same as LDA. Liu and Ravens (2007) reinforced the arguments of Barker and Ravens (2003) by providing illustrations on real data sets. Brereton and Lloyd (2014) provided a tutorial on PLS-DA, motivating it by using a somewhat questionable version of the bilinear calibration model described by Martens and Næs (1989, p. 91). They compared PLS-DA to LDA and to other methods of discrimination. PLS-DA is now available in most software for PLS regressions.

In this chapter, we revisit PLS-DA but differ from the literature by connecting it with principal fitted components (PFC), envelopes and reduced-rank (RR) regression. Results in previous chapters help to explain why it might be expected to be serviceable when the number of features is larger than the sample size. Familiarity with classical normal-theory linear and quadratic discriminant analysis is assumed here and in Chapter 8. (See for example McLachlan, 2004).

DOI: 10.1201/9781003482475-7

7.1 Envelope discriminant subspace

Given an observed vector $X \in \mathbb{R}^p$ of features associated with an object, the problem of estimating the class C to which it belongs from among $r+1$ possible classes indexed by the set $\chi = \{0, \ldots, r\}$ is frequently addressed by using Bayes' rule to assign the object to the class $\phi(X)$ with the maximum probability:

$$\phi(X) = \arg\max_{k \in \chi} \Pr(C = k \mid X).$$

It is widely recognized that we might lower substantially the chance of misclassification if we can reduce the dimension of X without loss of information on $\phi(X)$. With that goal in mind, Cook and Yin (2001) considered reducing, without loss of information, a vector X of continuous features by projecting it onto a subspace $\mathcal{S} \subseteq \mathbb{R}^p$ and then employing Bayes' rule on the reduced features $\phi_{\mathcal{S}}(X) := \arg\max_{k \in \chi} \Pr(C = k \mid P_{\mathcal{S}}X)$:

Definition 7.1. *If*

$$\phi(X) = \phi_{\mathcal{S}}(X) \tag{7.1}$$

then $\mathcal{S}$ is called a discriminant subspace. If the intersection of all discriminant subspaces is itself a discriminant subspace, it is called the central discriminant subspace and denoted as $\mathcal{D}_{C|X}$.

In this way the central discriminant subspace captures all of the classification information that X has about class C and thus has the potential to reduce the dimension of X without loss of information on $\phi(X)$. For example, suppose that there are three classes, $\chi = \{0, 1, 2\}$, two normal features $X = (X_1, X_2)^T$ and that $\Pr(C = k \mid X_1, X_2) = \Pr(C = k \mid X_1 + X_2)$. Then only the sum of the two predictors is relevant for classification and (7.1) holds with $\mathcal{S} = \text{span}((1,1)^T)$. This reflects the kind of setting we have in mind when envelopes are employed a bit later.

To illustrate the potential importance of discriminant subspaces, consider a stylized problem adapted from Cook and Yin (2001), still using three classes and two normal features. Define the conditional distribution of $C \mid X$

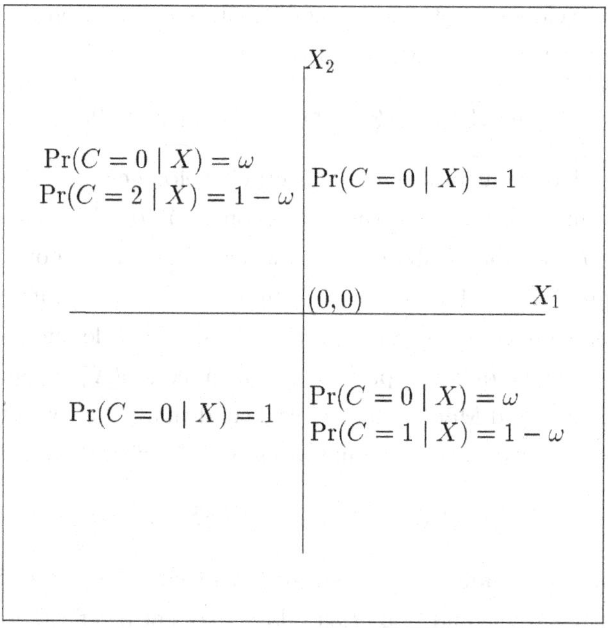

FIGURE 7.1
Illustration of the importance of the central discriminant subspace.

according to the following:

$$\begin{aligned}
\Pr(C=0 \mid X_1<0, X_2<0) &= \Pr(C=0 \mid X_1>0, X_2>0)=1, \\
\Pr(C=0 \mid X_1>0, X_2<0) &= \Pr(C=0 \mid X_1<0, X_2>0)=\omega, \\
\Pr(C=1 \mid X_1>0, X_2<0) &= 1-\omega \\
\Pr(C=2 \mid X_1<0, X_2>0) &= 1-\omega,
\end{aligned}$$

where $0<\omega<1$. These conditional distributions are depicted in Figure 7.1. The central subspace $\mathcal{S}_{C|X}=\mathbb{R}^2$ and no linear dimension reduction is possible. However, dimension reduction is possible for $\phi(X)$. Although we would not normally expect applications to be so intricate, the example does illustrate the potential relevance of discriminant subspaces.

If $\omega>1/2$ then $\phi(X)=0$ for all X, $\mathcal{D}_{C|X}=\text{span}((0,0)^T)$ is a discriminant subspace and consequently the two features provide no information to aid in classification. In this case we are able to discard X completely. If $\omega<1/2$ then $\phi(X)$ depends non-trivially on both predictors, $\mathcal{D}_{C|X}=\mathbb{R}^2$ and so no dimension reduction is possible.

To bring envelopes into the discussion, recall that conditions (2.3), restated here with Y replaced by C for ease of reference

$$(a)\ C \perp\!\!\!\perp X \mid P_{\mathcal{S}}X \text{ and (b) } P_{\mathcal{S}}\Sigma_X Q_{\mathcal{S}} = 0, \tag{7.2}$$

are operational versions of the more general envelope conditions (2.2) for reducing the dimension of X in the regression of Y on X. Condition (7.2a) requires that $P_{\mathcal{S}}X$ capture all of ways in which X can inform on the distribution of C. Dimension reduction for discriminant analysis is different since we are not interested in capturing all aspects of X that provide information about C. Rather we pursue only the part of X that affects $\phi(X)$. Adapting to this distinction, Zhang and Mai (2019) replaced condition (7.2a) with requirement (7.1), leading to consideration of subspaces $\mathcal{S} \subseteq \mathbb{R}^p$ with the properties

$$(a)\ \phi(X) = \phi_{\mathcal{S}}(X) \text{ and (b) } P_{\mathcal{S}}\Sigma_X Q_{\mathcal{S}} = 0. \tag{7.3}$$

The rationale for including (7.3b) is similar to that discussed in Section 2.1: methodology based on condition (7.3a) alone may not be effective when $p > n$ or the features are highly collinear because then it is hard in application to distinguish the material part $P_{\mathcal{S}}X$ of X that is required for $\phi(X)$ from the complementary part $Q_{\mathcal{S}}X$ that is immaterial. The role of condition (7.3b) is then to induce a measure of clarity in the separation of X into parts that are material and immaterial to $\phi(X)$. Zhang and Mai (2019) formalized the notion of the smallest subspace that satisfies (7.3) as follows:

Definition 7.2. *If the intersection of all subspaces that satisfy (7.3) is itself a subspace that satisfies (7.3), then it is called the envelope discriminant subspace.*

A variety of methods have been proposed for implementing aspects of these general ideas. Most focus on modeling $\Pr(C = k \mid X = x)$ directly, or indirectly via the conditional density f of X given $(C = k)$ (e.g. Cook and Yin, 2001, eq. (5)):

$$\Pr(C = k \mid X = x) = \frac{\pi_k f(x \mid C = k)}{\sum_{j=0}^{r} \pi_j f(x \mid C = j)}, \tag{7.4}$$

where $\pi_k = \Pr(C = k)$ is the marginal prior probability of class $k \in \chi$. Beginning with a model for $X \mid C$ and specification of prior probabilities π_k, this represents the Bayes posterior probability of class k. However, if $\Pr(C = k \mid X = x)$ is modeled directly, say with logistic regression when there are two classes, this Bayes interpretation may not play a useful role. PLS discriminant analysis (e.g. Brereton and Lloyd, 2014) arises by modeling $X \mid (C = k)$ as a conditional normal, leading to methodology closely associated with Fisherian LDA. While this may seem simple relative to the range of methods available, Hand (2006) argued that simple methods can outperform more intricate methods that do not adapt well to changing circumstances between classifier development and application.

We next turn to LDA where we bring in PLS methods. Quadratic discriminant analysis is considered in Chapter 8.

7.2 Linear discriminant analysis

The nominal stochastic structure underlying LDA for predicting the class $C \in \chi$ based on a vector of features $X \in \mathbb{R}^p$ is

$$X \mid (C = k) \sim N(\mu_k, \Sigma_{X|(C=k)}), \ \Pr(C = k) = \pi_k, \ k \in \chi, \qquad (7.5)$$

where $\pi_k > 0$, $\sum_{k=0}^{r} \pi_k = 1$ and the intra-class covariance matrices $\Sigma_{X|(C=k)}$ are assumed to be equal over the classes, $\Sigma_{X|(C=k)} = \Sigma_{X|(C=j)} := \Sigma_{X|C}$. Under this model, the Bayes' rule is to classify a new unit with feature vector X onto the class with maximum posterior probability, giving the classification function from (7.4)

$$\begin{aligned}
\phi_{\text{lda}}(X) &= \arg\max_{k \in \chi} \Pr(C = k \mid X) \\
&= \arg\max_{k \in \chi} \{\log \pi_k + \mu_k^T \Sigma_{X|C}^{-1}(X - \mu_k/2)\} \qquad (7.6) \\
&= \arg\max_{k \in \chi} \{\log(\pi_k/\pi_0) + \beta_k^T \Sigma_{X|C}^{-1}(X - (\mu_k + \mu_0)/2)\} \\
&= \arg\max_{k \in \chi} \{\log(\pi_k/\pi_0) + \beta_k^T \Sigma_{X|C}^{-1}(X - (\beta_k + 2\mu_0)/2)\}, \qquad (7.7)
\end{aligned}$$

where the vectors $\beta_k = (\mu_k - \mu_0) \in \mathbb{R}^p$, $k = 1, \ldots, r$, are the mean deviations relative to the reference population μ_0. The prior probabilities π_k are assumed to be known, perhaps $\pi_k = (1 + r)^{-1}$ or, when a training sample is available, π_k is set equal to the observed fractions of cases in class $C = k$. Given the stochastic setup (7.5), the success of a prediction rule in application depends on having good estimators of the β_k's and $\Sigma_{X|C}$. These parameters can be well-estimated by using a multivariate regression model when $n \gg p$.

Let $\beta = (\beta_1, \ldots, \beta_r)^T \in \mathbb{R}^{r \times p}$ and recall that $Y = (0, \ldots, 0, 1_j, 0, \ldots 0)^T$ denotes the $r \times 1$ vector indicating class $j \in \chi$. Then μ_0, β, and $\Sigma_{X|C}$ can be estimated by fitting the multivariate linear model implied by (7.5),

$$X_i = \mu_0 + \beta^T Y_i + \varepsilon_i, \; i = 1, \ldots, n, \tag{7.8}$$

where the errors are independent copies of $\varepsilon \sim N(0, \Sigma_{X|C})$. Maximum likelihood estimators of the parameters in this model were reviewed in Section 1.2.2. Since the roles of X and Y are different in (7.8) we restate the estimators for ease of reference:

$$\begin{aligned} \widehat{\beta}_{\text{ols}} &= S_Y^{-1} S_{Y,X} \\ \widehat{\Sigma}_{X|C} &= S_{X|Y} \\ \widehat{\mu}_0 &= \bar{X} - \widehat{\beta}_{\text{ols}}^T \bar{Y}. \end{aligned}$$

These estimators are then substituted into $\phi_{\text{lda}}(X)$ to estimate the class with the maximum posterior probability, a process that characterizes classical LDA. However, this estimation procedure can be inefficient if β has less than full row rank or if only a part of X is informative about C. In the extreme, if it were known that $\text{rank}(\beta) = 1$ then it may be possible to improve the classical method considerably. Incorporating the central discriminant subspace allows for the possibility that β is rank deficient, and envelopes were designed to deal the possibility that only a portion of X informs on C.

7.3 Principal fitted components

As used previously, let $\mathcal{B}' = \text{span}(\beta^T) \subseteq \mathbb{R}^p$. Zhang and Mai (2019) noted, as may be clear from (7.7), that under model (7.5),

$$\mathcal{D}_{C|X} = \text{span}(\Sigma_{X|C}^{-1} \beta^T) = \Sigma_{X|C}^{-1} \mathcal{B}', \tag{7.9}$$

so estimating the central discriminant subspace reduces to estimation of $\text{span}(\Sigma_{X|C}^{-1} \beta^T)$ for model (7.8), including its dimension. To incorporate this into model (7.8), let $d = \dim(\mathcal{B}') \leq \min(p, r)$, let $B \in \mathbb{R}^{p \times d}$ be a semi-orthogonal basis matrix for $\mathcal{B}'$ and define coordinate vectors b_k so that $\beta_k = B b_k$, $k = 1, \ldots, r$. Let $b = (b_1, \ldots, b_r) \in \mathbb{R}^{d \times r}$, which has rank d.

Then $\beta^T = Bb$ and model (7.8) becomes

$$X_i = \mu_0 + BbY_i + \varepsilon_i,\ i = 1, \ldots, n. \tag{7.10}$$

If $d = r$ then β^T has full column rank and this model reduces to model (7.8). If $d < r$ this becomes an instance of the general model for PFC (Cook, 2007; Cook and Forzani, 2008b), which is an extension to regression of the Tipping-Bishop model that yields probabilistic principal components (Tipping and Bishop, 1999). The general PFC model allows the Y-vector to be any user-specified function of a response, but in discriminant analysis Y is properly an indicator vector as defined previously. A key characteristic of model (7.10) is summarized in the following proposition (Cook (2007, Prop 6); Cook and Forzani (2008b, Thm 2.1)). In preparation, define a $p \times d$ basis matrix for the central discriminant subspace as

$$\Phi = \Sigma_{X|C}^{-1} B \in \mathbb{R}^{p \times d}.$$

Proposition 7.1. *Under model (7.10),*

$$X \mid (Y, \Phi^T X) \sim X \mid \Phi^T X.$$

Proof. Let (B, B_0) be an orthogonal matrix. We suppress the subscript i in this proof for notational simplicity. Transforming X in model (7.10) with $(\Phi, B_0)^T$, the model becomes

$$\begin{pmatrix} \Phi^T X \\ B_0^T X \end{pmatrix} = \begin{pmatrix} \Phi^T \mu_0 \\ B_0^T \mu_0 \end{pmatrix} + \begin{pmatrix} \Phi^T BbY \\ 0 \end{pmatrix} + \begin{pmatrix} \Phi^T \varepsilon \\ B_0^T \varepsilon \end{pmatrix}.$$

We see from this model that $B_0^T X \perp\!\!\!\perp Y$, so marginally $B_0^T X$ carries no information on the class C. Further,

$$\begin{aligned} \operatorname{cov}(\Phi^T \varepsilon, B_0^T \varepsilon) &= \Phi^T \operatorname{var}(\varepsilon) B_0 \\ &= \Phi^T \Sigma_{X|C} B_0 = B^T \Sigma_{X|C}^{-1} \Sigma_{X|C} B_0 \\ &= 0. \end{aligned}$$

Since ε is normally distributed, it follows that $\Phi^T X \perp\!\!\!\perp B_0^T X \mid Y$. This plus the previous conclusion $B_0^T X \perp\!\!\!\perp Y$ implies that $B_0^T X \perp\!\!\!\perp (\Phi^T X, Y)$ and thus that $B_0^T X \perp\!\!\!\perp Y \mid \Phi^T X$ (Cook, 1998, Proposition 4.6). Since $(B_0, \Phi)^T X$ is a full rank linear transformation of X, this last conclusion implies that $X \perp\!\!\!\perp Y \mid \Phi^T X$. The desired conclusion – $X \mid (Y, \Phi^T X) \sim X \mid \Phi^T X$ – follows.

□

In consequence of Proposition 7.1, $Y \perp\!\!\!\perp X \mid \Phi^T X$, which implies that $\Phi^T X$ holds all of the information that X has about Y and thus, together with (7.9),

$$\mathcal{S}_{Y|X} = \mathcal{D}_{C|X} = \text{span}(\Phi). \tag{7.11}$$

Accordingly, we lose no information when basing LDA classifications on the reduced features $\Phi^T X$ instead of the full features, and may gain by reducing substantially the probability of misclassification. In terms of the reduced features, the classification function is

$$\begin{aligned}
\phi_{\text{lda}}(\Phi^T X) &= \arg\max_{k \in \chi} \Pr(C = k \mid \Phi^T X) \\
&= \arg\max_{k \in \chi} \{\log(\pi_k/\pi_0) \\
&\quad + \beta_k^T \Phi(\Phi^T \Sigma_{X|C} \Phi)^{-1} \Phi^T (X - (\beta_k + 2\mu_0)/2)\}, \\
&= \arg\max_{k \in \chi} \{\log(\pi_k/\pi_0) \\
&\quad + (P_{B(\Sigma_{X|C}^{-1})} \beta_k)^T \Sigma_{X|C}^{-1} P_{B(\Sigma_{X|C}^{-1})} (X - (\beta_k + 2\mu_0)/2)\}, \\
&= \arg\max_{k \in \chi} \{\log(\pi_k/\pi_0) + b_k^T \Phi^T (X - (Bb_k + 2\mu_0)/2)\},
\end{aligned} \tag{7.12}$$

where still $\beta_k = (\mu_k - \mu_0) = Bb_k$, $k = 1, \ldots, r$. Result (7.12) was obtained by substituting

$$\Phi(\Phi^T \Sigma_{X|C} \Phi)^{-1} \Phi^T = \Sigma_{X|C}^{-1} P_{B(\Sigma_{X|C}^{-1})} = P_{B(\Sigma_{X|C}^{-1})}^T \Sigma_{X|C}^{-1} P_{B(\Sigma_{X|C}^{-1})}.$$

It might be concluded from (7.12) that $\mathcal{D}_{C|X} = \text{span}(\Phi b)$. But $\text{span}(\Phi b) = \text{span}(\Phi)$ since b has full row rank, and so we again arrive at (7.11).

Representation (7.12) expresses the classification in terms of the reduced features $\Phi^T X$, while (7.7) expresses it terms of the full features. Comparing (7.7) to the steps leading to (7.12) we see that in the classification with reduced features, the features have been projected onto $\text{span}(B)$ in the $\Sigma_{X|C}^{-1}$ inner product. In consequence, we would expect the reduced predictors to yield more reliable categorizations, at least in large samples, because directions of extraneous variation have been removed. There are several methods that can be used to obtain estimators of the unknown quantities in $\phi_{\text{lda}}(\Phi^T X)$. These estimators are based on variations of model (7.10) that all have the same mean structure but differ on the error structures. We discuss some of these in the following sections, leading to connections with envelopes and PLS methods in Sections 7.3.2.

7.3.1 Discrimination via PFC with Isotropic errors

Isotropic errors – $\Sigma_{X|C} = \sigma^2 I_p$ – may be appropriate when measurement error dominates model (7.10) and, given the correct classification, the features are conditionally independent and have the same variance. They may also be useful as an approximation to avoid estimating a large variance-covariance matrix with limited sample size. In particular, this isotropic classifier might be serviceable in some applications when $n \ll p$.

With isotropic errors, $Y \perp\!\!\!\perp X \mid B^T X$, $P_{B(\Sigma^{-1}_{X|C})} = P_B$, and the classification function reduces to

$$
\begin{aligned}
\phi_{\text{lda}}(\Phi^T X) &= \phi_{\text{lda}}(B^T X) \\
&= \arg\max_{k \in \chi} \{\log(\pi_k/\pi_0) + \sigma^{-2}\beta_k^T P_B (X - (\beta_k + 2\mu_0)/2)\} \\
&= \arg\max_{k \in \chi} \{\log(\pi_k/\pi_0) + \sigma^{-2} b_k^T B^T (X - (Bb_k + 2\mu_0)/2)\}.
\end{aligned}
$$

The maximum likelihood estimator of $\phi_{\text{lda}}(B^T X)$ can be constructed as follows. Let $\mathbb{Y}$ denote the $n \times r$ matrix with rows $(Y_i - \bar{Y})^T$, let $\mathbb{X}$ denote the $n \times p$ matrix with rows $(X_i - \bar{X})^T$, let $\widehat{\mathbb{X}} = P_{\mathbb{Y}}\mathbb{X}$ denote the $n \times p$ matrix of centered fitted values from the fit of the full feature model (7.8) and let $\hat{\lambda}_1, \ldots, \hat{\lambda}_d$ and $\widehat{\phi}_1, \ldots, \widehat{\phi}_d$ denote the first d eigenvalues and corresponding eigenvectors of $S_{\text{fit}} = \widehat{\mathbb{X}}^T \widehat{\mathbb{X}}/n$, the sample covariance matrix of the fitted vectors. The following estimators arise from model (7.10) with d specified and isotropic errors (Cook, 2007):

$$
\begin{aligned}
\widehat{B} &= (\widehat{\phi}_1, \ldots, \widehat{\phi}_d) \\
\widehat{b} &= \widehat{B}^T \mathbb{X}^T \mathbb{Y} (\mathbb{Y}^T \mathbb{Y})^{-1} \\
\widehat{\mu}_0 &= \bar{X} - \widehat{B}\widehat{b}\bar{Y} \\
\widehat{\beta} &= (\mathbb{Y}^T \mathbb{Y})^{-1} \mathbb{Y}^T \mathbb{X} P_{\widehat{B}} = (\widehat{\beta}_1, \ldots, \widehat{\beta}_r) \\
\widehat{\sigma}^2 &= p^{-1} \left\{ \operatorname{tr}(\mathbb{X}^T \mathbb{X}/n) - \sum_{i=1}^{d} \hat{\lambda}_i \right\}.
\end{aligned}
$$

These estimators, which require $n > r$ and $n > d$ but not $n > p$, can now we substituted into $\phi_{\text{lda}}(B^T X)$ for a sample classifier. If classes have equal prior probabilities, $\pi_k = \pi_j$, then the classification rule simplifies a bit and an estimator of σ^2 is no longer necessary:

$$
\phi_{\text{lda}}(B^T X) = \arg\max_{k \in \chi} \{b_k^T B^T (X - (Bb_k + 2\mu_0)/2)\}.
$$

The dimension d of $\mathcal{D}_{C|X} = \text{span}(B)$ can be estimated straightforwardly by using cross validation , with a holdout sample reserved for estimation of the classification error of the final classification function.

7.3.2 Envelope and PLS discrimination under model (7.8)

As suggested previously, methodology based on model (7.8) alone may not be effective when the features are highly collinear because then it is hard in application to distinguish the material part of X that is required for classification from the complementary part of X that does nothing more than induce extraneous variation. This is where envelopes play a key role via condition (7.3b), which induces a measure of clarity in the separation of X into its parts that are material and immaterial to $\phi(X)$. Operationally, an envelope parametrization of model (7.8) leads to a model with a structure for $\Sigma_{X|C}$ that can adapt to changing circumstances through the choice of dimension.

The maximum likelihood estimators based on model (7.8) are not serviceable for classification when $p > n$ because then $S_{X|Y}$ is singular and direct estimation of the classification function (7.7) is not possible. This is where PLS classification may be particularly useful, as it provides a method of estimating the material part of the feature vector when $p > n$. It may be useful also when $n \gg p$ and the features are collinear, although here likelihood-based envelopes may dominate.

Since $\mathcal{D}_{C|X} = \text{span}(\Phi)$ under model (7.10), it follows from Definition 7.2 and model (7.10) that the envelope discriminant subspace is the $\Sigma_{X|C}$-envelope of span(Φ), which leads to the following envelope equality after applying (1.26),

$$\mathcal{E}_{\Sigma_{X|C}}(\mathcal{D}_{C|X}) = \mathcal{E}_{\Sigma_X}(\mathcal{B}').$$

Let $u = \dim(\mathcal{E}_{\Sigma_X}(\mathcal{B}'))$. As in previous chapters, we can now parameterize model (7.8) in terms of a semi-orthogonal basis matrix $\Gamma \in \mathbb{R}^{p\times u}$ for $\mathcal{E}_{\Sigma_X}(\mathcal{B}')$. Let (Γ, Γ_0) be an orthogonal matrix. Then the PFC model becomes

$$\begin{aligned} X_i &= \mu_0 + \Gamma\eta Y_i + \varepsilon_i,\ i = 1, \ldots, n, \\ \Sigma_{X|C} &= \Gamma\Omega\Gamma^T + \Gamma_0\Omega_0\Gamma_0^T. \end{aligned} \tag{7.13}$$

This response envelope model (See Sections 2.4 and 2.5.3) was anticipated by Cook (2007), who proposed it as an extension (EPFC) of PFC model (7.10), but without the key understanding that can derive from the envelope structure. It was subsequently proposed specifically for discrimination by Zhang and Mai (2019), who called it the envelope discriminate subspace (ENDS) model. Based on model (7.13) we determine class membership by using (7.7) in combination with the multivariate model for the reduced features

$$\begin{aligned} \Gamma^T X_i &= \alpha_0 + \eta Y_i + \epsilon_i,\ i = 1, \dots, n, \\ \Sigma_{\Gamma^T X|C} &= \Omega, \\ \eta &= (\eta_1, \dots, \eta_r) \\ \alpha_0 &= \Gamma^T \mu_0. \end{aligned} \tag{7.14}$$

This results in the central discriminant subspace $\mathcal{D}_{C|\Gamma^T X} = \text{span}(\Omega^{-1}\eta)$ and corresponding classification function (7.12) expressed in terms of the reduced features

$$\phi_{\text{lda}}(\Gamma^T X) = \arg\max_{k \in \chi} \left\{ \log(\pi_k/\pi_0) + \eta_k^T \Omega^{-1}(\Gamma^T X - (\eta_k + 2\Gamma^T \mu_0)/2) \right\}. \tag{7.15}$$

7.3.2.1 Maximum likelihood envelope estimators

Maximum likelihood estimators of the unknown parameters in (7.15) can be constructed by adapting the estimators given in Section 2.5.3 using the features X as the response and the class indicator vector Y as the predictor:

$$\begin{aligned} \widehat{\Gamma} &= \arg\min_G \left\{ \log|G^T S_{X|Y} G| + \log|G^T S_X^{-1} G| \right\} \\ \widehat{\eta} &= \widehat{\Gamma}^T S_{X,Y} S_Y^{-1} = (\widehat{\eta}_1, \dots, \widehat{\eta}_r) \\ \widehat{\Gamma}\widehat{\eta} &= P_{\widehat{\Gamma}} S_{X,Y} S_Y^{-1}, \\ \widehat{\mu}_0 &= \bar{X} - \widehat{\Gamma}\widehat{\eta}\bar{Y} \\ \widehat{\Omega} &= \widehat{\Gamma}^T S_{X|Y} \widehat{\Gamma}, \\ \widehat{\Omega}_0 &= \widehat{\Gamma}_0^T S_X \widehat{\Gamma}_0, \\ \widehat{\Sigma}_{X|C} &= \widehat{\Gamma}\widehat{\Omega}\widehat{\Gamma}^T + \widehat{\Gamma}_0 \widehat{\Omega}_0 \widehat{\Gamma}_0^T, \end{aligned} \tag{7.16}$$

where $\min_G$ is over all semi-orthogonal matrices $G \in \mathbb{R}^{r \times u}$, $(\widehat{\Gamma}, \widehat{\Gamma}_0)$ is an orthogonal matrix. The fully maximized log-likelihood for fixed u is then

$$\begin{aligned} \widehat{L}_u &= -(nr/2)\log(2\pi) - nr/2 - (n/2)\log|S_X| \\ &\quad -(n/2)\log|\widehat{\Gamma}^T S_{X|Y} \widehat{\Gamma}| - (n/2)\log|\widehat{\Gamma}^T S_X^{-1} \widehat{\Gamma}|. \end{aligned}$$

The estimators $\widehat{\Gamma}$, $\widehat{\eta}_k$, $\widehat{\mu}_0$ and $\widehat{\Omega}$ are now substituted into (7.15) to determine the class with the maximum estimated probability.

7.3.2.2 Discrimination via PLS

Recall from the discussion of Chapter 3 that NIPALS and SIMPLS estimate the predictor envelope $\mathcal{E}_{\Sigma_X}(\mathcal{B})$ in the regression of Y on X. However, we also know from the discussions of Sections 2.4 and 3.11 that this envelope is the same as the response envelope in the regression of X on Y, which is exactly what appears in model (7.13). In other words, beginning with classification indicator vectors Y_i and corresponding features X_i we can use either NIPALS or SIMPLS weight matrices as estimates of a basis $\widehat{\Gamma}_{\text{pls}}$ for $\mathcal{E}_{\Sigma_X}(\mathcal{B})$, which is then used in place of $\widehat{\Gamma}$ from (7.16) to construct the remaining estimators for substitution into (7.15). The adaptation of PLS algorithms to discrimination problems is then seen to be straightforward, the methodology for constructing $\widehat{\Gamma}_{\text{pls}}$ being covered by the general discussions in previous chapters. In particular, the PLS algorithms are not hindered by the requirement that $n > p$ and their asymptotic behavior in high dimensions is governed by the results summarized in Chapter 4.

To be clear, a procedure for computing the PLS discrimination function is outlined as follows.

1. Run a NIPALS or SIMPLS algorithm with response Y and predictor X and extract the weight matrix W normalized if necessary so that $W^T W = I_p$. Set $\widehat{\Gamma}_{\text{pls}} = W$.

2. Construct the required estimators based on the estimators at (7.16) as follows.

$$\begin{aligned}
\widehat{\eta}_{\text{pls}} &= \widehat{\Gamma}^T_{\text{pls}} S_{X,Y} S_Y^{-1} = (\widehat{\eta}_{1,\text{pls}}, \ldots, \widehat{\eta}_{r,\text{pls}}) \\
\widehat{\mu}_{0,\text{pls}} &= \bar{X} - \widehat{\Gamma}_{\text{pls}} \widehat{\eta}_{\text{pls}} \bar{Y} \\
\widehat{\Omega}_{\text{pls}} &= \widehat{\Gamma}^T_{\text{pls}} S_{X|Y} \widehat{\Gamma}_{\text{pls}}.
\end{aligned}$$

3. Substitute these estimators into (7.15) to get the estimated PLS discriminant function:

$$\begin{aligned}
\widehat{\phi}_{\text{lda}}(\Gamma^T X) &= \arg\max_{k \in \chi} \\
&\times \left\{ \log(\pi_k/\pi_0) + \widehat{\eta}^T_{k,\text{pls}} \widehat{\Omega}^{-1}_{\text{pls}} (\widehat{\Gamma}^T_{\text{pls}} X - (\widehat{\eta}_{k,\text{pls}} + 2\widehat{\Gamma}^T_{\text{pls}} \widehat{\mu}_{0,\text{pls}})/2) \right\}.
\end{aligned}$$

7.4 Discrimination via PFC with $\Sigma_{X|C} > 0$

Discrimination via PFC with isotropic errors $\Sigma_{X|C} = \sigma^2 I_p$ was discussed in Section 7.3.1. In this section we discuss PFC with anisotropic errors $\Sigma_{X|C} > 0$.

In the context of model (7.10), which has anisotropic errors $\Sigma_{X|C} > 0$, PFC and RR regression models (Anderson, 1999; Izenman, 1975; Reinsel and Velu, 1998) are the same and produce the same maximum likelihood estimators of μ_0, β and $\Sigma_{X|C}$. Aspects of these models are related to our discussion of bilinear PLS models in Section 5.12. Estimators, including methods for inferring about d, were described by Cook and Forzani (2008b, Sec 3.1) in the context of PFC and by Cook, Forzani, and Zhang (2015, Sec 2.1) in the context of RR regression. The estimators can be obtained from either source, although those given by Cook et al. (2015) are more succinct. The estimators were also summarized by (Cook, 2018, Sec. 9.10.2).

On the other hand, PFC regressions include subspace estimation, while RR regressions normally do not. The PFC family of methods covers possibilities, like the isotropic and extended PFC models, that are natural extension of probabilistic principal components (Tipping and Bishop, 1999) to regression problems and that are not related to RR regression. Nevertheless, the methods are equivalent under model (7.10) and so we use the reference acronym PFC-RR to denote model (7.10) and quantities derived therefrom.

Define

$$C_{X,Y} = S_X^{-1/2} S_{X,Y} S_Y^{-1/2}$$

to be the matrix of sample correlations between the elements of the standardized vectors $S_X^{-1/2}X$ and $S_Y^{-1/2}Y$, with singular value decomposition $C_{X,Y} = UDV^T$. Extending this notation, let $C_{X,Y}^{(d)} = U_d D_d V_d^T$, where $U_d \in \mathbb{R}^{r\times d}$ and $V_d \in \mathbb{R}^{p\times d}$ consist of the first d columns of U and V, and D_d is the diagonal matrix consisting of the first d singular values of $C_{X,Y}$. We also use $C_{Y,X} = C_{X,Y}^T$. Then, assuming normal errors, that d is given and that $r < p$, the maximum likelihood estimators of the parameters in model (7.10) are (Cook et al., 2015, Sec 2.1)

$$\widehat{\mu}_0 = \bar{X} - \widehat{\beta}^T \bar{Y} \tag{7.17}$$

$$\widehat{\beta}^T = (\widehat{\beta}_1, \ldots, \widehat{\beta}_r) = S_X^{1/2} C_{X,Y}^{(d)} S_Y^{-1/2} \tag{7.18}$$

$$\begin{aligned}\widehat{\Sigma}_{X|C} &= S_X - \widehat{\beta}^T S_{Y,X} \\ &= S_X^{1/2}(I_r - C_{X,Y}^{(d)} C_{Y,X}^{(d)}) S_X^{1/2}. \end{aligned} \tag{7.19}$$

The usual OLS estimators are obtained by setting $d = r$ so there is no rank reduction. Once the PFC-RR estimators given in (7.17) – (7.19) are available they can be substituted into (7.7) to obtain the estimated classification function. The dimension d can then be selected by cross validation using the error rate as a criterion.

General estimation methods like that based on PFC-RR model (7.10) and the corresponding PFC discriminant function require that $n \gg p$ to insure that $\Sigma_{X|C}^{-1}$ can be well estimated. The same is true of the envelope discrimination discussed in Section 7.3.2. But PLS discrimination discussed at the end of Section 7.3.2 and PFC discrimination with isotropic errors may be serviceable without requiring $n \gg p$. Of these two, PLS fitting is surely the more versatile since it does not require isotropic errors.

7.5 Discrimination via PLS and PFC-RR

7.5.1 Envelopes and PFC-RR model (7.10) combined

Discrimination based on PFC-RR model (7.10) is designed to exploit aspects of regression of X on Y that are different from those exploited by envelope model (7.13). In PFC-RR model (7.10) there is no modeled connection between $\beta^T = Bb$ and the error covariance matrix $\Sigma_{X|C}$ as there is in (7.13). Envelope model (7.13) makes no direct allowance for the possibility that β is rank deficient, $\text{rank}(\beta^T) = \dim(\mathcal{B}') = d < \min(p, r)$. These distinctions have consequences. In binary classification, $r = 1$ and $p > 1$, so $\beta^T \in \mathbb{R}^p$ and its only possible ranks are 0 and 1. In this case PFC-RR is not useful, while an envelope can still lead to substantial gains if $\text{span}(\beta^T)$ is contained in a low-dimensional reducing subspace of $\Sigma_{X|C}$. More generally, PFC-RR offers no gain when β is full rank, while envelopes can still produce substantial gain. On the other hand, it is also possible to have situations where envelopes offer no gain, while PFC-RR provides notable gain. For instance, if $r > 1$, $p > 1$ and $d = 1$, so PFC-RR gives maximal gain, it is still possible that $\mathcal{E}_{\Sigma_{X|C}}(\mathcal{B}') = \mathbb{R}^p$ so that envelopes produce no gain.

There are also notable differences between the ways in which gains are produced by envelope and by PFC-RR regressions. The gain from PFC-RR regression results primarily from the reduction in the number of real parameters need to specify β (Cook, 2018, Sec. 9.2.1). On the other hand, the gain from a response envelope is due to the reduction in the number of parameters and to the structure of $\Sigma_{X|C} = \Gamma\Omega\Gamma^T + \Gamma_0\Omega_0\Gamma_0^T$, with massive gains possible depending on the relationship between $\|\Omega\|$ and $\|\Omega_0\|$.

These contrasts lead to the conclusion that envelope and PFC-RR regressions (7.10) are distinctly different methods of dimension reduction with different operating characteristics. Reasoning in the equivalent context of reduced-rank regression, Cook et al. (2015) combined PFC-RR and response envelopes, leading to a new dimension reduction paradigm that can automatically choose the better of the two methods and, if appropriate, can also give an estimator that does better than both of them.

When formulating envelope model (7.13) no explicit accommodation was included for the dimension d of $\mathcal{B}'$. The PFC-RR -envelope model includes such an accommodation by starting with model (7.10) and then incorporating a semi-orthogonal basis matrix $\Gamma \in \mathbb{R}^{r\times u}$ for $\mathcal{E}_{\Sigma_{X|C}}(\text{span}(B))$:

$$\begin{aligned} X_i &= \mu_0 + \Gamma\eta b Y_i + \varepsilon_i,\ i = 1,\ldots,n, \\ \Sigma_{X|C} &= \Gamma\Omega\Gamma^T + \Gamma_0\Omega_0\Gamma_0^T, \end{aligned} \tag{7.20}$$

where $\beta^T = \Gamma\eta b$, $\eta \in \mathbb{R}^{u\times d}$, $b \in \mathbb{R}^{d\times r}$ as defined for (7.10) and the remaining parameters are as defined in (7.13). This model contains two tuning dimension, u and d, that need to be determined subject to the constraints $0 \le d \le u \le \min(p, r)$. Maximum likelihood estimators of the parameters in this model as well as suggestions for determining the tuning parameters are available from Cook et al. (2015). These estimators can then be substituted into classification function (7.7).

7.5.2 Bringing in PLS

Cook et al. (2015) showed that, in large samples, PFC-RR envelope estimation converges to a two-stage procedure with the first stage based on envelope model (7.13) and the second based on a PFC-RR regression model starting from envelope-reduced model (7.14). Recognizing the advantages of using PLS instead of maximum likelihood estimation in the first stage when n is not large relative to $\min(p, r)$, leads to the following outline of the possibilities.

First, neglect d and estimate the envelope basis $\Gamma \in \mathbb{R}^{p\times u}$ using either

(i) maximum likelihood for envelope model (7.13) if $n \gg \min(p, r)$, or

(ii) PLS from the regression of Y on X, as discussed at the end of Section 7.3.2.

Standard methods based on cross validation or a holdout sample can be used to determine its dimension u. Let $\widehat{\Gamma}$ denote the estimated basis from either PLS or MLE. Selecting from the list following (7.16), the estimators needed from this fit are $\widehat{\Gamma}$, $\widehat{\mu}_0$ and $\widehat{\Sigma}_{X|C}$.

Second, substitute $\widehat{\Gamma}$ into model (7.14) and then use that as the basis for a PFC-RR regression. This corresponds to fitting the working RR model

$$\begin{aligned} \widehat{\Gamma}^T X_i &= \alpha_0 + GbY_i + e_i,\ i = 1, \ldots, n, \\ \Sigma_{\widehat{\Gamma}^T X|C} &= \Omega, \end{aligned}$$

where $G \in \mathbb{R}^{u\times d}$, $b \in \mathbb{R}^{d\times r}$ and, in reference to model (7.14), $\eta = Gb$. Let $\tilde{\alpha}_0$, $\tilde{G}$, $\tilde{b}$, and $\tilde{\Omega}$ denote the estimators coming from the reduced rank fit of this working model. Then the corresponding estimator of η is $\tilde{\eta} = \tilde{G}\tilde{b}$. The dimension d can be determined by cross validation or a holdout sample based on this working RR model.

Recalling that in the population $\alpha_0 = \Gamma^T \mu_0$, the estimated classification function is then found by substituting $\tilde{\alpha}_0$, $\tilde{\eta}$, and $\tilde{\Omega}$ into (7.15), giving an estimated classification function based on the compressed features $\widehat{\Gamma}^T X$:

$$\phi(\widehat{\Gamma}^T X) = \arg\max_{k\in\chi} \left\{\log(\pi_k/\pi_0) + \tilde{\eta}_k^T \tilde{\Omega}^{-1}\{\widehat{\Gamma}^T X - (\tilde{\eta}_k + 2\tilde{\alpha}_0)/2\}\right\}. \tag{7.21}$$

7.6 Overview of LDA methods

We have now considered seven procedures for estimating the discriminant function ϕ. These are based on the five models listed below along with the number of real parameters in each:

1. Standard model (7.8): $p + pr + p(p+1)/2$

2. PFC-RR model (7.10): $p + d(p - d) + rd + p(p + 1)/2$
3. Isotrophic PFC model, Sec. 7.3.1: $p + d(p - d) + rd + 1$
4. MLE based on envelope model (7.13): $p + ru + p(p + 1)/2$
5. PLS based on envelope model (7.13): $p + ru + p(p + 1)/2$
6. MLE based on envelope PFC-RR model (7.20): $p+d(u-d)+rd+p(p+1)/2$
7. PLS based on envelope PFC-RR model (7.20): $p+d(u-d)+rd+p(p+1)/2$

Our personal preference would be to use envelope methods 4 and 6 if $n \gg \min(p, r)$ since these methods subsume the standard and PFC-RR models 1 and 2. We would use the isotropic PFC model when there is good reason to set $\Sigma_{X|C} = \sigma^2 I_p$. The PLS methods 5 and 7 are at their best when n is not large relative to $\min(p, r)$, perhaps $n < \min(p, r)$. In all cases we tend to prefer cross validation based on classification rates for dimension determination, although information criteria are available for the likelihood-based methods. We would normally use a holdout sample to estimate the rate of misclassification for the selected method. Finally, recalling the discussion of Chapter 4, the envelope and PLS methods discussed here should work well in abundant discrimination problems where many features contribute information on the classification. They are not recommended in sparse problems where p is large and yet few features inform on classification.

7.7 Illustrations

The likelihood-based classification methods listed as items 2, 4, and 6 in Section 7.6 will eventually dominate for a sufficiently large sample, since the methods inherit optimality properties from general likelihood theory. Our primary goal for this section is to provide some intuition into the methods that do not require a large sample size, principally isotropic PFC and PLS, methods 3, 5 and 7. We choose two data sets – Coffee data and Olive Oil data – from the literature because these data sets have been used in studies to compare classification methods for small samples. That enabled us to compare our methods with other methods without the need to implement them.

7.7.1 Coffee

Zheng, Fu, and Ying (2014), using data from Downey, Briandet, Wilson, and Kemsley (1997), compared five discrimination methods on their ability to predict one of two coffee varieties, Arabica and Robusta. The data consist of 29 Arabica and 27 Robusta species with corresponding Fourier transform infrared spectral features obtained by sampling at $p = 286$ wavelengths. The five methods they compared, including PLS discriminant analysis, are named in Table 7.1. Detailed descriptions of the methods are available from their article.

The last five entries in the second row of Table 7.1 gives the rates of correct classification, estimated by using leave-one-out cross validation, from Zheng et al. (2014, Table 1). The second and third entries in the second row are the rates based on leave-one-out cross validation that we observed by applying methods 3 (ISO) and 5 (PLS) listed in Section 7.6. Using 10-fold cross validation, we chose 3 components for PLS and 4 components for classification via isotropic PFC. For the fourth entry, PLS+PFC, we first applied PLS and then used PFC to further reduce the compressed PLS features. This resulted in 3 PLS components and one PFC component. The linear classification rule was then based on the PLS+PFC compressed feature. Our implementation of PLS did better than three of the five methods use by Zheng et al. (2014) and did the same as two of their methods at 100% accuracy.

These data are sensitive to particular partition used to conduct the 10-fold cross validation. Depending on the seed to start the pseudo-random sampling,

TABLE 7.1
Olive oil and Coffee data: Estimates of the correct classification rates (%) from leave one out cross validation.

				From Zheng et al. (2014)				
Dataset	PLS	ISO	PLS+PFC	KNN	LS-SVM	PLS-DA	BP-ANN	ELM
Coffee	100	91.1	100	82.2	97.5	100	94.8	100
Olive Oil	99.2	88.3	93.3	83.2	95.1	93.1	90.0	97.4

ISO and PLS refer to methods 3 and 5 as listed in Section 7.6. The remaining designations are those used by Zheng et al. (2014). KNN: k-nearest neighbor. LS-SVM: least-squares support vector machine. PLS-DA: partial least-squares discriminant analysis. BP-ANN: back propagation artificial neural network. ELM: extreme learning machine.

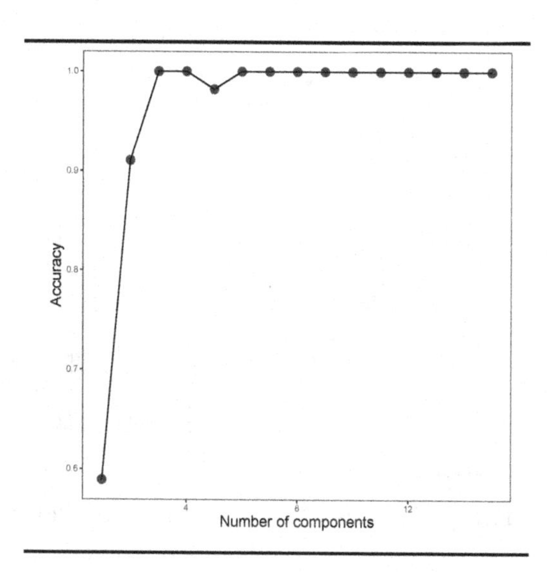

FIGURE 7.2
Coffee data: Plots of estimated classification rate (accuracy) versus the number of components

we estimated 3 or 5 components. With 3 components, the subsequent leave-one-out validation gave 100 percent accuracy, as reported in Table 7.1. With 5 component we observed one misclassification, resulting in a correct classification rate of $55/56 = 0.9821$. Shown in Figure 7.2 is a plot of the rate of correct classification versus the number of components. As long as one avoids $1, 2$ and 5 components, the percent of correct classification is 100.

The marked plot of the first two PLS components in Figure 7.3a shows good separation, although PLS required three components in total. The plot of the PLS+PFC direction in Figure 7.3b shows perfect separation. For contrast, Figure 7.3c shows a marked scatterplot of the first two ISO components.

As a benchmark, we randomly permuted the classification labels for the coffee data and re-ran our implementation of PLS. Again using 10-fold cross validation, we selected 4 component, which subsequently gave the leave-one-out classification rate of 48.21 arising from 27 correct classifications.

7.7.2 Olive oil

Zheng et al. (2014) employed also a spectral dataset from Tapp, Defernez, and Kemsley (2003) to see if their methods could distinguish olive oils by

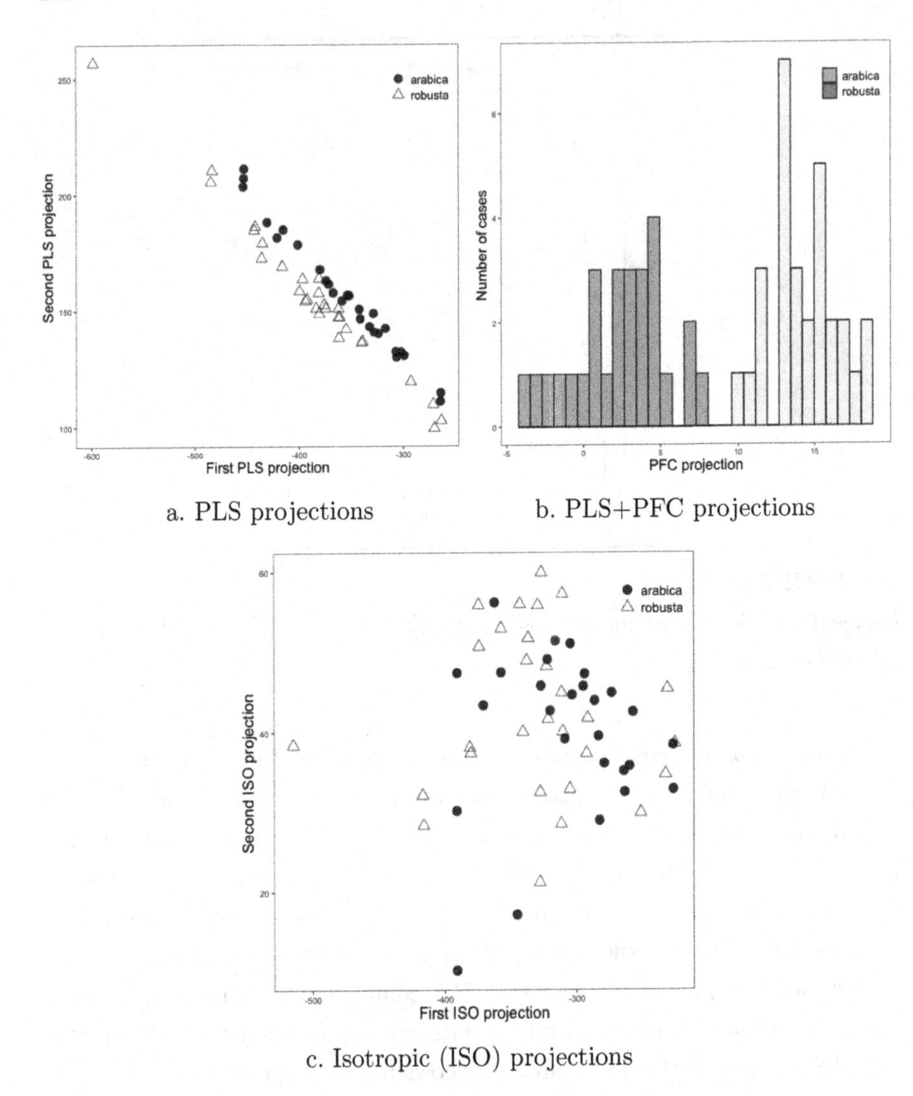

a. PLS projections

b. PLS+PFC projections

c. Isotropic (ISO) projections

FIGURE 7.3
Coffee data: Plots of PLS, PFC, and Isotropic projections.

their country of origin. There were 60 authenticated extra virgin olive oils from four countries: 10 from Greece, 17 from Italy, 8 from Portugal, and 25 from Spain. The $p = 570$ features were obtained from Fourier transform infrared spectroscopy of each of the 60 samples. The analyses of these data parallels those for the Coffee data in Section 7.7.1. The percentages of correct

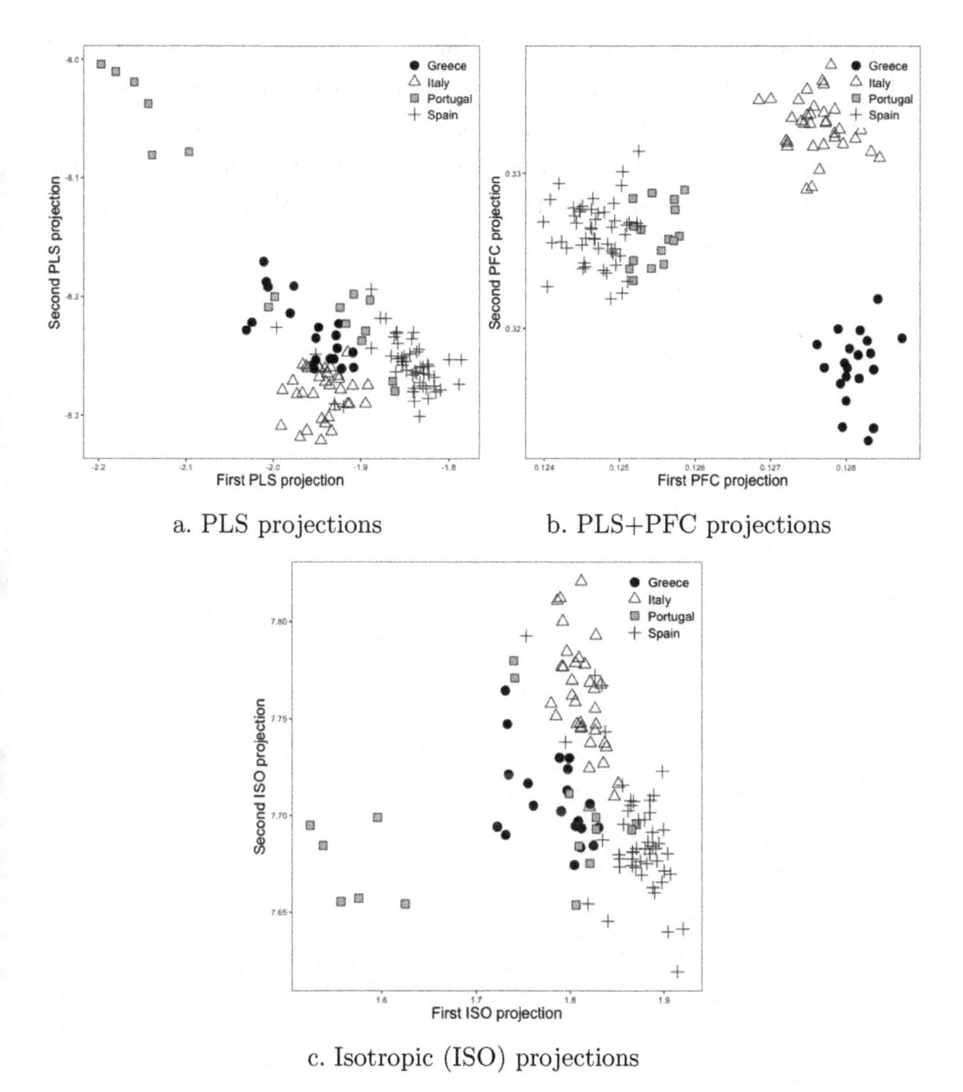

FIGURE 7.4
Olive oil data: Plots of PLS, PLS+PFC, and Isotropic projections.

classification are shown in the third row of Table 7.1 and the corresponding graphics are shown in Figure 7.4. Our implementation of PLS gave notably better results than the PLS-DA method of Zheng et al. (2014). We have no explanation for the difference.

Our implementation of 10-fold cross validation gave 28 components for PLS and 28 components for ISO. Marked plots of the first two PLS and ISO components are shown in panels a and c of Figures 7.4. The separation is not very clear, perhaps signaling the need for more components. Application of PLS+PFC resulted in 17 PLS components and 2 PFC component based on the PLS components. A marked plot of the two PLS components is shown in Figures 7.4b where we observe perfect separation of the four classes. One advantage of the PLS+PFC method is its ability to allow informative low dimensional plots, as illustrated here.

8

Quadratic Discriminant Analysis

Quadratic discriminant analysis (QDA) proceeds under the same model (7.5) as linear discriminant analysis, except that now the conditional covariance matrices are no longer assumed to be the same, so we may have $\Sigma_{X|C=k} \neq \Sigma_{X|C=j}$ for $k \neq j$. Let $\Sigma_k = \Sigma_{X|(C=k)}$. The nominal stochastic structure underlying QDA for predicting the class $C \in \chi$ based on a vector of continuous features $X \in \mathbb{R}^p$ is then

$$X \mid (C = k) \sim N(\mu_k, \Sigma_k), \; \Pr(C = k) = \pi_k, \; k \in \chi. \tag{8.1}$$

As in linear discriminant analysis, $\chi = \{0, 1, \ldots, r\}$, $\pi_k > 0$ and $\sum_{k=0}^{r} \pi_k = 1$, but here the intra-class covariance matrices Σ_k are permitted to be unequal over the classes. Under this model, the Bayes' rule is to classify a new unit with feature vector X onto the class with maximum posterior probability. From (7.4) the classification function is

$$\begin{aligned}
\phi_{\mathrm{qda}}(X) &= \arg\max_{k\in\chi} \Pr(C = k \mid X) \\
&= \arg\max_{k\in\chi} \left\{\log(\pi_k) + \tfrac{1}{2}\log|\Sigma_k| + \tfrac{1}{2}(X - \mu_k)^T \Sigma_k^{-1}(X - \mu_k)\right\} &(8.2)\\
&= \arg\max_{k\in\chi} \left\{a_k - X^T(\Sigma_k^{-1}\mu_k - \Sigma_0^{-1}\mu_0) + \tfrac{1}{2}X^T(\Sigma_k^{-1} - \Sigma_0^{-1})X\right\} \\
& &(8.3)\\
&= \arg\max_{k\in\chi} \Big\{a_k - X^T\left[(\Sigma_k^{-1} - \overline{\Sigma}^{-1})\mu_k + \overline{\Sigma}^{-1}(\mu_k - \overline{\mu})\right] \\
&\qquad + \tfrac{1}{2}X^T(\Sigma_k^{-1} - \overline{\Sigma}^{-1})X\Big\}, &(8.4)
\end{aligned}$$

DOI: 10.1201/9781003482475-8

where

$$\begin{aligned} a_k &= \log(\pi_k) + (1/2)\log|\Sigma_k| + (1/2)\mu_k^T\Sigma_k^{-1}\mu_k \\ \overline{\Sigma} &= \sum_{k\in\chi}\pi_k\Sigma_k = \mathrm{E}(\mathrm{var}(X \mid C)) \\ \overline{\mu} &= \sum_{k\in\chi}\pi_k\mu_k. \end{aligned}$$

Each of these three forms (8.2)–(8.4) may be useful depending on context. Form (8.2) arises directly from inspection of the posterior probability $\Pr(C = k \mid X)$. An expanded version of this form was used by Pardoe, Yin, and Cook (2007) in a study of graphical methods for quadratic discrimination. In form (8.3), which was used by Zhang and Mai (2019) in their study of quadratic discrimination, the terms are shifted by using the first category (μ_0, Σ_0) as a reference point. The average variance and average mean are used for centering in (8.4). Similar forms arise in the likelihood-based sufficient dimension reduction methodology developed by Cook and Forzani (2009). In developing (8.2)–(8.4), addends not depending on k were removed.

The maximum posterior probability for LDA contains only linear terms in X and the central discriminant subspace (Definition 7.1) $\mathcal{D}_{C|X} = \Sigma_{X|C}^{-1}\mathcal{B}'$, while $\phi_{\text{qda}}(X)$ contains both linear and quadratic terms in X. If the intraclass covariance matrices are equal, then $\Sigma_k = \overline{\Sigma} = \Sigma_{X|C}$, where $\Sigma_{X|C}$ is the notation used on Chapter 7 for the common intra-class covariance matrix. Using $\overline{\Sigma}$ to represent the common covariance matrix, the three forms (8.2)–(8.4) reduce straightforwardly to a form that is equivalent to (7.6):

$$\begin{aligned} \phi_{\text{qda}}(X) &= \arg\max_{k\in\chi}\left\{\log(\pi_k) + (1/2)(X-\mu_k)^T\overline{\Sigma}^{-1}(X-\mu_k)\right\} \\ &= \arg\max_{k\in\chi}\left\{a_k - X^T\overline{\Sigma}^{-1}(\mu_k-\mu_0)\right\} \\ &= \arg\max_{k\in\chi}\left\{a_k - X^T\overline{\Sigma}^{-1}(\mu_k-\overline{\mu})\right\}, \end{aligned}$$

where

$$\begin{aligned} a_k &= \log(\pi_k) + (1/2)\mu_k^T\overline{\Sigma}^{-1}\mu_k \\ \overline{\Sigma} &= \Sigma_{X|C}. \end{aligned}$$

If the intraclass means are equal, $\mu_k = \overline{\mu}$ where $\overline{\mu}$ represents the common mean, then the three forms (8.2)–(8.4) of the classification function reduce to

$$\begin{aligned}
\phi_{\text{qda}}(X) &= \arg\max_{k \in \chi} \left\{ \log(\pi_k) + \frac{1}{2}\log|\Sigma_k| + \frac{1}{2}(X-\overline{\mu})^T\Sigma_k^{-1}(X-\overline{\mu}) \right\} \\
&= \arg\max_{k \in \chi} \left\{ a_k - X^T(\Sigma_k^{-1} - \Sigma_0^{-1})\overline{\mu} + \frac{1}{2}X^T(\Sigma_k^{-1} - \Sigma_0^{-1})X \right\} \\
&= \arg\max_{k \in \chi} \left\{ a_k - X^T\left[(\Sigma_k^{-1} - \overline{\Sigma}^{-1})\overline{\mu}\right] + \frac{1}{2}X^T(\Sigma_k^{-1} - \overline{\Sigma}^{-1})X \right\},
\end{aligned}$$

where

$$\begin{aligned}
a_k &= \log(\pi_k) + (1/2)\log|\Sigma_k| + (1/2)\overline{\mu}^T\Sigma_k^{-1}\overline{\mu} \\
\overline{\Sigma} &= \sum_{k \in \chi} \pi_k \Sigma_k.
\end{aligned}$$

With sufficient observations per class, $\phi_{\text{qda}}(X)$ can be estimated consistently by substituting sample versions of Σ_k and μ_k, $k = 0, \ldots, r$. However, as with linear discriminant analysis, we strive to reduce the dimension of the feature vectors without loss of information and thereby reduce the rate of mis-classifications. It may be clear from the above forms that the feature vectors furnish classification information through the scaled mean deviations $\overline{\Sigma}^{-1}(\mu_k - \overline{\mu})$ and precision deviations $\Sigma_k^{-1} - \overline{\Sigma}^{-1}$. These deviations will play a key role when pursuing dimension reduction for quadratic discriminant analysis.

Beginning with a simple random sample (C_i, X_i), $i = 1, \ldots, n$, let n_k denote the number of observations in class $C = k$ so the total sample size can be represented as $n = \sum_{k=0}^{r} n_k$. In the remainder of this section, we frequently take $\pi_k = n_k/n$ for use in application as will often be appropriate. This also facilitates presentation, particularly connections with other methods.

8.1 Dimension reduction for QDA

Zhang and Mai (2019) showed that for model (8.1) the central discriminant subspace is again equal to the central subspace, $\mathcal{D}_{C|X} = \mathcal{S}_{C|X}$. Consequently we can pursue dimension reduction via the central subspace, which was introduced in Section 2.1. In their development of likelihood-based sufficient dimension reduction, Cook and Forzani (Likelihood acquired directions (LAD);

2009, Thm 1 and Prop 1) proved the key result shown in Proposition 8.1. In preparation, define

$$\begin{aligned}\beta_k &= \mu_k - \overline{\mu},\ k \in \chi \\ \beta^T &= (\beta_0, \beta_1, \ldots, \beta_r) \\ \mathcal{B}' &= \operatorname{span}(\beta^T).\end{aligned}$$

This definition of β_k differs from that used for linear discriminant analysis (see just below (7.7)). Here $\overline{\mu}$ is used for centering rather than μ_0. Recall also that, from Definition 2.1, a subspace $\mathcal{S} \subseteq \mathbb{R}^p$ with the property $C \perp\!\!\!\perp X \mid P_{\mathcal{S}}X$ is called a dimension reduction subspaces for the regression of C on X and that the central subspace $\mathcal{S}_{C|X}$ is the intersection of all dimension reduction subspaces.

Proposition 8.1. *Assume model (8.1), let $\mathcal{S}$ be a subspace of $\mathbb{R}^p$ with semi-orthogonal basis matrix Ψ and let $d = \dim(\mathcal{S})$. Then $\mathcal{S}$ is a dimension reduction subspace for the regression of C on X if and only if the following two conditions are satisfied*

(i) $\overline{\Sigma}^{-1}\mathcal{B}' \subseteq \mathcal{S}$

(ii) $\Sigma_k^{-1} = \overline{\Sigma}^{-1} + \Psi\Delta_k(\Psi)\Psi^T$ *for all* $k \in \chi$,

where

$$\begin{aligned}\Delta_k(\Psi) &= (\Psi^T\Sigma_k\Psi)^{-1} - (\Psi^T\overline{\Sigma}\Psi)^{-1} \\ &= \{\operatorname{var}(\Psi^TX \mid C = k)\}^{-1} - \{\operatorname{ave}_{j\in\chi}\operatorname{var}(\Psi^TX \mid C = j)\}^{-1}.\end{aligned}$$

Additionally, condition (ii) is equivalent to condition

(iii) $\Sigma_k = \overline{\Sigma} + P^T_{\Psi(\overline{\Sigma})}(\Sigma_k - \overline{\Sigma})P_{\Psi(\overline{\Sigma})}$ *or all* $k \in \chi$.

In Proposition 8.1 the dimension reduction subspace $\mathcal{S}$ need not be the smallest; that is, it need not be a basis for the central subspace. Later in this chapter we will be using this proposition for applications in which $\mathcal{S}$ is not necessarily minimal.

Condition (i) of Proposition 8.1 tells us that $\overline{\Sigma}^{-1}(\mu_k - \overline{\mu}) \in \mathcal{S}$ for all $k \in \chi$. Consequently, for each k there is a $b_k \in \mathbb{R}^d$ so that $\overline{\Sigma}^{-1}(\mu_k - \overline{\mu}) = \Psi b_k$. Substituting this and condition (ii) of Proposition 8.1 into $\phi_{\text{qda}}(X)$ given in (8.4) we obtain a reduced form of the quadratic discrepancy function. This reduced discrepancy function can be seen to arise by reducing $X \mapsto \Psi^T X \in \mathbb{R}^d$ and then rewriting (8.4) in terms of the moments of $\Psi^T X$:

$$\begin{aligned} \mathrm{E}(\Psi^T X \mid C = k) &= \Psi^T \mu_k \\ \mathrm{var}(\Psi^T X \mid C = k) &= \Psi^T \Sigma_k \Psi \\ \mathrm{E}\{\mathrm{var}(\Psi^T X \mid C)\} &= \Psi^T \overline{\Sigma} \Psi. \end{aligned}$$

Using these moments in conjunction with (8.4) gives the reduced classification function

$$\phi_{\text{qda}}(\Psi^T X) = \arg\max_{k \in \chi} \left\{a_k - L_k(\Psi^T X) + Q_k(\Psi^T X)\right\}, \tag{8.5}$$

where

$$a_k = \log(\pi_k) + (1/2)\log|\Psi^T \Sigma_k \Psi| + (1/2)\mu_k^T \Psi (\Psi^T \Sigma_k \Psi)^{-1} \Psi^T \mu_k,$$

the linear terms are represented by

$$\begin{aligned} L_k(\Psi^T X) &= X^T \Psi \\ &\quad \left[\left\{(\Psi^T \Sigma_k \Psi)^{-1} - (\Psi^T \overline{\Sigma} \Psi)^{-1}\right\} \Psi^T \mu_k + (\Psi^T \overline{\Sigma} \Psi)^{-1} \Psi^T (\mu_k - \overline{\mu})\right], \end{aligned}$$

the quadratic terms by

$$Q_k(\Psi^T X) = (1/2) X^T \Psi \left\{(\Psi^T \Sigma_k \Psi)^{-1} - (\Psi^T \overline{\Sigma} \Psi)^{-1}\right\} \Psi^T X,$$

and $\overline{\Sigma}$ and $\overline{\mu}$ are as defined previously. This classification function is computed in the same way as (8.4), except it is based on d features $\Psi^T X$ instead of the original p features X. For this to be useful as a basis for a method for QDA, we need to specify a dimension reduction subspace $\mathcal{S}$ and have available a method of estimating a basis. Once a basis estimator is available, a sample version of $\phi_{\text{qda}}(\Psi^T X)$ can be constructed by substituting $\Psi^T X$ and its sample moments into (8.5).

We discuss basis estimation methods in next sections. The first is based on taking $\mathcal{S}$ to be the central subspace $\mathcal{S} = \mathcal{S}_{C|X}$ and estimating a basis by maximum likelihood under model (8.1). The second is based on taking $\mathcal{S}$ to be the $\overline{\Sigma}$-envelope of $\mathcal{S}_{C|X}$, $\mathcal{S} = \mathcal{E}_{\overline{\Sigma}}(\mathcal{S}_{C|X})$, and estimating a basis by using maximum likelihood or partial least squares. Choosing $\mathcal{S} = \mathcal{S}_{C|X}$ with maximum likelihood estimation is reasonable when the covariance matrices Σ_k are well-estimated by their sample versions, so $n_k \gg p$, $k \in \chi$, and intraclass collinearity is negligible to moderate, since likelihood estimation tends to degrade generally if either if these conditions is not met. Choosing $\mathcal{S} = \mathcal{E}_{\overline{\Sigma}}(\mathcal{S}_{C|X})$ with maximum likelihood estimation is advisable when the covariance matrices Σ_k are well-estimated by their sample versions and intraclass collinearity is high in some classes. Choosing $\mathcal{S} = \mathcal{E}_{\overline{\Sigma}}(\mathcal{S}_{C|X})$ with partial least squares estimation should be considered when for some groups Σ_k is not well-estimated or the intraclass collinearity is moderate to high.

Recall that we can represent the marginal covariance matrix of the features X as

$$\begin{aligned} \Sigma_X &= \mathrm{E}(\mathrm{var}(X \mid C)) + \mathrm{var}(\mathrm{E}(X \mid C)) \\ &= \overline{\Sigma} + \sum_{k \in \chi} \pi_k (\mu_k - \overline{\mu})(\mu_k - \overline{\mu})^T. \end{aligned}$$

The Σ_X-envelope of $\mathcal{S}_{C|X}$ is equal to the $\overline{\Sigma}$-envelope of $\mathcal{S}_{C|X}$, so either could be used for an envelope version of QDA. We chose $\mathcal{E}_{\overline{\Sigma}}(\mathcal{S}_{C|X})$ because it fits well with past work (e.g. Cook and Forzani, 2009).

8.2 Dimension reduction with $\mathcal{S} = \mathcal{S}_{C|X}$

The subspace

$$\mathcal{L} \;:=\; \overline{\Sigma}^{-1}\mathcal{B}' = \mathrm{span}\{\overline{\Sigma}^{-1}(\mu_k - \mu_0) \mid k \in \chi\} = \mathrm{span}\{\overline{\Sigma}^{-1}(\mu_k - \overline{\mu}) \mid k \in \chi\},$$

which is the focus of Proposition 8.1(i), captures the classification information available from the class means. The subspace

$$\mathcal{Q} \;:=\; \mathrm{span}\{\Sigma_k^{-1} - \Sigma_0^{-1} \mid k \in \chi\} = \mathrm{span}\{\Sigma_k^{-1} - \overline{\Sigma}^{-1} \mid k \in \chi\},$$

which is the subject of Proposition 8.1(ii), captures the classification information available from the class variances. The following corollary restates the conditions of Proposition 8.1 in terms of the subspaces $\mathcal{L}$ and $\mathcal{Q}$ and adds the conclusion that $\mathcal{S}_{C|X} = \mathcal{L} + \mathcal{Q}$.

Corollary 8.1. *Assume model (8.1) and let $\mathcal{S}$ be a subspace of $\mathbb{R}^p$. Then $\mathcal{S}$ is a dimension reduction subspace if and only if the following two conditions are satisfied* *(i)* $\mathcal{L} \subseteq \mathcal{S}$

(ii) $\mathcal{Q} \subseteq \mathcal{S}$.

In addition,

(iii) $\mathcal{S}_{C|X} = \mathcal{D}_{C|X} = \mathcal{L} + \mathcal{Q}$.

Conditions (i) and (ii) are restatements of conditions (i) and (ii) in Proposition 8.1. The conclusion that $\mathcal{S}_{C|X} = \mathcal{L} + \mathcal{Q}$ follows because $\mathcal{L} + \mathcal{Q}$ is a dimension reduction subspace that is contained in all dimension reduction subspaces. The conclusion that $\mathcal{D}_{C|X} = \mathcal{S}_{C|X}$ was demonstrated by Zhang and Mai (2019).

Cook and Forzani (2009, Thm 2) showed that the maximum likelihood estimator $\widehat{\Phi}$ of a basis Φ for $\mathcal{S}_{C|X}$ can be constructed as follows. Let S_X and S_k denote the sample versions of Σ_X, the marginal covariance matrix of X, and Σ_k, the covariance matrix of X restricted to class $C = k$. Also let $\overline{S} = \sum_{k=0}^{r}(n_k/n)S_k$ denote the sample version of the average covariance matrix $\overline{\Sigma}$. Then under model (8.1) with fixed dimension $d = \dim(\mathcal{S}_{C|X})$ and $S_k > 0$ for all $k \in \chi$, the maximum likelihood estimator of a basis matrix Φ for $\mathcal{S}_{C|X}$ is

$$\widehat{\Phi} = \arg\max_{H} \ell_d(H),$$

where the maximization is over all semi-orthogonal matrices $H \in \mathbb{R}^{p\times d}$ and $\ell_d(H)$ is the log likelihood function maximized over all parameters except Φ (Cook and Forzani, 2008a),

$$\begin{aligned} \ell_d(H) &= C + (n/2)\log|H^T S_X H| - (n/2)\sum_{k=0}^{r}(n_k/n)\log|H^T S_k H| \\ C &= -(np/2)(1+\log(2\pi)) - (n/2)\log|S_X|. \end{aligned}$$

Once $\widehat{\Phi}$ is available, the sample version of the reduced quadratic classification function is constructed as

$$\phi_{\text{qda}}(\widehat{\Phi}^T X) = \arg\max_{k\in\chi}\left\{a_k - L_k(\widehat{\Phi}^T X) + Q_k(\widehat{\Phi}^T X)\right\}, \tag{8.6}$$

where

$$a_k = \log(\pi_k) + (1/2)\log|\widehat{\Phi}^T S_k \widehat{\Phi}| + (1/2)\bar{X}_k^T \widehat{\Phi}(\widehat{\Phi}^T S_k \widehat{\Phi})^{-1}\widehat{\Phi}^T \bar{X}_k,$$

and the linear terms are represented by

$$L_k(\widehat{\Phi}^T X) = X^T\widehat{\Phi}\left[\left\{(\widehat{\Phi}^T S_k\widehat{\Phi})^{-1} - (\widehat{\Phi}^T\overline{S}\widehat{\Phi})^{-1}\right\}\widehat{\Phi}^T\bar{X}_k + (\widehat{\Phi}^T\overline{S}\widehat{\Phi})^{-1}\widehat{\Phi}^T(\bar{X}_k - \bar{X})\right].$$

The quadratic terms in (8.6) are

$$Q_k(\widehat{\Phi}^T X) = (1/2) X^T \widehat{\Phi} \left\{ (\widehat{\Phi}^T S_k \widehat{\Phi})^{-1} - (\widehat{\Phi}^T \overline{S} \widehat{\Phi})^{-1} \right\} \widehat{\Phi}^T X,$$

where $\bar{X}_k$ is the average feature vector in class $C = k$ and $\bar{X}$ is the overall average. This reduced classification function may be reasonable when the intraclass sample sizes are relatively large $n_k \gg p$ for all $k \in \chi$ and intraclass collinearity is negligible to moderate.

8.3 Dimension reduction with $\mathcal{S} = \mathcal{E}_{\overline{\Sigma}}(\mathcal{S}_{C|X})$

We begin this section by using envelopes to enhance the structure of model (8.1) .

8.3.1 Envelope structure of model (8.1)

Let Γ be a semi-orthogonal basis matrix for $\mathcal{E}_{\overline{\Sigma}}(\mathcal{S}_{C|X})$, the $\overline{\Sigma}$-envelope of $\mathcal{S}_{C|X}$ with dimension u. The population classification function is then $\phi_{\text{qda}}(\Gamma^T X)$, so only $\Gamma^T X$ is material to classification. Select $\Gamma_0 \in \mathbb{R}^{p \times p-u}$ so that (Γ, Γ_0) is an orthogonal matrix. It follows from Proposition 1.2 and Corollary 1.1 that,

$$\begin{aligned} \overline{\Sigma} &= \Gamma \Omega \Gamma^T + \Gamma_0 \Omega_0 \Gamma_0^T \\ \overline{\Sigma}^{-1} &= \Gamma \Omega^{-1} \Gamma^T + \Gamma_0 \Omega_0^{-1} \Gamma_0^T. \end{aligned} \tag{8.7}$$

We see from this that one role of the envelope is to separate $\overline{\Sigma}$ into a part $\Gamma\Omega\Gamma^T$ that is material to classification and a complementary part $\Gamma_0\Omega_0\Gamma_0^T$ that is immaterial. Since $\mathcal{S}_{C|X} \subseteq \mathcal{E}_{\overline{\Sigma}}(\mathcal{S}_{C|X})$, the envelope space $\mathcal{E}_{\overline{\Sigma}}(\mathcal{S}_{C|X})$ is a dimension reduction subspace for the regression of C on X and, from Corollary 8.1, $\mathcal{L} \subseteq \mathcal{E}_{\overline{\Sigma}}(\mathcal{S}_{C|X})$ and $\mathcal{Q} \subseteq \mathcal{E}_{\overline{\Sigma}}(\mathcal{S}_{C|X})$. The following corollary to Proposition 8.1 describes key relationships that will guide the methodology in this section.

Corollary 8.2. *Assume the hypotheses in Proposition 8.1. Then*

(I) $\mathcal{B}' \subseteq \overline{\Sigma}\mathcal{S}_{C|X} \subseteq \mathcal{E}_{\overline{\Sigma}}(\overline{\Sigma}\mathcal{S}_{C|X}) = \mathcal{E}_{\overline{\Sigma}}(\mathcal{S}_{C|X})$,

(II) $\text{span}(\overline{\Sigma} - \Sigma_k) \subseteq \overline{\Sigma}\mathcal{S}_{C|X} \subseteq \mathcal{E}_{\overline{\Sigma}}(\overline{\Sigma}\mathcal{S}_{C|X}) = \mathcal{E}_{\overline{\Sigma}}(\mathcal{S}_{C|X})$,

(III) $\text{span}(\overline{\Sigma}^{-1} - \Sigma_k^{-1}) \subseteq \mathcal{E}_{\overline{\Sigma}}(\overline{\Sigma}\mathcal{S}_{C|X}) = \mathcal{E}_{\overline{\Sigma}}(\mathcal{S}_{C|X})$.

Proof. Since $\mathcal{S}_{C|X}$ is a dimension reduction subspace by construction and since $\mathcal{S}_{C|X} \subseteq \mathcal{E}_{\overline{\Sigma}}(\mathcal{S}_{C|X})$, it follows that $\mathcal{E}_{\overline{\Sigma}}(\mathcal{S}_{C|X})$ is also a dimension reduction subspace.

To show conclusion (I), we have that Proposition 8.1(i) implies that

$$\mathcal{B}' \subseteq \overline{\Sigma}\mathcal{S}_{C|X} \subseteq \overline{\Sigma}\,\mathcal{E}_{\overline{\Sigma}}(\mathcal{S}_{C|X}).$$

Since $\overline{\Sigma}$ trivially commutes with itself, Proposition 1.5 implies that

$$\overline{\Sigma}\,\mathcal{E}_{\overline{\Sigma}}(\mathcal{S}_{C|X}) = \mathcal{E}_{\overline{\Sigma}}(\overline{\Sigma}\,\mathcal{S}_{C|X}) = \mathcal{E}_{\overline{\Sigma}}(\mathcal{S}_{C|X}),$$

which implies conclusion (I).

Turning to conclusion (II), Proposition 8.1(iii) implies that for each $k \in \chi$

$$\operatorname{span}(\overline{\Sigma} - \Sigma_k) \subseteq \overline{\Sigma}\mathcal{S}_{C|X} \subseteq \overline{\Sigma}\,\mathcal{E}_{\overline{\Sigma}}(\mathcal{S}_{C|X}).$$

Conclusion (II) now follows by using the argument in the justification of conclusion (I).

Conclusion (III) follows immediately from Proposition 8.1(ii). □

To describe the structure of Σ_X, we have from Corollary 8.2(I) that there is a positive definite matrix U so that

$$\operatorname{var}(\mathrm{E}(X \mid C)) = \sum_{k=0}^{r} \pi_k(\mu_k - \overline{\mu})(\mu_k - \overline{\mu})^T = \sum_{k=0}^{r} \pi_k \beta_k \beta_k^T = \Gamma U \Gamma^T, \tag{8.8}$$

where Γ is a semi-orthogonal basis matrix for $\mathcal{E}_{\overline{\Sigma}}(\mathcal{S}_{C|X})$ as defined at the outset of this section. Using this result and (8.7) we have

$$\begin{aligned} \Sigma_X &= \mathrm{E}(\operatorname{var}(X \mid C)) + \operatorname{var}(\mathrm{E}(X \mid C)) && (8.9)\\ &= \overline{\Sigma} + \Gamma U \Gamma^T \\ &= \Gamma(\Omega + U)\Gamma^T + \Gamma_0 \Omega_0 \Gamma_0^T. && (8.10) \end{aligned}$$

We next discuss two ways in which this structure can be used to estimate the corresponding reduced classification function $\phi_{\text{qda}}(\Gamma^T X)$. The first is likelihood-based, requiring relatively large sample settings where $n_k \gg p$ and so $S_k > 0$ for all $k \in \chi$. The second is when n_k is not large relative to p for some k. This is where PLS comes into play since then the likelihood-based classification function may be unserviceable or not sufficiently reliable.

8.3.2 Likelihood-based envelope estimation

Su and Cook (2013, eq. (2.3)) and later Zhang and Mai (2019) showed that, under the QDA model (8.1) with $S_k > 0$ for all $k \in \chi$, a maximum likelihood estimator of a basis Γ for $\mathcal{E}_{\overline{\Sigma}}(\mathcal{S}_{C|X})$ is

$$\widehat{\Gamma} = \arg\min_{G} \left\{ \log |G^T S_X^{-1} G| + \sum_{k=0}^{r} (n_k/n) \log |G^T S_k G| \right\}. \tag{8.11}$$

This objective function is similar in structure to $\ell_d(H)$ for estimating $\mathcal{S}_{C|X}$, as described in Section 8.2. There are consequential differences, however, because $\ell_d(H)$ is designed for estimation of the central discriminant subspace, while $\widehat{\Gamma}$ is an estimated basis for an upper bound $\mathcal{E}_{\overline{\Sigma}}(\mathcal{S}_{C|X})$ on that subspace. As discussed previously, the bound is intended to accommodate collinearity among the features.

The maximum likelihood estimators of other key parameters associated with model (8.1) are (Su and Cook, 2013)

$$\begin{aligned}
\widehat{\bar{\mu}} &= \bar{X} \\
\widehat{\mu}_k &= \bar{X} + P_{\widehat{\Gamma}}(\bar{X}_k - \bar{X}) \\
\widehat{\Sigma}_k &= P_{\widehat{\Gamma}} S_k P_{\widehat{\Gamma}} + Q_{\widehat{\Gamma}} S_X Q_{\widehat{\Gamma}} \\
\widehat{\overline{\Sigma}} &= P_{\widehat{\Gamma}} \overline{S} P_{\widehat{\Gamma}} + Q_{\widehat{\Gamma}} S_X Q_{\widehat{\Gamma}} \\
\widehat{\Sigma}_X &= P_{\widehat{\Gamma}} S_X P_{\widehat{\Gamma}} + Q_{\widehat{\Gamma}} S_X Q_{\widehat{\Gamma}},
\end{aligned}$$

where $\overline{S} = \sum_{k \in \chi} (n_k/n) S_k$. Substituting these estimators into (8.5) gives the classification function under the ENDS-QDA model of Zhang and Mai (2019):

$$\phi_{\text{qda}}(\widehat{\Gamma}^T X) = \arg\max_{k \in \chi} \left\{ a_k - L_k(\widehat{\Gamma}^T X) + Q_k(\widehat{\Gamma}^T X) \right\} \tag{8.12}$$

where

$$a_k = \log(\pi_k) + (1/2) \log |\widehat{\Gamma}^T S_k \widehat{\Gamma}| + (1/2) \bar{X}_k^T \widehat{\Gamma} (\widehat{\Gamma}^T S_k \widehat{\Gamma})^{-1} \widehat{\Gamma}^T \bar{X}_k,$$

the linear terms are represented by

$$L_k(\widehat{\Gamma}^T X) = X^T \widehat{\Gamma} \left[\left\{ (\widehat{\Gamma}^T S_k \widehat{\Gamma})^{-1} - (\widehat{\Gamma}^T \overline{S} \widehat{\Gamma})^{-1} \right\} \widehat{\Gamma}^T \bar{X}_k + (\widehat{\Gamma}^T \overline{S} \widehat{\Gamma})^{-1} \widehat{\Gamma}^T (\bar{X}_k - \bar{X}) \right],$$

the quadratic terms by

$$Q_k(\widehat{\Gamma}^T X) = (1/2) X^T \widehat{\Gamma} \left\{ (\widehat{\Gamma}^T S_k \widehat{\Gamma})^{-1} - (\widehat{\Gamma}^T \overline{S} \widehat{\Gamma})^{-1} \right\} \widehat{\Gamma}^T X,$$

where $\bar{X}_k$ is the average feature vector in class $C = k$ and $\bar{X}$ is the overall average. This reduced classification function may be advisable when the intraclass sample sizes are relatively large $n_k \gg p$ for all $k \in \chi$ and intraclass collinearity is high for some classes.

Rounding out the discussion, Su and Cook (2013) developed envelope estimation of multivariate means from populations with different covariance matrices. The basic model that they used to develop envelope estimation is the same as (8.1), although they were interested only in estimation of population means and not subsequent classification. They based their methodology on the following definition of a generalized envelope.

Definition 8.1. *Let* $\mathcal{M}$ *be a collection of real* $t \times t$ *symmetric matrices and let* $\mathcal{V} \in \mathrm{span}(M)$ *for all* $M \in \mathcal{M}$*. The* $\mathcal{M}$*-envelope of* $\mathcal{V}$*, denoted* $\mathcal{E}_{\mathcal{M}}(\mathcal{V})$*, is the intersection of all subspaces of* $\mathbb{R}^t$ *that contain* $\mathcal{V}$ *and reduce each member of* $\mathcal{M}$*.*

They applied this definition with $\mathcal{M} = \{\Sigma_k \mid k \in \chi\}$ and $\mathcal{V} = \mathcal{B}'$, and showed that the maximum likelihood estimator of a basis for $\mathcal{E}_{\mathcal{M}}(\mathcal{B}')$ is given by (8.11), suggesting that there is an intrinsic connection between the envelopes $\mathcal{E}_{\mathcal{M}}(\mathcal{B}')$ and $\mathcal{E}_{\overline{\Sigma}}(\mathcal{S}_{C|X})$.

8.3.3 Quadratic discriminant analysis via algorithms $\mathbb{N}$ and $\mathbb{S}$

When the intraclass sample sizes are relatively small, perhaps $n_k < p$ for some $k \in \chi$ or intraclass collinearity is high for some classes, it is wise to consider estimating $\mathcal{E}_{\overline{\Sigma}}(\mathcal{S}_{C|X})$ by using Algorithms $\mathbb{N}$ and $\mathbb{S}$ as introduced in Section 1.5 since these methods do not require nonsingular empirical class covariance matrices. In reference to those general algorithms, the first selections might be $M = \overline{\Sigma}$ and $\mathcal{A} = \mathcal{S}_{C|X}$. To use these choices in applications, we need to have an estimator of $\overline{\Sigma}$ and an estimator of (a basis for) $\mathcal{S}_{C|X}$. The average covariance matrix $\overline{\Sigma}$ can be estimated straightforwardly by using $\overline{S}$. However, carrying this ideas to fruition, we find that $\overline{S}$ needs to be nonsingular for the estimation of $\mathcal{S}_{C|X}$, sending us back to consider likelihood-based methods.

Corollary 8.2 tells us that $\overline{\Sigma}\,\mathcal{S}_{C|X} \subseteq \mathcal{E}_{\overline{\Sigma}}(\overline{\Sigma}\,\mathcal{S}_{C|X}) = \mathcal{E}_{\overline{\Sigma}}(\mathcal{S}_{C|X})$, which provides an alternative choice: $M = \overline{\Sigma}$ and $\mathcal{A} = \overline{\Sigma}\mathcal{S}_{C|X}$, which does not require $\overline{S}$ to be nonsingular when used as an estimator of $\overline{\Sigma}$. The next corollary describes the population foundation for an estimator of $\overline{\Sigma}\mathcal{S}_{C|X}$.

Corollary 8.3. *Assume the hypotheses in Proposition 8.1. Then*

$$\operatorname{span}\left\{\sum_{k\in\chi} \pi_k(\Sigma_X - \Sigma_k)^2\right\} = \overline{\Sigma}\,\mathcal{S}_{C|X}.$$

Proof. Let $\Pi = \operatorname{diag}(\pi_0, \pi_1, \ldots, \pi_p)$, and recall from (8.8) that

$$\operatorname{var}(\mathrm{E}(X \mid C)) = \sum_{k=0}^{r} \pi_k \beta_k \beta_k^T.$$

Then it follows from Corollary 8.2(I) that

$$\operatorname{span}\{\operatorname{var}(\mathrm{E}(X \mid C))\} = \operatorname{span}(\beta^T \Pi \beta) \subseteq \overline{\Sigma}\mathcal{S}_{C|X},$$

and from Corollary 8.2(II) that

$$\operatorname{span}(\overline{\Sigma} - \Sigma_k) \subseteq \overline{\Sigma}\mathcal{S}_{C|X}.$$

Combining these we have that

$$\operatorname{span}(\overline{\Sigma} - \Sigma_k + \beta^T \Pi \beta) \subseteq \overline{\Sigma}\mathcal{S}_{C|X}.$$

But using (8.9)

$$\begin{aligned} \overline{\Sigma} + \beta^T \Pi \beta - \Sigma_k &= \mathrm{E}(\operatorname{var}(X \mid C)) + \operatorname{var}(\mathrm{E}(X \mid C)) - \Sigma_k \\ &= \Sigma_X - \Sigma_k, \end{aligned}$$

which implies that $\operatorname{span}(\Sigma_X - \Sigma_k) \subseteq \overline{\Sigma}\mathcal{S}_{C|X}$ and thus that

$$\operatorname{span}\left\{\sum_{k\in\chi} \pi_k(\Sigma_X - \Sigma_k)^2\right\} \subseteq \overline{\Sigma}\,\mathcal{S}_{C|X}. \tag{8.13}$$

□

The left-hand side of (8.13) is the span of the kernel function for sliced average variance estimation (SAVE) as proposed by Cook and Weisberg (1991) and developed further by Cook (2000) and Shao, Cook, and Weisberg (2007, 2009). A comprehensive treatment is available from Li (2018, Ch. 5). Equality follows from Cook and Forzani (2009, Discussion of Prop. 3).

While normality guarantees equality in (8.13), containment holds under much weaker conditions. In particular, if $\operatorname{var}(X \mid P_{\mathcal{S}_{C|X}} X)$ is a non-random matrix and if the linearity condition as given in Definition 9.3 holds then the containment represented in (8.13) is assured. See Li (2018, Ch. 5) for further discussion.

Algorithms $\mathbb{N}$ and $\mathbb{S}$ can now be used in applications by setting

$$M = \overline{S} \text{ and } A = \sum_{k \in \chi} (n_k/n)(S_X - S_k)^2, \tag{8.14}$$

and using predictive cross validation to determine the dimension of the envelope.

The methodology implied by Corollary 8.3 implicitly treats the mean and variance components – $\beta^T \Pi \beta$ and $\overline{\Sigma} - \Sigma_k$ – of Corollary 8.2 equally. In some applications, it may be useful to differentially weight these components, particularly if they contribute unequally to classification. Let $0 \leq a \leq 1$ and combine the weighted components $(1-a)(\overline{\Sigma} - \Sigma_k)$ and $a\beta^T \pi \beta$ as

$$\sum_{k \in \chi} \pi_k \{(1-a)(\overline{\Sigma} - \Sigma_k) + a\beta^T \pi \beta\}^2 = (1-a)^2 \sum_{k \in \chi} \pi_k \left\{(\overline{\Sigma} - \Sigma_k)^2\right\} + a^2 (\beta^T \pi \beta)^2,$$

where the equality holds because the cross product terms sum to zero. Without loss of generality, we can rescale the right-hand side to give

$$\text{span}\left\{(1-\lambda) \sum_{k \in \chi} \pi_k (\overline{\Sigma} - \Sigma_k)^2 + \lambda (\beta^T \pi \beta)^2 \right\} \subseteq \overline{\Sigma} \mathcal{S}_{C|X},$$

where $\lambda = a^2 / \{(1-a)^2 + a^2\}$. This then implies that we use the sample version of Algorithm $\mathbb{N}$ or $\mathbb{S}$ with $M = \overline{S}$ and

$$A_\lambda = (1-\lambda) \sum_{k \in \chi} (n_k/n)(\overline{S} - S_k)^2 + \lambda (S_X - \overline{S})^2. \tag{8.15}$$

The coefficient λ and the number of PLS components can be determined by cross validation with the error rate of the final choices estimated from a hold-out sample. The resulting weight matrix is then substituted for $\widehat{\Gamma}$ in (8.12) to give the PLS classification function.

When $\lambda = 0$,

$$\text{span}\left\{\sum_{k \in \chi} \pi_k (\overline{\Sigma} - \Sigma_k)^2 \right\} \subseteq \overline{\Sigma} \mathcal{S}_{C|X},$$

which is closely related to the methods for studying covariance matrices proposed by Cook and Forzani (2008a).

8.4 Overview of QDA methods

All methods summarized begin with the QDA model (8.1) and then use various classification functions and estimation methods depending on the dimension reduction paradigm. The primary methods are listed below along with the number of parameters N and notes on application.

1. QDA: Classification function (8.4) with no dimension reduction:

$$N = (r+1)\{p + p(p+1)/2\}.$$

   This method is most appropriate when μ_k and Σ_k are well estimated, $k \in \chi$. Even in this setting, dimension reduction may still reduce the misclassification error materially. We illustrate this method in Section 8.5.

2. Classification function (8.6) based on reduction via the central subspace $\mathcal{S}_{C|X}$ of dimension d estimated using LAD (Cook and Forzani, 2009):

$$N = p + rd + p(p+1)/2 + d(p-d) + rd(d+1)/2.$$

   This method is appropriate when μ_k and Σ_k are well estimated, $k \in \chi$, and dimension reduction is effective. The dimension d of the central subspace can be estimated using likelihood testing, an information or cross validation (Cook and Forzani, 2009). Cross validation tends to be preferred as it is less dependent on underlying distribution requirements.

   In addition to LAD, a variety of other methods are available to estimate the central subspace. We use LAD, sliced inverse regression (SIR; Li, 1991) and SAVE (Cook and Weisberg, 1991; Cook, 2000) in Section 8.5 to illustrate classification via the central subspace.

3. Classification function (8.12) based on dimension reduction via the enveloped central subspace $\mathcal{S}_{C|X}$ of dimension u estimated by using the maximum likelihood function from Zhang and Mai (2019):

$$N = p + ru + p(p+1)/2 + ru(u+1)/2.$$

   This method may be useful when collinearity is present. Cross validation can be used to determine the dimension u of the envelope. It is illustrated in Section 8.5.

4. AN-Q1: Classification function (8.12) based on dimension reduction via the enveloped central subspace $\mathcal{S}_{C|X}$ of dimension u estimated by using PLS:

$$N = p + ru + p(p+1)/2 + ru(u+1)/2.$$

   The weight matrix U is obtained by using Algorithm $\mathbb{N}$ or $\mathbb{S}$ of Section 1.5 with $M = \overline{S}$ and $A = \sum_{k \in \chi}(n_k/n)(S_X - S_k)^2$, and using predictive cross validation to determine the dimension of the envelope. The function (8.12) is then used for classification after replacing $\widehat{\Gamma}$ with U in the function itself and in the estimators leading to that function.

   This method may be particularly useful when μ_k is well-estimated but estimation of Σ_k is questionable because of collinearity, $k \in \chi$. It is illustrated in Section 8.5 There it is called AN-Q1, indicating Algorithm $\mathbb{N}$, first quadratic method.

5. AN-Q2: Classification function (8.12) based on dimension reduction via the enveloped central subspace $\mathcal{S}_{C|X}$ of dimension u estimated by using PLS:

$$N = p + ru + p(p+1)/2 + ru(u+1)/2.$$

   The weight matrix U is obtained by using Algorithm $\mathbb{N}$ or $\mathbb{S}$ of Section 1.5 with $M = \overline{S}$ and A_λ as given in (8.15). Predictive cross validation can be used to determine the dimension of the envelope and the mixing parameter λ. The function (8.12) is then used for classification after replacing $\widehat{\Gamma}$ with U in the function itself and in the estimators leading to that function.

   This method may be particularly useful for assessing classification errors as they depend on the relative contributions λ of the means and variances. It is illustrated in Section 8.5 using Algorithm $\mathbb{N}$. There it is called AN-Q2, indicating Algorithm $\mathbb{N}$, second quadratic method.

Methods 3–5 all depend on the same classification function but differ on the method of estimating the compressed predictors.

8.5 Illustrations

In this section we compare the PLS-type methods AN-Q1 and AN-Q2 to other classification methods studied in the literature. We confine attention mostly

TABLE 8.1
Birds-planes-cars data: Results from the application of three classification methods. Accuracy is the percent correct classification based on leave-one-out cross validation.

Dim. Reduction Method	u	ϕ_{qda}	Accuracy (%)
AN-Q2 ($\lambda = 0.6$)	4	(8.15)	98.8
AN-Q1	3	(8.14)	97.6
QDA, no reduction	13	(8.4)	92.1

to methods that require and thus benefit from having nonsingular class covariance matrices S_k. We see this as a rather stringent test for AN-Q1 and AN-Q2, which do not require the class covariance matrices S_k to be nonsingular.

8.5.1 Birds, planes, and cars

This illustration is from a study to distinguish birds, planes and cars by the sounds they make. A two-hour recording was made in the city of Ermont, France, and then 5-second snippets of sounds were selected. This resulted in 58 recordings identified as birds, 43 as cars and 64 as planes. Each recording was processed and ultimately represented by 13 Scale Dependent Mel-Frequency Cepstrum Coefficients, which constitute the features for distinguishing birds, planes and cars. This classification problem then has $p = 13$ feature vectors and $n = 165$ total observations. These data were used by Cook and Forzani (2009) to illustrate dimension reduction via LAD and by Zhang and Mai (2019) in their development of classification methodology based on the envelope discriminant subspace (ENDS) for QDA.

We applied the three methods shown in Table 8.1. The number u of condensed features and λ were selected by 10-fold cross validation for AN-Q1 and AN-Q2. The estimated classification functions were assessed using leave-one-out cross validation. Plots of the first two compressed features for AN-Q1 and AN-Q2 are shown in Figure 8.1.

Zhang and Mai (2019, Table 4) compared their proposed method ENDS to 11 other classification methods, including classifiers based on Bayes' rule, the support vector machine, SIR and SAVE. Their results are listed in columns 5–11 of Table 8.2 along with those from Table 8.1. For QDA without reduction they reported a misclassification rate of 7.7% which corresponds to an

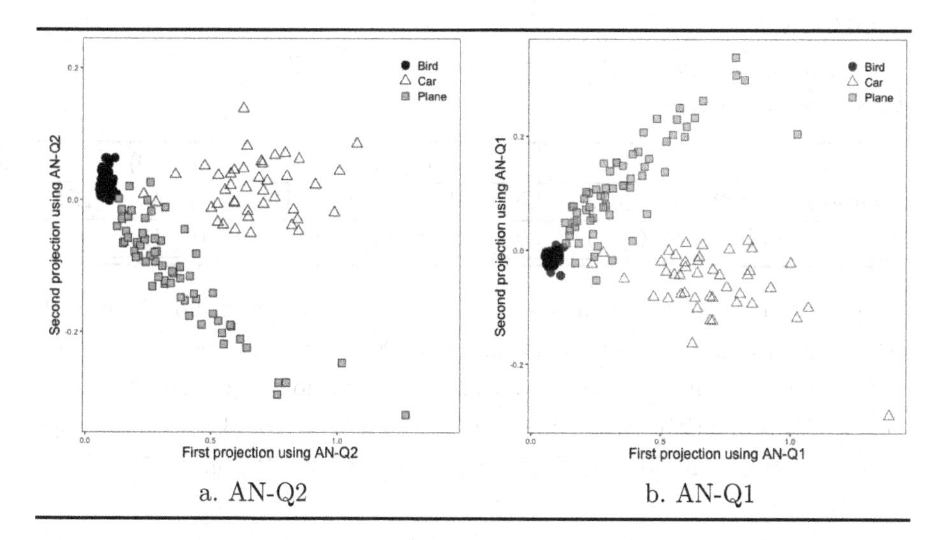

a. AN-Q2

b. AN-Q1

FIGURE 8.1
Bird-plane-cars data: Plots of the first two projected features for AN-Q1 and AN-Q2.

accuracy rate of 92.3%. This agrees well with our estimate of 92.1% in Table 8.1 and in the fourth column of Table 8.2. Due to this agreement, we feel comfortable comparing our results based on Algorithm $\mathbb{N}$ with their results for 12 other methods. Of those 12 methods, their new method ENDS had the best accuracy at 96.2%, which is less than our estimated accuracy rates for the two methods based on Algorithm $\mathbb{N}$. Zhang and Mai (2019, Figure 3) gave a plot of the first two feature vectors compressed by using ENDS. Their plot is similar to those in Figure 8.1 but the point separation does not appear as crisp.

All of the methods studied by Zhang and Mai (2019) require that the intraclass covariance matrices S_k be nonsingular, except perhaps for their implementation of SVM. However, being based on Algorithm $\mathbb{N}$, methods AN-Q2 and AN-Q1 do not require the S_k's to be nonsingular. We take this as a general indication of the effectiveness of the methods based on Algorithm $\mathbb{N}$ since they are able to perform on par with or better than methods that rely on relatively large sample sized. A similar conclusion arises from the fruit example of the next section.

TABLE 8.2

Birds-planes-cars data: Estimates of the correct classification rates (%) from leave one out cross validation.

				From Zhang and Mai (2019)						
Dataset	AN-Q2	AN-Q1	QDA	NB	SVM	QDA	ENDS	SIR	SAVE	DR
Birds+	98.2	97.6	92.1	94.0	87.4	92.3	96.2	93.0	71.8	91.1

Classification rates for columns 5–11 were taken from Zhang and Mai (2019). NB: Naive Bayes classifier (Hand and Yu, 2001). SVM: support vector machine. QDA: quadratic discriminant analysis with no reduction. ENDS: envelope discrimination. SIR: sliced inverse regression. SAVE: sliced average variance estimation. DR: directional regression (Li and Wang, 2007).

8.5.2 Fruit

This dataset contains a collection of 983 infrared spectra collected from strawberry, 351 samples, and non-strawberry, 632 samples, fruit purees. The spectral range was restricted to 554–11,123 nm, and each spectrum contained 235 variables. This then is a classification problem with $p = 235$ features for classifying fruit purees as strawberry or non-strawberry. It was used by Zheng, Fu, and Ying (2014) to compare several classification methods.

Percentages of correct classifications are shown in Table 8.3. The number of compressed features, determined by 10-fold cross validation , for AN-Q1 and AN-Q2 were 14 and 18. The value of λ for AN-Q2 was estimated similarly to be 1. The rates of correct classification shown in Table 8.3 were determined by leave-one-out cross validation. Except for the relatively poor performances by KNN and PLS-DA, there is little to distinguish between the methods. Plots of the first two compressed feature vectors are shown in Figure 8.2, although the results of Zheng et al. (2014) show that more than two compressed features are needed for each method.

8.6 Coffee and olive oil data

Recall that the sizes of the coffee and olive oil data discussed in Sections 7.7.1 and 7.7.2 are $n = 56, p = 256$ and $n = 60, p = 570$. Recall also from Table 7.1

TABLE 8.3

Fruit data: Estimates of the correct classification rates (%) from leave one out cross validation.

			From Zheng et al. (2014)				
Dataset	AN-Q1	AN-Q2	KNN	LS-SVM	PLS-DA	BP-ANN	ELM
Fruit	95.63	95.73	67.01	96.01	85.07	95.33	95.05

Classification rates for columns 4–8 were taken from Zheng et al. (2014). KNN: k-nearest neighbor. LS-SVM: least-squares support vector machine. PLS-DA: partial least-squares discriminant analysis. BP-ANN: back propagation artificial neural network. ELM: extreme learning machine.

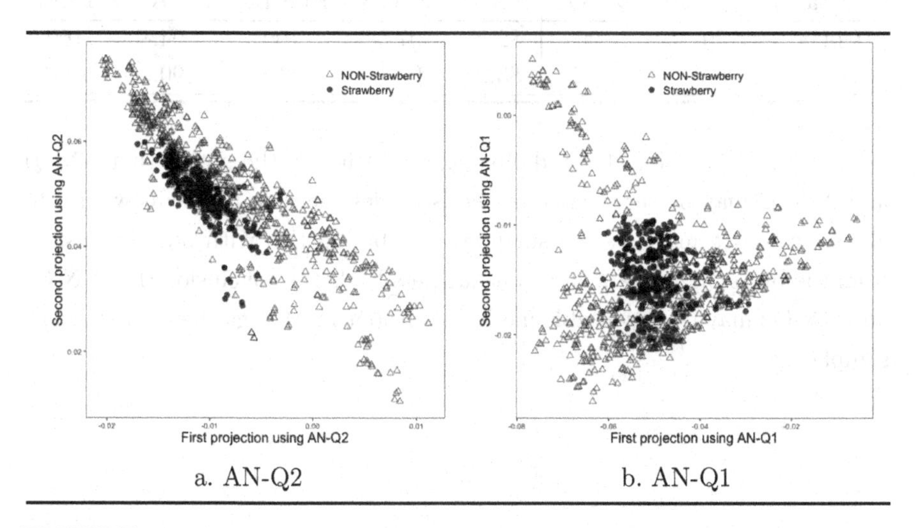

a. AN-Q2 b. AN-Q1

FIGURE 8.2

Fruit data: Plots of the first two projected features for AN-Q1 and AN-Q2.

that the PLS methods discussed in Chapter 7 did well for these data when compared to the methods of Zheng et al. (2014).

The coffee and olive oil datasets may not be large enough to give compelling evidence that the intraclass covariance matrices are unequal and consequently the PLS classification methods of Chapter 7 would be a natural first choice. It could also be argued reasonably that, to error on the safe side, quadratic methods should be tried as well. Shown in columns 2 and 3 of Table 8.4 are the rates of correct classification from applying AN-Q1 and AN-Q2 to the coffee and olive oil data. For comparison, we again listed the results from Zheng et al. (2014). For the coffee data, these quadratic methods did as well as the best method studied by Zheng et al. (2014) and, from Table 7.1, as well as the best PLS method. Viewing the results for the olive oil data, the quadratic methods did reasonably, but not as well as the PLS or PLS+PFC results in Table 7.1.

TABLE 8.4
Olive oil and Coffee data: Estimates of the correct classification rates (%) from leave one out cross validation. Results in columns 4–8 are as described in Table 7.1.

			From Zheng et al. (2014)				
Dataset	AN-Q1	AN-Q2	KNN	LS-SVM	PLS-DA	BP-ANN	ELM
Coffee	100	100	82.2	97.5	100	94.8	100
Olive oil	95.0	94.2	83.2	95.1	93.1	90.0	97.4

The result for the coffee and olive oil data support the notion that AN-Q1 and AN-Q2 may be serviceable methods in classification problems where the class covariance matrices are singular. Combining this with our conclusions from the birds-planes-cars and fruit datasets leads the conclusion that AN-Q1 and AN-Q2 may be serviceable methods without much regardless for the class sample sizes.

9

Non-linear PLS

The behavior of predictions stemming from the PLS regression algorithms discussed in Chapter 3 depends in part on the accuracy of the multivariate linear model (1.1). In particular, the usefulness of those algorithms may be called into question if the conditional mean $\mathrm{E}(Y \mid X)$ is a non-linear function of X, leading to a possibly non-linear model that we represent as

$$Y_i = \mathrm{E}(Y \mid X_i) + \varepsilon_i, \; i = 1, \ldots, n, \tag{9.1}$$

where the errors ε are independent and identically distributed random vectors with mean 0 and variance $\Sigma_{Y|X}$. Model (9.1) is intended to be the same as model (1.1) except for the possibility that the mean is a non-linear function of X. If $\mathrm{E}(Y \mid X) = \beta_0 + \beta^T X$ then model (9.1) reduces to model (1.1). It is recognized that predictions from PLS algorithms based on (1.1) are not serviceable when the mean function $\mathrm{E}(Y \mid X)$ has significant non-linear characteristics (e.g. Shan et al., 2014).

Following Cook and Forzani (2021), in this chapter we study the behavior of the PLS regression algorithms under model (9.1), without necessarily specifying a functional form for the mean. We restrict attention to univariate responses ($r = 1$) starting in Section 9.5, but until that point the response may be multivariate. Our discussion is based mainly on the NIPALS algorithm (Table 3.1) although our conclusions apply equally to SIMPLS (Table 3.4) and to Helland's algorithm (Table 3.5).

DOI: 10.1201/9781003482475-9

9.1 Synopsis

We bring in two new ideas to facilitate our discussion of PLS algorithms under non-linearity (9.1). The first is a construction – the *central mean subspace (CMS)* – that is used to characterize the mean function $E(Y \mid X)$ in a way that is compatible with envelopes (Section 9.2). The second is a *linearity condition* that is used to constrain the marginal distribution of the predictors (Section 9.3). This condition is common in sufficient dimension reduction (e.g. Cook, 1998; Li, 2018) and is used to rule out anomalous predictor behavior by requiring that certain regressions among the predictors themselves are all linear. Using these ideas we conclude in this chapter that

1. Plots of the responses against NIPALS fitted values derived from the algorithm in Table 3.1 can be used to diagnose non-linearity in the mean function $\mathrm{E}(Y \mid X)$. Linearity in such plots supports linear model (1.1), while a clearly non-linear trend contradicts model (1.1) in support of model (9.1) (see Proposition 9.4).

2. Under the linearity condition, NIPALS may be serviceable for dimension reduction under non-linear model (9.1). However, while linear model (1.1) provides for straightforward predictions based on PLS weights, non-linear model (9.1) does not. This means that additional analysis is required to formulate predictions under non-linear model (9.1). Possibilities for prediction are discussed in Section 9.6.

3. By introducing a generalization of the logic leading to Krylov subspaces, we argue in Section 9.5 that PLS algorithms can be generalized to yield methodology that can adapt to non-linearities without prespecifying a parametric form for $E(Y \mid X)$.

9.2 Central mean subspace

Cook and Li (2002) introduced the CMS, which is designed for dimension reduction targeted at the conditional mean $\mathrm{E}(Y \mid X)$ without requiring a pre-specified parametric model. The CMS plays a central role in characterizing the behavior of PLS regression algorithms under (9.1). We first introduce the idea of a mean dimension-reduction subspace, which is also from Cook and Li (2002):

Definition 9.1. *A subspace $\mathcal{S} \subseteq \mathbb{R}^p$ is a mean dimension-reduction subspace for the regression of Y on X if $Y \perp\!\!\!\perp \mathrm{E}(Y \mid X) \mid P_{\mathcal{S}}X$.*

The intuition behind this definition is that the projection $P_{\mathcal{S}}X$ carries all of the information that X has about the conditional mean $\mathrm{E}(Y \mid X)$. Let $\alpha \in \mathbb{R}^{p\times \dim(\mathcal{S})}$ be a basis for a mean dimension reduction subspace $\mathcal{S}$. Then if $\mathcal{S}$ were known, we might expect that $\mathrm{E}(Y \mid X) = \mathrm{E}(Y \mid \alpha^T X)$, thus reducing the dimension of X for the purpose of estimating the conditional mean. This expectation is confirmed by the following proposition (Cook and Li, 2002) whose proof is sketched in Appendix A.7.1.

Proposition 9.1. *The following statements are equivalent.*

(i) $Y \perp\!\!\!\perp \mathrm{E}(Y \mid X) \mid \alpha^T X$,

(ii) $\mathrm{cov}\{(Y, \mathrm{E}(Y \mid X)) \mid \alpha^T X\} = 0$,

(iii) $\mathrm{E}(Y \mid X)$ *is a function of* $\alpha^T X$.

Statement (i) is equivalent to Definition 9.1, although here it is stated in terms of a basis α. Statement (ii) says that the conditional covariance between Y and $\mathrm{E}(Y \mid X)$ is 0, and statement (iii) tells us that $\mathrm{E}(Y \mid X) = \mathrm{E}(Y \mid \alpha^T X)$. As in our discussion of the central subspace, the most parsimonious reduction is provided by the smallest mean dimension-reduction subspace:

Definition 9.2. *If the intersection of all mean dimension-reduction subspaces is itself a mean dimension-reduction subspace then it is called the CMS and denoted as $\mathcal{S}_{\mathrm{E}(Y|X)}$ with dimension $q = \dim(\mathcal{S}_{\mathrm{E}(Y|X)})$.*

The CMS does not always exist, but it does exist under mild conditions that should not be worrisome in practice (Cook and Li, 2002). We assume existence of the CMS throughout this chapter.

Single index model.

Suppose $r = 1$ and that the regression of Y on X follows the single index model

$$Y = f(\beta_1^T X) + \varepsilon, \tag{9.2}$$

where $X \perp\!\!\!\perp \varepsilon$, $\beta_1 \in \mathbb{R}^p$ and the function f is unknown. Any subspace $\mathcal{S} \subseteq \mathbb{R}^p$ that contains $\text{span}(\beta_1)$ is a mean dimension-reduction subspace because $\text{E}(Y \mid X)$ is constant conditional on $P_{\mathcal{S}}X$ and then trivially $Y \perp\!\!\!\perp \text{E}(Y \mid X) \mid P_{\mathcal{S}}X$. Conclusion (ii) of Proposition 9.1 holds also because $\text{E}(Y \mid X)$ is constant conditional on $\alpha^T X$, where α is still a basis for $\mathcal{S}$. When referring to this model in subsequent developments, we will always require that $\Sigma_{X,Y} \neq 0$, unless stated otherwise.

Double index model.

If $r = 1$, then the regression of Y on X follows a double index model if

$$Y = f(\beta_1^T X, \beta_2^T X) + \varepsilon, \tag{9.3}$$

where $X \perp\!\!\!\perp \varepsilon$, $\beta_j \in \mathbb{R}^p$, $j = 1, 2$ and the function f is unknown. For example, we might have

$$Y = \gamma_0 + \beta_1^T X + \gamma_1 e^{-\beta_2^T X} + \varepsilon.$$

Any subspace $\mathcal{S} \subseteq \mathbb{R}^p$ that contains $\text{span}(\beta_1, \beta_2)$ is a mean dimension-reduction subspace because $\text{E}(Y \mid X)$ is constant conditional on $P_{\mathcal{S}}X$ and then trivially $Y \perp\!\!\!\perp \text{E}(Y \mid X) \mid P_{\mathcal{S}}X$. Conclusion (ii) of Proposition 9.1 holds also because $\text{E}(Y \mid X)$ is constant conditional on $\alpha^T X$, where α is a basis for $\mathcal{S}$.

We defined in Section 2.1 the central subspace $\mathcal{S}_{Y|X}$ as the intersection of all subspaces $\mathcal{S}$ such that $Y \perp\!\!\!\perp X \mid P_{\mathcal{S}}X$, while the CMS $\mathcal{S}_{\text{E}(Y|X)}$ is defined as the intersection of all subspaces $\mathcal{S}$ such that $Y \perp\!\!\!\perp \text{E}(Y \mid X) \mid P_{\mathcal{S}}X$. The central subspace captures all of the ways in which X affects the distribution of Y, while the CMS captures all the ways in which just $\text{E}(Y \mid X)$ depends on X. Clearly, we must have $\mathcal{S}_{\text{E}(Y|X)} \subseteq \mathcal{S}_{Y|X}$. Under models (9.2) and (9.3) Y depends on X only via its mean function and in those cases $\mathcal{S}_{\text{E}(Y|X)} = \mathcal{S}_{Y|X}$. In consequence, the model-free predictor envelope described in Definition 2.2 reduces to $\mathcal{E}_{\Sigma_X}(\mathcal{S}_{Y|X}) = \mathcal{E}_{\Sigma_X}(\mathcal{S}_{\text{E}(Y|X)})$, which becomes one tool for our study of PLS regression algorithms under (9.1). If linear model (1.1) holds, then the coefficient matrix β is well-defined and $\mathcal{S}_{\text{E}(Y|X)} = \mathcal{S}_{Y|X} = \text{span}(\beta)$. If non-linear model (9.1) holds, then there is no model-based β defined and, as mentioned previously, we have only $\mathcal{S}_{\text{E}(Y|X)} = \mathcal{S}_{Y|X}$.

The ability to estimate vectors in the CMS or its Σ_X-envelope $\mathcal{E}_{\Sigma_X}(\mathcal{S}_{\mathrm{E}(Y|X)})$ is crucial for understanding PLS regression algorithms under model (9.1). Estimation is facilitated when the predictors satisfy a linearity condition on their marginal distribution.

9.3 Linearity conditions

We first introduce the linearity condition in terms of an essentially arbitrary random vector and then show how it is applied in the regression context.

Definition 9.3. *Let $W \in \mathbb{R}^p$ be a random vector with mean 0 and covariance matrix $\Sigma_W > 0$. Let $\alpha \in \mathbb{R}^{p\times q}$, $q \leq p$, be any basis matrix for a subspace $\mathcal{S} \subseteq \mathbb{R}^p$. We say that W satisfies the linearity condition relative to $\mathcal{S}$ if there exists a non-stochastic $p \times q$ matrix M such that $\mathrm{E}(W \mid \alpha^T W = u) = Mu$ for all u.*

The essence of the linearity condition is that $\mathrm{E}(W \mid \alpha^T W) = \mathrm{E}(W \mid P_{\mathcal{S}} W)$ must be a linear function of $\alpha^T W$. If the linearity condition holds for a subspace $\mathcal{S}$ it need not necessarily hold for any other distinct subspace. For instance, the linearity conditions holds trivially for $\mathcal{S} = \mathbb{R}^p$, but that does not imply that it holds for any proper subspace of $\mathbb{R}^p$. The following proposition, which is adapted from Cook (1998, Prop. 9.1), gives properties of M under Definition 9.3. Its proof is sketched in Appendix A.7.2. The notion of a subspace reducing a matrix was described in Definition 1.1.

Proposition 9.2. *Assume that $W \in \mathbb{R}^p$, a random vector with mean 0 and covariance matrix $\Sigma_W > 0$, satisfies the linearity condition relative to $\mathcal{S}$ under Definition 9.3. Let α be a basis for $\mathcal{S}$. Then*

1. $M = \Sigma_W \alpha(\alpha^T \Sigma_W \alpha)^{-1}$.

2. *M^T is a generalized inverse of α.*

3. *αM^T is the orthogonal projection operator for $\mathcal{S}$ relative to the Σ_W inner product (see equation 1.3).*

4.
$$\mathrm{E}(W \mid \alpha^T W) - \mu_W = P^T_{\mathcal{S}(\Sigma_W)}(W - \mu_W),$$
where for completeness we have allowed $\mu_W = \mathrm{E}(W)$ to be non-zero.

5. *If $\mathcal{S}$ reduces Σ_W then $P_{\mathcal{S}(\Sigma_W)} = P_{\mathcal{S}}$.*

If $W \in \mathbb{R}^p$ is elliptically contoured with mean μ_W and variance Σ_W, which includes the multivariate normal family, then it is known that W satisfies the linearity condition for any subspace $\mathcal{S} \subseteq \mathbb{R}^p$ with basis matrix α (Cook, 1998, Section 7.3.2),

$$\mathrm{E}(W \mid \alpha^T W) = \mu_W + P^T_{\mathcal{S}(\Sigma_W)}(W - \mu_W).$$

Moreover, Eaton (1986) has shown that a random vector $W \in \mathbb{R}^p$ satisfies the linearity condition for *all* subspace $\mathcal{S} \subseteq \mathbb{R}^p$ if and only if W is elliptically contoured. Being elliptically contoured is then sufficient to imply the linearity condition under Definition 9.3, although it is not necessary since the definition requires linearity for only the selected subspace and not all subspaces. Diaconis and Freedman (1984) and Hall and Li (1993) showed that almost all projections of high-dimensional data are approximately normal and thus satisfy the linearity condition. For all of these reasons, the linearity condition is largely seen as mild when $p \gg q$. It should hold well in the analysis of spectral data in chemometrics, since then p is often in the hundreds and q is relatively small. See Li and Wang (2007) and Cook (1998, 2018) for further discussion.

We will be requiring the linearity condition of the predictors relative to subspaces associated with the regression of Y on X. For these applications, it may be sufficient to have the distribution of X given Y satisfy the linearity condition relative to the central subspace. The proof of the following proposition is available in Appendix A.7.3.

Proposition 9.3. *Assume that the distribution of $X \mid Y$ satisfies the linearity condition relative to $\mathcal{S}_{Y|X}$ for each value of Y. Then the marginal distribution of X also satisfies the linearity condition relative to $\mathcal{S}_{Y|X}$.*

The next corollary and lemma describe special settings for the linearity condition.

Corollary 9.1. *(I) If model (9.1) holds and if $X \mid Y$ satisfies the linearity condition relative to $\mathcal{S}_{\mathrm{E}(Y|X)}$ for each value of Y, then X satisfies the linearity condition relative to $\mathcal{S}_{\mathrm{E}(Y|X)}$. (II) If $X \mid Y$ is elliptically contoured, then X satisfies the linearity condition relative to $\mathcal{S}_{Y|X}$.*

Proof. (I) The result follows because under model (9.1), $\mathcal{S}_{\mathrm{E}(Y|X)} = \mathcal{S}_{Y|X}$. (II) If $X \mid Y$ is elliptically contoured, then it satisfies the linearity condition for all values of Y relative to any subspace $\mathcal{S} \subseteq \mathbb{R}^p$. The conclusion follows from Proposition 9.3. □

9.4 NIPALS under the non-linear model

As a first use of the linearity condition, consider the NIPALS algorithm under model (9.1). Let η be a basis matrix for $\mathcal{S}_{\mathrm{E}(Y|X)}$, the CMS, and define $\beta = \Sigma_X^{-1}\Sigma_{X,Y}$, the OLS coefficients from the linear population regression of Y on X, and correspondingly let $\mathcal{B} = \mathrm{span}(\beta)$. In this use of β we are not assuming that linear model (1.1) holds. Rather, β is simply a parameter vector constructed from Σ_X and $\Sigma_{X,Y}$. The following proposition connects β with the CMS.

Proposition 9.4. *Assume the linearity condition for X relative to $\mathcal{S}_{\mathrm{E}(Y|X)}$ and that non-linear model (9.1) holds. Then* $\mathrm{span}(\Sigma_{X,Y}) \subseteq \Sigma_X \mathcal{S}_{\mathrm{E}(Y|X)}$ *and consequently* $\mathcal{B} \subseteq \mathcal{S}_{\mathrm{E}(Y|X)}$.

Proof. We step through the proof so the role of various structures can be seen. Recall that η is a basis matrix for $\mathcal{S}_{\mathrm{E}(Y|X)}$. Expanding the expectation operator, we first have

$$\begin{aligned}\Sigma_{X,Y} &= \mathrm{E}_{X,Y}\{(X-\mu_X)(Y-\mu_Y)^T\} \\ &= \mathrm{E}_X \mathrm{E}_{Y|X}\{(X-\mu_X)(Y-\mu_Y)^T\} \\ &= \mathrm{E}_X\{(X-\mu_X)(\mathrm{E}(Y \mid X)-\mu_Y)^T\}.\end{aligned}$$

From Proposition 9.1, $\mathrm{E}(Y \mid X) = \mathrm{E}(Y \mid \eta^T X)$ and so

$$\begin{aligned}\Sigma_{X,Y} &= \mathrm{E}_X\{(X-\mu_X)(\mathrm{E}(Y \mid \eta^T X)-\mu_Y)^T\} \\ &= \mathrm{E}_{\eta^T X}\mathrm{E}_{X|\eta^T X}\{(X-\mu_X)(\mathrm{E}(Y \mid \eta^T X)-\mu_Y)^T\} \\ &= \mathrm{E}_{\eta^T X}\{(\mathrm{E}(X \mid \eta^T X)-\mu_X)(\mathrm{E}(Y \mid \eta^T X)-\mu_Y)^T\}.\end{aligned}$$

From the assumed linearity condition and conclusion 4 of Proposition 9.2 we have

$$\mathrm{E}(X \mid \eta^T X) - \mu_X = P^T_{\eta(\Sigma_X)}(X-\mu_X).$$

Substituting back, we get

$$\begin{aligned}\Sigma_{X,Y} &= \mathrm{E}_{\eta^T X}\{P^T_{\eta(\Sigma_X)}(X-\mu_X)(\mathrm{E}(Y \mid \eta^T X)-\mu_Y)^T\} \\ &= P^T_{\eta(\Sigma_X)}\Sigma_{X,Y} \qquad (9.4)\\ \mathrm{span}(\beta) &\subseteq \mathcal{S}_{\mathrm{E}(Y|X)},\end{aligned}$$

and our claim follows. □

In the proof of Proposition 9.4, the linearity condition is applied only to the marginal distribution of the predictors; it does not involve the response. This will also be true for all subsequent applications of the linearity condition.

The following corollary gives Proposition 9.4 when restricted to single index models (9.2) with non-zero β. Its proof is straightforward and omitted.

Corollary 9.2. *Assume the linearity condition for X relative to $\mathcal{S}_{\mathrm{E}(Y|X)}$. Assume also that the regression of Y on X follows the single index model (9.2) and that $\Sigma_{X,Y} \neq 0$. Then $\mathcal{B} = \mathcal{S}_{\mathrm{E}(Y|X)}$.*

The next proposition provides a formal connection between the predictor envelope $\mathcal{E}_{\Sigma_X}(\mathcal{B})$ and the envelope of the CMS $\mathcal{E}_{\Sigma_X}(\mathcal{S}_{\mathrm{E}(Y|X)})$. Its proof is in Appendix A.7.4.

Proposition 9.5. *Assume the linearity condition for X relative to $\mathcal{S}_{\mathrm{E}(Y|X)}$. Then under the non-linear model (9.1), we have*

$$\mathcal{E}_{\Sigma_X}(\mathcal{B}) \subseteq \mathcal{E}_{\Sigma_X}(\mathcal{S}_{\mathrm{E}(Y|X)}).$$

Moreover, if the single index model (9.2) holds and $\Sigma_{X,Y} \neq 0$ then $\mathcal{E}_{\Sigma_X}(\mathcal{B}) = \mathcal{E}_{\Sigma_X}(\mathcal{S}_{\mathrm{E}(Y|X)})$.

We know from the discussion in Section 3.2 that the weight matrix from the NIPALS algorithm provides an estimator of a basis for the predictor envelope $\mathcal{E}_{\Sigma_X}(\mathcal{B})$, which by Proposition 9.5 is contained in the corresponding envelope of the CMS. It follows that the compressed predictors $W^T X$ are as relevant for the non-linear regression model (9.1) as they are for the linear model (1.1), provided that the linearity condition holds. In some regressions it may happen that $\mathcal{E}_{\Sigma_X}(\mathcal{B})$ is a proper subset of $\mathcal{E}_{\Sigma_X}(\mathcal{S}_{\mathrm{E}(Y|X)})$, in which case the NIPALS compression may miss relevant directions in which the mean function is non-linear. However, if the single index model (9.2) holds then

$\mathcal{E}_{\Sigma_X}(\mathcal{B}) = \mathcal{E}_{\Sigma_X}(\mathcal{S}_{\mathrm{E}(Y|X)})$ and NIPALS will not miss any relevant directions in the population.

These results have implications for graphical diagnostics to detect non-linearities in a regressions. The following steps may be helpful in regressions with a real response.

1. Fit the data with NIPALS, SIMPLS or a PLS regression variation thereof and get the estimated coefficient matrix $\widehat{\beta}_{\mathrm{pls}}$.

2. Plot the response Y versus the fitted values $\widehat{\beta}^T_{\mathrm{pls}}X$. If the plot shows clear curvature, then the serviceability of the envelope model (2.5) is questionable. If no worrisome non-linearities are observed, then the model is sustained.

   The conclusion $\mathcal{E}_{\Sigma_X}(\mathcal{B}) \subseteq \mathcal{E}_{\Sigma_X}(\mathcal{S}_{\mathrm{E}(Y|X)})$ from Proposition 9.5 implies that $\mathcal{E}_{\Sigma_X}(\mathcal{B})$ could be a proper subset of $\mathcal{E}_{\Sigma_X}(\mathcal{S}_{\mathrm{E}(Y|X)})$ and thus that some directions in the CMS envelope are missed. However, if the underlying regression is a single index model, then the plot of the response versus the fitted values is sufficient for detecting non-linearity.

3. A closer inspection may be useful for some regressions. In these cases, plots of the response versus selected linear combinations, $a^T W^T X$ with $a \in \mathbb{R}^q$, of the compressed predictors $W^T X$ may be informative. Since the only role of the weight matrix is to provide an estimate of a basis for the envelope, the linear combinations to plot may be unclear. Plotting the response versus the principal components of $W^T X$ may be a useful default choice. Another potentially useful possibility is to plot the combinations $w_j^T X$ against Y, where w_j is the j-th weight vector generated by the NIPALS algorithm. The first few linear combinations will likely be most important. Clear non-linearity in a plot of Y versus $w_j^T X$ is then an indication that the mean function is non-linear.

If non-linearities are detected, the usual predictive procedures for selecting the number of components q may not be useful here because they rely on a linear model, and predictions based on the linear model may no longer be serviceable. Different methods may be needed in such cases. Several possibilities are illustrated in Section 9.6.

9.5 Extended PLS algorithms for dimension reduction

It is known from our discussion in Section 3.5.1 that for single-response linear regressions in the context of model (1.1) the population weight matrix W from the NIPALS algorithm arises as Gram-Schmidt orthogonalization of the Krylov sequence $K_q(\Sigma_X, \sigma_{X,Y})$ (Helland, 1990; Manne, 1987). We also know from Proposition 1.10 that the Krylov subspace $\mathcal{K}_q(\Sigma_X, \sigma_{X,Y}) = \mathcal{E}_{\Sigma_X}(\mathcal{B})$, so

$$\text{span}(\beta) \subseteq \mathcal{E}_{\Sigma_X}(\mathcal{B}) = \text{span}(W) = \mathcal{K}_q(\Sigma_X, \sigma_{X,Y}), \tag{9.5}$$

which is the fundamental population justification for using the NIPALS algorithm for dimension reduction in conjunction with using linear model (1.1) for prediction. In particular, it implies that there is a vector $\eta \in \mathbb{R}^q$ so that $\beta = W\eta$, leading to the reduced linear model $Y = \alpha + \eta^T W^T X + \varepsilon$.

We propose in Section 9.5.1 a procedure for generalizing PLS algorithms for dimension reduction in regressions covered by non-linear model (9.1). These generalizations effect dimension reduction but not prediction. Methods for prediction are discussed in Section 9.6. We show in Section 9.5.2 that the generalized procedure can produce the Krylov sequence (9.5) and in Section 9.5.3 we show how to use it to remove linear trends to facilitate detecting non-linearities. We expect that there will be many other applications of the generalized method in the future.

We confine discussion to regressions with a real response, $r = 1$, in the remainder of this chapter.

9.5.1 A generalization of the logic leading to Krylov spaces

Recall the following from our discussion of invariant subspaces and envelopes in Section 1.3. If $\mathcal{S}$ is an invariant subspace of Σ_X and if $x \in \mathcal{S}$, then we can iteratively transform x by Σ_X to find additional vectors in $\mathcal{S}$, eventually collecting enough vectors to span $\mathcal{S}$. The NIPALS algorithm for a univariate response operates with $\mathcal{S} = \mathcal{E}_{\Sigma_X}(\mathcal{B})$ and $x = \sigma_{X,Y}$, which gives rise to the Krylov sequence $K_q(\Sigma_X, \sigma_{X,Y})$. This idea of iteratively transforming a vector in a selected invariant subspace to collect more vector in that subspace is the fundamental idea behind NIPALS and our generalization. In the following proposition, a version of which was first proven by Cook and Li (2002, Theorem 3), our subspace is $\mathcal{S} = \mathcal{E}_{\Sigma_X}(\mathcal{S}_{\mathrm{E}(Y|X)})$.

Proposition 9.6. *Assume model (9.1) with a real response. Let* Γ *be a basis matrix for* $\mathcal{E}_{\Sigma_X}(\mathcal{S}_{\mathrm{E}(Y|X)})$ *and let* U *and* V *be real-valued functions of* $\Gamma^T X$*. Assume also that* X *satisfies the linearity condition relative to* $\mathcal{E}_{\Sigma_X}(\mathcal{S}_{\mathrm{E}(Y|X)})$*. Then*

$$\mathrm{E}\{(UY+V)(X-\mu_X)\} \in \mathcal{E}_{\Sigma_X}(\mathcal{S}_{\mathrm{E}(Y|X)}).$$

Proof. Assume without loss of generality that $\mu_X = 0$. Then

$$\mathrm{E}((UY+V)X) = \mathrm{E}_X\{(U\mathrm{E}(Y \mid X)+V)X\}.$$

Let η be basis matrix for $\mathcal{S}_{\mathrm{E}(Y|X)}$. By Proposition 9.1, $\mathrm{E}(Y \mid X) = \mathrm{E}(Y \mid \eta^T X)$. Since $\mathcal{S}_{\mathrm{E}(Y|X)} \subseteq \mathcal{E}_{\Sigma_X}(\mathcal{S}_{\mathrm{E}(Y|X)})$, it follows that $\mathrm{E}(Y \mid X) = \mathrm{E}(Y \mid \eta^T X) = \mathrm{E}(Y \mid \Gamma^T X)$. Consequently,

$$\begin{aligned}\mathrm{E}((UY+V)X) &= \mathrm{E}_X\{(U\mathrm{E}(Y \mid \Gamma^T X)+V)X\} \\ &= \mathrm{E}_{\Gamma^T X}\{(U\mathrm{E}(Y|\Gamma^T X)+V)\mathrm{E}(X \mid \Gamma^T X)\}.\end{aligned}$$

Since X satisfies the linearity condition relative to $\mathcal{E}_{\Sigma_X}(\mathcal{S}_{\mathrm{E}(Y|X)}) = \mathrm{span}(\Gamma)$, we know from conclusions 4 and 5 of Proposition 9.2 that $\mathrm{E}(X \mid \Gamma^T X) = P_\Gamma X$. Using the condition that U and V are real-valued functions of $\Gamma^T X$, we have

$$\begin{aligned}\mathrm{E}((UY+V)X) &= \mathrm{E}_{\Gamma^T X}\{(U\mathrm{E}(Y|\Gamma^T X)+V)\mathrm{E}(X \mid \Gamma^T X)\} \\ &= \mathrm{E}_{\Gamma^T X}\{(U\mathrm{E}(Y|\Gamma^T X)+V)P_\Gamma X\} \\ &= P_\Gamma \mathrm{E}\{(UY+V)X\} \\ &\in \mathcal{E}_{\Sigma_X}(\mathcal{S}_{\mathrm{E}(Y|X)}).\end{aligned}$$

□

The importance of this proposition is as follows. Suppose we know one vector ν_0 in the envelope $\mathcal{E}_{\Sigma_X}(\mathcal{S}_{\mathrm{E}(Y|X)})$. We can find another vector in $\mathcal{E}_{\Sigma_X}(\mathcal{S}_{\mathrm{E}(Y|X)})$ by defining functions $u : \mathbb{R} \mapsto \mathbb{R}$ and $v : \mathbb{R} \mapsto \mathbb{R}$, which are then used to form a constructed response $Y^* = u(\nu_0^T X)Y + v(\nu_0^T X)$. The covariance between the constructed response and X is in the envelope: $\nu_1 = \mathrm{cov}(Y^*, X) \in \mathcal{E}_{\Sigma_X}(\mathcal{S}_{\mathrm{E}(Y|X)})$. This can then be iterated to find additional vectors in the envelope:

$$\nu_j = \mathrm{cov}\{u(\nu_{j-1}^T X)Y + v(\nu_{j-1}^T X), X\} \in \mathcal{E}_{\Sigma_X}(\mathcal{S}_{\mathrm{E}(Y|X)}),\ j = 1, 2, \ldots .$$

9.5.2 Generating NIPALS vectors

In this section we show how to use Proposition 9.6 to generate the Krylov sequence. This then is another way to reach Proposition 9.5 and the conclusions that follow it. This argument shows how to achieve generalizations of NIPALS dimension reduction, which may be useful in some regressions.

Setting $U = 1$, $V = 0$ we have a *first vector*

$$\begin{aligned} \mathrm{E}((UY+V)X) &= \sigma_{X,Y} \\ \mathrm{span}(\sigma_{X,Y}) &\subseteq \mathcal{E}_{\Sigma_X}(\mathcal{S}_{\mathrm{E}(Y|X)}). \end{aligned}$$

There are many ways to construct second and subsequent vectors in $\mathcal{E}_{\Sigma_X}(\mathcal{S}_{\mathrm{E}(Y|X)})$ depending on the choices for U and V. Our selections here were chosen to generate the Krylov sequence $K_q(\Sigma_X, \sigma_{X,Y})$ associated with the NIPALS algorithm.

Take $U = 0$ and $V = v(\sigma_{X,Y}^T X) = \sigma_{X,Y}^T X$. Then

$$\mathrm{E}((UY+V)X) = \mathrm{E}(XX^T\sigma_{X,Y}) = \Sigma_X\sigma_{X,Y} \in \mathcal{E}_{\Sigma_X}(\mathcal{S}_{\mathrm{E}(Y|X)})$$

is a second vector in the envelope. To get a third vector, we use the second vector since $\Sigma_X\sigma_{X,Y} \in \mathcal{E}_{\Sigma_X}(\mathcal{S}_{\mathrm{E}(Y|X)})$. Then with $U = 0$ and $V = v\{(\Sigma_X\sigma_{X,Y})^T X\} = (\Sigma_X\sigma_{X,Y})^T X$, we have

$$\mathrm{E}((UY+V)X) = \Sigma_X^2\sigma_{X,Y} \subseteq \mathcal{E}_{\Sigma_X}(\mathcal{S}_{\mathrm{E}(Y|X)}).$$

Continuing by induction, we generate the Krylov subspace,

$$\mathrm{span}(\sigma_{X,Y}, \Sigma_X\sigma_{X,Y}, \Sigma_X^2\sigma_{X,Y}, \ldots) \subseteq \mathcal{E}_{\Sigma_X}(\mathcal{S}_{\mathrm{E}(Y|X)}).$$

Under the non-linear model we still have strict monotone containment of the Krylov subspaces following (1.23) with equality after reaching q components is reached. However, here $\mathcal{K}_q(\Sigma_X, \sigma_{X,Y})$ is contained in $\mathcal{E}_{\Sigma_X}(\mathcal{S}_{\mathrm{E}(Y|X)})$ without necessarily being equal:

$$\mathcal{K}_1(\Sigma_X, \sigma_{X,Y}) \subset \cdots \subset \mathcal{K}_q(\Sigma_X, \sigma_{X,Y}) = \mathcal{K}_{q+1}(\Sigma_X, \sigma_{X,Y}) \cdots \subseteq \mathcal{E}_{\Sigma_X}(\mathcal{S}_{\mathrm{E}(Y|X)}).$$

As we have used in the past, q is the number of NIPALS components. However, in his context, q is not necessarily equal to the dimension of the corresponding envelope, $q \leq \dim\{\mathcal{E}_{\Sigma_X}(\mathcal{S}_{\mathrm{E}(Y|X)})\}$. Putting it all together we have the non-linear model counterpart to (9.5),

Corollary 9.3. *Under the hypotheses of Proposition 9.6,*

$$\mathcal{K}_q(\Sigma_X, \sigma_{X,Y}) = \mathrm{span}(W) = \mathcal{E}_{\Sigma_X}(\mathcal{B}) \subseteq \mathcal{E}_{\Sigma_X}(\mathcal{S}_{\mathrm{E}(Y|X)}), \tag{9.6}$$

with equality $\mathcal{E}_{\Sigma_X}(\mathcal{B}) = \mathcal{E}_{\Sigma_X}(\mathcal{S}_{\mathrm{E}(Y|X)})$ under the single index model.

This result reinforces the conclusions of Proposition 9.5 and the discussion that follows it: the NIPALS algorithm can be used to diagnose non-linearity in the mean function. In fact, the mild linearity condition for the envelope is the only novel condition needed.

This graphical diagnostic procedure discussed at the end of Section 9.4 will display linear as well as non-linear trends. For the purpose of diagnosing non-linearity in the mean function, it may be desirable to remove linear trends first and use the NIPALS residuals

$$R \;=\; Y - \mathrm{E}(Y) - \beta_{\mathrm{npls}}^T(X - \mu_X)$$

as a response, where β_{npls} denotes the population NIPALS coefficient vector. However, using the residuals to generate a Krylov sequence $K_{p-1}(\Sigma_X, \sigma_{X,r})$ may not be effective because the residual covariance $\sigma_{X,r}$ will likely be small and uninformative; $\sigma_{X,r} = 0$ when the OLS residuals are used. In the next sections we propose a different diagnostic method based on an application of Proposition 9.6 with residuals.

9.5.3 Removing linear trends in NIPALS

Diagnosing non-linearity in the mean function is most effectively done by first removing linear trends and then dealing with the residuals. To accomplish that we define the third-moment matrix

$$\Sigma_{RXX} \;=\; \mathrm{E}\{R(X - \mathrm{E}(X))(X - \mathrm{E}(X))^T\},$$

where the expectation is with respect to the joint distribution of (R, X), and then apply Proposition 9.6 with first vector $\beta_{\mathrm{npls}} \in \mathcal{E}_{\Sigma_X}(\mathcal{S}_{\mathrm{E}(Y|X)})$. To get the second vector, assume without loss of generality that $\mathrm{E}(X) = 0$ and use $U = u(\beta_{\mathrm{npls}}^T X) = \beta_{\mathrm{npls}}^T X$ and $V = v(\beta_{\mathrm{npls}}^T X) = -\beta_{\mathrm{npls}}^T X \mathrm{E}(Y) - (\beta_{\mathrm{npls}}^T X)^2$. This gives

$$\begin{aligned}
\mathrm{E}((UY + V)X) &= \mathrm{E}(\{YX^T\beta_{\mathrm{npls}} - \mathrm{E}(Y)X^T\beta_{\mathrm{npls}} - (X^T\beta_{\mathrm{npls}})^2\}X) \\
&= \mathrm{E}(\{Y - \mathrm{E}(Y) - X^T\beta_{\mathrm{npls}}\}XX^T\beta_{\mathrm{npls}}) \\
&= \mathrm{E}(rXX^T)\beta_{\mathrm{npls}} = \Sigma_{RXX}\beta_{\mathrm{npls}} \in \mathcal{E}_{\Sigma_X}(\mathcal{S}_{\mathrm{E}(Y|X)}),
\end{aligned}$$

where the final containment follows from Proposition 9.6. For the third vector we operationally replace β_{npls} in u and v with $\Sigma_{RXX}\beta_{\text{npls}}$ to get

$$\begin{aligned} U &= u((\Sigma_{RXX}\beta_{\text{npls}})^T X) = (\Sigma_{RXX}\beta_{\text{npls}})^T X \\ V &= v((\Sigma_{RXX}\beta_{\text{npls}})^T X) = -(\Sigma_{RXX}\beta_{\text{npls}})^T X \mathrm{E}(Y) - ((\Sigma_{RXX}\beta_{\text{npls}})^T X)^2. \end{aligned}$$

This gives the third vector

$$\mathrm{E}((UY + V)X) = \Sigma_{RXX}^2 \beta_{\text{npls}} \in \mathcal{E}_{\Sigma_X}(\mathcal{S}_{\mathrm{E}(Y|X)}).$$

Repeating this operation iteratively, we get by induction

$$\operatorname{span}(\beta_{\text{npls}}, \Sigma_{RXX}\beta_{\text{npls}}, \ldots, \Sigma_{RXX}^j \beta_{\text{npls}}, \ldots) \subseteq \mathcal{E}_{\Sigma_X}(\mathcal{S}_{\mathrm{E}(Y|X)}), \tag{9.7}$$

which implies that we run NIPALS with $\widehat{\Sigma}_{RXX}$ and $\widehat{\beta}_{\text{npls}}$. Using residuals in $\widehat{\Sigma}_{RXX}$ has the advantage that the linear trend is removed making non-linearities easier to identify.

The graphical diagnostic for non-linearity in the mean function can be summarized as follows:

1. Run NIPALS$(\widehat{\Sigma}_X, \widehat{\sigma}_{X,Y})$, obtain $\widehat{\beta}_{\text{npls}}$ and then form the sample residuals

$$\widehat{R}_i = Y_i - \bar{Y} - \widehat{\beta}_{\text{npls}}^T (X_i - \bar{X}),\ i = 1, \ldots, n.$$

2. Construct the third moment matrix

$$\widehat{\Sigma}_{RXX} = n^{-1} \sum_{i=1}^{n} \widehat{R}_i (X_i - \bar{X})(X_i - \bar{X})^T.$$

3. Run NIPALS$(\widehat{\Sigma}_{RXX}, \widehat{\beta}_{\text{npls}})$ and extract the weights h_j, $j = 1, \ldots, p-1$. Here we use h_j to denote the vectors of weights to distinguish them from the usual weights w_j defined in Table 3.1.

4. Plot Y or residuals versus $h_j^T X$ for $j = 1, 2, \ldots$ and look for clear non-linear trends. Since β_{npls} is the first vector listed on the left-hand side of (9.7), $h_1 = \widehat{\beta}_{\text{npls}} / \|\widehat{\beta}_{\text{npls}}\|$, where the normalization arises from the eigenvector computation in Table 3.1. In consequence, a plot of Y or residuals against $h_1^T X$ is effectively the same as a plot of Y or residuals against the NIPALS fitted values $\widehat{\beta}_{\text{npls}}^T X_i$, $i = 1, \ldots, n$.

If the linear model is incorrect, we expect that $\mathrm{E}(R \mid X)$ will exhibit quadratic or high-order behavior, giving a clear signal that the model is somehow deficient. Further interpretation of non-linear trends can be problematic because it is possible that $\dim(\mathcal{S}_{\mathrm{E}(R|X)}) > \dim(\mathcal{S}_{\mathrm{E}(Y|X)})$, implying that the residual regression $R \mid X$ is more complicated than the original regression $Y \mid X$ because the residual regression can introduce extraneous directions into the picture. If non-linearity is detected in residual plots, it may be wise to use the plots at the end of Section 9.4 to guide model development.

9.6 Prediction

The graphical methods described in Sections 9.5.2 and 9.5.3 are used to assess the presence of non-linearity in the mean function. If no notable non-linearity is detected then the linear methods described in Chapter 3 may be serviceable for prediction. If non-linearity is detected, the compressed predictors $w_1^T X, w_2^T X, \ldots$ can still serve for dimension reduction under Corollary 9.3, but there is no model or rule for prediction and consequently there is no way to determine the number of components. To complete the analysis we need a method or model for predicting Y from $w_1^T X, \ldots, w_d^T X$, $d = 1, 2, \ldots, \min\{p, n-1\}$, along with the associated estimated predictive mean squared error used in selecting the number of components.

9.6.1 Available approaches to prediction

A variety of methods have been proposed to handle non-linear PLS applications, ranging from models that describe relationships between latent variables in a non-linear way (Wold et al., 2001), to quadratic PLS (Baffi et al., 1999; Li et al., 2005), to spline PLS (Wold, 1992), to neural network PLS (Baffi, Martin, and Morris, 1999; Chiappini et al., 2020; Chiappini, Goicoechea, and Olivieri, 2020; Olivieri, 2018). A tutorial on the use of neural networks for multivariate calibration is available from Despagne and Luc Massart (1998). Shan et al. (2014) reviewed these and other methods, pointing out their relative strengths and weaknesses. They then proposed a new method based on slicing the response into non-overlapping regions within which the relationship between Y and X follows linear model (1.1), and they demonstrated good performance against a variety of other methods. Nevertheless, being the basis for sliced inverse regression, the slicing operation is well understood in sufficient dimension reduction. In particular, overall efficiency tends to drop off as the number of slices increases because the number of observations per slice decreases resulting in a loss of information for each intra-slice linear model fit.

The available non-linear methods seem to either ignore dimension reduction or to have a new dimension reduction paradigm built in. However, we have seen as a consequence of Corollary 9.3 that standard PLS regression algorithms like NIPALS and SIMPLS can be used for dimension reduction even when the mean function $\mathrm{E}(Y \mid X)$ is an unknown non-linear function of X. Thus, as indicated previously in Section 9.5.2, we need a method or model for predicting Y from $w_1^T X, \ldots, w_d^T X$, $d = 1, 2, \ldots, \min\{p, n-1\}$, along with the associated estimated predictive mean squared error for determining the number of components. Dimension reduction is no longer an issue per se.

Since dimension reduction is no longer an issue, the regression of Y on $w_1^T X, \ldots, w_d^T X$ might be developed by using classical methods. In some settings a simple transformation of the response may be enough to induce linearity in the mean function sufficient to give good predictions. The Box-Cox (Box and Cox, 1964) family of power transformations (see also Cook and Weisberg, 1999) should be sufficient for positive responses, while the relatively recent method by Hawkins and Weisberg (2017) can be used for non-positive responses. In other cases adding all quadratic $(w_k^T X)^2$ and cross product $w_k^T X w_j^T X$ terms may be sufficient, provided d is not too large since there will be a total of $d(d+1)/2$ terms of these forms. Another possibility is to use classical non-parametric regression (e.g. Green and Silverman, 1994), but again tuning is required and problems may arise if d is too large. Methods based on decision trees may be useful in some settings.

We present in the next section a novel method of prediction adapted from the general proposal of Adragni and Cook (2009). Our method gives good predictions for continuous predictors, does not require an entirely new analysis for each value of d, is not hindered by large values of d and, based on our case studies, performs as well as or better than existing methods.

9.6.2 Prediction via inverse regression

For notational convenience, let $Z_d = (w_1^T X, \ldots, w_d^T X)^T$. The premise underlying the methodology of this section is that it is generally easier to adequately model the inverse regression of Z_d on Y than it is to model the forward regression of Y on Z_d since the regression of Z_d on Y consists of d simple regressions. The inverse regression of Z_d on Y can itself be inverted to provide a method for estimating the forward mean function $\mathrm{E}(Y \mid X)$ without specifying explicitly a model for the forward regression. We denote the density of $Z_d \mid Y$ by $g(z_d \mid Y)$. Then

$$\mathrm{E}(Y \mid Z_d = z_d) = \frac{\mathrm{E}\{Yg(z_d \mid Y)\}}{\mathrm{E}\{g(z_d \mid Y)\}}, \tag{9.8}$$

where all right-hand side expectations are with respect to the marginal distribution of Y. A sample version of (9.8) provides an estimator of the forward mean function:

$$\begin{aligned} \widehat{\mathrm{E}}(Y \mid Z_d = z_d) &= \sum_{i=1}^{n} \omega_i(z_d) Y_i \\ \omega_i(z_d) &= \frac{\widehat{g}(z_d \mid Y_i)}{\sum_{i=1}^{n} \widehat{g}(z_d \mid Y_i)}, \end{aligned} \tag{9.9}$$

where $\widehat{g}$ denotes an estimated density. This estimator may be reminiscent of a non-parametric kernel estimator (e.g. Simonoff, 1996, Ch. 4), but there are important differences. The weights in a kernel estimator do not depend on the response, while the weights ω_i here do. Multivariate kernels are usually taken to be the product of univariate kernels, corresponding here to constraining the components of Z_d to be independent, but (9.9) requires no such constraint. Finally, there is no explicit bandwidth in our weights as they are determined entirely from $\widehat{g}$, which eliminates the need for bandwidth estimation by, for example, cross validation .

To get an operational version of (9.9), we need to decide how to get the estimated densities. One straightforward method is based on the multivariate normal model

$$Z_d = \mu_d + \Theta_d f(Y) + \varepsilon_d, \tag{9.10}$$

where $f \in \mathbb{R}^s$ is a known user-specified vector-valued function of the response, Θ_d is a $d \times s$ matrix of regression coefficients and $\varepsilon_d \sim N_d(0, \Sigma_{Z_d|Y})$ is independent of Y.

For instance, suppose that by inspecting plots of the $d = 10$ components of Z_d versus Y we conclude correctly that each element of $\mathrm{E}(Z_d \mid Y)$ is at most a quadratic function of Y, implying that $f(Y) = (Y, Y^2)^T$ and that

$$Z_{10} = \mu + \Theta_{10\times 2} \begin{pmatrix} Y \\ Y^2 \end{pmatrix} + \varepsilon.$$

Let $S_{Z_d|Y}$ denote the covariance matrix of the residuals and let

$$\widehat{Z}_{d,i} = \widehat{\mu}_d + \widehat{\Theta}_d f(Y_i),\ i = 1, \ldots, n$$

denote the fitted values from a fit of (9.10). Then based on model (9.10), we have

$$\omega_i(z_d) \propto \exp\left\{-\frac{1}{2}(z_d - \widehat{Z}_{d,i})^T S_{Z_d|Y}^{-1}(z_d - \widehat{Z}_{d,i})\right\}.$$

These weights are then substituted into (9.9) to obtain the predictions $\widehat{Y}(z_d) = \widehat{E}(Y \mid Z_d = z_d)$, which can be used with a holdout sample or cross validation to estimate the predictive mean squared error.

The performance of this predictive methodology depends on the approximate normality of Z_d and the choice of f. Since low-dimensional projections of high-dimensional data are approximately normal (Diaconis and Freedman, 1984; Hall and Li, 1993), it is reasonable to rely on approximate normality of Z_d. Consistency and the asymptotic distribution for prediction were established by Forzani et al. (2019) for the case of p fixed and $n \to \infty$. There are several generic possibilities for choice of f, perhaps guided by graphics as implied earlier. Fractional polynomials (Royston and Sauerbrei, 2008) or polynomials deriving from a Taylor approximation may be useful for single-response regressions,

$$f(Y) = \{Y, Y^2, Y^3, \ldots, Y^s\}^T,$$

are one possibility. Periodic behavior could be modeled using a Fourier series form

$$f(Y) = \{\cos(2\pi Y), \sin(2\pi Y), \ldots, \cos(2\pi k Y), \sin(2\pi k Y)\}^T$$

as perhaps in signal processing applications. Here, k is a user-selected integer and $s = 2k$. Splines and other types of non-parametric constructions could also be used to form a suitable f. A variety of basis functions are available from Adragni (2009).

Another option with a single continuous response consists of "slicing" the observed values of Y into H bins C_h, $h = 1, \ldots, H$. We can then set $s = H - 1$ and specify the h-th element of f to be $J_h(Y)$, where J is the indicator function, $J_h(Y) = 1$ if Y is in bin h and 0 otherwise. This has the effect of approximating each component $\mathrm{E}(w_j^T X \mid Y)$ of $\mathrm{E}(Z_d \mid Y)$ as a step function of Y with H steps,

$$\mathrm{E}(Z_d \mid Y) \approx \mu + \xi J,$$

where $J = (J_1(Y), J_2(Y), \ldots, J_{H-1}(Y)^T)$.

Cook and Forzani (2008b) studied a generalized version of model (9.10) that leads to principal fitted components. Their large sample results apply to our present case and help us understand the importance of normality and the choice of f. Consider model (9.10) but with possibly non-normal errors that are independent of Y and have finite fourth moments. The fitted model has mean function $\mu_d + \Theta_d f(Y)$ but we no longer assume that $\mathrm{E}(Z_d \mid Y) = \mu_d + \Theta_d f(Y)$ and thus allowing for the possibility that f is incorrect. It follows from (Cook and Forzani, 2008b, Section 3.2) that our estimators are $\sqrt{n}$ consistent with p fixed and $n \to \infty$ if and only if the matrix of correlations between $\mathrm{E}(Z_d \mid Y)$ and $f(Y)$ has rank q. This indicates that we can still expect useful results when f is misspecified provided there is sufficient correlation with the true mean function.

9.7 Tecator data

The Tecator data is a well-known chemometrics dataset that consists of 215 NIR absorbance spectra of meat samples, recorded on a Tecator Infratec Food and Feed Analyzer (Boggaard and Thodberg, 1992). The three response variables are the percent moisture, fat, and protein content in each meat sample as determined by analytic chemistry. The spectra ranged from 850 to 1050 nm, discretized into $p = 100$ wavelength values which are the predictors. Training and testing datasets contain 172 and 43 samples. A plot of the absorbance versus wavelength is available from Shan et al. (2014).

Figure 9.1 shows four diagnostic plots for the response fat. Plot (a) corresponds to the PLS1 graphical procedure described in Section 9.3. Under the linearity condition for the CMS, we have from (9.4) that $\beta \subseteq \mathcal{S}_{\mathrm{E}(Y|X)}$. Accordingly, the trend in a plot of fat vs. $\widehat{\beta}^T_{\mathrm{npls}} X$ should appear linear if model (2.5) holds and non-linear under model (9.1). Because the plot is clearly curved as seen by the fitted quadratic, we conclude that there is evidence for model (9.1). This is supported by plot (b) which is a scatterplot of the residuals $r_i = Y_i - \bar{Y} - \widehat{\beta}^T_{\mathrm{npls}} X_i$ vs. $\widehat{\beta}^T_{\mathrm{npls}} X_i$. In plots (a) and (b), the horizontal axis can also be interpreted as $R_1^T X_i$, as discussed under step 4 of the graphical procedure described in Section 9.5.3. Plots (c) and (d) correspond to the second and third plots of step 4. These plots can show if there is non-linearity present

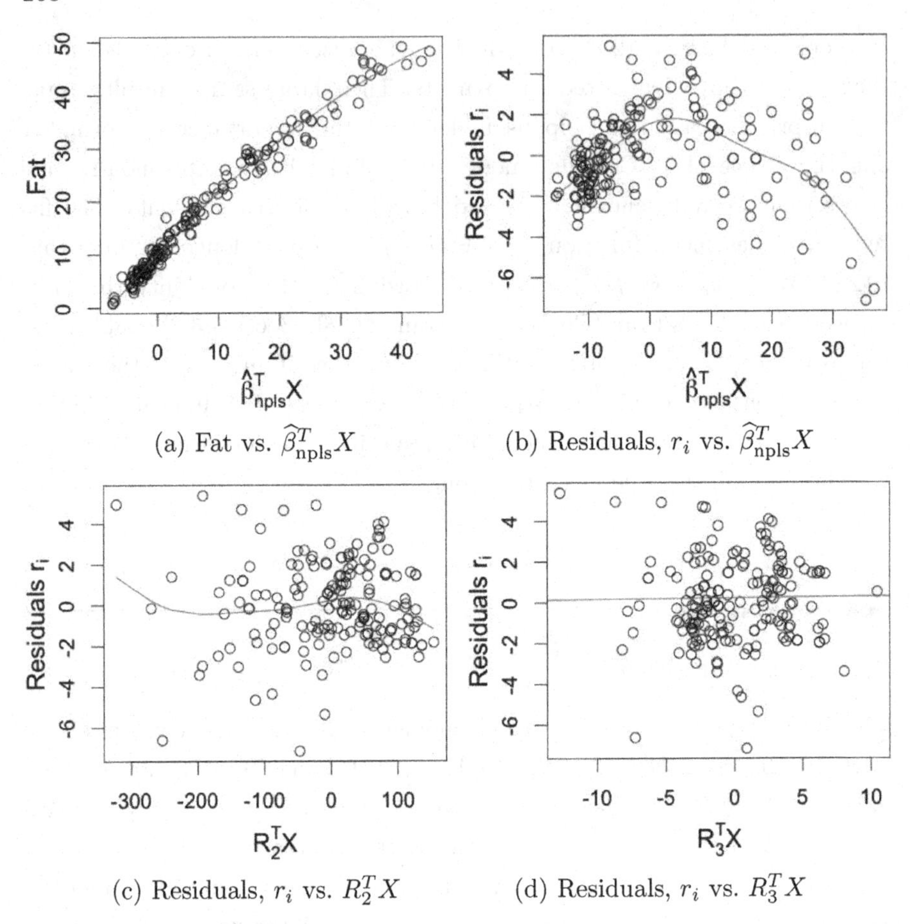

(a) Fat vs. $\widehat{\beta}^T_{\text{npls}}X$

(b) Residuals, r_i vs. $\widehat{\beta}^T_{\text{npls}}X$

(c) Residuals, r_i vs. $R_2^T X$

(d) Residuals, r_i vs. $R_3^T X$

FIGURE 9.1
Diagnostic plots for fat in the Tecator data. Plot (a) was smoothed with a quadratic polynomial. The other three plots were smoothed with splines of order 5. (From Fig. 1 of Cook and Forzani (2021) with permission.)

beyond that displayed in plots (a) and (b). In this case there does not seem to be any notable non-linearity remaining and consequently we could proceed with modeling based on plot (a). In particular, a reasonable model would have $\mathrm{E}(\text{fat} \mid X) = f(\beta^T X)$ for some scalar-valued function f that is likely close to a quadratic.

Figures 9.2 and 9.3 show for protein and water content the same constructions as Figure 9.1 does for fat content. Our interpretation of these plots is qualitatively similar to that for Figure 9.1: In this case there does not seem

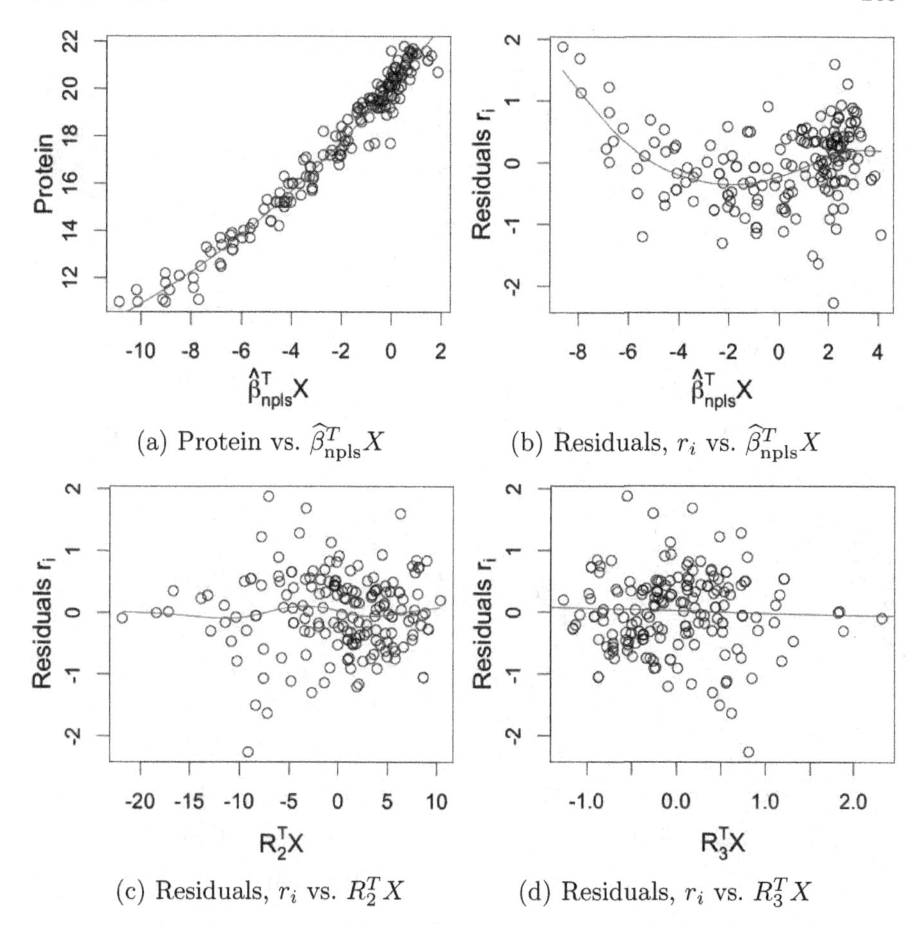

(a) Protein vs. $\widehat{\beta}^T_{\text{npls}}X$

(b) Residuals, r_i vs. $\widehat{\beta}^T_{\text{npls}}X$

(c) Residuals, r_i vs. $R_2^T X$

(d) Residuals, r_i vs. $R_3^T X$

FIGURE 9.2
Diagnostic plots for protein in the Tecator data. Plot (a) was smoothed with a quadratic polynomial. The other three plots were smoothed with splines of order 5.

to be any notable non-linearity remaining and consequently we could proceed in each case with modeling based on plot (a). Since each plot (a) seems well-described by a quadratic, the overall three-component model, with one component for each response, suggested by these plots is

$$Y = \begin{pmatrix} \text{Fat} \\ \text{Protein} \\ \text{Water} \end{pmatrix} = \begin{pmatrix} \beta_{10} + \beta_{11}(\alpha_1^T X) + \beta_{12}(\alpha_1^T X)^2 \\ \beta_{20} + \beta_{21}(\alpha_2^T X) + \beta_{22}(\alpha_2^T X)^2 \\ \beta_{30} + \beta_{31}(\alpha_3^T X) + \beta_{32}(\alpha_3^T X)^2 \end{pmatrix} + \varepsilon, \tag{9.11}$$

where $\|\alpha_j\| = 1$, $j = 1, 2, 3$. Sample size permitting, this model can now

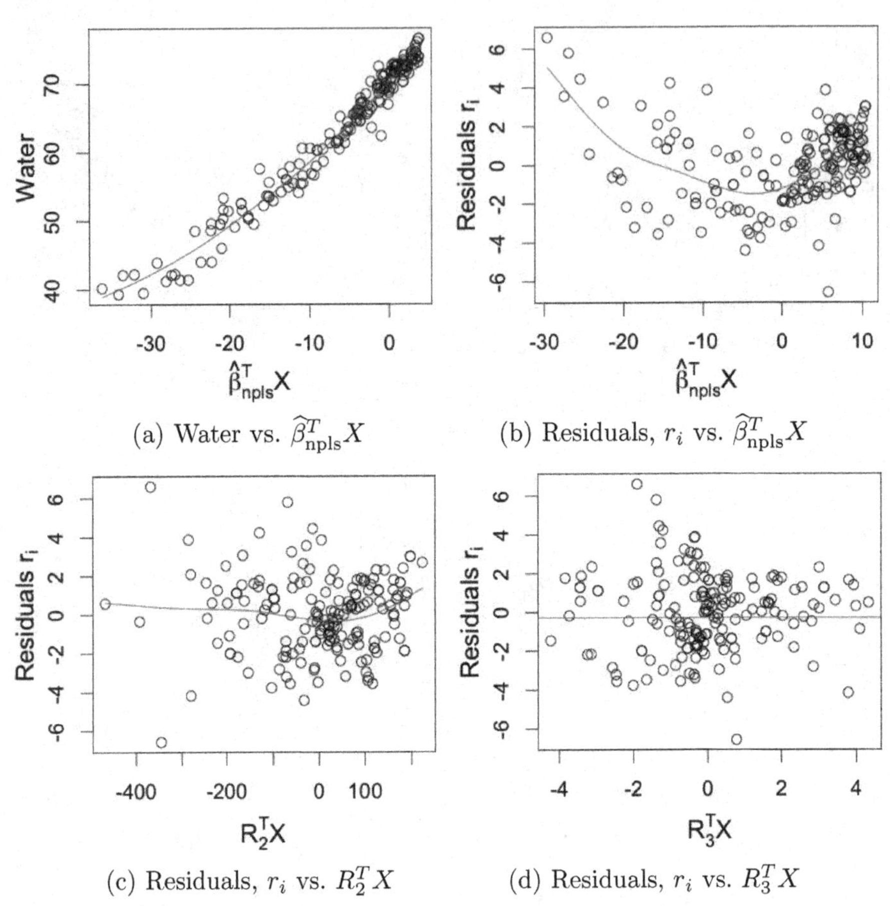

(a) Water vs. $\widehat{\beta}^T_{\text{npls}}X$

(b) Residuals, r_i vs. $\widehat{\beta}^T_{\text{npls}}X$

(c) Residuals, r_i vs. R_2^TX

(d) Residuals, r_i vs. R_3^TX

FIGURE 9.3

Diagnostic plots for water in the Tecator data. Plot (a) was smoothed with a quadratic polynomial. The other three plots were smoothed with splines of order 5.

be fitted and studied using traditional methods for non-linear regression. When the sample size is deficient, the α_j's can be replaced by their estimates $\widehat{\alpha}_j = \widehat{\beta}_{\text{npls}}/\|\widehat{\beta}_{\text{npls}}\|$ coming from a PLS algorithm, yielding a model in one compressed predictor, $\widehat{\alpha}_j^TX$, $j = 1, 2, 3$, for each response.

$$Y = \begin{pmatrix} \text{Fat} \\ \text{Protein} \\ \text{Water} \end{pmatrix} = \begin{pmatrix} \beta_{10} + \beta_{11}(\widehat{\alpha}_1^TX) + \beta_{12}(\widehat{\alpha}_1^TX)^2 \\ \beta_{20} + \beta_{21}(\widehat{\alpha}_2^TX) + \beta_{22}(\widehat{\alpha}_2^TX)^2 \\ \beta_{30} + \beta_{31}(\widehat{\alpha}_3^TX) + \beta_{32}(\widehat{\alpha}_3^TX)^2 \end{pmatrix} + \varepsilon. \quad (9.12)$$

Standard inferences based on this model will be optimistic because they do not take the variation in the $\widehat{\alpha}_j$'s into account, although assessing predictive performance using cross validation or a holdout sample is still feasible.

In some regressions it may be worthwhile to entertain the notion that each response is responding to the same linear combination of the predictors, so $\alpha_1 = \alpha_2 = \alpha_3 := \alpha$. This leads to the model

$$Y = \beta_0 + \beta^T \begin{pmatrix} (\alpha^T X) \\ (\alpha^T X)^2 \end{pmatrix},$$

where β_0 is a 3×1 vector of intercepts and β is a 2×3 matrix regression coefficients. For the Tecator data, the correlations among the fitted vectors $\widehat{\beta}^T_{\text{npls}} X$ are

$$\begin{matrix} \text{Fat} \\ \text{Water} \\ \text{Protein} \end{matrix} \begin{pmatrix} 1.000 & -0.989 & -0.862 \\ -0.989 & 1.000 & 0.820 \\ -0.862 & 0.820 & 1.000 \end{pmatrix}.$$

Because the absolute correlation between Fat and Water fits is high, it may be reasonable to conjecture that they are controlled by the same linear combination of the predictors. This conjecture in terms of model (9.11) is that $\alpha_1 = \alpha_2$. Contrasting the fits of model (9.11) with the version in which $\alpha_1 = \alpha_2 \neq \alpha_3$ would then be a natural next step.

We next describe five methods that we used to illustrate how prediction might be carried out in a non-linear setting:

1. Linear forward PLS.

Here we used NIPALS as summarized in Table 3.1 to fit each response individually. The number components was determined by using 5-fold predictive cross validation on the training data. This gave $q = 13, 13$, and 17 for fat, protein and water. In view of the curvature present in panel (a) of Figures 9.1–9.3 we would expect non-linear methods to give smaller mean squared prediction errors.

2. Quadratic forward PLS.

The fitted model for prediction is as shown in (9.12) with the coefficients determined by using linear PLS. The number of components was again determined by using 5-fold predictive cross validation on the training data, giving $q = 12, 13$, and 16 for fat, protein and water. This method will perform best when the CMS has dimension 1, as suggested by the plots in Figures 9.1–9.3, and the linearity condition holds. If the dimension of the CMS is greater than 1, this method could still give prediction mean squared errors that are smaller than those from linear PLS, provided that the response surface is well approximated by a quadratic.

3. Non-parametric forward PLS.

This method is similar to quadratic PLS, except the regression of Y on $\widehat{\beta}_{\text{npls}}^T X$ was fitted using a non-parametric method instead of a quadratic. We used the `R` package `np` that uses kernel regression estimates. Details of this method are available in the literature that accompanies the package. Our choice here was based mainly on convenience; many other non-parametric fitting methods are available.

The number of components was again determined by using 5-fold predictive cross validation on the training data, giving $q = 13, 14$, and 17 for fat, protein and water. Like quadratic PLS, this method will perform best under the single index model; that is, when the CMS has dimension 1 and the linearity condition holds, as described in Corollary 9.2 and Proposition 9.5. If the dimension of the CMS is greater than 1, this method could still give prediction mean squared errors that are smaller than those from linear PLS, without requiring that the response surface be well approximated by a quadratic. Asymptotic results, including consistency, are available from Forzani, Rodriguez, and Sued (2023).

4. Non-parametric inverse PLS (NP-I-PLS) using $\widehat{\beta}_{\text{npls}}^T X$.

This is a version of the method described in Section 9.6.2 where the weights (9.9) were determined using $Z_d = \widehat{\beta}_{\text{npls}}^T X$ and $f(Y) = (Y, Y^2, Y^{1/2})$ was selected by graphical inspection. The number components for $\widehat{\beta}_{\text{npls}}$ was again selected by using 5-fold predictive cross validation on the training data, giving $q = 13, 18$, and 18 for fat, protein and water. Like quadratic PLS, and non-parametric forward PLS, this method will perform best when the CMS has dimension 1 and the linearity condition holds but may still be useful otherwise.

5. Non-parametric inverse PLS (NP-I-PLS) using $W^T X$.

This is the method described in Section 9.6.2 using $f(Y) = (Y, Y^2, Y^{1/2})$. The number components for W was again determined by using 5-fold predictive

TABLE 9.1

Tecator Data: Number of components based on 5-fold cross validation using the training data for each response. NP-PLS denotes non-parametric PLS and NP-I-PLS denotes non-parametric inverse PLS.

Method	Fat	Protein	Water
1. Linear PLS	13	13	17
2. Quadratic PLS	12	13	16
3. NP-PLS	13	14	17
4. NP-I- PLS with $\widehat{\beta}^T_{\text{npls}}X$	13	18	18
5. NP-I-PLS with W^TX	15	17	22

cross validation on the training data, giving $q = 15, 17$, and 22 for fat, protein and water. The previous four methods will all be at their best when the dimension of the CMS is 1. This method does not have that restriction and should work well even when the dimension of the CMS is greater than one.

The number of components for each method-response combination is summarized in Table 9.1 and the root mean squared prediction errors

$$\left((1/43)\sum_{i=1}^{43}\left(\widehat{Y}_{\text{test},i} - Y_{\text{test},i}\right)^2\right)^{1/2}$$

determined from the 43 observations in the testing dataset are given in Table 9.2(a). The root mean squared prediction errors typically changed relatively little when the number of components was varied by, say, ± 2. This is illustrated in Figure 9.4, which shows the cross validation root mean squared prediction error for linear PLS, method 1, and non-parametric inverse PLS with $\widehat{\beta}^T_{\text{npls}}X$, method 4. The prediction errors are much larger for q outside the range of the horizontal axis shown in Figure 9.4.

The number of training cases in this dataset $n = 172$ is greater than the number of predictors, $p = 100$, but none of the prediction methods used here requires that $n > p$ and all are serviceable when $n < p$.

The first method, linear PLS, is the same as the standard NIPALS algorithm outlined in Table 3.1. From Table 9.2(a) it seems clear that in this example any of methods 2–4 is preferable to the NIPALS algorithm. This is perhaps not surprising in view of the curvature shown in panel (a) of Figures 9.1–9.3. Methods 2–4 are all based on the NIPALS fitted values $\widehat{\beta}^T_{\text{npls}}X$. As mentioned

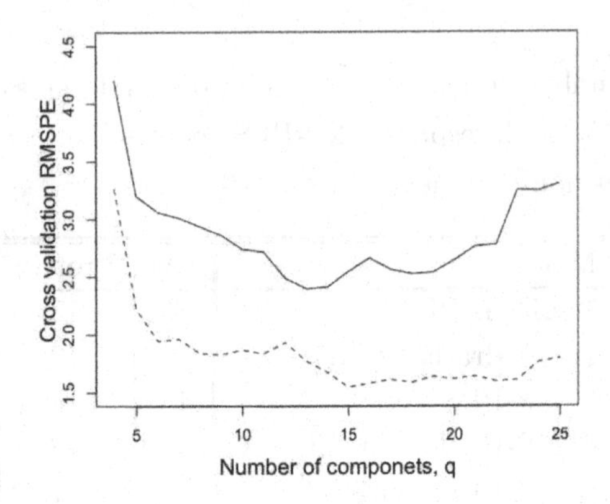

FIGURE 9.4
Tecator data: root mean squared prediction error versus number of components for fat. Upper curve is for linear PLS, method 1 in Table 9.1; lower curve is for inverse PLS prediction with $\widehat{\beta}^T_{\text{npls}}X$, method 4. (From Fig. 2 of Cook and Forzani (2021) with permission.)

previously, these methods will do their best when the CMS has dimension 1. We judge their relative behavior in this example to be similar, except it might be argued that non-parametric PLS demonstrates a slight advantage. Method 5, non-parametric inverse PLS with W^TX, has the smallest root mean squared prediction errors in this example and is clearly the winner. This method does not rely on the CMS having one dimension, and we expect that it will dominate in any analysis where linear PLS regression is judged to be serviceable apart from the presence of non-linearity in the conditional mean $\text{E}(Y \mid X)$.

As discussed in Section 9.6.1, Shan et al. (2014) propose a new method, called PLS-SLT, of non-linear PLS based on slicing the response into non-overlapping bins within which the relationship between Y and X is assumed to follow linear model (1.1). In their analysis of the Tecator data, PLT-SLT showed superior predictive performance against linear PLS and five competing non-linear methods. Table 9.2(b) gives the root mean squared prediction error reported by (Shan et al., 2014, Table 6) for linear PLS and their new method PLS-SLT. They used 12 components for each of the six scenarios shown in Table 9.2(b). Although we used 13 components for fat and protein, the root mean squared prediction errors for linear PLS shown in Table 9.2 are nearly identical. We used 17 components for the linear PLS analysis of water, while Shan et al. (2014) used 12. This may account for the discrepancy

TABLE 9.2

Tecator Data:(a) Root mean squared training and testing prediction errors for five methods of prediction.

	Fat		Protein		Water	
(a) Method	Train	Test	Train	Test	Train	Test
1. Linear PLS	2.04	2.10	0.61	0.62	1.48	1.72
2. Quadratic PLS	1.69	1.85	0.56	0.60	1.30	1.51
3. NP-PLS	1.29	1.90	0.51	0.58	1.19	1.32
4. NP-I-PLS with $\widehat{\beta}^T_{\text{npls}}X$	1.66	1.91	0.54	0.62	1.20	1.41
5. NP-I-PLS with W^TX	1.40	1.49	0.47	0.54	0.88	1.32
(b) Method						
I. Linear PLS	–	2.10	–	0.63	–	2.03
II. PLS-SLT	–	1.66	–	0.60	–	1.36

The three responses were analyzed separately as in PLS1. NP-PLS denotes non-parametric PLS and NP-I-PLS denotes non-parametric inverse PLS. (b) Root mean squared testing prediction error for two methods of prediction from Shan et al. (2014). PLS-SLT is the slice transform version of PLS proposed by Shan et al. (2014).

in linear PLS prediction errors for water, 1.72 and 2.03 in parts (a) and (b) of Table 9.2. Most importantly, the root mean squared prediction error for PLS-SLT is comparable to or smaller than the prediction errors for methods 1–4 in Table 9.2 (a), but its predictor errors are larger than those for method 5, non-parametric inverse PLS with W^TX.

The methods discussed here all require a number of user-selected tuning parameters or specifications. Our method 5, non-parametric inverse regression with $Z_d = W^TX$, requires $f(y)$ and the number of components q represented in W. Recall that we selected $f(y) = (y, y^2, \sqrt{y})^T$ based on graphical inspection of inverse regression functions, and we selected q by using 5-fold cross validation on the training data. PLS-SLT requires selecting the number of bins and the number of components, which we still represented as q. Shan et al. (2014) restricted q to be at most 15 and the number of bins to be at most 10. For each number of bins, they determined the optimal value of q by using a testing procedure proposed by Haaland and Thomas (1988). The number of bins giving the best predictive performance was then selected. We conclude from these descriptions that the procedures for selecting the tuning parameters are an integral part of the methods; changing those procedures in effect changes the method. Thus, although based on different numbers of components, it is fair to compare the root mean squared prediction errors in Table 9.2.

9.8 Etanercept data

Chiappini et al. (2020) used a multi-layer perceptron (MLP) artificial neural network to develop a novel multivariate calibration strategy for the online prediction of etanercept, a protein used in the treatment of autoimmune diseases like rheumatoid arthritis. An artificial neural network was used instead of a traditional PLS algorithm because of the expected non-linear relationship between etanercept concentrations and the spectral predictors.

Etanercept samples were studied using fluorescence excitation-emission data generated in the spectral ranges 225.0 and 495.0 nm and 250.0 and 599.5 nm for excitation and emission modes. Their experimental protocol resulted in $38,500$ data points for the vectorized set of excitation and emission wavelengths. After deletion of one outlier, their full data set consisted of $n = 35$ independent samples, each consisting of $r = 1$ etanercept concentration and $p = 38,500$ spectral readings as predictors. However, it is difficult if not impossible to adequately train an artificial neural network with so many predictors (Despagne and Luc Massart, 1998) and, in consequence, Chiappini et al. (2020) first reduced the number of predictors by systematically removing wavelengths to yield a more manageable p = 9,623 predictors. This number is still too large and so the authors further reduced these predictors to between 4 and 12 principal components, which then constituted the input layer for their network. They trained their MLP network on a subset of $n = 26$ training samples, reserving the remaining 9 samples for testing and estimating the root mean squared prediction error. Their training protocol consisted of running a Box-Benken design with 5 center points, 4, 8, and 12 input nodes (the principal components computed from the $p = 9,623$ spectral readings); 2, 6, and 10 nodes for layer 1; 0, 4, and 8 nodes for layer 2 and one output node. They used a desirability function to select the optimum settings for their MLP network. After considerable analysis, their protocol resulted in 12 input variables, two nodes at layer 1 and 0 nodes at layer 2. For their final MLP

TABLE 9.3

Etanercept Data: Numbering of the methods corresponds to that in Table 9.2. Part (a): Number of components was based on leave-one-out cross validation using the training data with $p = 9{,}623$ predictors. RMSEP is the root mean squared error of prediction for the testing data. Part (b): The number of components for MLP is the number of principal components selected for the input layer of the network. Part (c): Number of components was based on leave-one-out cross validation using all 35 data points. CVRPE is the cross validation root mean square error of prediction based on the selected 6 components.

(a) Method, $p = 9{,}623$	Components	RMSEP
1. Linear PLS	7	15.28
3. NP-PLS	4	15.10
4. NP-I-PLS with $\widehat{\beta}^T_{\text{npls}}X$	5	12.43
5. NP-I-PLS with W^TX	6	9.20
(b) MLP Network, $p = 12$	12	8.60
(c) Method, $p = 38{,}500$	Components	CVRPE
NP-I-PLS with W^TX	6	8.46

network, the root mean squared prediction error for the testing data was 8.6, which is considerably smaller than the corresponding error of 15.28 from their implementation of PLS.

We applied four of the methods used for Table 9.2 to the same training and testing data with $p = 9,623$. Quadratic PLS (method 2 in Table 9.2) was not considered because the response surface was noticeably non-quadratic. The fitting function $f(Y) = (Y, Y^2, Y^3, Y^{1/2})^T$ for inverse PLS with W^TX (see Section 9.6.2) was selected by smoothing a few plots of predictors versus the response from the training data. Using the testing data, the root mean squared prediction errors for these four methods are shown in Table 9.3(a), along with the prediction error for the final MLP network in part (b). The prediction error of 9.20, which is the best of those considered, is a bit larger than the MLP error found by Chiappini et al. (2020) but still considerably smaller than the PLS error.

Wold et al. (2001) recommended that "A good nonlinear regression technique has to be simple to use." We agree subject to the further condition that a simple technique must produce competitive or useful results. The general approach behind the MLP network requires considerable analysis and many subjective decisions, including the initial reduction to 9,623 spectral measurements followed by reduction to 12 principal components, outlier detection methodology, the type of experimental design used to train the network, the desirability function, specific characteristics of the network construction, division of the data into testing and training sets, and so on. Because of all these required data-analytic decisions, it is not clear to us that the relative performance of the neural network approach displayed in the etanercept data would hold in future analyses. In contrast, the inverse PLS approach with $W^T X$ proved best in the Tecator data and had a strong showing in the analysis of the etanercept data with many fewer data analytic decisions required.

To gain intuition into what might be achieved by minimizing the number of data-analytic decisions required, we conducted another analysis using NP-I-PLS with $W^T X$ based on the original 38,500 predictors and $n = 35$ data points. For each fixed number of components, the root mean square error of prediction was estimated by using leave-one-out cross validation with the fitting function used previously. For each subset of 34 observations, the number of components minimizing the prediction error was always 6. The cross validation root mean square error of prediction was determined to be 8.46 as shown in part (c) of Table 9.3. This relatively simple method requires only two essential data-analytic choices: the number of components and the fitting function $f(Y)$, and yet it produced a root mean square error of prediction that is smaller than that found by the carefully tuned MLP network.

9.9 Solvent data

Lavoie, Muteki, and Gosselin (2019) reaffirmed the general inappropriateness of linear PLS in the presence of non-linear relationships and proposed a new non-linear PLS algorithm. They used constrained piecewise cubic splines to model non-linear relationships iteratively, and calculated variable weights using a methods originally proposed by Indahl (2005). While we find their method to be somewhat elusive, it performed quite well against four other proposed non-linear PLS methods in the three case studies that they reported. One of their case studies used a dataset on properties of different solvents. We

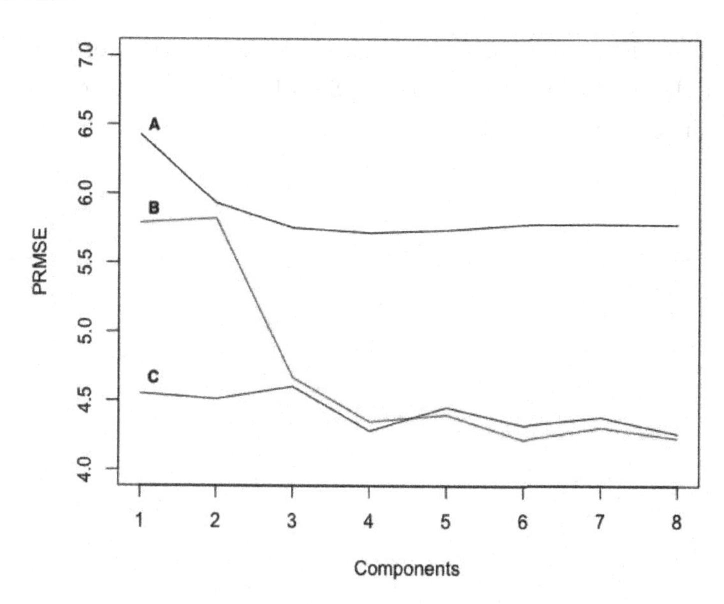

FIGURE 9.5
Solvent data: Predictive root mean squared error PRMSE versus number of components for three methods of fitting. A: linear PLS. B: non-parametric inverse PLS with W^TX, as discussed in Section 9.7. C: The non-linear PLS method proposed by Lavoie et al. (2019).

used the same solvent data to compare the method proposed by Lavoie et al. (2019) to linear PLS and to non-parametric inverse PLS with W^TX.

We studied the solvent data following the steps described by Lavoie et al. (2019). For instance, we removed the 5 observations that Lavoie et al. (2019) flagged outliers, leaving $n = 98$ observations on $p = 8$ chemical properties of the solvents as predictors and one response, the dielectric constant. Ten-fold cross validation was used to measure the predictive root mean squared error, PRMSE.

The results of our comparative study are shown in Figure 9.5. Curve A gives the PRMSE for linear PLS and it closely matches the results of Lavoie et al. (2019, Fig. 6(b), curve A). Curve C gives the PRMSE for the Lavoie et al. method read directly from their Figure 6. Curve B gives the PRMSE of our proposed NP-I-PLS with W^TX using the fractional polynomial (e.g. Royston and Sauerbrei, 2008) fitting function $f(Y) = (Y^{1/3}, Y^{1/2}, Y, Y^2, Y^3, \log Y)^T$. Figure 9.5 reinforces our previous conclusions that the dimension reduction step of linear PLS algorithms is serviceable in the presence of non-linearity and that the non-parametric inverse PLS with W^TX method competes well against all other methods.

10

The Role of PLS in Social Science Path Analyses

In this chapter, we describe the current and potential future roles for partial least squares (PLS) algorithms in path analyses. After reviewing the present debate on the value of PLS for studying path models in the social sciences and establishing a context, we conclude that, depending on specific objectives, PLS methods have considerable promise, but that the present social science method identified as PLS is only weakly related to PLS and is perhaps more akin to maximum likelihood estimation. Developments necessary for integrating proper PLS into the social sciences are described. A critique of covariance-based structural equation modeling (CB|SEM), as it relates to PLS, is given as well. The discussion in this chapter follows Cook and Forzani (2023).

10.1 Introduction

Path modeling is a standard way of representing social science theories. It often involves concepts like "customer satisfaction" or "competitiveness" for which there are no objective measurement scales. Since such concepts cannot be measured directly, multiple surrogates, which may be called indicators, observed or manifest variables, are used to gain information about them indirectly. One role of a path diagram is to provide a representation of the relationships between the concepts, which are represented by latent variables, and the indicators. A fully executed path diagram is in effect a model that can be used to guide subsequent analysis, rather like an algebraic model in statistics.

DOI: 10.1201/9781003482475-10

A path model is commonly translated to a structural equation model (SEM) for analysis. There are two common paradigms for studying a path diagram after translation to an SEM, called partial least squares (PLS|SEM) and covariance-based analysis (CB|SEM). The acronyms just given are used here to distinguish between the PLS methods used to study SEMs and PLS methods more generally. These paradigms have for decades been used for estimation and inference, but their advocates now seem at loggerheads over relative merit, due in part to publications that are highly critical of PLS|SEM methods.

10.1.1 Background on the PLS|SEM vs CB|SEM debate

Much discussion has taken place in the social sciences literature about the role of PLS|SEM in path modeling, including Akter, Wamba, and Dewan (2017), Evermann and Rönkkö (2021), Goodhue, Lewis, and Thompson (2023), Henseler et al. (2014), Rönkkö and Evermann (2013), Russo and Stol (2023), Sarstedt et al. (2016), Sharma et al. (2022), and Rönkkö et al. (2016b).

Rönkkö et al. (2016b) asserted that "...the majority of claims made in PLS literature should be categorized as statistical and methodological myths and urban legends," and recommended that the use of "...PLS should be discontinued until the methodological problems explained in [their] article have been fully addressed." In response, proponents of PLS|SEM appealed for a more equitable assessment. Henseler and his nine coauthors (Henseler et al., 2014) tried to reestablish a constructive discussion on the role PLS|SEM in path analysis while arguing that, rather than moving the debate forward, Rönkkö and Evermann (2013) created new myths. McIntosh, Edwards, and Antonakis (2014) attempted to resolve the issues separating Rönkkö and Evermann (2013) and Henseler et al. (2014) in an "even-handed manner," concluding by lending their support to Rigdon's recommendation that steps be taken to divorce PLS|SEM path analysis from its competitors by developing methodology from a purely PLS perspective (Rigdon, 2012). The appeal for balance was evidently found wanting by the Editors of *The Journal of Operations Management*, who established a policy of desk-rejecting papers that use PLS|SEM (Guide and Ketokivi, 2015), and by Rönkkö et al. (2016b), who restated and reinforced the views of Rönkkö and Evermann (2013).

Later, Evermann and Rönkkö (2021) tempered their ardent criticism of PLS|SEM when writing "...to ensure that [Information Systems (IS)] researchers have up-to-date methodological knowledge of PLS if they decide to use it." Although dressed up a bit to highlight new developments, their criticism are essentially the same as those they leveled in previous writings. There was much give-and-take in the subsequent discussion, both for and against PLS|SEM. Goodhue et al. (2023) recommended against using PLS in path analysis, arguing that it "... violates accepted norms for statistical inference." They further recommended that key journals convene a task force to assess the advisability of accepting PLS-based work. Russo and Stol (2023) took exception to some of Evermann and Rönkkö (2021) conclusions, while commending their efforts. Sharma et al. (2022) expressed matter-of-factly that the PLS|SEM claims leveled by Evermann and Rönkkö (2021) are misleading, extraordinary and questionable. They then set about "... to bring a positive perspective to this debate and highlight the recent developments in PLS that make it an increasingly valuable technique in IS and management research in general".

There is a substantial literature that bears on this debate (Rönkkö et al., 2016b, cites about 150 references). The preponderance of articles rely mostly on intuition and simulations to support sweeping statements about intrinsically mathematical/statistical issues without sufficient supporting theory. But adequately addressing the methodological issues in path analysis requires in part avoiding ambiguity by employing a degree of context-specific theoretical specificity.

In this chapter, we use the acronym PLS to designate the partial least squares methods that stem from the theoretical foundations by Cook, Helland, and Su (2013). These link with early work by Wold (Geladi, 1988) and cover chemometrics applications (Cook and Forzani, 2020, 2021; Martens and Næs, 1989), as well as a host of subsequent methods, particularly the PLS methods for simultaneous reduction of predictors and responses discussed in Chapter 5, which are relevant to the analysis of the path models. We rely on these foundations in this chapter.

10.1.2 Chapter goals

We elucidate PLS|SEM by describing what it can and cannot achieve and by comparing it to widely accepted methods like maximum likelihood. We also describe the potential of PLS methods in path analyses and discuss CB|SEM as a necessary backdrop for PLS|SEM. We restrict discussion to reflective path models so the context is clear, as it seems to us that many claims are made without adequate foundations. By appealing to specific and relatively straightforward path models, we are able to state clearly results that carry over qualitatively to more intricate settings. And we are able to avoid terms and phrases that do not seem to be understood in the same way across the community of path modelers (e.g. McIntosh et al., 2014; Sarstedt et al., 2016). We took the articles by Wold (1982), Dijkstra (1983, 2010), and Dijkstra and Henseler (2015a,b) as the gold standard for technical details on PLS|SEM. Output from our implementation of their description of a PLS|SEM estimator agreed with output from an algorithm by Rönkkö et al. (2016a), which supports our assessment of the method. We confine our discussion largely to issues encountered in Wold's first-stage algorithm (Wold, 1982, Section 1.4.1).

10.1.3 Outline

To establish a degree of context-specific theoretical specificity, we cast our development in the framework of common reflective path models that are stated in Section 10.2 along with the context and goals. These models cover simulation results by Rönkkö et al. (2016b, Figure 1), which is one reason for their adoption. We show in Section 10.3 that the apparently novel observation that this setting implies a reduced rank regression (RRR) model for the observed variables (Anderson, 1951; Cook, Forzani, and Zhang, 2015; Izenman, 1975; Reinsel and Velu, 1998). From here we address identifiability and estimation using RRR. Estimators are discussed in Sections 10.4 and 10.5. The approach we take and the broad conclusions that we reach should be applicable to other perhaps more complicated contexts.

Some of the concerns expressed by Rönkkö et al. (2016b) involve contrasts with CB|SEM methodology. Building on RRR, we bring CB|SEM methodology into the discussion in Section 10.5 and show that under certain key assumptions additional parameters are identifiable under the SEM model. In Section 10.6 we quantify the bias and revisit the notion of consistency at large. The chapter concluded with simulation results and a general discussion.

10.2 A reflective path model

10.2.1 Path diagrams used in this chapter

The upper and lower path diagrams presented in Figure 10.1 are typical reflective path models. In each diagram elements of $Y = (Y_j) \in \mathbb{R}^r$, $X = (X_j) \in \mathbb{R}^p$

denote observable random variables, which are called *indicators*, that are assumed to reflect information about underlying real latent *constructs* $\xi, \eta \in \mathbb{R}$. This is indicated by the arrows in Figure 10.1 leading from the constructs to the indicators, which imply that the indicators reflect the construct. Boxed variables are observable while circled variables are not. This restriction to univariate constructs is common practice in the social sciences, but is not required. An important path modeling goal in this setting is to infer about the association between the latents η and ξ, as indicated by the double-headed curved paths between the constructs. We take this to be a general objective of both PLS|SEM and CB|SEM, and it is one focal point of this chapter. Each indicator is affected by an unobservable error ϵ. The absence of paths between these errors indicates that, in the upper path diagram, they are independent conditional on the latent variables. They are allowed to be conditionally dependent in the lower path diagram, as indicated by the double-arrowed paths that join them. We refer to the lower model as the correlated error (CORE) model and to the upper model as the uncorrelated error (UNCORE) model.

The models of Figure 10.1 occur frequently in the literature on path models (e.g. Lohmöller, 1989). The UNCORE model is essentially the same as that described by Wold (1982, Fig. 1) and it is the basis for Dijkstra and Henseler's studies (Dijkstra, 1983; Dijkstra and Henseler, 2015a,b) of the asymptotic properties of PLS|SEM. We have not seen a theoretical analysis of PLS|SEM under the CORE model.

For instance, Bollen (1989, pp. 228–235, Fig. 7.3) described a case study on the evolution of political democracy in developing countries. The variables $X_1, \ldots, X_p$ were indicators of specific aspects of a political democracy in developing countries in 1960, like measures of the freedom of the press and the fairness of elections. The latent construct ξ was seen as a real latent construct representing the level of democracy in 1960. The variables $Y_1, \ldots, Y_r$ were the same indicator from the same source in 1965, and correspondingly η was interpreted as a real latent construct that measures the level of democracy in 1965. One goal was to estimate the correlation between the levels of democracy in 1960 and 1965.

The path diagrams in Figure 10.1 are uncomplicated relative to those that may be used in sociological studies. For example, Vinzi, Trinchera, and Amato (2010) described a study of fashion in which an UNCORE model was imbedded in a larger path diagram with the latent constructs "Image" and "Character"

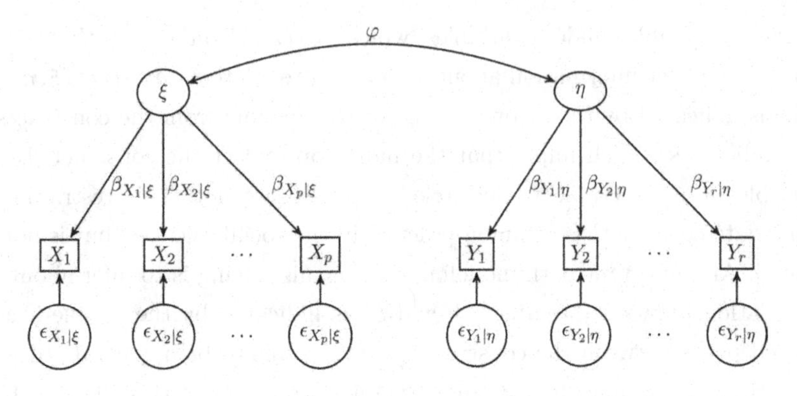

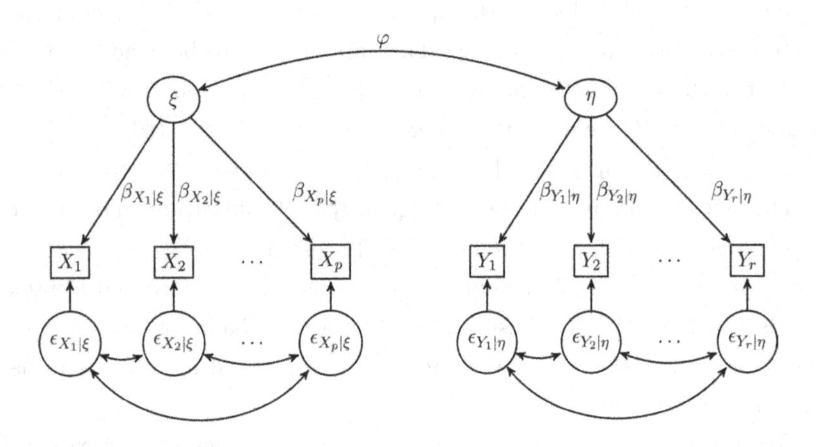

FIGURE 10.1
Reflective path diagrams relating two latent constructs ξ and η with their respective univariate indicators, $X_1, \ldots, X_p$ and $Y_1, \ldots, Y_r$. φ denotes either $\text{cor}(\xi, \eta)$ or $\Psi = \text{cor}\{E(\xi \mid X), E(\eta \mid Y)\}$. Upper and lower diagrams are the UNCORE and CORE models. (From Fig. 2.1 of Cook and Forzani (2023) with permission.)

serving as reflective indicators of a third latent construct "Brand Preference" with its own indicators. Our discussion centers largely on the path diagrams of Figure 10.1 since these are sufficient for the objectives of this chapter.

The models of Figure 10.1 are called *reflective* because the indicators reflect properties of the constructs, as indicated by the arrows pointing to the constructs from the indicators. These can also be viewed as regressions where the constructs represent a single predictor and the indicators collectively represent a multivariate response. If the direction of the arrows is reversed, so they point from the indicators to the constructs, then the path models are called *formative* because the indicators form the constructs. From a regression view, the indicators now become the predictors and the construct is the response.

10.2.2 Implied model

Assuming that all effects are additive, the upper and lower path diagrams in Figure 10.1 can be represented as a common multivariate (multi-response) regression model with the following stipulations:

$$\begin{pmatrix} \xi \\ \eta \end{pmatrix} = \begin{pmatrix} \mu_\xi \\ \mu_\eta \end{pmatrix} + \epsilon_{\xi,\eta}, \text{ where } \epsilon_{\xi,\eta} \sim N_2(0, \Sigma_{(\xi,\eta)}). \tag{10.1}$$

and

$$\left.\begin{aligned} Y &= \mu_Y + \beta_{Y|\eta}(\eta - \mu_\eta) + \epsilon_{Y|\eta}, \text{ where } \epsilon_{Y|\eta} \sim N_r(0, \Sigma_{Y|\eta}) \\ X &= \mu_X + \beta_{X|\xi}(\xi - \mu_\xi) + \epsilon_{X|\xi}, \text{ where } \epsilon_{X|\xi} \sim N_p(0, \Sigma_{X|\xi}). \end{aligned}\right\} \tag{10.2}$$

We use σ_ξ^2, $\sigma_{\xi,\eta}$ and σ_η^2 to denote the elements of $\Sigma_{(\xi,\eta)} \in \mathbb{R}^{2\times 2}$, and we further assume that $\epsilon_{\xi,\eta}$, $\epsilon_{X|\xi}$ and $\epsilon_{Y|\eta}$ are mutually independent so jointly X, Y, ξ, and η follow a multivariate normal distribution. Normality per se is not necessary. However, certain implications of normality like linear regressions with constant variances are needed in this chapter. Assuming normality overall avoids the need to provide a list of assumptions, which might obscure the overarching points about the role of PLS. As defined in Chapter 1, we use the notation $\Sigma_{U,V}$ to denote the matrix of covariances between the elements of the random vectors U and V, and we use $\beta_{U|V} = \Sigma_{U,V}\Sigma_V^{-1}$ to indicate the matrix of population coefficients from multi-response linear regression of U on V, where $\Sigma_V = \text{var}(V)$. Lemma A.5 in Appendix A.8.1 gives the mean and variance of the joint multivariate normal distribution of X, Y, ξ, and η. Since only X, Y are observable, all estimation and inference must be based on the joint multivariate distribution of (X, Y).

10.2.3 Measuring association between η and ξ

The association between η and ξ has been measured in the path modeling literature via the conditional correlation $\Psi := \text{cor}\{\text{E}(\eta \mid Y), \text{E}(\xi \mid X)\}$ and via the marginal correlation $\mathrm{cor}(\eta, \xi)$. These measures can reflect quite different views of a reflective setting. The marginal correlation $\mathrm{cor}(\eta, \xi)$ implies that η and ξ represent concepts that are uniquely identified regardless of any subjective choices made regarding the variables Y and X that are reasoned to reflect their properties up to linear transformations. Two investigators studying the same constructs would naturally be estimating the same correlation even if they selected a different Y or a different X. In contrast, $\Psi = \mathrm{cor}\{\mathrm{E}(\eta \mid Y), \mathrm{E}(\xi \mid X)\}$, the correlation between population regressions, suggests a conditional view: η and ξ exist only by virtue of the variables that are selected to reflect their properties, and so attributes of η and ξ cannot be claimed without citing the corresponding (Y, X). Two investigators studying the same concepts could understandably be estimating different correlations if they selected a different Y or a different X. For instance, the latent construct "happiness" might reflect in part a combination of happiness at home and happiness at work. Two sets of indicators X and X^* that reflect these happiness subtypes differently might yield different correlations with a second latent construct, say "generosity." The use of Ψ as a measure of association can be motivated also by an appeal to dimension reduction. Under the models (10.2), $X \perp\!\!\!\perp \xi \mid \mathrm{E}(\xi \mid X)$ and $Y \perp\!\!\!\perp \eta \mid \mathrm{E}(\eta \mid Y)$. In consequence, the constructs affect their respective indicators only through the linear combinations given by the conditional means, $\mathrm{E}(\xi \mid X) = \Sigma_{X,\xi}^T \Sigma_X^{-1}(X - \mu_X)$ and $\mathrm{E}(\eta \mid Y) = \Sigma_{Y,\eta}^T \Sigma_Y^{-1}(Y - \mu_Y)$.

10.2.4 Construct interpretation

Rigdon (2012) questioned whether constructs like ξ and η are necessarily the same as the theoretical concepts they are intended to capture or whether they are approximations of the theoretical concepts allowed by the selected indicators. If the latter view is taken, then an understanding of the constructs cannot be divorced from the specific indicators selected for the study.

Further, Rigdon, Becker, and Sarstedt (2019) argued that there is *always* a gap between a targeted concept like consumer satisfaction or trustworthiness and its representation as a construct via selected indicators. Attempts to reify a concept through a specific set of indicators must naturally reflect the researchers' vision, insights and ideology, with the unavoidable implication that $\mathrm{cor}(\eta, \xi)$ is a researcher-specific parameter. Two independent researcher groups studying the same concepts, but using different indicators, will inherently be dealing with distinct correlations $\mathrm{cor}(\eta, \xi)$. Whether they arrive at the same substantive conclusions is a related issue that can also be relevant, but is beyond the scope of this book. The correlation Ψ brings these issues to the fore; its explicit dependence on the indicators manifests important transparency and it facilitates the reification of a concept, although it does not directly address discrepancies between the target concept and construct.

The diagrams in Figure 10.1 can be well described as instances of operational measurement theory in which concepts are defined in terms of the operations used to measure them (e.g. Bridgman, 1927; Hand, 2006). In operationalism there is no assumption of an underlying objective reality; the concept being pursued is precisely that which is being measured, so concept and construct are the same. In the alternate measurement theory of representationalism, the measurements are distinct from the underlying reality. There is a universal reality underlying the concept of length regardless of how it is measured. However, a universal understanding of "happiness" must rely on unanimity over the measurement process.

10.2.5 Constraints on ξ and η

The constructs η and ξ are not well-defined since any non-degenerate linear transformations of them lead to an equivalent model and, in consequence, it is useful to introduce harmless constraints to facilitate estimation, inference and interpretation. To this end, we consider two sets of constraints:

$$\begin{aligned}\text{Regression constraints}: &\quad \mu_\xi = \mu_\eta = 0 \text{ and } \mathrm{var}\{\mathrm{E}(\xi \mid X)\} = \mathrm{var}\{\mathrm{E}(\eta \mid Y)\} = 1.\\ \text{Marginal constraints}: &\quad \mu_\xi = \mu_\eta = 0 \text{ and } \mathrm{var}(\xi) = \mathrm{var}(\eta) = 1.\end{aligned}$$

The regression constraints fix the variances of the conditional means $\mathrm{E}(\xi \mid X)$ and $\mathrm{E}(\eta \mid Y)$ at 1, while the marginal constraints fix the marginal variances of ξ and η at 1. Under the regression constraints, $\Psi = \mathrm{cor}(\mathrm{E}(\eta \mid Y), \mathrm{E}(\xi \mid X)) = \mathrm{cov}(\mathrm{E}(\eta \mid Y), \mathrm{E}(\xi \mid X))$. Under the marginal constraints, $\mathrm{cor}(\xi, \eta) = \mathrm{cov}(\xi, \eta)$. The regression and marginal constraints are related via the variance decompositions

$$\begin{aligned}\mathrm{var}(\xi) &= \mathrm{var}\{\mathrm{E}(\xi \mid X)\} + \mathrm{E}\{\mathrm{var}(\xi \mid X)\}\\ \mathrm{var}(\eta) &= \mathrm{var}\{\mathrm{E}(\eta \mid Y)\} + \mathrm{E}\{\mathrm{var}(\eta \mid Y)\}.\end{aligned}$$

It seems that the choice of constraints is mostly a matter of taste: We show later in Lemma 10.2 that $\mathrm{cov}(\xi, \eta)$ is unaffected by the choice of constraint.

10.2.6 Synopsis of estimation results

CB|SEM gives the maximum likelihood estimator of $\text{cov}(\xi, \eta)$, provided that there are identified off-diagonal elements of the conditional covariance matrices $\Sigma_{X|\xi}$ and $\Sigma_{Y|\eta}$ that are known to be zero and set to zero in the model. It seems common in application to assume that $\Sigma_{X|\xi}$ and $\Sigma_{Y|\eta}$ are diagonal matrices, in effect assuming that ξ and η account for all of the dependence between the indicators. If there are no identified off-diagonal elements that are known to be zero, then $\text{cov}(\xi, \eta)$ is not identified. In the models of Figure 10.1, CB|SEM can be used to estimate $\text{cov}(\xi, \eta)$ in the UNCORE model, but not in the CORE model. Schönemann and Wang (1972), Henseler et al. (2014, Fig. 1) and others have argued that the UNCORE model rarely holds in practice. Nevertheless, it still seems to be used frequently in the social sciences.

The maximum likelihood estimator $|\widehat{\Psi}_{\text{mle}}|$ of $|\Psi|$ can be derived from the RRR of Y on X and X on Y, as described in Section 10.3. We will see in Section 10.4.1 that $|\widehat{\Psi}_{\text{mle}}|$ is simply the first canonical correlation between X and Y. A corresponding moment estimator $\widehat{\Psi}_{\text{mt}}$ is introduced in Section 10.4.2.

PLS methods are moment-based envelope methods that give estimators of Ψ. There are no restrictions on $\Sigma_{X|\xi}$ and $\Sigma_{Y|\eta}$, other than the general restriction that they are positive definite.

CB|SEM and the PLS methods have two distinct targets, $|\text{cov}(\xi, \eta)|$ and $|\Psi|$. PLS methods do not provide generally serviceable estimators of $|\text{cov}(\xi, \eta)|$ and CB|SEM does not provide a serviceable estimator of $|\Psi|$.

10.3 Reduced rank regression

PLS path algorithms gain information on the latent constructs from the multi-response linear regressions of Y on X and X on Y. Consequently, it is useful to provide a characterization of these regressions.

In the following lemma we state that for the model presented in (10.1) and (10.2) of Section 10.2.2, the regression of Y on X qualifies as a RRR. The essential implication of the lemma is that, under the reflective model described in Section 10.2, the coefficient matrix $\beta_{Y|X}$ in the multi-response linear regression of Y on X must have rank 1 because $\text{cov}(Y, X) = \Sigma_{Y,X}$ has rank 1. A similar conclusion applies to the regression of X on Y. Proofs for Lemma 10.1 and Proposition 10.1 are available in Appendix Sections A.8.3 and A.8.4.

Lemma 10.1. *For the* CORE *model presented in presented in (10.1) and (10.2) and without imposing either the regression or marginal constraints, we have* $E(Y \mid X) = \mu_Y + \beta_{Y|X}(X - \mu_X)$, *where* $\beta_{Y|X} = AB^T\Sigma_X^{-1}$, *with* $A \in \mathbb{R}^{r\times 1}$, $B \in \mathbb{R}^{p\times 1}$, *and* $\Sigma_{Y,X} = AB^T$.

It is known from the literature on RRR that the vectors A and B are not identifiable, while AB^T is identifiable (see, for example, Cook et al., 2015). As a consequence of being a RRR model, we are able to state in Proposition 10.1 which parameters are identifiable in the reflective model of (10.1) and (10.2).

Proposition 10.1. *The parameters* Σ_X, Σ_Y, $\Sigma_{Y,X} = AB^T$, $\beta_{Y|X}$, μ_Y, *and* μ_X *are identifiable in the reduced rank model of Lemma 10.1, but A and B are not. Additionally, under the regression constraints, the quantities* $\sigma_{\xi,\eta}/(\sigma_\eta^2\sigma_\xi^2)$, Ψ, $E(\eta|Y)$, $\Sigma_{X,\xi}$, *and* $\Sigma_{Y,\eta}$ *are identifiable in the reflective model except for sign, while* $\mathrm{cor}(\eta,\xi)$, σ_ξ^2, σ_η^2, $\sigma_{\xi,\eta}$, $\beta_{X|\xi}$, $\beta_{Y|\eta}$, $\Sigma_{Y|\eta}$, *and* $\Sigma_{X|\xi}$ *are not identifiable. Moreover,*

$$|\Psi| = |\sigma_{\xi,\eta}|/(\sigma_\eta^2\sigma_\xi^2) = \mathrm{tr}^{1/2}(\beta_{X|Y}\beta_{Y|X}) \tag{10.3}$$

$$= \mathrm{tr}^{1/2}(\Sigma_{X,Y}\Sigma_Y^{-1}\Sigma_{Y,X}\Sigma_X^{-1}). \tag{10.4}$$

10.4 Estimators of Ψ

10.4.1 Maximum likelihood estimator of Ψ

Although A and B are not identifiable, $|\Psi|$ is identifiable since from (10.4) it depends on the identifiable quantities Σ_X, Σ_Y, and on the rank 1 covariance matrix $\Sigma_{Y,X} = AB^T$ in the RRR of Y on X. Let $\widehat{\Sigma}_{Y,X}$ denote the maximum likelihood estimator of $\Sigma_{Y,X}$ from fitting the RRR model of Lemma 10.1, and recall that S_X and S_Y denote the sample versions of Σ_X and Σ_Y. Then the maximum likelihood estimator of $|\Psi|$ can be obtained by substituting these estimators into (10.4):

$$|\widehat{\Psi}_{\mathrm{mle}}| = \mathrm{tr}^{1/2}(\widehat{\Sigma}_{X,Y}S_Y^{-1}\widehat{\Sigma}_{Y,X}S_X^{-1}).$$

Let $\Sigma_{\tilde{Y},\tilde{X}} = \Sigma_Y^{-1/2}\Sigma_{Y,X}\Sigma_X^{-1/2}$ denote the standardized version of $\Sigma_{Y,X}$ that corresponds to the rank one regression of $\tilde{Y} = \Sigma_Y^{-1/2}Y$ on $\tilde{X} = \Sigma_X^{-1/2}X$.

Then $|\Psi|$ can be written also as

$$|\Psi| = \mathrm{tr}^{1/2}(\Sigma_{\tilde{X},\tilde{Y}}\Sigma_{\tilde{Y},\tilde{X}}) = \|\Sigma_{\tilde{Y},\tilde{X}}\|_F, \tag{10.5}$$

which can be seen as the Frobenius norm $\|\cdot\|_F$ of the standardized covariance matrix $\Sigma_{\tilde{Y},\tilde{X}}$.

The maximum likelihood estimator $|\widehat{\Psi}_{\mathrm{mle}}|$ can be computed in the following steps starting with the data (Y_i, X_i), $i = 1, \ldots, n$ (Cook et al., 2015):

1. Standardize $\tilde{Y}_i = S_Y^{-1/2}(Y_i - \bar{Y})$ and $\tilde{X}_i = S_X^{-1/2}(X_i - \bar{X})$, $i = 1, \ldots, n$.
2. Construct $\widehat{\Sigma}_{\tilde{Y},\tilde{X}}$, the matrix of sample correlations between the elements of the standardized vectors $\tilde{X}$ and $\tilde{Y}$.
3. Form the singular value decomposition $\widehat{\Sigma}_{\tilde{Y},\tilde{X}} = UDV^T$ and extract U_1 and V_1, the first columns of U and V, and D_1 is the corresponding (largest) singular value of $\widehat{\Sigma}_{\tilde{Y},\tilde{X}}$.
4. Then
$$|\widehat{\Psi}_{\mathrm{mle}}| = D_1.$$
In short, the maximum likelihood estimator $|\widehat{\Psi}_{\mathrm{mle}}|$ is the largest singular value of $\widehat{\Sigma}_{\tilde{Y},\tilde{X}}$. Equivalently it is the first sample canonical correlation between X and Y.

Wold (1982, Section 5.2), pointed out that the first canonical correlation arises under formative models, which established a connection between canonical correlations and the analysis of Wold's two-block, two-LV soft models.

This maximum likelihood estimator makes use of the reduced dimension that arises because $\Sigma_{X,Y}$ has rank 1. In a sense then, it is a dimension reduction estimator. However, it is not formally related to envelope or PLS estimators, which are also based on reducing dimensions. Envelope and reduced rank estimators are distinct methods that were combined by Cook et al. (2015) to yield a reduced-rank envelope estimator.

10.4.2 Moment estimator

Define

$$|\widehat{\Psi}_{\mathrm{mt}}| = \mathrm{tr}^{1/2}(S_{X,Y}S_Y^{-1}S_{Y,X}S_X^{-1}), \tag{10.6}$$

where tr denotes the trace operator. This moment estimator of $|\Psi|$, which is constructed by simply substituting moment estimators for the quantities in (10.4), may not be as efficient as the maximum likelihood estimator under the model of 10.2.2, but it might possess certain robustness properties. This estimator does not use reduced dimensions that arises from the rank of $\Sigma_{X,Y}$ or from envelope constructions and, for this reason, it is likely to be inferior.

10.4.3 Envelope and PLS estimators of Ψ

The results of Chapter 5 together with our starting models provide a framework for deriving the envelope and PLS estimators of Ψ. We begin by establishing notation for describing the response envelopes for model (10.1)–(10.2).

Let $\mathcal{B}_{Y|\eta} = \text{span}(\beta_{Y|\eta})$, let $\mathcal{B}_{X|\xi} = \text{span}(\beta_{X|\xi})$, let Γ denote a semi-orthogonal basis matrix of the response envelope $\mathcal{E}_{\Sigma_{Y|\eta}}(\mathcal{B}_{Y|\eta})$ for Y model in (10.2) and let Φ denote a semi-orthogonal basis matrix of the response envelope $\mathcal{E}_{\Sigma_{X|\xi}}(\mathcal{B}_{X|\xi})$ for X model in (10.2). Let u and q denote the dimensions of $\mathcal{E}_{\Sigma_{Y|\eta}}(\mathcal{B}_{Y|\eta})$ and $\mathcal{E}_{\Sigma_{X|\xi}}(\mathcal{B}_{X|\xi})$ and let $(\Gamma, \Gamma_0) \in \mathbb{R}^{r\times r}$ and $(\Phi, \Phi_0) \in \mathbb{R}^{p\times p}$ be orthogonal matrices.

The envelope versions of models (10.2) can now be written as

$$\left.\begin{array}{l} Y = \mu_Y + \Gamma\gamma(\eta - \mu_\eta) + \epsilon_{Y|\eta}; \quad \epsilon_{Y|\eta} \sim N_r(0, \Gamma\Omega\Gamma^T + \Gamma_0\Omega_0\Gamma_0^T) \\ X = \mu_X + \Phi\phi(\xi - \mu_\xi) + \epsilon_{X|\xi}; \quad \epsilon_{X|\xi} \sim N_p(0, \Phi\Delta\Phi^T + \Phi_0\Delta_0\Phi_0^T). \end{array}\right\} \quad (10.7)$$

Here $\gamma \in \mathbb{R}^u$ and $\phi \in \mathbb{R}^q$ give the coordinates of $\beta_{Y|\eta}$ and $\beta_{X|\xi}$ in terms of the basis matrices Γ and Φ, and Δ, Δ_0, Ω, and Ω_0 are positive definite matrices with dimensions that conform to the indicated matrix multiplications.

It follows from the envelope models given by (10.7) that

$$\begin{aligned} \Sigma_Y &= \Gamma(\gamma\gamma^T\sigma_\eta^2 + \Omega)\Gamma^T + \Gamma_0\Omega_0\Gamma_0^T \\ \Sigma_{Y,X} &= \Gamma\gamma\phi^T\Phi^T\sigma_{\xi\eta} \\ \Sigma_X &= \Phi(\phi\phi^T\sigma_\xi^2 + \Delta)\Phi^T + \Phi_0\Delta_0\Phi^T \\ |\Psi| &= |\text{cor}\{E(\xi|\Phi^T X), E(\eta|\Gamma^T Y)\}|. \end{aligned}$$

The essential implication of this last result is as follows. Once estimators $\widehat{\Gamma}$ and $\widehat{\Phi}$ are known, we can use the estimated composite indicators $\widehat{\Gamma}^T Y$ and $\widehat{\Phi}^T X$ in place of Y and X to estimate $|\Psi|$. Using these in combination with maximum likelihood estimation, we can again see that $|\Psi|$ is estimated as the first canonical correlation between $\widehat{\Gamma}^T Y$ and $\widehat{\Phi}^T X$. Moment estimation (10.6) proceeds similarly by replacing Y and X with $\widehat{\Gamma}^T Y$ and $\widehat{\Phi}^T X$.

If $\widehat{\Gamma}^T Y$ and $\widehat{\Phi}^T X$ are real, then the maximum likelihood and moment estimators both yield the absolute sample correlation between these composite indicators. To see this result, consider the four computational steps in Section 10.4: $\tilde{Y}_i = S^{-1/2}_{\widehat{\Gamma}^T Y}(\widehat{\Gamma}^T Y_i - \widehat{\Gamma}^T \bar{Y})$ and $\tilde{X}_i = S^{-1/2}_{\widehat{\Phi}^T X}(\widehat{\Phi}^T X_i - \widehat{\Phi}^T \bar{X})$ are both real and so $\widehat{\Sigma}_{\tilde{Y},\tilde{X}}$ is just the sample covariance between $\widehat{\Gamma}^T Y$ and $\widehat{\Phi}^T X$, which equals the usual sample correlation in view of the standardization. The moment estimator is evaluated in the same way.

Envelope and PLS methods are distinguished then by how they construct the estimated composite indicators:

Envelope estimators.

The envelope estimator $|\widehat{\Psi}_{\text{env}}|$ is obtained by using the algorithms of Section 5.2.2 to construct $\widehat{\Gamma}$ and $\widehat{\Phi}$.

PLS estimators.

The PLS estimator $|\widehat{\Psi}_{\text{pls}}|$ is obtained by using the algorithms of Table 5.1 to construct $\widehat{\Gamma}$ and $\widehat{\Phi}$, which are denoted as G and W in the table.

These are envelope and PLS estimators that do not take into account that $\Sigma_{Y,X}$ has rank 1. Nevertheless, they are still serviceable dimension reduction methods that can compete effectively with the maximum likelihood and moment estimators of Sections 10.4.1 and 10.4.2. This potential is demonstrated by the simulations of Section 10.7.

10.4.4 The PLS|SEM estimator used in structural equation modeling of path diagrams

Using Dijkstra (1983) and Dijkstra and Henseler (2015a) as guides, the method designated as PLS in the social sciences differs from those in Section 10.4.3 in two crucial aspects. First, the data are scaled by normalizing each real indicator with its sample standard deviation, so the resulting scaled indicator has sample mean 0 and sample variance 1. Second, the method is based on the condition that $u = q = 1$. Working in the scaled variables, the PLS|SEM method then effectively assumes that all reflective information is captured in one linear combination of the scaled indicators.

Specific steps to compute the PLS|SEM estimator are as follows:

1. Scale and center the indicators marginally, $X^{(s)} = \text{diag}^{-1/2}(S_X)(X - \bar{X})$ and $Y^{(s)} = \text{diag}^{-1/2}(S_Y)(Y - \bar{Y})$, where diag$(\cdot)$ denotes the diagonal matrix with diagonal elements the same as those of the argument.

2. Construct the first eigenvector ℓ_1 of $S_{Y^{(s)}X^{(s)}}S_{X^{(s)}Y^{(s)}}$ and the first eigenvector r_1 of $S_{X^{(s)}Y^{(s)}}S_{Y^{(s)}X^{(s)}}$.

3. Construct the proxy latent variables $\bar{\xi} = r_1^T X^{(s)}$ and $\bar{\eta} = \ell_1^T Y^{(s)}$.

4. Then the estimated covariance and correlation between the proxies $\bar{\eta}$ and $\bar{\xi}$ are

$$\begin{aligned} \widehat{\text{cov}}(\bar{\eta}, \bar{\xi}) &= \ell_1^T S_{Y^{(s)}X^{(s)}} r_1 \\ \widehat{\text{cor}}(\bar{\eta}, \bar{\xi}) &= \frac{\ell_1^T S_{Y^{(s)}X^{(s)}} r_1}{\{\ell_1^T S_{Y^{(s)}} \ell_1 r_1^T S_{X^{(s)}} r_1\}^{1/2}}. \end{aligned} \quad (10.8)$$

Aside from the scaling, the reduction to the proxy variables $\bar{\xi}$ and $\bar{\eta}$ is the same as that from the simultaneous PLS method developed by (Cook, Forzani, and Liu, 2023b, Section 4.3) and discussed in Section 5.3 when $u = q = 1$.

However, if we standardize jointly, then we recover the MLE. Rewriting the MLE algorithm given in Section 10.4.1 to reflect its connection with PLS|SEM, we have

1. Standardize the indicators jointly $\tilde{X} = S_X^{-1/2}(X - \bar{X})$, and $\tilde{Y} = S_Y^{-1/2}(Y - \bar{Y})$.

2. Construct the first eigenvector U_1 of $S_{\tilde{Y},\tilde{X}}S_{\tilde{X},\tilde{Y}}$ and the first eigenvector V_1 of $S_{\tilde{X},\tilde{Y}}S_{\tilde{Y},\tilde{X}}$.

3. Construct the proxy latent variables $\bar{\eta} = V_1^T \tilde{X}$ and $\bar{\xi} = U_1^T \tilde{Y}$.

4. Then the estimated correlation between the proxies $\bar{\eta}$ and $\bar{\xi}$ is

$$\widehat{\text{cor}}(\bar{\eta}, \bar{\xi}) = U_1^T S_{\tilde{Y},\tilde{X}} V_1 = D_1.$$

Consequently, we reach the following three essential conclusions:

I. If joint standardization is used for the indicator vectors, then the PLS|SEM estimator $\widehat{\text{cor}}(\bar{\eta}, \bar{\xi})$ is the same as the MLE of $|\Psi|$ under the model (10.1)–(10.2). This involves no dimension reduction beyond that arising from the reduced rank model of Section 10.3. It is unrelated to PLS.

We recommend that PLS|SEM be based on joint standardization and not marginal scaling, when permitted by the sample size.

II. If no standardization is used for the indicator vectors, then the PLS|SEM estimator $\widehat{\text{cor}}(\bar{\eta}, \bar{\xi})$ is the same as the simultaneous PLS estimator (see Section 5.3) when $u = q = 1$. But we may lose information if $u \neq 1$ or $q \neq 1$.

III. If marginal scaling is used, it's not clear what $\widehat{\text{cor}}(\bar{\eta}, \bar{\xi})$ is estimating. It might be viewed as an approximation of the MLE and simultaneous PLS estimators of Ψ, but this has not been studied to our knowledge.

We have not seen a compelling reason to use marginal scaling. Marginal scaling might be useful when the sample size is relatively small, but we are not aware of work on this possibility.

10.5 CB|SEM

As hinted in Proposition 10.1, the parameter $\text{cor}(\eta, \xi)$ is not identifiable without further assumptions. We give now sufficient conditions to have identification of $\text{cor}(\xi, \eta)$. The conditions needed are related to the identification of $\Sigma_{X|\xi}$ and $\Sigma_{Y|\eta}$. As seen in Lemma A.6 of Appendix A.8.4, under model (10.1)–(10.2),

$$\Sigma_X = \Sigma_{X|\xi} + \sigma_\xi^{-2}(B^T\Sigma_X^{-1}B)^{-1}BB^T \tag{10.9}$$

$$\Sigma_Y = \Sigma_{Y|\eta} + \sigma_\eta^{-2}(A^T\Sigma_Y^{-1}A)^{-1}AA^T. \tag{10.10}$$

The terms $(B^T\Sigma_X^{-1}B)^{-1}BB^T$ and $(A^T\Sigma_Y^{-1}A)^{-1}AA^T$ in (10.9) and (10.10) are identifiable. Since the goal of CB|SEM is to estimate $\text{cor}(\xi, \eta)$ and since we know from Proposition 10.1 that, under the regression constraints,

$$|\Psi| = |\text{cor}\{E(\xi \mid X), E(\eta \mid Y)\}| = |\sigma_{\xi,\eta}|/(\sigma_\xi^2\sigma_\eta^2) \tag{10.11}$$

is identifiable, we will be able to identify

$$|\text{cor}(\xi, \eta)| = |\sigma_{\xi,\eta}|/\sigma_\xi \sigma_\eta = \sigma_\xi \sigma_\eta |\Psi|$$

if we can identify σ_ξ^2 and σ_η^2. From (10.9) and (10.10) that is equivalent to identifying $\Sigma_{Y|\eta}$ and $\Sigma_{X|\eta}$. We show in Proposition A.4 of Appendix A.8.4 that, under the regression constraints, (a) if $\Sigma_{Y|\eta}$ and $\Sigma_{X|\xi}$ are identifiable, then σ_ξ^2, σ_η^2, $|\sigma_{\xi,\eta}|$, $\beta_{X|\xi}$, $\beta_{Y|\eta}$ are identifiable and that (b) $\Sigma_{Y|\eta}$ and $\Sigma_{X|\xi}$ are identifiable if and only if σ_ξ^2, σ_η^2 are so.

The next proposition gives conditions that are sufficient to ensure identifiability; its proof is given in Appendix A.8.5. Let $(M)_{ij}$ denote the ij-th element of the matrix M and $(V)_i$ the i-th element of the vector V.

Proposition 10.2. *Under the regression constraints, (I) if $\Sigma_{X|\xi}$ contains an off-diagonal element that is known to be zero, say $(\Sigma_{X|\xi})_{ij} = 0$, and if $(B)_i(B)_j \neq 0$, then $\Sigma_{X|\xi}$ and σ_ξ^2 are identifiable. (II) if $\Sigma_{Y|\eta}$ contains an off-diagonal element that is known to be zero, say $(\Sigma_{Y|\eta})_{ij} = 0$, and if $(A)_i(A)_j \neq 0$ then $\Sigma_{Y|\eta}$ and σ_η^2 are identifiable.*

Corollary 10.1. *Under the regression constraints, if $\Sigma_{Y|\eta}$ and $\Sigma_{X|\xi}$ are diagonal matrices and if A and B each contain at least two non-zero elements then $\Sigma_{Y|\eta}$, $\Sigma_{X|\xi}$, σ_ξ^2, and σ_η^2 are identifiable.*

The usual assumption in SEM is that $\Sigma_{X|\xi}$ and $\Sigma_{Y|\eta}$ are diagonal matrices (e.g. Henseler et al., 2014; Jöreskog, 1970). We see from (10.11) and Corollary 10.1 that this assumption along with the regression constraints is sufficient to guarantee that $|\text{cor}(\xi, \eta)|$ is identifiable provided B and A contain at least two non-zero elements. However, from Proposition 10.2, we also see that it is not necessary for $\Sigma_{Y|\eta}$ and $\Sigma_{X|\xi}$ to be diagonal. The assumption that $\Sigma_{X|\xi}$ and $\Sigma_{Y|\eta}$ are diagonal matrices means that, given ξ and η, the elements of Y and X must be independent. In consequence, elements of X and Y are correlated only by virtue of their association with η and ξ. The presence of any residual correlations after accounting for ξ and η would negate the model and possibly lead to spurious conclusions. See Henseler et al. (2014) for a related discussion.

In full, the usual SEM requires that $\Sigma_{Y|\eta}$ and $\Sigma_{X|\xi}$ are diagonal matrices, and it adopts the marginal constraints instead of the regression constraints. By Proposition 10.2, our ability to identify parameters is unaffected by the constraints adopted. However, we need also to be sure that the meaning of

$\text{cor}(\xi, \eta)$ is unaffected by the constraints adopted. The proof of the following proposition is in Appendix A.8.6.

Lemma 10.2. *Starting with (10.2),* $\text{cor}(\xi, \eta)$ *is unaffected by choice of constraint,* $\sigma_\xi^2 = \sigma_\eta^2 = 1$ *or* $\text{var}\{E(\xi|X)\} = \text{var}\{E(\eta|Y)\} = 1$.

Now, to estimate the identifiable parameters with the regression or marginal constraints, we take the joint distribution of (X, Y) and insert the conditions imposed. The maximum likelihood estimators can then be found under normality, as shown in Proposition A.3 of Appendix A.8.2. From this we conclude that the same type of maximization problem arises under either sets of constraints.

A connection between envelopes and $\text{cor}(\eta, \xi)$ similar to that between envelopes and Ψ is problematic. Since only $\text{span}(\Gamma) = \mathcal{E}_{\Sigma_{Y|\eta}}(\mathcal{B}_{Y|\eta})$ and $\text{span}(\Phi) = \mathcal{E}_{\Sigma_{X|\xi}}(\mathcal{B}_{X|\xi})$ are identifiable, arbitrary choices of bases for these subspaces will not likely result in $\Sigma_{\Gamma^T Y|\eta} = \Omega$ and $\Sigma_{\Phi^T X|\xi} = \Delta$ being diagonal matrices, as required by CB|SEM. Choosing arbitrary bases Γ and Φ and using the marginal constraints, the envelope composites have the following structure,

$$\begin{aligned} \Sigma_{\Gamma^T Y} &= \gamma\gamma^T + \Omega \\ \Sigma_{\Gamma^T Y, \Phi^T X} &= \gamma\phi^T \text{cor}(\xi, \eta) \\ \Sigma_{\Phi^T X} &= \phi\phi^T + \Delta, \end{aligned}$$

where the notation is as used for model (10.7). From this we see that the joint distribution of the envelope composites $(\Gamma^T Y, \Phi^T X)$ has the same structure as the SEM shown in equation (A.31) of Appendix A.8.6, except that assuming Ω and Δ to be diagonal matrices is untenable from this structure alone. Additionally, Ω and Δ are not identifiable because they are confounded with $\gamma\gamma^T$ and $\phi\phi^T$.

In short, it does not appear that there is a single method that can provide estimators of both of the parameters $|\Psi|$ and $\text{cov}(\xi, \eta)$. However, an estimator of $|\Psi|$ can provide an estimated lower bound on $|\text{cov}(\xi, \eta)|$, as discussed in Section 10.6.

10.6 Bias

Bias is a malleable concept, depending on the context, the estimator and the quantity being estimated.

If the goal is to estimate $|\text{cor}(\xi, \eta)|$ via maximum likelihood while assuming that $\Sigma_{Y|\eta}$ and $\Sigma_{X|\xi}$ are diagonal matrices, bias might not be a worrisome issue. Although maximum likelihood estimators are generally biased, the bias typically vanishes at a fast rate as the sample size increases. Bias may be an issue when the sample size is not large relative to the number of parameters to be estimated while S_X and S_Y are still nonsingular. This issue is outside the scope of this chapter.

If the goal is to estimate $|\Psi|$ without assuming diagonal covariance matrices and $n > \min(p, r) + 1$ then its maximum likelihood estimator, the first canonical correlation between X and Y, can be used and again bias might not be a worrisome issue.

In some settings we may wish to use an estimator of $|\Psi|$ also as an estimator of $|\text{cor}(\xi, \eta)|$ without assuming diagonal covariance matrices. The implications of doing so are a consequence of the next proposition. Its proof is in Appendix A.8.7.

Proposition 10.3. *Under the model that stems from (10.1) and (10.2),*

$$\Psi = [\text{var}\{\text{E}(\xi \mid X)\}\text{var}\{\text{E}(\eta \mid Y)\}]^{1/2}\frac{\text{cor}(\xi, \eta)}{\sigma_\xi \sigma_\eta}.$$

Under either the regression constraints or the marginal constraints

$$|\Psi| \leq |\text{cor}(\eta, \xi)|. \tag{10.12}$$

Relationship (10.12) enables us to define a population bias in this case as the difference

$$0 \leq |\text{cor}(\xi, \eta)| - |\Psi| \leq 1,$$

which agrees with Dijkstra's (Dijkstra, 1983, Section 4.3) conclusion that PLS will underestimate $|\text{cor}(\xi, \eta)|$. Under the marginal constraints,

$$\begin{aligned}
\sigma_\xi^2 &= 1 = \text{E}\{\text{var}(\xi \mid X)\} + \text{var}\{\text{E}(\xi \mid X)\} \\
\sigma_\eta^2 &= 1 = \text{E}\{\text{var}(\eta \mid Y)\} + \text{var}\{\text{E}(\eta \mid Y)\},
\end{aligned}$$

this bias will be small when $\mathrm{E}\{\mathrm{var}(\xi \mid X)\}$ and $\mathrm{E}\{\mathrm{var}(\eta \mid Y)\}$ are small, so ξ and η are well predicted by X and Y. This may happen with a few highly informative indicators. It may also happen as the number of informative indicators increases, a scenario that is referred to as an *abundant* regression in statistics (Cook, Forzani, and Rothman, 2012, 2013). On the other extreme, if ξ and η are not well predicted by X and Y, then it is possible to have $|\Psi|$ close to 0 while $|\mathrm{cor}(\xi, \eta)|$ is close to 1, in which case the bias is close to 1. Like the assumption that $\Sigma_{Y|\eta}$ and $\Sigma_{X|\xi}$ are diagonal matrices, it may be effectively impossible to gain from the data support for claiming that $\mathrm{E}(\mathrm{var}(\xi \mid X))$ and $\mathrm{E}(\mathrm{var}(\eta \mid Y))$ are small.

Under the regression constraints,

$$
\begin{aligned}
\sigma_\xi^2 &= \mathrm{E}\{\mathrm{var}(\xi \mid X)\} + 1 \\
\sigma_\eta^2 &= \mathrm{E}\{\mathrm{var}(\eta \mid Y)\} + 1,
\end{aligned}
$$

this bias will again be small when $\mathrm{E}\{\mathrm{var}(\xi \mid X)\}$ and $\mathrm{E}\{\mathrm{var}(\eta \mid Y)\}$ are small. In short, an estimator of $|\Psi|$ is also an estimator of a lower bound on $|\mathrm{cor}(\xi, \eta)|$. Under either the marginal or regression constraints, the bias $|\mathrm{cor}(\xi, \eta)| - |\Psi|$ will be small when the indicators are good predictors of the constructs.

10.7 Simulation results

In this section, we provide simulation results to support our general conclusions and to demonstrate what can happen. In all simulations, the constructs η and ξ are real random variables with $0 \leq \mathrm{cor}(\eta, \xi) \leq 2/3$. In Section 10.7.1 we use simulations in the manner of Rönkkö et al. (2016b) to illustrate that the PLS methods in the social sciences may overestimate both $\mathrm{cor}(\xi, \eta)$ and Ψ, while envelope methods, including PLS, perform notably better for estimating Ψ. The simulation results in Section 10.7.2 are intended to underscore the observation that CB|SEM and PLS|SEM methods may materially underestimate Ψ and $\mathrm{cor}(\xi, \eta)$ when $\Sigma_{X|\xi}$ and $\Sigma_{Y|\eta}$ are not diagonal matrices, while the envelope-based methods continue to do well. We illustrate in Section 10.7.3 that CB|SEM and PLS|SEM methods may again exhibit substantial bias when there are multiple reflective composites.

10.7.1 Simulation results with $\Sigma_{X|\xi} = \Sigma_{Y|\eta} = I_3$

Rönkkö et al. (2016b) illustrated some of their concerns via simulations with $p = r = 3$ and sample size $n = 100$ under the marginal constraints with $0 \leq \text{cor}(\eta, \xi) \leq 2/3$ and $\Sigma_{X|\xi} = \Sigma_{Y|\eta} = I_3$.

We follow their general setup by simulating data with coefficient vectors for the linear regressions of Y on η and X on ξ set equal to $L = (8/10, -7/10, 6/10)^T$. Their choice for the covariance matrices implies that $u = q = 1$, so there is only one material composite of the indicators, which agrees with the conditions for PLS|SEM. It follows from Proposition 10.3 that

$$\Psi = 0.5984 \times \text{cor}(\eta, \xi), \tag{10.13}$$

which agrees with our general conclusion (10.12). Following Rönkkö et al. (2016b), we simulated data with sample sizes $N = 100$ and $N = 1000$ observations on (X, Y) according to these settings with various values for $\text{cor}(\eta, \xi)$. The estimators that we used are as follows:

MLE: This is the estimator described in Section 10.4.1. Recall that it is the same as the PLS|SEM estimator with joint standardization, as developed in Section 10.4.4.

Matrixpls: This R program, which was used to compute a version of PLS|SEM, was provided to us by Mikko Rönkkö (Rönkkö et al., 2016b, Footnote 1).

ENV: This envelope estimator was computed using the methods described in Cook and Zhang (2015b). It is as discussed in Section 10.4.3.

CB|SEM: This estimator was discussed in Section 10.5. It was computed using the *lavaan* package based on Rosseel (2012).

PLS|SEM: This estimator is as discussed in Section 10.4.4. We computed it using our own code, but found the estimates to be quite close to those from Matrixpls.

PLS: This designates the PLS estimator discussed in Section 10.4.3. It was computed using the methods described by Cook, Forzani, and Liu (2023b, Table 2).

We can see from the results shown in Figure 10.2, which are the averages over 10 replications, that at $N = 100$, Matrixpls tends to underestimate Ψ, while the other five estimators of Ψ clearly overestimate for all

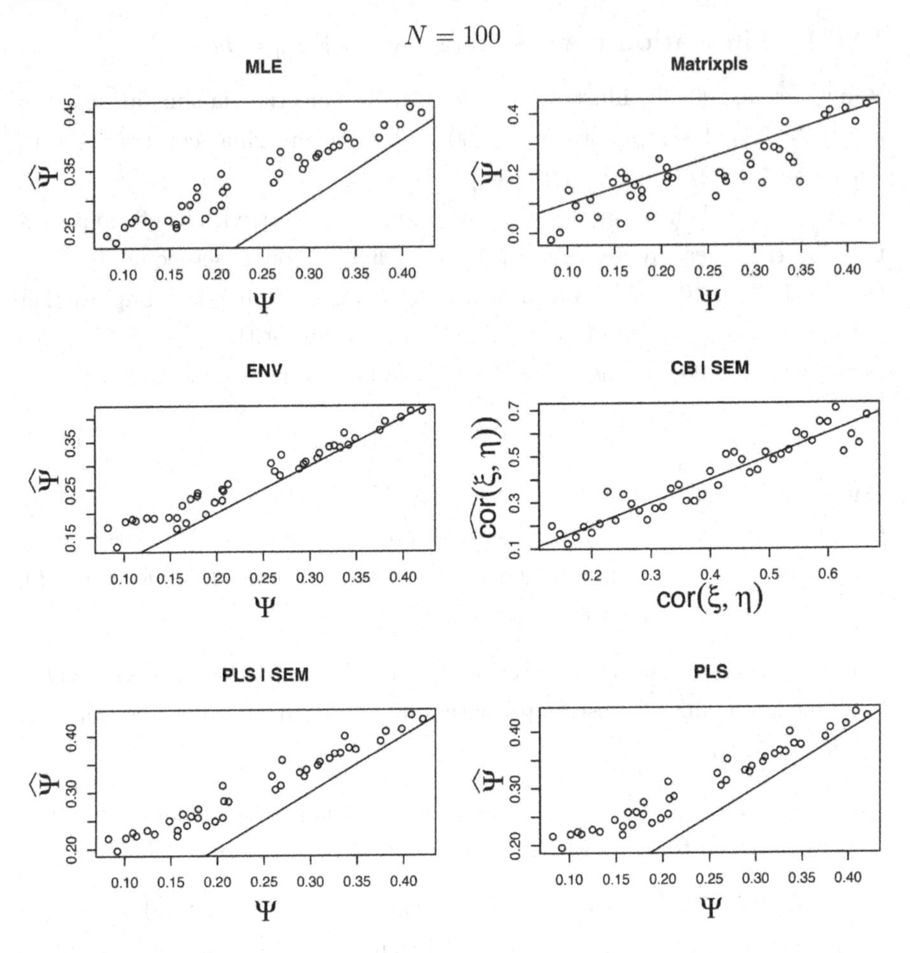

FIGURE 10.2
Simulations with $N = 100$, $\Sigma_{X|\xi} = \Sigma_{Y|\eta} = I_3$. Horizontal axes give the true correlations; vertical axes give the estimated correlations. Lines $x = y$ represented equality.

$\text{cor}(\eta, \xi) \in (0, 2/3)$, with ENV doing a bit better than the others. Because the indicators are independent with constant variances conditional on the constructs, we did not expect much difference between MLE and PLS|SEM. Other than CB|SEM, none of these estimators made use of the fact that $\Sigma_{X|\xi}$ and $\Sigma_{Y|\eta}$ are diagonal matrices. At $N = 1000$ shown in Figure 10.3 the performance of ENV is essentially flawless, while the other PLS-type estimators show a slight propensity to overestimate Ψ.

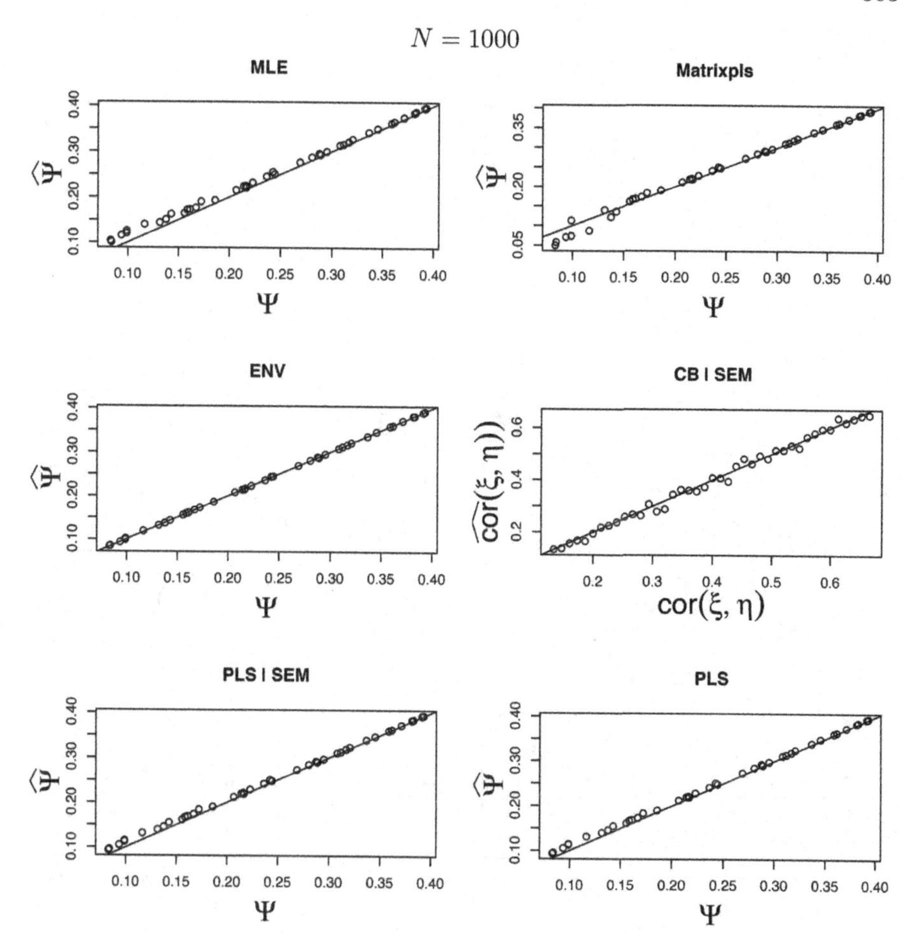

FIGURE 10.3
Simulations with $N = 1000$, $\Sigma_{X|\xi} = \Sigma_{Y|\eta} = I_3$. Horizontal axes give the true correlations; vertical axes give the estimated correlations. Lines $x = y$ represented equality.

10.7.2 Non-diagonal $\Sigma_{X|\xi}$ and $\Sigma_{Y|\eta}$

Simulating with $\Sigma_{X|\xi} = \Sigma_{Y|\eta} = I_3$, as we did in the construction of Figures 10.2 and 10.3, conforms to the standard assumption that these matrices be diagonal and partially accounts for the success of CB|SEM in Figures 10.2 and 10.3. To illustrate the importance of these assumptions, we next took $\Sigma_{X|\xi} = \Sigma_{Y|\eta} = L(L^TL)^{-1}L^T + 8L_0L_0^T$ with $L_0^TL_0 = I_2$ and $L^TL_0 = 0$.

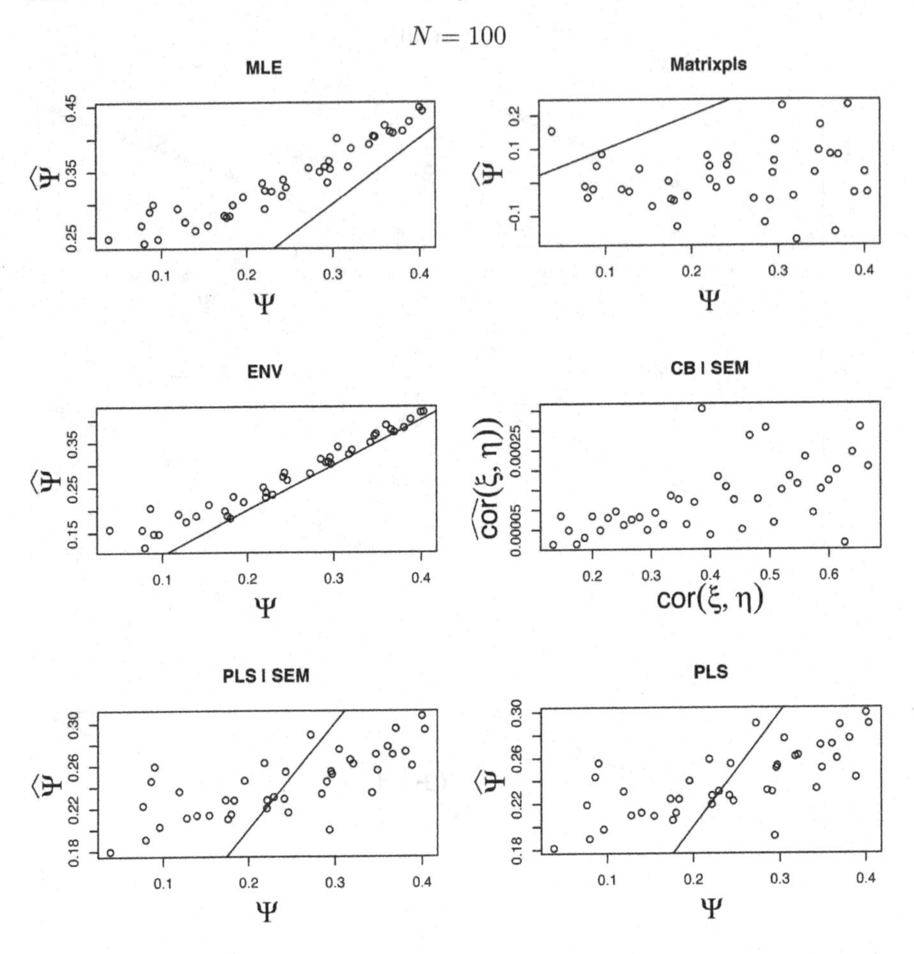

FIGURE 10.4
Simulations with $N = 100$, $\Sigma_{X|\xi} = \Sigma_{Y|\eta} = L(L^TL)^{-1}L^T + 3L_0L_0^T$. Horizontal axes give the true correlations; vertical axes give the estimated correlations. The lines represent $x = y$.

Otherwise, the settings are as described in Section 10.7.1, including the relationship between Ψ and $\text{cor}(\xi, \eta)$ given at (10.14). With this structure, the usual PLS assumption that $q = u = 1$ is still correct, the material information being L^TX and L^TY. At $N = 100$ shown in Figure 10.4, ENV is still the best with the MLE coming in second. The performance of the other four estimators is terrible, as there seems to be little relationship between the estimator and the estimand. At $N = 1000$ shown in Figure 10.4, the performance of ENV

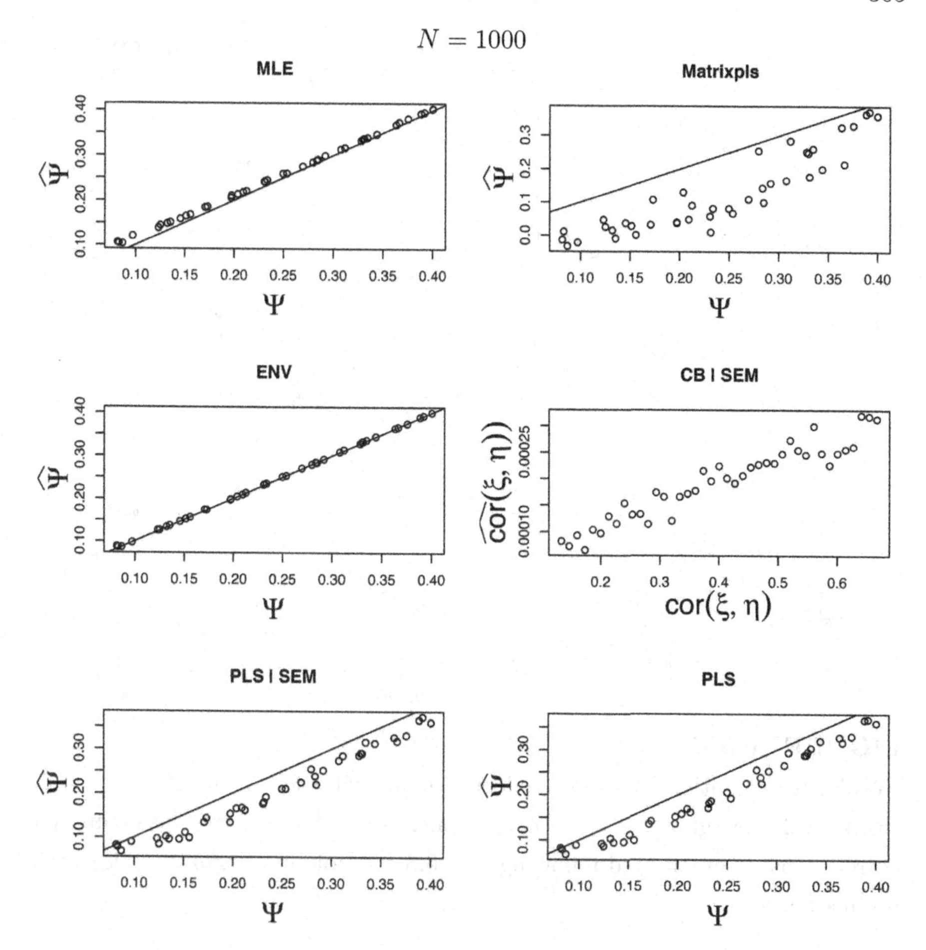

FIGURE 10.5
Results of simulations with $N = 1000$, $\Sigma_{X|\xi} = \Sigma_{Y|\eta} = L(L^TL)^{-1}L^T + 3L_0L_0^T$. Horizontal axes give the true correlations; vertical axes give the estimated correlations. The lines represent $x = y$.

is essentially flawless, while the MLE does well. The other four estimators all have a marked tendency toward underestimation, particularly CB|SEM.

10.7.3 Multiple reflective composites

Figure 10.6 gives the results of a simulation with $q = 2$ reflective composites for X and $u = 2$ reflective composites for Y. The simulation was structured

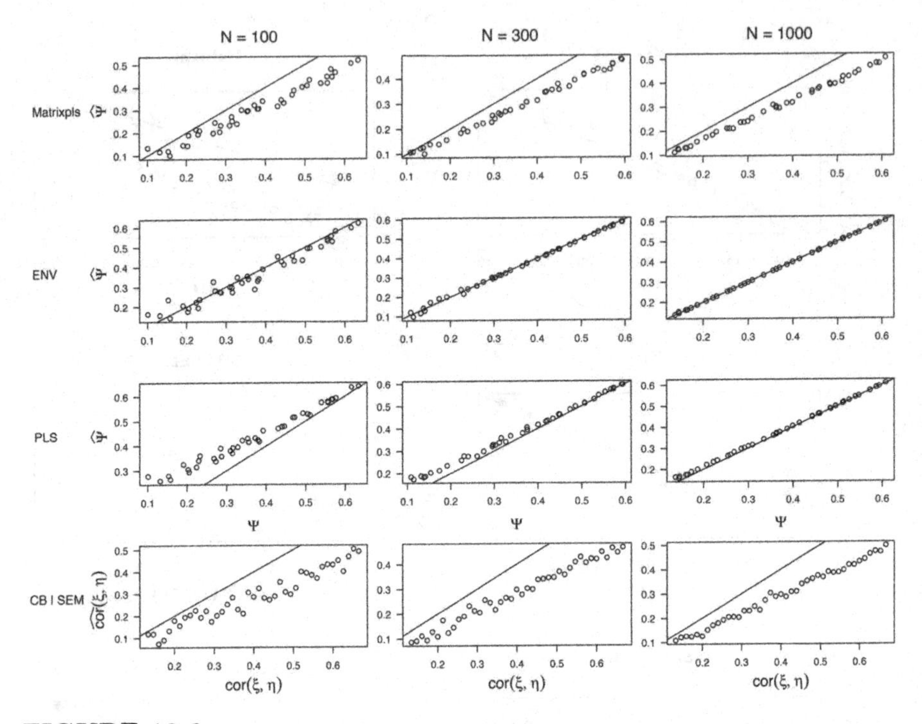

FIGURE 10.6
Results of simulations with two reflective composites for X and Y, $q = u = 2$. Horizontal axes give the true correlations; vertical axes give the estimated correlations. (Constructed following Fig. 6.3 of Cook and Forzani (2023) with permission.)

like that described in Section 10.7.2, except now $p = r = 21$, L is a 21×2 matrix and

$$\Psi \quad = \quad 0.9997224 \times \mathrm{cor}(\eta, \xi). \tag{10.14}$$

Specifically, let 1_k be a $k \times 1$ vector of ones, let $L_1 = (8, -0.7, 60, 1_9^T, -1_9^T)^T$, $L_2 = (1, 0.3, -0.59/60, -1_9^T, 1_9^T)^T$ and $L = (L_1, L_2)$. The conditional means and variances were generated as $\mathrm{E}(X \mid \xi) = 0.1(L_1 + 0.9L_2)\xi$, $\Sigma_{X|\xi} = 5L(L^TL)^{-1}L^T + 0.1L_0L_0^T$ and $E(Y \mid \eta) = 0.1(L_1 + 0.9L_2)\eta$, $\Sigma_{Y|\eta} = 5L(L^TL)^{-1}L^T + 0.1L_0L_0^T$. From this structure, we have, $\Sigma_{Y,\eta} = \Sigma_{X,\xi} = 0.1(L_1 + L_2)$, and

$$\Sigma_{Y,X} = 0.01(L_1 + L_2)(L_1 + L_2)^T = 0.01L1_21_2^TL^T.$$

It follows from this structure that $\Phi = \Gamma = L(L^TL)^{-1/2}$ and so $q = u = 2$.

We see from Figure 10.6 that the ENV estimator does the best at all sample sizes, while PLS does well at the larger sample sizes. The Matrixpls estimator from Rönkkö et al. (2016b) underestimates the true correlation Ψ at all displayed sample sizes because it implicitly assumes that $q = u = 1$. The CB|SEM estimator also underestimates its target $\mathrm{cor}(\xi, \eta)$ because it cannot deal with non-diagonal covariance matrices.

10.8 Discussion

This chapter is based on the relatively simple path diagram of Figure 10.1. Phrased in terms of a construct ξ and a corresponding vector of indicators X, the following overarching conclusions about the role of PLS in path analyses apply regardless of the complexity of the path model. The relationship between the indicators and the construct can be reflective or formative. The same conclusions hold with ξ and X replaced with η and Y.

Path modeling.

At its core, path modeling hinges on the ability of the investigators to identify sets of indicators that are related to the constructs in the manner specified. Our view is that an understanding of a construct should not be divorced from the specific indicators selected for its study. This view leads naturally to using $E(\xi \mid X)$ as a means of construct reification. Here, PLS and ENV can have a useful role in reducing the dimension of X without loss of information on ξ, which allows X to be replaced by reduced predictors X_R so that $E(\xi \mid X) = E(\xi \mid X_R)$.

On the other hand, if marginal characteristics of the constructs like $\mathrm{cov}(\xi, \eta)$ are of sole interest, then we see PLS as having little or no relevance to the analysis, unless the lower bound $|\Psi| \leq |\mathrm{cov}(\xi, \eta)|$ is useful.

PLS|SEM

The success of this method for estimating Ψ depends critically on the standardization/scaling used.

No scaling or standardization works best when one real composite of the indicators, say $\Phi^T X$, extracts all of the available information from X about ξ.

That is, once the real composite $\Phi^T X$ is known and fixed, ξ and X are independent or at least uncorrelated. However, we found no rational for adopting this one-composite framework, which we see as tying the hands of PLS|SEM. Envelopes and their descendent PLS methods include methodology for estimating the number of composites needed to extract all of the available information. Expanding the one-composite framework now used by PLS|SEM has the potential to increase its efficacy considerably.

Standardization using sample covariance matrices produces the maximum likelihood estimator under the CORE model. We expect that it will also be called for in more complicated model, but further investigation is needed to affirm this. We see no compelling reason to use marginal scaling.

CB|SEM

This likelihood-based method for estimating $\text{cor}(\eta, \xi)$ requires constraints on covariance matrices for identifiability. We find the common assumption of diagonal covariance matrices to be quite restrictive.

PLS versus unit-weighted summed composites

The unit-weighted summed composite is the sum of the individual indicators, $X_{\text{sum}} = \sum_{j=1}^{p} X_j$. The effectiveness of X_{sum} depends on its relationship to the material composites. In one-component path models, if the multiple correlation between X_{sum} and $\Phi^T X$is high then X_{sum} may be a useful composite. On the other extreme, if the multiple correlation between X_{sum} and the immaterial composites $\Phi_0^T X$ is large then X_{sum} is effectively an immaterial composite and no useful results based on X_{sum} should be expected. In short, the usefulness of summed composites depends on the anatomy of the paths. In some path analyses they might be quite useful, while useless in others.

Bias

We do not see bias as playing a dominant role in the debate over methodology. If the goal is to estimate $\text{cor}(\eta, \xi)$ using CB|SEM, then estimation bias may, depending on the sample size, play a notable role in CB|SEM, as maximum likelihood estimators are biased but unbiased asymptotically. If the goal is to estimate $\text{cor}(\eta, \xi)$ using PLS, then structural bias is relevant. But if the goal is to estimate Ψ, then structural bias has no special relevance if PLS is used.

11

Ancillary Topics

In this chapter we present various sidelights and some extensions to enrich our discussions from previous chapters. In Section 11.1 we discuss the NIPALS and SIMPLS algorithms as instances of general algorithms $\mathbb{N}$ and $\mathbb{S}$ introduced in Section 1.5. In Section 11.2 we discuss bilinear models that have been used to motivate PLS algorithms, particularly the simultaneous reduction of responses and predictors, and show that they rely on an underlying envelope structure. This connects with the discussion in Chapter 5 on simultaneous reduction. The relationship between NIPALS, SIMPLS, and conjugate gradient algorithms is discussed in Section 11.3. Sparse PLS is discussed briefly in Section 11.4, and Section 11.5 has an introductory discussion of PLS for multi-way (tensor-valued) predictors. A PLS algorithm for generalized linear regression is proposed in Section 11.6.

11.1 General application of the algorithms

In this section, we discuss issues related to using algorithms $\mathbb{N}$ and $\mathbb{S}$ for the construction of an estimator of a general envelope $\mathcal{E}_M(\mathcal{A})$. As described in Sections 1.5.3 and 1.5.4, these algorithms require symmetric positive semi-definite matrices A and M and their estimators. Let $\widehat{M}$ be a $\sqrt{n}$-consistent estimator of M and let $\widehat{A}$ be a symmetric $\sqrt{n}$-consistent estimator of a basis matrix for $\mathcal{A}$. If we start with a non-symmetric basis estimator $\widehat{U}$, then we simply set $\widehat{A} = \widehat{U}\widehat{U}^T$. For a particular sample application, we denote these algorithms as $\mathbb{N}(\widehat{A}, \widehat{M})$ and $\mathbb{S}(\widehat{A}, \widehat{M})$, which provide for $\sqrt{n}$-consistent estimators of a projection onto $\mathcal{E}_M(\mathcal{A})$ when $M > 0$. Application of these algorithms in this general context depends on the problem and the goals, and may suggest different versions of the algorithms even for predictor reduction.

DOI: 10.1201/9781003482475-11

We know from Chapter 3 that the NIPALS and SIMPLS algorithms for predictor reduction provide estimators of the envelope $\mathcal{E}_{\Sigma_X}(\mathcal{B})$ and that they depend on the data only via $S_{X,Y}$ and S_X. To emphasize this aspect of the algorithms, we denote them as NIPALS$(S_{X,Y}, S_X)$ and SIMPLS$(S_{X,Y}, S_X)$. With $\widehat{U} = S_{X,Y}$ and $\widehat{M} = S_X$ we have the following connection between these algorithms for predictor reduction in linear models:

$$\begin{aligned} \text{NIPALS}(S_{X,Y}, S_X) &= \mathbb{N}(S_{X,Y}S_{X,Y}^T, S_X) \\ \text{SIMPLS}(S_{X,Y}, S_X) &= \mathbb{S}(S_{X,Y}S_{X,Y}^T, S_X). \end{aligned}$$

However, the underlying theory allows for many other options for using $\mathbb{N}$ and $\mathbb{S}$ to estimate $\mathcal{E}_{\Sigma_X}(\mathcal{B})$. Recall from Proposition 1.6 that, for all k, $\mathcal{E}_M(M^k\mathcal{A}) = \mathcal{E}_M(\mathcal{A})$ and, for $k \neq 0$, $\mathcal{E}_{M^k}(\mathcal{A}) = \mathcal{E}_M(\mathcal{A})$. In particular, $\mathcal{E}_{\Sigma_X}(\mathcal{B}) = \mathcal{E}_{\Sigma_X}(\mathcal{C}_{X,Y})$. This suggests that when $n \gg p$, we could also use the algorithms $\mathbb{N}(\widehat{\beta}_{\text{ols}}\widehat{\beta}_{\text{ols}}^T, S_X)$ and $\mathbb{S}(\widehat{\beta}_{\text{ols}}\widehat{\beta}_{\text{ols}}^T, S_X)$ to estimate $\mathcal{E}_{\Sigma_X}(\mathcal{B})$. As a second instance, it follows also that

$$\mathcal{E}_{\Sigma_X}(\mathcal{B}) = \mathcal{E}_{\Sigma_X^2}(\mathcal{C}_{X,Y}) = \mathcal{E}_{\Sigma_X}(\text{span}(\Sigma_X^2\Sigma_{X,Y})),$$

which indicates that we could use also $\mathbb{N}(S_{X,Y}S_{X,Y}^T, S_X^2)$, $\mathbb{N}(S_X^2 S_{X,Y}S_{X,Y}^T S_X^2, S_X)$, or the corresponding versions from algorithm $\mathbb{S}$ to estimate $\mathcal{E}_{\Sigma_X}(\mathcal{B})$. The essential point here is that there are many choices for $\widehat{M}$ and $\widehat{A}$, and consequently many different versions of NIPALS and SIMPLS, that give the same envelope in the population but can produce different estimates in application. Likelihood-based approaches like that for predictor envelopes discussed in Section 2.3 can help alleviate this ambiguity.

To illustrate the associated reasoning, recall that the likelihood-based development in Section 2.3.1 was put forth as a basis for estimating $\mathcal{E}_{\Sigma_X}(\mathcal{B})$, but further reasoning is needed to see its implications for PLS algorithms. We start by rewriting the partially maximized likelihood function from (2.10),

$$\begin{aligned} L_q(G) &= \log|G^T S_{X|Y} G| + \log|G^T S_X^{-1} G| \\ &= \log|G^T S_{X|Y} G| + \log|G^T \{S_{X|Y} + S_{X\circ Y}\}^{-1} G| \\ &= \log|G^T S_{X|Y} G| + \log|G^T \{S_{X|Y} + S_{X,Y}S_Y^{-1}S_{Y,X}\}^{-1} G|. \end{aligned}$$

Comparing this to the kernel of algorithm $\mathbb{L}$ in Proposition 1.14, we see that in the population this function forms the basis for estimating $\mathcal{E}_M(\mathcal{A})$, where

$M = \Sigma_{X|Y}$ and $A = \Sigma_{X,Y}\Sigma_Y^{-1}\Sigma_{Y,X}$. From Proposition 1.8 and the discussion of (1.26) we have the following equivalences

$$\mathcal{E}_M(\mathcal{A}) = \mathcal{E}_{\Sigma_{X|Y}}(\Sigma_{X,Y}\Sigma_Y^{-1/2}) = \mathcal{E}_{\Sigma_X}(\Sigma_{X,Y}\Sigma_Y^{-1/2}) = \mathcal{E}_{\Sigma_X}(\mathcal{B})\Sigma_Y^{-1/2} = \mathcal{E}_{\Sigma_X}(\mathcal{B}).$$

As pointed out in Section 3.9, we again see that the likelihood is based on the standardized response vector $Z = S_Y^{-1/2}Y$, which suggests that NIPALS and SIMPLS algorithms also be based on the standardized responses: NIPALS$(S_{X,Y}S_Y^{-1/2}, S_X)$ and SIMPLS$(S_{X,Y}S_Y^{-1/2}, S_X)$, with corresponding adaptations to algorithms $\mathbb{N}$ and $\mathbb{S}$. This idea was introduced in Section 3.9, but here it is used as an illustration of the general recommendation that a likelihood can be used to guide the implementation of a PLS algorithm. The algorithms NIPALS$(S_{X,Y}, S_X)$ and SIMPLS$(S_{X,Y}, S_X)$ with the original unstandardized responses might be considered when S_Y is singular.

11.2 Bilinear models

Bilinear models have long been used to motivate PLS algorithms, particularly NIPALS. The idea of using bilinear models originated around 1975 in the early work of H. Wold (Martens and Næs, 1989, p. 118). We introduced bilinear models for simultaneous reduction of the predictors and the responses in Section 5.4.1, but that is not the only version of a bilinear model that has been proposed as a basis for PLS. In this section, we continue our discussion of bilinear models, starting with that proposed by Martens and Næs (1989).

11.2.1 Bilinear model of Martin and Næs

The bilinear calibration model described by Martens and Næs (1989, p. 91), which they used in part to motivate the NIPALS algorithms for dimension reduction in X, is formulated in terms of the centered data matrices $\mathbb{X} \in \mathbb{R}^{n\times p}$ and $\mathbb{Y} \in \mathbb{R}^{n\times r}$:

$$\begin{aligned} \mathbb{X}_{n\times p} &= TR_{p\times u}^T + E_{n\times p} \\ \mathbb{Y}_{n\times r} &= TU_{r\times u}^T + F_{n\times r} \\ T_{n\times u} &= \mathbb{X}W_{p\times u} \end{aligned}$$

where $\min(p, r) \geq u$ and the matrix W of weights has full column rank. The rows of R and U represent the loadings and the rows of T represent the scores. Descriptions of this bilinear model in the PLS literature rarely mentioning any stochastic properties of E and F, regarding them generally as unstructured residual or error matrices. Martens and Næs (1989), as well as others, seem to treat the bilinear model as a data description rather than as a statistical model per se. To develop the connection with envelopes, it is helpful to reformulate the model in terms of uncentered random vectors X_i^T and Y_i^T that correspond to the rows of $\mathbb{X}$ and $\mathbb{Y}$, and in terms of independent zero mean error vectors e_i^T and f_i^T, representing the rows of E and F. Then written in vector form the bilinear model becomes for $i = 1, \ldots, n$

$$\left.\begin{aligned} X_i &= \alpha_X + Rt_i + e_i \\ Y_i &= \alpha_Y + Ut_i + f_i \\ t_i &= W^T X_i, \end{aligned}\right\} \tag{11.1}$$

where α_X and α_Y are intercept vectors that are needed because the model is in terms of uncentered data, and $e \perp\!\!\!\perp f$. In the bilinear model (5.13) for simultaneous reduction of responses and predictors, the latent vectors v_i and t_i are assumed to be independent, while in (11.1) the corresponding latent vector are the same and this common latent vector is a linear function of X. As shown in the following discussion, the condition $t_i = W^T X_i$ has negligible impact on the X model but it does lead to envelopes in terms of the Y model.

It follows from (11.1) that we can take W to be semi-orthogonal without any loss of generality, so we assume that in the following. Substituting for t in the Y equation,

$$Y_i = \alpha_Y + UW^T X_i + f_i. \tag{11.2}$$

Thinking of this Y-model in the form of the multivariate linear model (1.1), we must have $\mathcal{B} \subseteq \text{span}(W)$. Without structure beyond assuming that the error vectors f_i are independent copies of a random vector f with mean 0 and positive definite variance, this is a reduced rank multivariate regression model (Cook, Forzani, and Zhang, 2015; Izenman, 1975). With W regarded as known, the estimator for β is

$$W\widehat{U}_W^T = W(W^T S_X W)^{-1} W^T S_{X,Y},$$

which is the same form as given in Tables 3.1 and 3.4. The issue then is how we estimate W, the basis for which must come from the X-model.

Substituting t_i into the X-model, we have

$$X_i = \alpha_X + R(W^T X_i) + e_i. \tag{11.3}$$

This represents a model for the conditional distribution of $X \mid W^T X$. Multiplying both sides of the X-model (11.3) by W^T, we then have,

$$W^T X_i = W^T \alpha_X + W^T R(W^T X_i) + W^T e_i.$$

Since this equality holds for all values of $W^T X$ and W has full column rank, we must have $W^T \alpha_X = 0$, $W^T R = I_u$, and $W^T e \equiv 0$. In consequence, we can take $R = W$ without loss of generality. Recalling that W is semi-orthogonal, the X-model (11.3) then reduces to $X_i = \alpha_X + P_W X_i + Q_W e_i$. This implies that $Q_W X = Q_W \alpha_X + Q_W e$ and so the X-model reduces further to simply

$$\begin{aligned} X &= \alpha_X + P_W X + e \\ &= P_W \alpha_X + Q_W \alpha_X + P_W X + Q_W e \\ &= P_W \alpha_X + P_W X + Q_W X \\ &= P_W X + Q_W X, \end{aligned} \tag{11.4}$$

since $W^T \alpha_X = 0$. This holds for any W, so the X-model doesn't really add any restrictions to the problem. The only restriction on W arises from the Y-model, which implies that span(W) must contain span(β). This line of reasoning does not give rise to envelopes directly because (11.4) holds for any span(W) that contains span(β). The final step to reach envelopes is to require that $\text{cov}(P_W X, Q_W X) = \text{cov}(P_W X, Q_W e) = 0$. With this and the previous conclusion that $\mathcal{B} \subseteq \text{span}(W)$, we see that span$(W)$ is a reducing subspace of Σ_X that contains $\mathcal{B}$, and then $u = q$ becomes the number of components. As we have seen previously in this chapter, PLS algorithms NIPALS and SIMPLS require this condition in the population, although we have not seen it stated in the literature as part of a bilinear model.

11.2.2 Bilinear probabilistic PLS model

Recall from the discussion following (2.5) that the constituent parameters of envelope models or PLS formulations are not identifiable, while key parameters are identifiable. el Bouhaddani et al. (2018) proposed a bilinear PLS model in which the parameters are identifiable up to a sign, and they showed how to use the EM algorithm to estimate those parameters. We describe their bilinear PLS model in this section.

The el Bouhaddani et al. (2018) model, which was motivated by the work of Tipping and Bishop (1999) on probabilistic principal component analysis, begins with the requirement that the response and predictor vectors $Y \in \mathbb{R}^r$ and $X \in \mathbb{R}^p$ are related via $q \times 1$ latent vectors $t \sim N(0, \mathrm{diag}(\sigma_{t_1}^2, \ldots, \sigma_{t_q}^2))$ and u by the following bilinear probabilistic PLS model.

$$\left.\begin{array}{l} X = \mu_X + Wt + e, \text{ with } e \sim N(0, \sigma_e^2 I_p) \\ Y = \mu_Y + Vu + f, \text{ with } f \sim N(0, \sigma_f^2 I_r) \\ u = Bt + h, \text{ with } h \sim N(0, \sigma_h^2 I_q), \end{array}\right\} \tag{11.5}$$

where $\mu_{(\cdot)}$'s are unknown intercept vectors, $W \in \mathbb{R}^{p\times q}$, $V \in \mathbb{R}^{r\times q}$, $B = \mathrm{diag}(b_1, \ldots, b_q)$ and the error vectors e, f and h are mutually independent. Additionally, W and V have full column ranks, t is independent of $\{e, f, h\}$, and u is independent of $\{e, f\}$. As defined in Section 2.3.1, we let $C = (X^T, Y^T)^T$ denote the random vector constructed by concatenating X and Y. It is also required that $q \leq \min(p, r)$; otherwise, simultaneous reduction of X and Y dimension reduction may have no benefits in this context.

It then follows from the model that

$$\Sigma_C = \begin{pmatrix} \Sigma_X & \Sigma_{X,Y} \\ \Sigma_{Y,X} & \Sigma_Y \end{pmatrix} = \begin{pmatrix} W\Sigma_t W^T + \sigma_e^2 I_p & W\Sigma_t B^T V^T \\ VB\Sigma_t W^T & V\{B\Sigma_t B + \sigma_h^2 I_q\}V^T + \sigma_f^2 I_r \end{pmatrix}.$$

In consequence,

$$\beta = \Sigma_X^{-1}\Sigma_{X,Y} = (W\Sigma_t W^T + \sigma_e^2 I_p)^{-1} W\Sigma_t B^T V^T.$$

We next demonstrate that $\mathrm{span}(W)$ is an envelope; that is, it is the smallest reducing subspace of Σ_X that contains $\mathcal{B} = \mathrm{span}(\beta) = \mathrm{span}(W)$. First we show that $\mathrm{span}(W)$ reduces Σ_X and contains $\mathcal{B} = \mathrm{span}(W)$.

To show that $\mathrm{span}(W)$ reduces Σ_X, define a $q \times q$ nonsingular matrix A so that $\tilde{W} = WA$ is semi-orthogonal, and choose $\tilde{W}_0$ so that $(\tilde{W}, \tilde{W}_0)$ is orthogonal. Then

$$\Sigma_X = \tilde{W}(A^{-1}\Sigma_t A^{-1} + \sigma_e^2 I_q)\tilde{W}^T + \sigma_e^2 \tilde{W}_0 \tilde{W}_0^T. \tag{11.6}$$

It follows from Proposition 1.24 that $\mathrm{span}(W)$ reduces Σ_X.

Using (11.6), the following calculations show that $\text{span}(\beta) \subseteq \text{span}(W)$.

$$\begin{aligned}
\beta &= \Sigma_X^{-1}\Sigma_{X,Y} \\
&= (\tilde{W}(A^{-1}\Sigma_t A^{-1} + \sigma_e^2 I_q)\tilde{W}^T + \sigma_e^2 \tilde{W}_0 \tilde{W}_0^T)^{-1} W \Sigma_t B^T V^T \\
&= \left\{\tilde{W}(A^{-1}\Sigma_t A^{-1} + \sigma_e^2 I_q)^{-1}\tilde{W}^T + \sigma_e^{-2}\tilde{W}_0\tilde{W}_0^T\right\} W\Sigma_t B^T V^T \\
&= \left\{\tilde{W}(A^{-1}\Sigma_t A^{-1} + \sigma_e^2 I_q)^{-1}\tilde{W}^T\right\} \tilde{W} A^{-1}\Sigma_t B^T V^T \\
&= \tilde{W}(A^{-1}\Sigma_t A^{-1} + \sigma_e^2 I_q)^{-1} A^{-1}\Sigma_t B^T V^T.
\end{aligned}$$

Since $(A^{-1}\Sigma_t A^{-1} + \sigma_e^2 I_q)^{-1} A^{-1}\Sigma_t B^T$ has rank q, $r > q$ and V has full column rank, we have that

$$\text{span}((A^{-1}\Sigma_t A^{-1} + \sigma_e^2 I_q)^{-1} A \Sigma_t B^T V^T) = \mathbb{R}^q$$

and, in consequence, $\mathcal{B} = \text{span}(W)$ and $\mathcal{E}_{\Sigma_X}(\mathcal{B}) = \text{span}(\beta) = \text{span}(W)$. It follows also from these results that $\text{span}(W)$ is equal to the span of the first q eigenvectors of Σ_X, and thus that the material predictors $\tilde{W}^T X$ correspond to the first q principal components. Under this bilinear model then, principal component regression coincides with partial least squares in the population.

The bilinear probabilistic PLS model may seem similar to the Martin-Næs model described in Section 11.2.1, but there are important differences. Most notably, the Martin-Næs model is not formulated in terms of latent variables, but is instead described directly in terms of loadings and scores. As indicated previously, the requirement that the errors e and f be isotropic in the probabilistic PLS model links it to the probabilistic principal component model PPCA proposed by Tipping and Bishop (1999). As a result, the latent variables t and u must account *all* variation in X and Y up to isotropic errors. One consequence of this can be seen in the covariance matrices for the material and immaterial predictors: From (11.6), the variation in the material predictors $\text{var}(\tilde{W}^T X) = A^{-1}\Sigma_t A^{-1} + \sigma_e^2 I_q$ is unconstrained, while the variation in the immaterial predictors is modeled as isotropic, $\text{var}(\tilde{W}_0^T X) = \sigma_e^2 I_{p-q}$. This is much more restrictive than the envelope model for predictor reduction (2.5). Zhang and Chen (2020) discuss the limitations of having isotropic errors as a key ingredient of probabilistic principal component analysis. In view of the isotropic errors required by the probabilistic PLS model (11.5), their criticism are applicable here as well. In sum, it seems that this probabilistic PLS model solved a paltry issue with an overly restrictive model, resulting in methodology that is essentially equivalent to reduction by principal components. Etiévant and Viallon (2022) raised additional issues and proposed a generalization that addresses some of them.

Multiple versions of the bilinear model have been used to motivate PLS. Our assessment is that they can be confusing and more of hindrance rather than a help, particularly since PLS can be motivated fully using envelopes.

11.3 Conjugate gradient, NIPALS, and SIMPLS

11.3.1 Synopsis

There are many techniques available for solving the linear systems of equations (e.g. Björck, 1966; Elman, 1994),

$$A\omega = b, \tag{11.7}$$

where $b \in \mathbb{R}^r$ is known and $A \in \mathbb{S}^{r\times r}$ is positive definite and known. The solution $\tilde{\omega}$ can of course be represented as $\tilde{\omega} = A^{-1}b$. However, there are situations in which computing this naive solution directly may not be the best course, depending on the relationship between b and the reducing subspaces of A. Let $G \in \mathbb{R}^{r\times s}$ be a semi-orthogonal basis matrix for $\mathcal{E}_A(\text{span}(b))$ and let $(G, G_0) \in \mathbb{R}^{r\times r}$ be an orthogonal matrix. Then, from Proposition 1.2 and Corollary 1.1, A and A^{-1} can be represented as

$$\begin{aligned} A &= GHG^T + G_0H_0^{-1}G_0^T \\ A^{-1} &= GH^{-1}G^T + G_0H_0^{-1}G_0^T \end{aligned}$$

where $H = G^TAG$. It follows that we can compute $\tilde{\omega}$ as

$$\begin{aligned} \tilde{\omega} &= GH^{-1}G^Tb \\ &= G(G^TAG)^{-1}G^Tb, \end{aligned}$$

which serves also to highlight the fact that only $\text{span}(G) = \mathcal{E}_A(\text{span}(b))$ matters. This form for $\tilde{\omega}$ could be particularly advantageous if G were known and its column dimension $s = \dim\{\mathcal{E}_A(\text{span}(b))\}$ were small relative to r, or if the eigenvalues of H_0 were small enough to cause numerical difficulties. Of course, implementations of this idea require an accurate numerical approximation of a suitable G. We might consider developing approximations of G by using the general PLS-type algorithms $\mathbb{N}(b, A)$ or $\mathbb{S}(b, A)$, but in this discussion there is not necessarily a statistical context associated with (11.7), so it may be unclear how to use cross validation or a holdout sample to aid in selecting a suitable dimension s. Nevertheless, we demonstrate in this section that the highly regarded conjugate gradient method for solving (11.7) is in fact an envelope method that relies on NIPALS and SIMPLS for an approximation of G (e.g. Phatak and de Hoog, 2002; Stocchero, de Nardi, and Scarpa, 2020).

In keeping with the theme of this book, we now consider (11.7) in the context of model (1.1) with a univariate response and our standard notation $A = \text{var}(X) = \Sigma_X$, $\omega = \beta$ and $b = \text{cov}(X, Y) = \sigma_{X,Y}$. In this context, (11.7) becomes the normal equations for the population, $\Sigma_X \beta = \sigma_{X,Y}$. Sample versions are discussed later in this section.

Table 11.1 gives the conjugate gradient algorithm in the context of solving the normal equations $\Sigma_X \beta = \sigma_{X,Y}$ for a regression with a real response. It was adapted from Elman (1994) and it applies to solving any linear system $A\omega = b$ for A symmetric and positive definite.

11.3.2 Details for the conjugate gradient algorithm of Table 11.1

To gain an understanding of this algorithm, we consider the first two iterations, as we have done for other algorithms. This then will lead via induction to a general conclusion about the relationship between the conjugate gradient algorithm (CGA), PLS, and envelopes. Recall from Tables 3.1 and 3.4 that $W_q = (w_1, \ldots, w_q)$ and $V_q = (v_1, \ldots, v_q)$ refer to the NIPALS and SIMPLS weight matrices with q components.

At $k = 0$, $p_0 = r_0 = \sigma_{X,Y}$, $\alpha_0 = \sigma_{X,Y}^T \sigma_{X,Y} (\sigma_{X,Y}^T \Sigma_X \sigma_{X,Y})^{-1}$ and

$$
\begin{aligned}
\beta_1 &= \alpha_0 p_0 = p_0 (p_0^T \Sigma_X p_0)^{-1} p_0^T \sigma_{X,Y} \\
&= \{\sigma_{X,Y}^T \sigma_{X,Y} (\sigma_{X,Y}^T \Sigma_X \sigma_{X,Y})^{-1}\} \sigma_{X,Y} \\
&= \{\sigma_{X,Y} (\sigma_{X,Y}^T \Sigma_X \sigma_{X,Y})^{-1} \sigma_{X,Y}^T\} \sigma_{X,Y} \\
r_1 &= r_0 - \alpha_0 \Sigma_X p_0 \\
&= \sigma_{X,Y} - \{\sigma_{X,Y}^T \sigma_{X,Y} (\sigma_{X,Y}^T \Sigma_X \sigma_{X,Y})^{-1}\} \Sigma_X \sigma_{X,Y} \\
&= \{I - \Sigma_X \sigma_{X,Y} (\sigma_{X,Y}^T \Sigma_X \sigma_{X,Y})^{-1} \sigma_{X,Y}^T\} \sigma_{X,Y} \\
&= Q_{\sigma_{X,Y}(\Sigma_X)}^T \sigma_{X,Y}.
\end{aligned}
$$

From this we see that $\text{span}(p_0) = \text{span}(r_0) = \text{span}(w_1) = \text{span}(v_1)$ and

TABLE 11.1

Conjugate gradient algorithm: Population version of the conjugate gradient method for solving the linear system of equations $\Sigma_X \beta = \Sigma_{X,Y}$ for β.

(a) *Population version*	
Initialize	$\beta_0 = 0$, $r_0 = \sigma_{X,Y}$, $p_0 = r_0$ and tolerance ϵ.
	$R_0 = (r_0)$, $\bar{P}_0 = (p_0)$
For $k = 0, 1, 2 \ldots$	
	$\alpha_k = r_k^T r_k / p_k^T \Sigma_X p_k$
	$\beta_{k+1} = \beta_k + \alpha_k p_k$
	$r_{k+1} = r_k - \alpha_k \Sigma_X p_k$
Append	$R_{k+1} = (R_k, r_{k+1})$
End when first	$\lVert r_{k+1} \rVert < \epsilon \lVert \sigma_{X,Y} \rVert$ and then set $\beta_{\text{cg}} = \beta_{k+1}$ and $q = k+1$
	$B_k = r_{k+1}^T r_{k+1} / r_k^T r_k$
	$p_{k+1} = r_{k+1} + B_k p_k$
Append	$\bar{P}_{k+1} = (\bar{P}_k, p_{k+1})$
(b) *Notes*:	w_{k+1} and v_{k+1} denote NIPALS and SIMPLS weights.
Weights	$\operatorname{span}(r_k) = \operatorname{span}(w_{k+1})$, $\operatorname{span}(p_k) = \operatorname{span}(v_{k+1})$
	for all $j < k$, $r_j^T r_k = 0$, $p_j^T \Sigma_X p_k = 0$, $p_j^T r_k = 0$.
Envelope connection	$\operatorname{span}(R_q) = \mathcal{E}_{\Sigma_X}(\mathcal{B})$, the Σ_X-envelope of $\mathcal{B} = \operatorname{span}(\beta)$ when $r_q = 0$.

$\beta_1 = \beta_{\text{npls}} = \beta_{\text{spls}}$ when the number of components is $q = 1$ (cf. Tables 3.1 and 3.4). The estimators are thus identical when the respective stopping criteria are met. For CGA to stop at β_1 we need $\lVert Q_{\sigma_{X,Y}(\Sigma_X)}^T \sigma_{X,Y} \rVert < \epsilon$. For NIPALS to stop with $q = 1$ we need, from Table 3.1, $Q_{\sigma_{X,Y}(\Sigma_X)}^T \sigma_{X,Y} = 0$. Thus aside from a relatively minor difference in the population stopping criterion, the CGA and NIPALS are so far identical.

Assuming that the stopping criterion is not met, the next part of CGA is to compute

$$
\begin{aligned}
B_0 &= r_1^T r_1 / r_0^T r_0 = \sigma_{X,Y}^T Q_{\sigma_{X,Y}(\Sigma_X)} Q_{\sigma_{X,Y}(\Sigma_X)}^T \sigma_{X,Y} / \sigma_{X,Y}^T \sigma_{X,Y} \\
p_1 &= r_1 + B_0 p_0 \\
&= Q_{\sigma_{X,Y}(\Sigma_X)}^T \sigma_{X,Y} + P_{\sigma_{X,Y}} Q_{\sigma_{X,Y}(\Sigma_X)} Q_{\sigma_{X,Y}(\Sigma_X)}^T \sigma_{X,Y}.
\end{aligned}
$$

Key characteristics of this second step are, as shown in Appendix A.9.1,

$\text{span}(p_1) = \text{span}(v_2)$. Since $\text{span}(p_0) = \text{span}(v_1)$, $p_0^T \Sigma_X p_1 = 0$. Additionally, since $r_1^T \sigma_{X,Y} = 0$,

$$\begin{aligned} r_1^T \sigma_{X,Y} &= \sigma_{X,Y}^T Q_{\sigma_{X,Y}(\Sigma_X)} \sigma_{X,Y} = 0 \\ p_1^T \sigma_{X,Y} &= r_1^T \sigma_{X,Y} + \sigma_{X,Y}^T Q_{\sigma_{X,Y}(\Sigma_X)} Q_{\sigma_{X,Y}(\Sigma_X)}^T \sigma_{X,Y} \\ &= r_1^T r_1. \end{aligned}$$

Using these results, we get

$$\begin{aligned} \beta_2 &= \beta_1 + \frac{r_1^T r_1}{p_1^T \Sigma_X p_1} p_1 \\ &= p_0 (p_0^T \Sigma_X p_0)^{-1} p_0^T \sigma_{X,Y} + p_1 (p_1^T \Sigma_X p_1)^{-1} p_1^T \sigma_{X,Y} \\ &= V_2 (V_2^T \Sigma_X V_2)^{-1} V_2^T \sigma_{X,Y} \\ &= \beta_{\text{spls}} \text{ when } q = 2, \end{aligned}$$

where V_2 is the second SIMPLS weight matrix.

The following proposition and its proof covers these initial observations and casts them into a general result connecting NIPALS, SIMPLS and CGA. The notation is that used in Tables 3.1, 3.4, and 11.1. The index i in the proposition is offset by 1 for NIPALS and SIMPLS. This arises because the PLS algorithms are indexed from $i = 1$, while CGA is indexed from $i = 0$.

Proposition 11.1. *For a single-response regression with $\Sigma_X > 0$ and dimension q envelope $\mathcal{E}_{\Sigma_X}(\mathcal{B})$, we have*

1. $\text{span}(r_i) = \text{span}(w_{i+1})$, $i = 0, 1, \ldots, q-1$
2. $\text{span}(p_i) = \text{span}(v_{i+1})$, $i = 0, 1, \ldots, q-1$
3. $\text{span}(r_0, \ldots, r_{q-1}) = \text{span}(W_q) = \text{span}(V_q) = \mathcal{E}_{\Sigma_X}(\mathcal{B})$
4. $\beta_{\text{cg}} = \beta_{\text{npls}} = \beta_{\text{spls}}$.

Proof. The proof in Appendix A.9.1 is by induction on q. □

This proposition tells us that in effect CGA uses NIPALS and SIMPLS to pursue an envelope solution to the linear system of equations $\Sigma_X \beta = \sigma_{X,Y}$. All three algorithms – CGA, NIPALS, and SIMPLS – can be regarded as methods for estimating β based on the predictor envelope $\mathcal{E}_{\Sigma_X}(\mathcal{B})$. The conjugate gradient method (Hentenes and Stiefel, 1952) preceded NIPALS and SIMPLS and, being based on a positive definite sample covariance matrix S_X, it proceeds by updating the estimate of β. NIPALS and SIMPLS do not require S_X to be positive definite and they proceed by generating the weights w_j and v_j reserving estimation of β until the end. There are relatively recent studies of the conjugate gradient method when A is singular (see, for example, Hayami, 2020) and those might be closer to the PLS algorithms.

Proposition 11.1 also holds in the sample, provided $S_X > 0$, and we assume that all three algorithms stop at the same point. In that case, r_i, p_i, w_{i+1} and v_{i+1} are based on the sample, the last equality in item 3 becomes $\text{span}(V_q) = \widehat{\mathcal{E}}_{\Sigma_X}(\mathcal{B})$, the estimated envelope, and the fourth item becomes $\widehat{\beta}_{\text{cg}} = \widehat{\beta}_{\text{npls}} = \widehat{\beta}_{\text{spls}}$. In application, CGA may differ appreciably from the PLS methods because of the nature of their stopping criteria: CGA uses a sample version of the PLS population stopping criteria described in Tables 3.1 and 3.4, while the stopping criterion for the PLS methods is typically based on cross validation or a holdout sample. Of course, cross validation could be used in conjunction with CGA as well.

Recall from Tables 3.1 and 3.4 that at termination NIPALS and SIMPLS employ matrices in the computation of the their estimates of β. In contrast, the CGA estimator $\widehat{\beta}_{\text{cg}}$ is a simple additive update of the previous iterate; no matrix valuation or storage is required. In consequence, CGA may be computationally more stable than NIPALS or SIMPLS.

A version of Proposition 11.1 holds also when S_X is singular. It is well-known that the linear system $S_X\widehat{\beta} = \widehat{\sigma}_{X,Y}$ always has at least one solution; it has infinitely many solutions when S_X is singular. To find a solution that fits with Proposition 11.1 when S_X is singular, we can use the NIPALS method discussed in Section 3.1.3. Let H be a semi-orthogonal basis matrix for $\text{span}(S_X)$. Then $S_{H^TX} > 0$ and the linear system $S_{H^TX}\widehat{\alpha} = \widehat{\sigma}_{H^TX,Y}$ is covered by Proposition 11.1. The solution to the original equations is then $\widehat{\beta}_{\text{cg}} = H\widehat{\alpha}$. Since the PLS methods use the same technique for dealing with rank deficiency, the connections between CGA and the PLS methods described in Proposition 11.1 still hold.

11.3.3 Origins of CGA

In the previous section we described the connection between CGA, NIPALS, SIMPLS, and envelopes. However, the origins of CGA are quite different, tracing back to the basic notion of steepest decent (See Björck, 1966, Ch. 7, for an historical overview.) The numerical goal is to find an approximate solution to $\Sigma_X \beta = \sigma_{X,Y} \in \mathbb{R}^p$ by minimizing $\phi(\beta) = ||\sigma_{X,Y} - \Sigma_X \beta||_2^2$. The general idea is to start with a possible solution β_0 and then iterate to a final solution, moving from step k to step $k+1$ in the direction of maximum descend $d_k := -\nabla\phi(\beta_k) = \sigma_{X,Y} - \Sigma_X \beta_k$. The $(k+1)$-st iterate is then defined as $\beta_{k+1} = \beta_k + \alpha d_k$, with α chosen to minimize $\phi(\beta_k + \alpha d_k)$. This gives $\alpha_k = d_k^T d_k / d_k^T \Sigma_X d_k$. The algorithm stops when the update d_k is sufficiently small.

An alternative version of steepest decent optimizes overall α coefficients simultaneously at each step:

1. Choose β_1 to minimize ϕ over $\{\beta_0 + \alpha d_0, \alpha \in \mathbb{R}\}$. We get $\beta_1 = \beta_0 + \delta_0 d_0$. Define $d_1 = \sigma_{X,Y} - \Sigma_X \beta_1$.

2. Choose β_2 to minimize ϕ over $\{\beta_0 + \alpha_0 d_0 + \alpha_1 d_1, \alpha_0, \alpha_1 \in \mathbb{R}\}$, we find $\beta_2 = \beta_0 + \delta_0 d_0 + \delta_1 d_1$ and define $d_2 = \sigma_{X,Y} - \Sigma_X \beta_2$.

3. Choose β_{k+1} to minimize ϕ over $\{\beta_0 + \alpha_0 d_0 + \cdots + \alpha_k d_k, \alpha_i \in \mathbb{R}\}$, and define d_{k+1} accordingly.

4. Stop when $\|d_{k+1}\| < \epsilon \|\sigma_{X,Y}\|$.

This alternative version of steepest decent is in principle better than the basic version because at each step it optimizes simultaneously over the coefficients α_j if all directions d_j, but it is relatively complicated and rather unwieldily in application. However, there is an equivalent algorithm that gives the same solution at the end (but not in the intermediate steps) and updates only from the last step. The key in this variation is to find a single direction that gives the same iterate at each step. This leads then to CGA (Björck, 1966; Elman, 1994):

1. Initialize $\beta_0 = 0$, $d_0 = \sigma_{X,Y} - \Sigma_X \beta_0$, and $p_0 = d_0$. At initialization the new direction p_0 is the same as the original direction.

2. Set $\beta_1 = \beta_0 + \delta_0 p_0$, where

$$\delta_0 = \arg\min_{\delta} \phi(\beta_0 + \delta p_0) = \frac{d_0^T d_0}{p_0^T \Sigma_X p_0},$$

   $d_1 = d_0 - \delta_0 \Sigma_X p_0$ and $p_1 = d_1 + \gamma_0 p_0$ with $\gamma_0 = d_1^T d_1 / d_0^T d_0$.

3. At the $(k+1)$-st step, update β_k in the direction of p_k: $\beta_{k+1} = \beta_k + \delta_k p_k$, where

$$\delta_k = \arg\min_{\delta} \phi(\beta_k + \delta p_k) = \frac{d_k^T d_k}{p_k^T \Sigma_X p_k}$$

$d_{k+1} = d_k - \delta_k \Sigma_X p_k$ and $p_{k+1} = d_{k+1} + \gamma_k p_k$ with

$$\gamma_k = \frac{d_{k+1}^T d_{k+1}}{d_k^T d_k}.$$

4. Terminate when $||d_{k+1}|| < \epsilon ||\sigma_{X,Y}||$.

We find it rather remarkable that the CGA turns out to be an envelope method that relies on NIPALS and SIMPLS for iteration improvement since the rationale for the algorithm makes no direct appeal to reducing subspaces.

11.4 Sparse PLS

As mentioned in Chapter 4, sparse versions of PLS regression have been proposed by Chun and Keleş (2010), Liland et al. (2013), and Zhu and Su (2020). From our experience, we conclude that most chemometric regressions are closer to abundant than sparse. Nevertheless, assuming sparsity can be a sound step if there is prior supporting information, but it should not be taken for granted just because one may be dealing with high dimensions. Chun and Keleş (2010) used penalization to induce a sparse version of the PLS algorithm. The method is relatively popular in statistics but its theoretical properties are largely unknown. Zhu and Su (2020) developed a sparse version of envelope methodology that performs much better than the proposal by Chun and Keleş (2010). They referred to their sparse methods as envelope-based sparse PLS.

To introduce sparsity into envelope model (2.5), Zhu and Su (2020) first partitioned X into active predictors $X_{\mathcal{A}} \in \mathbb{R}^{p_{\mathcal{A}}}$ and inactive predictor $X_{\mathcal{I}} \in \mathbb{R}^{p_{\mathcal{I}}}$, with $X = (X_{\mathcal{A}}^T, X_{\mathcal{I}}^T)^T$, $p = p_{\mathcal{A}} + p_{\mathcal{I}}$. Under sparsity, the rows of ϕ corresponding to the inactive predictors are all zero,

$$\phi = \begin{pmatrix} \phi_{\mathcal{A}} \\ 0 \end{pmatrix},$$

which, from model (2.5), leads to

$$\beta = \phi\eta = \begin{pmatrix} \phi_{\mathcal{A}}\eta \\ 0 \end{pmatrix} := \begin{pmatrix} \beta_{\mathcal{A}} \\ 0 \end{pmatrix}.$$

Recall from (2.11) that the predictor envelope estimator of ϕ in model (2.5) is based on minimizing over semi-orthogonal matrices $G \in \mathbb{R}^{p \times q}$ the objective function

$$L_q(G) = \log|G^T S_{X|Y} G| + \log|G^T S_X^{-1} G|.$$

Zhu and Su (2020) based their development of envelope-based sparse PLS on a functionally equivalent objective function, say L^*, developed by Cook, Forzani, and Su (2016) that can be solved using unconstrained minimization over $\mathbb{R}^{(p-q)\times q}$ rather than constrained minimization in $\mathbb{R}^{p\times q}$. Then the Zhu-Su envelope-based sparse PLS estimator of ϕ is found by optimizing a constrained version of L^* with the sparse parameterization. Their method is more akin to maximum likelihood and is not based on a PLS-type algorithm. In consequence, we see their label "envelope-based sparse PLS" as a misnomer. They proved that their method has desirable theoretical properties and demonstrated via simulation that it can perform much better than the method by Chun and Keleş (2010).

11.5 PLS for multi-way predictors

In this book we have dealt with regression analysis in which the predictors take the form of vectors. In chemometrics, these are typically vectorial spectral data, like those captured through near-infrared (NIR) sample absorption. With increasingly sophisticated instrumentation, predictor may take on higher dimensional forms, like matrices or three-dimensional arrays.

Imagine that we aim to determine the concentration of a substance in water using single excitation and emission wavelength fluorescence measurements, where the sole emitter is the substance in question. Classical calibration involves using a range of substance standards of different known concentrations and constructing a regression for substance detection. The same methodology is not feasible when gauging the concentration of a substance in samples containing other fluorescent components. However, we could measure fluorescence spectra for a set of calibration samples containing both our substance and possible interferences that might occur in new samples. This is a first-order multivariate calibration method that allows quantification of the substance in mixtures. But the original question remains: Can we do this using standards of the pure substance? Generally, the answer is expected to be "no."

However, if we take measurements of matrix excitation-emission fluorescence data, we can indeed predict the concentrations of substances in new samples, having calibrated with pure substance standards. This highlights the primary advantage of multiway calibration. The goal of the earliest paper on this topic, published in 1978, was to determine the polycyclic aromatic hydrocarbon perylene in mixtures with anthracene, calibrated solely with perylene solutions, through appropriate processing of multiway fluorescence data. Multiway calibration can therefore be seen as relatively recent in the field of analytical chemistry. It results in models where Y is a real number but the predictors X are matrices. See Olivieri and Escandar (2014) for further explanation of these kind of data.

There are more intricate experiments where the predictors may be of higher dimensions, but for simplicity we will now focus on matrix-valued predictors $X \in \mathbb{R}^{p \times s}$ and real responses Y. The model will follow this structure.

Vectorizing X and adopting a linear model in $\text{vec}(X)$,

$$Y = \beta_0 + \beta^T \text{vec}(X) + \epsilon, \text{ with } \epsilon \sim N(0, \sigma^2_{Y|X}), \tag{11.8}$$

results in a setting that is covered by the dimension reduction methods in this book. This includes PLS regression methods of Tables 3.1 and 3.4 with X replaced by $\text{vec}(X)$. This type of analysis is commonly called U-PLS in analytic chemistry. Recall that the NIPALS and SIMPLS algorithms in the population give consistent estimators for the envelope model:

$$\begin{aligned} \beta &= \Gamma\eta \\ \Sigma_{\text{vec}(X)} &= \Gamma\Omega\Gamma^T + \Gamma_0\Omega_0\Gamma_0^T \end{aligned}$$

for some $\Gamma \in \mathbb{R}^{ps \times q}$, semi-orthogonal, $\eta \in \mathbb{R}^q$ and positive definite matrices Ω and Ω_0. PLS analyses based on this vectorization are quite flexible, but unfortunately they muddle the two-dimensional structure of the predictor and may lead to a loss in chemical interpretability.

In Bro (1998) a multiway regression method called N-way partial least squares (N-PLS) was presented. According to the authors, the developed algorithm is superior to the vectorization methods, primarily owing to a stabilization of the decomposition. Nevertheless, this method does not work as expected. As explained in Olivieri and Escandar (2014), this is due to the fact that the bilinear structure is only partially used in the estimation process.

Following the ideas for response envelopes given by Ding and Cook (2018), we propose here a new algorithm that uses the structure of bilinearity from the beginning. Specifically, consider the bilinear model

$$Y = \beta_0 + \beta_1^T X \beta_2 + \epsilon \text{ with } \epsilon \sim N(0, \sigma^2_{Y|X}), \tag{11.9}$$

with $X \in \mathbb{R}^{p\times s}$, $\beta_1 \in \mathbb{R}^p$, $\beta_2 \in \mathbb{R}^s$, $\Sigma_{\text{vec}(X)} = \Sigma_2 \otimes \Sigma_1$, $\Sigma_1 \in \mathbb{R}^{p\times p}$ and $\Sigma_2 \in \mathbb{R}^{s\times s}$. Also, assuming that $E(X) = 0$ without loss of generality, it follows that

$$\begin{aligned} \text{cov}(\text{vec}(X)Y) &= E(\text{vec}(X)Y) = E(\text{vec}(X)\text{vec}^T(X)(\beta_2 \otimes \beta_1)) \\ &= \Sigma_2\beta_2 \otimes \Sigma_1\beta_1 \\ \text{unvec}(\text{cov}(\text{vec}(X)Y)) &= \Sigma_1\beta_1\beta_2^T\Sigma_2. \end{aligned}$$

As a stepping stone to PLS methods, we define a *Kronecker envelope* as $\text{span}(W_{q_1} \otimes V_{q_2})$, where $\text{span}(W_{q_1})$ and $\text{span}(W_{q_2})$ are the smallest subspace of $\mathbb{R}^p$ and $\mathbb{R}^s$, respectively, whose semi-orthogonal basis matrices W_{q_1} and W_{q_2} satisfy

$$\begin{aligned} \beta_1 &= W_{q_1}A_1 \\ \beta_2 &= V_{q_2}A_2 \\ \Sigma_1 &= W_{q_1}\Omega_1W_{q_1}^T + W_{q_1,0}\Omega_{1,0}W_{q_1,0}^T \\ \Sigma_2 &= V_{q_1}\Omega_2V_{q_1}^T + V_{q_1,0}\Omega_{2,0}V_{q_1,0}^T. \end{aligned}$$

As a consequence, a PLS estimator of the envelope for model (11.9) is given in Table 11.2 and will be denoted as K-PLS (Kronecker-PLS). Further work is needed to compare the estimators and study its asymptotic behavior.

In this section we have restricted attention to matrix-values predictors. The ideas extend straightforwardly to general tensor-valued predictors, although the associated algebra is notably more changeling. In the statistics literature Zhang and Li (2017) developed a PLS estimator for tensor predictors that requires the inverse of the covariance matrix. As we know, PLS methods are generally applicable because they do not require inverting large matrices. More work is needed to determine if these algorithms can be implemented without inverting such matrices.

TABLE 11.2
NIPALS K-PLS algorithm:

Population model (11.9)	$\Sigma_{\text{vec}(X)} = \Sigma_2 \otimes \Sigma_1$, $E(\text{vec}(X)Y) = \Sigma_{2Y} \otimes \Sigma_{1Y}$, $\text{unvec}(E(\text{vec}(XY))) = \Sigma_{1Y}\Sigma_{2Y}^T$
(a) *Population Version*	
Initialize	$w_1 \in \mathbb{R}^p \propto \Sigma_{1Y}$, $v_1 \in \mathbb{R}^s \propto \Sigma_{2Y}$ $W_1 = (w_1)$, $V_1 = (v_1)$
For $d = 1, 2 \ldots,$	
and $l = 1, 2, \ldots,$	
End for W if	$Q_{W_d(\Sigma_1)}^T \Sigma_{1Y} = 0$ and set $q_1 = d$
End for V if	$Q_{V_d(\Sigma_2)}^T \Sigma_{2Y} = 0$ and set $q_2 = d$
compute weights	$w_{d+1} \propto Q_{W_d(\Sigma_1)}^T \Sigma_{1Y}$
compute weights	$v_{d+1} \propto Q_{V_d(\Sigma_2)}^T \Sigma_{2Y}$
Append	$W_{d+1} = (W_d, w_{d+1})$ and $V_{d+1} = (V_d, v_{d+1})$
End when first	$Q_{W_{d+1}(\Sigma_X)}^T \Sigma_{X,Y} = 0$ and then set $q = d + 1$
Regression coefficients	$\beta_{\text{npls}} = W_{q_1}(W_{q_1}^T \Sigma_1 W_{q_1})^{-1} W_{q_1}^T \Sigma_{1,Y} \otimes V_{q_2}(W_{q_2}^T \Sigma_2 W_{q_2})^{-1} W_{q_2}^T \Sigma_{2,Y}$
(b) *Sample Version.*	
Substitute for the population versions	$\widehat{\Sigma}_{1Y} = \left(\frac{1}{sn}\sum_{k=1}^{s}\sum_{i=1}^{n} X_{ijk}(Y_i - \bar{Y})\right)_{j=1,p}$
	$\widehat{\Sigma}_{2Y} = \left(\frac{1}{pn}\sum_{j=1}^{p}\sum_{i=1}^{n} X_{ijk}(Y_i - \bar{Y})\right)_{j=1,p}$
	$\widehat{\Sigma}_1 = \left(\frac{1}{ns^2}\sum_{k=1}^{s}\sum_{t=1}^{s}\sum_{i=1}^{n}(X_{ijk} - \bar{X}_{jk})(X_{iht} - \bar{X}_{ht})\right)_{j,h=1,p}$
	$\widehat{\Sigma}_2 = \left(\frac{1}{np^2}\sum_{j=1}^{s}\sum_{h=1}^{s}\sum_{i=1}^{n}(X_{ijk} - \bar{X}_{jk})(X_{iht} - \bar{X}_{ht})\right)_{t,k=1,s}$
(c) Notes	
Orthogonal weights	$W_{q_1}^T W_{q_1} = I_{q_1}$ and $V_{q_2}^T V_{q_2} = I_{q_2}$
Envelope connection	$\text{span}(W_{q_1}) = \mathcal{E}_{\Sigma_1}(\mathcal{B}_1)$, the Σ_1-envelope of $\mathcal{B}_1 := \text{span}(\beta_1)$
	$\text{span}(W_{q_2}) = \mathcal{E}_{\Sigma_2}(\mathcal{B}_2)$, the Σ_2-envelope of $\mathcal{B}_2 := \text{span}(\beta_2)$

11.6 PLS for generalized linear models

As described at the outsets of Chapters 1 and 2, the goal of envelope methodology is to separate with clarity information in the data that is material to the goals of the analysis from that which is immaterial. Pursuing this notion in the case of the multivariate linear regression model (1.1) led to serviceable dimension reduction paradigms and close connections with PLS. Adaptations for linear and quadratic discriminant analysis and for non-linear regression were discussed in Chapters 7–9.

Cook and Zhang (2015a) developed a foundation for extending envelopes to relatively general settings, including generalized linear models (GLMs). In this section, we briefly review their foundation and its implications for GLMs. This will then lead to PLS for predictor reduction in GLMs. To facilitate making connections with literature, we use the notation of Cook and Zhang (2015a) provided that it does not conflict with notation that we used previously. We assume familiarity with GLMs, providing background only to set the stage for envelopes and PLS.

11.6.1 Foundations

Instead of proposing a specific modeling environment for envelopes, Cook and Zhang (2015a) started with an asymptotically normal estimator. Let $\theta \in \Theta \subseteq \mathbb{R}^m$ denote a parameter vector, which we decompose into a vector $\phi \in \mathbb{R}^p$, $p \leq m$, of targeted parameters and a vector $\psi \in \mathbb{R}^{m-p}$ nuisance parameters. We require that $\sqrt{n}(\widehat{\phi} - \phi)$ converge in distribution to a normal random vector with mean 0 and covariance matrix $V_{\phi\phi}(\theta) > 0$ as $n \to \infty$. Allowing $V_{\phi\phi}(\theta)$ to depend on the full parameter vector θ means that the variation in $\widehat{\phi}$ is can depend on the parameters of interest ϕ in addition to the nuisance parameters ψ. In many problems we may construct ϕ and ψ to be orthogonal parameters in the sense of Cox and Reid (1987). In the remainder of this section, we suppress notation indicating that $V_{\phi\phi}(\theta)$ may depend on θ and write instead $V_{\phi\phi}$ in place of $V_{\phi\phi}(\theta)$.

In the context of linear model (1.1),

$$\theta = \begin{pmatrix} \alpha \\ \text{vec}(\beta) \\ \text{vech}(\Sigma_{Y|X}) \\ \text{vech}(\Sigma_X) \end{pmatrix}; \ \psi = \begin{pmatrix} \alpha \\ \text{vech}(\Sigma_{Y|X}) \\ \text{vech}(\Sigma_X) \end{pmatrix}; \ \phi = \text{vec}(\beta).$$

When targeting the OLS estimator $\widehat{\phi} = \text{vec}(\widehat{\beta}_{\text{ols}})$ for improvement, we have from (1.18),

$$V_{\phi\phi} = \text{avar}(\sqrt{n}\text{vec}(\widehat{\beta}_{\text{ols}}) \mid \mathbb{X}_0) = \Sigma_{Y|X} \otimes \Sigma_X^{-1},$$

which depends on the nuisance parameters but not on β.

Let $\mathcal{F} = \text{span}(\phi)$. Then we construct an envelope for improving $\widehat{\phi}$ as follows (Cook and Zhang, 2015a) .

Definition 11.1. *The envelope for the parameter $\phi \in \mathbb{R}^p$ is defined as the smallest reducing subspace of $V_{\phi\phi}$ that contains $\mathcal{F}$, which is represented as $\mathcal{E}_{V_{\phi\phi}}(\mathcal{F}) \subseteq \mathbb{R}^p$.*

This definition expands our previous approaches in three fundamental ways. First, it links the envelope to a particular pre-specified method of estimation through the covariance matrix $V_{\phi\phi}$; a model is not required. Second, the matrix to be reduced – here $V_{\phi\phi}$ – is dictated by the method of estimation. Third, the matrix to be reduced can now depend on the parameter being estimated, in addition to perhaps other parameters. Definition 11.1 reproduces all of the envelope methods discussed in this book (Cook and Zhang, 2015a).

The potential improvement from using envelopes in this general context arises in much the same way as we discussed in Section 2.2 for predictor reduction. For this illustration, we assume that a basis matrix Γ for the envelope $\mathcal{E}_{V_{\phi\phi}}(\mathcal{F})$ is known and we construct the envelope estimator $\widehat{\phi}_{\text{env}}$ of ϕ by projecting $\widehat{\phi}$ onto span(Γ), $\widehat{\phi}_{\text{env}} = P_\Gamma\widehat{\phi}$. Then, since span($\Gamma$) reduces $V_{\phi\phi}$, we have that

$$\begin{aligned} V_{\phi\phi} &= P_\Gamma V_{\phi\phi} P_\Gamma + Q_\Gamma V_{\phi\phi} Q_\Gamma \\ \text{avar}(\sqrt{n}\widehat{\phi}_{\text{env}}) &= \text{avar}(\sqrt{n}P_\Gamma\widehat{\phi}) = P_\Gamma V_{\phi\phi} P_\Gamma \leq V_{\phi\phi}. \end{aligned}$$

See Cook and Zhang (2015a) for methods of estimation in the general setting.

11.6.2 Generalized linear models

We restrict discussion to a GLM in which $Y \mid X$ follows a one-parameter exponential family with probability mass or density function $f(y \mid \vartheta) = \exp\{y\vartheta - b(\vartheta) + c(y)\}$, where ϑ is the canonical parameter and $y \in \mathbb{R}^1$. We consider only regressions based on a one-parameter family with the canonical link function, $\vartheta(\alpha, \beta) = \alpha + \beta^T X$, which gives $\theta = (\alpha, \beta^T)^T$, $\mu(\vartheta) = E(Y \mid \vartheta) = b'(\vartheta)$ and $\text{var}(Y \mid \vartheta) = b''(\vartheta)$, where $'$ and $''$ denote first and second derivatives with respect to the argument evaluated at its population value. The overarching objective is to reduce the dimension of $X \in \mathbb{R}^p$ with the goal of improving the estimation of β, which is the parameter vector of interest; α is the nuisance parameter.

For a sample (y_i, X_i), $i = 1, \ldots, n$, the log likelihood for the i-th observation, which varies for different exponential family distributions of $Y|X$, is

$$C(\vartheta_i \mid y_i) := \log f(y_i|\vartheta_i) = y_i\vartheta_i - b(\vartheta_i) + c(y_i) = C(\vartheta_i) + c(y_i),$$

where $\vartheta_i = \alpha + \beta^T X_i$ and $C(\vartheta_i) = y_i\vartheta_i - b(\vartheta_i)$ is the kernel of the log likelihood. The full log-likelihood can be written as

$$C_n(\alpha, \beta) = \sum_{i=1}^{n} C(\vartheta_i \mid y_i) = \sum_{i=1}^{n} C(\vartheta_i) + \sum_{i=1}^{n} c(y_i).$$

Different log likelihood functions are summarized in Table 11.3 via the kernel.

We next briefly review Fisher scoring, which is the standard iterative method for maximizing $C_n(\alpha, \beta)$. At each iteration of the Fisher scoring method, the update step for $\widehat{\beta}$ can be summarized in the form of a

TABLE 11.3

A summary of one-parameter exponential families. For the normal, $\sigma = 1$. $A(\vartheta) = 1 + \exp(\vartheta)$. $C'(\vartheta)$ and $C''(\vartheta)$ are the first and second derivatives of $C(\vartheta)$ evaluated at the true value.

	E$(Y\|X)$	$C(\vartheta)$	$C'(\vartheta)$	$-C''(\vartheta)$ $(\propto \omega)$
Normal	ϑ	$Y\vartheta - \vartheta^2/2$	$Y - \vartheta$	1
Poisson	$\exp(\vartheta)$	$Y\vartheta - \exp(\vartheta)$	$Y - \exp(\vartheta)$	$\exp(\vartheta)$
Logistic	$\exp(\vartheta)/A(\vartheta)$	$Y\vartheta - \log A(\vartheta)$	$Y - \exp(\vartheta)/A(\vartheta)$	$\exp(\vartheta)/A^2(\vartheta)$
Exponential	$-\vartheta^{-1} > 0$	$Y\vartheta - \log(-\vartheta)$	$(Y - 1/\vartheta)$	ϑ^{-2}

weighted least squares (WLS) estimator where the weights are defined as $\omega(\vartheta) = -C''(\vartheta)$. With the canonical link, as we are assuming, $-C''(\vartheta) = b''(\vartheta) = \text{var}(Y \mid \vartheta)$. For a sample of size n, we define the population weights as

$$\omega_i = \omega(\vartheta_i) = \frac{\text{var}(Y \mid \vartheta_i)}{\sum_{j=1}^n \text{var}(Y \mid \vartheta_j)}, \quad i = 1, \ldots, n,$$

which are normalized so that $\sum_{i=1}^n \omega_i = 1$. Estimated weights are obtained by simply substituting estimates for the ϑ_i's. In keeping with our convention, we use the same notation for population and estimated weights, which should be clear from context. Let $\Omega = \text{diag}(\omega_1, \ldots, \omega_n)$ and define the weighted sample estimators, which use sample weights,

$$\begin{aligned} \bar{X}_{(\Omega)} &= \sum_{i=1}^n \omega_i X_i \\ S_{X(\Omega)} &= \sum_{i=1}^n \{\omega_i [X_i - \bar{X}_{(\Omega)}][X_i - \bar{X}_{(\Omega)}]^T\} \\ S_{X,\widehat{Z}(\Omega)} &= \sum_{i=1}^n \left\{\omega_i [X_i - \bar{X}_{(\Omega)}][\widehat{Z}_i - \bar{Z}_{(\Omega)}]^T\right\}, \end{aligned}$$

where $\widehat{Z}_i = \widehat{\vartheta}_i + \{Y_i - \mu(\widehat{\vartheta}_i)\}/\omega_i$ is a pseudo-response variable at the current iteration. The weighted covariance $S_{X(\Omega)}$ is the sample version of the population-weighted covariance matrix

$$\Sigma_{X(\omega)} = \text{E}\left\{\omega[X - \text{E}(\omega X)][X - \text{E}(\omega X)]^T\right\},$$

and the weighted cross-covariance $S_{X\widehat{Z}(\omega)}$ is the sample estimator of the weighted cross-covariance matrix

$$\Sigma_{X,Z(\omega)} = \text{E}\left\{\omega[X - \text{E}(\omega X)][Z - \text{E}(\omega Z)]^T\right\},$$

where Z is the population version of $\widehat{Z}$.

Then the updated estimator of β at the k-th iteration can now be represented as a WLS estimator

$$\widehat{\beta} \longleftarrow S_{X(\Omega)}^{-1} S_{X,\widehat{Z}(\Omega)}. \tag{11.10}$$

The corresponding updated estimator of α can be determined as $\widehat{\alpha} = \arg\max_\alpha C_n(\alpha, \widehat{\beta})$. Upon convergence of the Fisher scoring process, the final WLS estimator, denoted $\widehat{\beta}_{\text{wls}}$. is a function of $\widehat{\alpha}$, $\widehat{\beta}$, and ω. This iterative form allows β to be represented in the population using a weighted construction,

$$\beta = \Sigma_{X(\omega)}^{-1} \Sigma_{X,Z(\omega)}.$$

The asymptotic covariance matrix of $\widehat{\beta}_{\text{wls}}$ is (e.g. Cook and Zhang, 2015a)

$$\text{avar}(\sqrt{n}\widehat{\beta}) = V_{\beta\beta}(\theta) = \left\{\text{E}(-C'') \cdot \Sigma_{X(\omega)}\right\}^{-1},$$

while $\mathcal{E}_{V_{\beta\beta}}(\mathcal{B})$ is the corresponding envelope for improving estimation of β, where still $\mathcal{B} = \text{span}(\beta)$. Writing $\mathcal{E}_{V_{\beta\beta}}(\mathcal{B})$ in a bit more detail and using Proposition 1.6 and the discussion that immediately follows, we have

$$\mathcal{E}_{V_{\beta\beta}}(\mathcal{B}) = \mathcal{E}_{\{\text{E}(-C'')\cdot\Sigma_{X(\omega)}\}^{-1}}(\mathcal{B}) = \mathcal{E}_{\Sigma_{X(\omega)}}(\text{span}(\Sigma_{X,Z(\omega)})).$$

From this we see that the envelope for improving β in a GLM has the same form as the envelope $\mathcal{E}_{\Sigma_X}(\mathcal{B}) = \mathcal{E}_{\Sigma_X}(\mathcal{C}_{X,Y})$ for linear predictor reduction. This implies that we can construct PLS-type estimators of β in GLM's by first performing dimension reduction using a NIPALS or SIMPLS algorithm. Specifically, implement a sample version of the NIPALS algorithm in Table 3.1b or the SIMPLS algorithm in Table 3.4b, substituting $S_{X(\Omega)}$ for Σ_X and $S_{X,\widehat{Z}(\Omega)}$ for $\Sigma_{X,Y}$. Following reduction, prediction can be based on the GLM regression of Y on the reduced predictors W^TX. Let $\widehat{\nu}$ denote the estimator of the coefficient vector from this regression. The corresponding PLS estimator of β is $W\widehat{\nu}$.

Following these ideas, Table 11.4 gives an algorithm for using a PLS to fit a one-parameter family GLM. The algorithm has two levels of iteration. The outer level, which is shown in the table, are the score-based iterations for fitting a GLM. The inner iterations, which are not shown explicitly, occur when calling a PLS algorithm during each outer GLM iteration. The "For $k = 1, 2, \ldots,$" instruction indexes the GLM iterations. For each value of k there is an instruction to "call PLS algorithm" with the current parameter values. The calls to a PLS algorithm all have the same number of components q and so these PLS iterations terminate after q stages. The overall algorithm stops when the coefficient estimates no longer change materially. The algorithm does not require $n > p$.

11.6.3 Illustration

In this section we use the relatively straightforward simulation scenario of Cook and Zhang (2015a, Sec. 5.1) to support the algorithm of Table 11.4. We generated $n = 150$ observations according to a logistic regression $Y \mid X \sim \text{Bernoulli}(\text{logit}(\beta^TX))$, where $\beta = (0.25, 0.25)^T$ and X follows a

TABLE 11.4

A PLS algorithm for predictor reduction in GLMs.

(a) *Sample Version*	
Select	Number of components, q
Initialize	$\omega_i^{(1)} = n^{-1}, \Omega^{(1)} = \text{diag}(\omega_1^{(1)}, \ldots, \omega_n^{(1)})$
	$\widehat{\beta}^{(1)}$ = estimator of β from a PLS fit of Y on X, $\alpha^{(1)} = \bar{y} + \widehat{\beta}^{(1)^T}\bar{X}$.
For $k = 1, 2 \ldots,$ compute	
weighted means	$\bar{X}^{(k)} = \sum_{i=1}^n \omega_i^{(k)} X_i,$
weight-centered data	$\mathbb{X}^{(k)}$ is $n \times p$ with rows $(X_i - \bar{X}^{(k)})^T$
weighted variance	$S_{X(\Omega)}^{(k)} = \mathbb{X}^{(k)^T}\Omega^{(k)}\mathbb{X}^{(k)}$
pseudo responses	$Z_i^{(k)} = \widehat{\alpha}^{(k)} + \widehat{\beta}^{(k)^T} X_i + [y_i - \mu(\widehat{\vartheta}_i)]/\omega_i^{(k)}, i = 1, \ldots, n$
sample covariance	$S_{X,Z(\Omega)}^{(k)} = \sum_{i=1}^n \left\{\omega_i^{(k)}[X_i - \bar{X}^{(k)}][Z_i^{(k)} - \bar{Z}^{(k)}]\right\}$
call PLS algorithm	Send $S_{X(\Omega)}^{(k)}$ and $S_{X,Z(\Omega)}^{(k)}$ to a PLS algorithm and return the $p \times q$ matrix $W^{(k+1)}$ of PLS weights.
GLM regression	Fit the GLM regression of Y on $W^{(k+1)^T}X$, giving coefficient vector $\widehat{\nu}^{(k+1)}$ and $\widehat{\beta}^{(k+1)} = W^{(k+1)}\widehat{\nu}^{(k+1)}$.
End if convergence	$\lVert\widehat{\beta}^{(k)} - \widehat{\beta}^{(k+1)}\rVert \leq \epsilon$
	then set $\widehat{\beta}_{\text{pls}} = \widehat{\beta}^{(k+1)}$ and $\widehat{\alpha}_{\text{pls}} = \arg\max_\alpha C_n(\alpha, \widehat{\beta}_{\text{pls}})$
Otherwise	
estimate $\widehat{\alpha}$	$\widehat{\alpha}^{(k+1)} = \arg\max_\alpha C_n\left(\alpha, \widehat{\beta}^{(k+1)}\right)$
re-compute weights	$\omega_i^{(k+1)} = \omega_i(\widehat{\vartheta}^{(k+1)}),\ i = 1, \ldots, n$
(b) Notes	
Weights	All ω weights are sample versions, and are distinct from those that occur in the PLS algorithm called during iteration.
$n < p$	The algorithm does not require that $S_{X(\Omega)}^{(k)}$ be non-singular.
Components	The call to a PLS algorithm requires that the number of components q be specified. This should be the same for all calls. q can be estimated by using predictive cross validation appropriate for the distribution of $Y \mid X$.

bivariate normal distribution with mean 0 and variance

$$\Sigma_X = (10/\lVert\beta\rVert^2)\beta\beta^T + 0.1\beta_0\beta_0^T,$$

where β_0 is a 2×1 vector of length 1 that is orthogonal to β. It follows that the Σ_X-envelope of span(β) is span(β) itself, since $\beta/\lVert\beta\rVert$ is an eigenvector of Σ_X. The correlation between the two predictors is about 0.98, so estimation of β

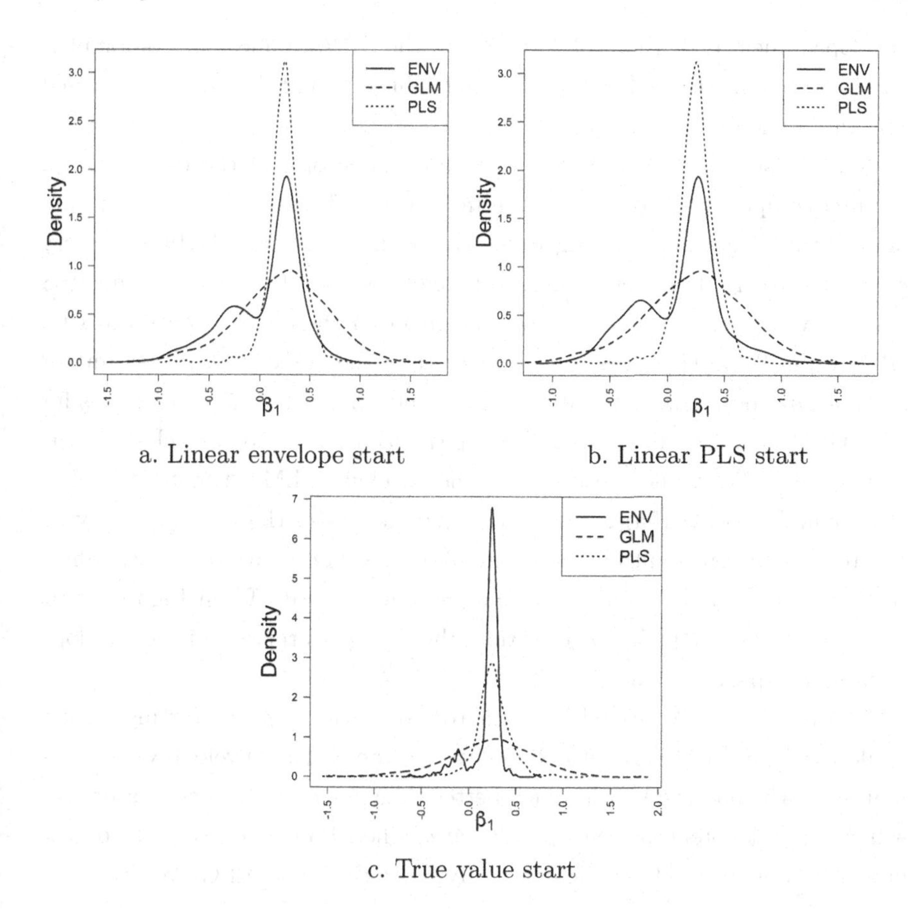

FIGURE 11.1
PLS-GLM data: Estimates of the densities of the estimators of the first component β_1 in the simulation to illustrate the PLS algorithm for GLMs in Table 11.4. Linear envelope and PLS starts refer to starting iteration at the envelope and PLS estimators from a fit of the linear model, ignoring the GLM structure.

may be a challange for standard likelihood-based methods. We repeated this simulation 500 times and for each dataset we estimated β using the standard GLM estimator, the envelope estimator of Cook and Zhang (2015a) and the PLS estimator from Table 11.4, the latter estimators using $q = 1$. To get some intuition on the importance of starting values, where necessary we ignored the logistic structure and started iterations using the estimators from (a) the enveloped linear regressions of Y on X, (b) the NIPALS linear regression of Y on X and (c) the true value of β. Starting values (a) and (b) were as discussed in Chapters 2 and 3.

The densities, estimated from the 500 simulations, of the estimators of the first component β_1 of β are given in Figure 11.1. Several observations are noteworthy. Regarding starting value, the results in Figures 11.1a,b are very close, suggesting that there is little difference between the envelope and PLS starting values based on the linear model. From these starting values, PLS GLM from Table 11.4 does much better than the standard GLM estimator and the envelope GLM estimator of Cook and Zhang (2015a). The means for PLS GLM (0.26) and the standard GLM estimator (0.23) are close to the true value (0.25), while the mean for the envelope GLM estimates (0.07) of Cook and Zhang (2015a) is not. The apparent bias of the estimator may be due to the inherent multimodal nature of envelope objective functions, which tends to be less important as the sample size increases. From Figure 11.1c, we see that, when starting at the true value, the performance of the envelope estimator improves greatly.

In short, the envelope GLM estimator seems sensitive to starting values, while the PLS GLM estimator is relatively stable. The envelope GLM estimator has the potential to do much better than the PLS GLM estimator, but achieving that potential seems problematic. The PLS GLM estimator, on the other hand, is relatively stable and easily beats the standard GLM estimator.

A

Proofs of Selected Results

A.1 Proofs for Chapter 1

A.1.1 Justification for (1.7)

$$\widehat{\beta}_{\text{ols}} = (\mathbb{X}^T\mathbb{X})^{-1}\mathbb{X}^T\mathbb{Y} = (\mathbb{X}^T\mathbb{X})^{-1}\mathbb{X}_0^T\mathbb{Y} = S_X^{-1}S_{Y,X}.$$

Let $\widehat{Y}_i = \bar{Y} + \widehat{\beta}_{\text{ols}}X_i$ and $r_i = Y_i - \widehat{Y}_i$ denote the i-th vectors of fitted values and residuals, $i = 1, \ldots, n$, and let $D = \beta - \widehat{\beta}_{\text{ols}}$. Then after substituting $\bar{Y}$ for α, the remaining log likelihood $L(\beta, \Sigma_{Y|X})$ to be maximized can be expressed as

$$\begin{aligned}
(2/n)L(\beta, \Sigma_{Y|X}) &= c - \log|\Sigma_{Y|X}| - n^{-1}\sum_{i=1}^{n}(Y_i - \bar{Y} - \beta X_i)^T\Sigma_{Y|X}^{-1} \\
&\quad \times (Y_i - \bar{Y} - \beta X_i) \\
&= c - \log|\Sigma_{Y|X}| - n^{-1}\sum_{i=1}^{n}(r_i - DX_i)^T\Sigma_{Y|X}^{-1}(r_i - DX_i) \\
&= c - \log|\Sigma_{Y|X}| - n^{-1}\text{tr}\left(\sum_{i=1}^{n} r_i r_i^T\Sigma_{Y|X}^{-1}\right) \\
&\quad - n^{-1}\text{tr}\left(D\sum_{i=1}^{n} X_iX_i^T D^T\Sigma_{Y|X}^{-1}\right) \\
&= c - \log|\Sigma_{Y|X}| - n^{-1}\text{tr}\left(\sum_{i=1}^{n} r_i r_i^T\Sigma_{Y|X}^{-1}\right) \\
&\quad - \text{tr}(DS_XD^T\Sigma_{Y|X}^{-1}),
\end{aligned}$$

where $c = -r\log(2\pi)$ and the penultimate step follows because $\sum_{i=1}^{n} r_iX_i^T = 0$. Consequently, $L(\beta, \Sigma_{Y|X})$ is maximized over β by setting $\beta = \widehat{\beta}_{\text{ols}}$ so $D = 0$, leaving the partially maximized log likelihood

$$(2/n)L(\Sigma_{Y|X}) = -r\log(2\pi) - \log|\Sigma_{Y|X}| - n^{-1}\mathrm{tr}\left(\sum_{i=1}^{n} r_i r_i^T \Sigma_{Y|X}^{-1}\right).$$

It follows that the maximum likelihood estimator of $\Sigma_{Y|X}$ is $S_{Y|X} := n^{-1}\sum_{i=1}^{n} r_i r_i^T$ and that the fully maximized log likelihood is

$$\widehat{L} = -(nr/2)\log(2\pi) - nr/2 - (n/2)\log|S_{Y|X}|.$$

A.1.2 Lemma 1.1

RESTATEMENT. Let $\mathcal{R}$ be a u dimensional subspace of $\mathbb{R}^r$ and let $M \in \mathbb{R}^{r\times r}$. Then $\mathcal{R}$ is an invariant subspace of M if and only if, for any $A \in \mathbb{R}^{r\times s}$ with $\mathrm{span}(A) = \mathcal{R}$, there exists a $B \in \mathbb{R}^{s\times s}$ such that $MA = AB$.

Proof. Suppose there is a B that satisfies $MA = AB$. For every $v \in \mathcal{R}$ there is a $t \in \mathbb{R}^s$ so that $v = At$. Consequently, $Mv = MAt = ABt \in \mathcal{R}$, which implies that $\mathcal{R}$ is an invariant subspace of M.

Suppose that $\mathcal{R}$ is an invariant subspace of M, and let a_j, $j = 1,\dots,s$ denote the columns of A. Then $Ma_j \in \mathcal{R}$, $j = 1,\dots,s$. Consequently, $\mathrm{span}(MA) \subseteq \mathcal{R}$, which implies there is a $B \in \mathbb{R}^{s\times s}$ such that $MA = AB$. □

A.1.3 Proposition 1.2

RESTATEMENT. $\mathcal{R}$ reduces $M \in \mathbb{R}^{r\times r}$ if and only if M can be written as

$$M = P_{\mathcal{R}} M P_{\mathcal{R}} + Q_{\mathcal{R}} M Q_{\mathcal{R}}. \tag{A.1}$$

Proof. Assume that M can be written as in (A.1). Then for any $v \in \mathcal{R}$, $Mv \in \mathcal{R}$, and for $v \in \mathcal{R}^{\perp}$, $Mv \in \mathcal{R}^{\perp}$. Consequently, $\mathcal{R}$ reduces M.

Next, assume that $\mathcal{R}$ reduces M. We must show that M satisfies (A.1). Let $u = \dim(\mathcal{R})$. It follows from Lemma 1.1 that there is a $B \in \mathbb{R}^{u\times u}$ that satisfies $MA = AB$, where $A \in \mathbb{R}^{r\times u}$ and $\mathrm{span}(A) = \mathcal{R}$. This implies $Q_{\mathcal{R}} MA = 0$ which is equivalent to $Q_{\mathcal{R}} M P_{\mathcal{R}} = 0$. By the same logic applied to $\mathcal{R}^{\perp}$, $P_{\mathcal{R}} M Q_{\mathcal{R}} = 0$. Consequently,

$$M = (P_{\mathcal{R}} + Q_{\mathcal{R}})M(P_{\mathcal{R}} + Q_{\mathcal{R}}) = P_{\mathcal{R}} M P_{\mathcal{R}} + Q_{\mathcal{R}} M Q_{\mathcal{R}}.$$

□

With this we have the following alternate definition of a reducing subspace

Definition A.1. *A subspace $\mathcal{R} \subseteq \mathbb{R}^r$ is said to be a reducing subspace of the real symmetric $r \times r$ matrix M if $\mathcal{R}$ decomposes M as $M = P_{\mathcal{R}} M P_{\mathcal{R}} + Q_{\mathcal{R}} M Q_{\mathcal{R}}$. If $\mathcal{R}$ is a reducing subspace of M, we say that $\mathcal{R}$ reduces M.*

A.1.4 Corollary 1.1

Restatement. Let $\mathcal{R}$ reduce $M \in \mathbb{R}^{r\times r}$, let $A \in \mathbb{R}^{r\times u}$ be a semi-orthogonal basis matrix for $\mathcal{R}$, and let A_0 be a semi-orthogonal basis matrix for $\mathcal{R}^{\perp}$. Then

1. M and $P_{\mathcal{R}}$, and M and $Q_{\mathcal{R}}$ commute.
2. $\mathcal{R} \subseteq \mathrm{span}(M)$ if and only if $A^T M A$ is full rank.
3. $|M| = |A^T M A| \times |A_0^T M A_0|$.
4. If M is full rank then

$$\begin{aligned} M^{-1} &= A(A^T M A)^{-1} A^T + A_0 (A_0^T M A_0)^{-1} A_0^T && \text{(A.2)} \\ &= P_{\mathcal{R}} M^{-1} P_{\mathcal{R}} + Q_{\mathcal{R}} M^{-1} Q_{\mathcal{R}}. && \text{(A.3)} \end{aligned}$$

5. If $\mathcal{R} \subseteq \mathrm{span}(M)$ then

$$M^{\dagger} = A(A^T M A)^{-1} A^T + A_0 (A_0^T M A_0)^{\dagger} A_0^T.$$

Proof. The first conclusion follows immediately from Proposition 1.2.

To show the second conclusion, first assume that $A^T M A$ is full rank. Then, from Lemma 1.1, B must be full rank in the representation $MA = AB$. Consequently, any vector in $\mathcal{R}$ can be written as a linear combination of the columns of M and thus $\mathcal{R} \subseteq \mathrm{span}(M)$. Next, assume that $\mathcal{R} \subseteq \mathrm{span}(M)$. Then there is a full rank matrix $V \in \mathbb{R}^{r\times u}$ such that $MV = A$ and thus that $A^T M V = I_u$. Substituting M from Proposition 1.2, we have $(A^T M A)(A^T V) = I_u$. It follows that $A^T M A$ is of full rank.

To demonstrate the third conclusion, write $P_{\mathcal{R}} = AA^T$, $Q_{\mathcal{R}} = A_0 A_0^T$ and

$$\begin{aligned} M &= AA^T M A A^T + A_0 A_0^T M A_0 A_0^T && \text{(A.4)} \\ &= (A, A_0) \begin{pmatrix} A^T M A & 0 \\ 0 & A_0^T M A_0 \end{pmatrix} (A, A_0)^T. \end{aligned}$$

The conclusion follows since (A, A_0) is an orthogonal matrix and thus has determinant 1.

For the fourth conclusion, since M is full rank $\mathcal{R} \subseteq \text{span}(M)$ and $\mathcal{R}^{\perp} \subseteq \text{span}(M)$. Consequently, both $A^T M A$ and $A_0^T M A_0$ are full rank. Thus both addends on the right hand side of (A.2) are defined. Multiplying (A.4) and the right hand side of (A.2) completes the proof of (A.2).

We use the first conclusion to prove (A.3). Since M and $P_{\mathcal{R}}$ commute, M^{-1} and $P_{\mathcal{R}}$ must also commute. Thus, $P_{\mathcal{R}} M^{-1} P_{\mathcal{R}} = M^{-1} P_{\mathcal{R}}$. Similarly, $Q_{\mathcal{R}} M^{-1} Q_{\mathcal{R}} = M^{-1} Q_{\mathcal{R}}$, which gives

$$P_{\mathcal{R}} M^{-1} P_{\mathcal{R}} + Q_{\mathcal{R}} M^{-1} Q_{\mathcal{R}} = M^{-1} P_{\mathcal{R}} + M^{-1} Q_{\mathcal{R}} = M^{-1}.$$

The final conclusion follows similarly: Since $\mathcal{R} \subseteq \text{span}(M)$, $A^T M A$ is full rank. The conclusion follows by checking the conditions for the Moore-Penrose inverse. □

A.1.5 Lemma 1.3

RESTATEMENT. Suppose that $M \in \mathbb{S}^{r \times r}$ is positive definite and that the column-partitioned matrix $(A, A_0) \in \mathbb{R}^{r \times r}$ is orthogonal with $A \in \mathbb{R}^{r \times u}$. Then

I. $|A_0^T M A_0| = |M| \times |A^T M^{-1} A|$

II. $\log |A^T M A| + \log |A_0^T M A_0| \geq \log |M|$

III. $\log |A^T M A| + \log |A^T M^{-1} A| \geq 0.$

Proof. Part I. Define the $r \times r$ matrix

$$K = \begin{pmatrix} I_u, A^T M A_0 \\ 0, A_0^T M A_0 \end{pmatrix}.$$

Since (A, A_0) is an orthogonal matrix and $|K| = |A_0^T M A_0|$,

$$\begin{aligned} |A_0^T M A_0| &= |(A, A_0) K (A, A_0)^T| = |AA^T + AA^T M A_0 A_0^T + A_0 A_0^T M A_0 A_0^T| \\ &= |AA^T + (AA^T + A_0 A_0^T) M A_0 A_0^T| \\ &= |AA^T + M A_0 A_0^T| \\ &= |M - (M - I_p) AA^T| = |M||I_u - A^T (I_r - M^{-1}) A| \\ &= |M||A^T M^{-1} A|. \end{aligned}$$

Part II. Let $O = (A, A_0)$.

$$\begin{aligned} |M| &= |O^T M O| = \begin{vmatrix} A^T M A & A^T M A_0 \\ A_0^T M A & A_0^T M A_0 \end{vmatrix} \\ &= |A^T M A| \times |A_0^T M A_0 - A_0^T M A (A^T M A)^{-1} A^T M A_0| \\ &\leq |A^T M A| \times |A_0^T M A_0|. \end{aligned}$$

Part III: This conclusion follows straightforwardly by combining the results of parts I and II.

□

A.1.6 Proposition 1.3

RESTATEMENT. If $\mathcal{R}$ reduces $M \in \mathbb{S}^{r \times r}$ and $\mathcal{R} \subseteq \text{span}(M)$ then $P_{\mathcal{R}(M)} = P_{\mathcal{R}}$.

Proof. Since $\mathcal{R}$ reduces M we have by Proposition 1.2

$$\begin{aligned} M &= P_{\mathcal{R}} M P_{\mathcal{R}} + Q_{\mathcal{R}} M Q_{\mathcal{R}} \\ &= R(R^T M R) R^T + Q_{\mathcal{R}} M Q_{\mathcal{R}} \\ &:= R \Omega R^T + Q_{\mathcal{R}} M Q_{\mathcal{R}}. \end{aligned}$$

Thus, $R^T M = \Omega R^T$. From conclusion 2 of Corollary 1.1, $(R^T M R)$ is full rank, and consequently we have

$$P_{\mathcal{R}(M)} = R(R^T M R)^{-1} R^T M = R \Omega^{-1} \Omega R^T = P_{\mathcal{R}}.$$

□

A.1.7 Proposition 1.4

RESTATEMENT. The intersection of any two reducing subspaces of $M \in \mathbb{R}^{r \times r}$ is also a reducing subspace of M.

Proof. Let $\mathcal{R}_1$ and $\mathcal{R}_2$ be reducing subspaces of M. Then by definition $M\mathcal{R}_1 \subseteq \mathcal{R}_1$ and $M\mathcal{R}_2 \subseteq \mathcal{R}_2$. Clearly, if $\mathbf{v} \in \mathcal{R}_1 \cap \mathcal{R}_2$ then $M\mathbf{v} \in \mathcal{R}_1 \cap \mathcal{R}_2$, and it follows that the intersection is an invariant subspace of M. The same argument shows that if $\mathbf{v} \in (\mathcal{R}_1 \cap \mathcal{R}_2)^\perp = \mathcal{R}_1^\perp + \mathcal{R}_2^\perp$ then $M\mathbf{v} \in \mathcal{R}_1^\perp + \mathcal{R}_2^\perp$:

If $\mathbf{v} \in \mathcal{R}_1^\perp + \mathcal{R}_2^\perp$ then it can be written as $\mathbf{v} = \mathbf{v}_1 + \mathbf{v}_2$, where $\mathbf{v} \in \mathcal{R}_1^\perp$ and $\mathbf{v} \in \mathcal{R}_2^\perp$. Then $M\mathbf{v} = M\mathbf{v}_1 + M\mathbf{v}_2 \in \mathcal{R}_1^\perp + \mathcal{R}_2^\perp$.

□

A.1.8 Lemma 1.4

RESTATEMENT. $\mathcal{E}_{M_1}(\mathcal{S}_1) \oplus \mathcal{E}_{M_2}(\mathcal{S}_2) = \mathcal{E}_{M_1 \oplus M_2}(\mathcal{S}_1 \oplus \mathcal{S}_2)$.

Proof. From Proposition 1.9, we have $\mathcal{E}_{M_i}(\mathcal{S}_i) = \sum_{j=1}^{q_i} P_{ij}\mathcal{S}_i$, $i = 1, 2$, where P_{ij} is the projection onto the j-th eigenspace of M_i. The eigen-projections of $M_1 \oplus M_2$, are of the forms $P_{1j} \oplus 0_{p_2}$, $j = 1, \ldots, q_1$, and $0_{p_1} \oplus P_{2k}$, $k = 1, \ldots, q_2$. Therefore, applying Proposition 1.9 again, we have

$$
\begin{aligned}
\mathcal{E}_M(\mathcal{S}) &= \left\{ \sum_{j=1}^{q_1} [(P_{1j} \oplus 0)(\mathcal{S}_1 \oplus \mathcal{S}_2)] \right\} + \left\{ \sum_{k=1}^{q_2} [(0 \oplus P_{2k})(\mathcal{S}_1 \oplus \mathcal{S}_2)] \right\} \\
&= \{\mathcal{E}_{M_1}(\mathcal{S}_1) \oplus 0\} + \{0 \oplus \mathcal{E}_{M_2}(\mathcal{S}_2)\} \\
&= \mathcal{E}_{M_1}(\mathcal{S}_1) \oplus \mathcal{E}_{M_2}(\mathcal{S}_2).
\end{aligned}
$$

□

A.1.9 Proposition 1.8

RESTATEMENT. Let $\Delta \in \mathbb{S}^{r \times r}$ be a positive definite matrix and let $\mathcal{S}$ be a u-dimensional subspace of $\mathbb{R}^r$. Let $G \in \mathbb{R}^{r \times u}$ be a semi-orthogonal basis matrix for $\mathcal{S}$ and let $V \in \mathbb{S}^{u \times u}$ be positive semi-definite. Define $\Psi = \Delta + GVG^T$. Then $\Delta^{-1}\mathcal{S} = \Psi^{-1}\mathcal{S}$ and

$$
\mathcal{E}_\Delta(\mathcal{S}) = \mathcal{E}_\Psi(\mathcal{S}) = \mathcal{E}_\Delta(\Delta^{-1}\mathcal{S}) = \mathcal{E}_\Psi(\Psi^{-1}\mathcal{S}) = \mathcal{E}_\Psi(\Delta^{-1}\mathcal{S}) = \mathcal{E}_\Delta(\Psi^{-1}\mathcal{S}).
$$

Proof. Using a variant of the Woodbury identity for matrix inverses we have

$$
\begin{aligned}
\Psi^{-1} &= \Delta^{-1} - \Delta^{-1}G(V^{-1} + G^T\Delta^{-1}G)^{-1}G^T\Delta^{-1}, \\
\Delta^{-1} &= \Psi^{-1} - \Psi^{-1}G(-V^{-1} + G^T\Psi^{-1}G)^{-1}G^T\Psi^{-1}.
\end{aligned}
$$

Multiplying both equations on the right by G, the first implies $\operatorname{span}(\Psi^{-1}G) \subseteq \operatorname{span}(\Delta^{-1}G)$; the second implies $\operatorname{span}(\Delta^{-1}G) \subseteq \operatorname{span}(\Psi^{-1}G)$. Hence $\Psi^{-1}\mathcal{S} = \Delta^{-1}\mathcal{S}$. From this we have also that $\mathcal{E}_\Psi(\Psi^{-1}\mathcal{S}) = \mathcal{E}_\Psi(\Delta^{-1}\mathcal{S})$ and $\mathcal{E}_\Delta(\Psi^{-1}\mathcal{S}) = \mathcal{E}_\Delta(\Delta^{-1}\mathcal{S})$.

We next show that $\mathcal{E}_\Delta(\mathcal{S}) = \mathcal{E}_\Psi(\mathcal{S})$ by demonstrating that $\mathcal{R} \subseteq \mathbb{R}^r$ is a reducing subspace of Δ that contains $\mathcal{S}$ if and only if it is a reducing subspace of Ψ that contains $\mathcal{S}$. Suppose $\mathcal{R}$ is a reducing subspace of Δ that contains $\mathcal{S}$. Let $\alpha \in \mathcal{R}$. Then $\Psi\alpha = \Delta\alpha + GVG^T\alpha$. $\Delta\alpha \in \mathcal{R}$ because $\mathcal{R}$ reduces Δ; the second term on the right is a vector in $\mathcal{R}$ because $\mathcal{S} \subseteq \mathcal{R}$. Thus, $\mathcal{R}$ is a reducing subspace of Ψ and by construction it contains $\mathcal{S}$. Next, suppose $\mathcal{R}$ is a reducing subspace of Ψ that contains $\mathcal{S}$. The reverse implication follows similarly by reasoning in terms of $\Delta\alpha = \Psi\alpha - GVG^T\alpha$. We have $\Psi\alpha \in \mathcal{R}$ because $\mathcal{R}$ reduces Ψ; the second term on the right is a vector in $\mathcal{R}$ because $\mathcal{S} \subseteq \mathcal{R}$. The remaining equalities follow immediately from Proposition 1.6.

□

A.2 Proofs for Chapter 2

A.2.1 Lemma 2.1

RESTATEMENT. Let $\mathcal{S} \subseteq \mathbb{R}^p$. Then $\mathcal{S}$ reduces Σ_X if and only if $\mathrm{cov}(P_\mathcal{S}X, Q_\mathcal{S}X) = 0$.

Proof. Since $P_\mathcal{S} + Q_\mathcal{S} = I_p$,

$$\begin{aligned} \Sigma_S &= (P_S + Q_S)\Sigma_S(P_S + Q_S) \\ &= P_S\Sigma_S P_S + Q_S\Sigma_S Q_S + P_S\Sigma_S Q_S + (P_S\Sigma_S Q_S)^T. \end{aligned}$$

The conclusion now follows from Proposition 1.2 which implies that $\mathcal{S}$ reduces Σ_X if and only if

$$\Sigma_X = P_\mathcal{S}\Sigma_X P_\mathcal{S} + Q_\mathcal{S}\Sigma_X Q_\mathcal{S}.$$

□

A.3 Proofs for Chapter 3

A.3.1 Lemma 3.1

RESTATEMENT. Following the notation from Table 3.1(a), for the sample version of the NIPALS algorithm

$$W_d^T W_d = I_d, d = 1, \ldots, q.$$

Proof. Since the columns of W_d are all eigenvectors with length one, the diagonal elements of $W_d^T W_d$ must all be 1. We show orthogonality by induction. For $d = 2$,

$$\begin{aligned} w_2 &= \ell_1(\mathbb{X}_2^T \mathbb{Y}_2 \mathbb{Y}_2^T \mathbb{X}_2) \\ &= \ell_1(Q_{w_1(S_{X_1})}^T \mathbb{X}_1^T \mathbb{Y}_2 \mathbb{Y}_2^T \mathbb{X}_2), \end{aligned}$$

where the second step follows by substituting (3.1) for $\mathbb{X}_2$. It follows that $w_1^T w_2 = 0$.

For $d = 3$, substituting (3.1) twice,

$$\begin{aligned} w_3 &= \ell_1(\mathbb{X}_3^T \mathbb{Y}_3 \mathbb{Y}_3^T \mathbb{X}_3) \\ &= \ell_1(Q_{w_2(S_{X_2})}^T \mathbb{X}_2^T \mathbb{Y}_3 \mathbb{Y}_3^T \mathbb{X}_3) \\ &= \ell_1(Q_{w_2(S_{X_2})}^T Q_{w_1(S_{X_1})}^T \mathbb{X}_1^T \mathbb{Y}_3 \mathbb{Y}_3^T \mathbb{X}_3). \end{aligned}$$

Clearly, $w_2^T w_3 = 0$. For w_1,

$$Q_{w_2(S_{X_2})} w_1 = w_1 - w_2(w_2^T S_{X_2} w_2)^{-1} w_2^T S_{X_2} w_1$$

But it follows from (3.4) that $S_{X_2} w_1 = 0$ and consequently $Q_{w_2(S_{X_2})} w_1 = w_1$ and it follows that $w_1^T w_3 = 0$. The rest of the justification follows straightforwardly by induction and is omitted.

A.3.2 Lemma 3.2

RESTATEMENT.

$$\mathbb{X}_{d+1}^T s_j = \mathbb{X}_{d+1}^T \mathbb{X}_j w_j = 0$$

for $d = 1, \ldots, q-1$, $j = 1, \ldots, d$.

Proof. The proof is by induction. For $d = 1$, substituting from Table 3.1 or using (3.2),

$$\begin{aligned} \mathbb{X}_2 &= \mathbb{X}_1 - s_1 l_1^T \\ &= \mathbb{X}_1 - \mathbb{X}_1 w_1 (w_1^T \mathbb{X}_1^T \mathbb{X}_1 w_1)^{-1} w_1^T \mathbb{X}_1^T \mathbb{X}_1 \\ &= Q_{\mathbb{X}_1 w_1} \mathbb{X}_1. \end{aligned}$$

Clearly,

$$\mathbb{X}_2^T s_1 = \mathbb{X}_2^T \mathbb{X}_1 w_1 = \mathbb{X}_1^T Q_{\mathbb{X}_1 w_1} \mathbb{X}_1 w_1 = 0$$

so the conclusion holds for $d = 1$.

For $d = 2$ and $j = 2$, $\mathbb{X}_3^T \mathbb{X}_2 w_2 = \mathbb{X}_2^T Q_{\mathbb{X}_2 w_2} \mathbb{X}_2 w_2 = 0$, where the first equality comes from substituting from(3.2). For $d = 2$ and $j = 1$,

$$\begin{aligned} \mathbb{X}_3^T \mathbb{X}_1 w_1 &= \mathbb{X}_1^T Q_{\mathbb{X}_1 w_1} Q_{\mathbb{X}_2 w_2} \mathbb{X}_1 w_1 \\ &= \mathbb{X}_1^T Q_{\mathbb{X}_1 w_1} (I - P_{\mathbb{X}_2 w_2}) \mathbb{X}_1 w_1 \\ &= \mathbb{X}_1^T Q_{\mathbb{X}_1 w_1} \mathbb{X}_1 w_1 \\ &= 0, \end{aligned}$$

where the third equality holds because $\mathbb{X}_2^T \mathbb{X}_1 w_1 = 0$.

Under the induction hypothesis, $\mathbb{X}_d^T \mathbb{X}_j w_j = 0$ for $j = 1, \ldots, d-1$. To prove the lemma we must show that $\mathbb{X}_{d+1}^T \mathbb{X}_j w_j = 0$ for $j = 1, \ldots, d$. Again using (3.2)

$$\begin{aligned} \mathbb{X}_{d+1}^T \mathbb{X}_j w_j &= \mathbb{X}_d^T Q_{\mathbb{X}_d w_d} \mathbb{X}_j w_j \\ &= \mathbb{X}_d^T (I - P_{\mathbb{X}_d w_d}) \mathbb{X}_j w_j. \end{aligned}$$

By the induction hypothesis, $\mathbb{X}_d^T \mathbb{X}_j w_j = 0$ for $j = 1, \ldots, d-1$ and so we have the desired conclusion for $j = 1, \ldots, d-1$: $\mathbb{X}_{d+1}^T \mathbb{X}_j w_j = 0$ for $j = 1, \ldots, d-1$. For $j = d$,

$$\mathbb{X}_{d+1}^T \mathbb{X}_d w_d = \mathbb{X}_d^T (I - P_{\mathbb{X}_d w_d}) \mathbb{X}_d w_d = 0,$$

and the conclusion follows. □

A.3.3 Lemma 3.3

RESTATEMENT. For $d = 1, \ldots, q$,

$$\begin{aligned} \mathbb{X}_{d+1} &= (I - P_{s_d} - P_{s_{d-1}} - \cdots - P_{s_2} - P_{s_1}) \mathbb{X}_1 \\ &= (I - P_{S_d}) \mathbb{X}_1 = Q_{S_d} \mathbb{X}_1 && \text{(A.5)} \\ \mathbb{Y}_{d+1} &= (I - P_{s_d} - P_{s_{d-1}} - \cdots - P_{s_2} - P_{s_1}) \mathbb{Y}_1 \\ &= (I - P_{S_d}) \mathbb{Y}_1 = Q_{S_d} \mathbb{Y}_1 && \text{(A.6)} \\ S_{X_d, Y_d} &= S_{X_d, Y_1} \\ l_d &= \mathbb{X}_1^T s_d / \|s_d\|^2 \\ m_d &= \mathbb{Y}_1^T s_d / \|s_d\|^2 \\ w_{d+1} &= \ell_1(\mathbb{X}_1^T Q_{S_d} \mathbb{Y}_1 \mathbb{Y}_1^T Q_{S_d} \mathbb{X}_1) && \text{(A.7)} \\ S_d &= \mathbb{X}_1 W_d \\ \widehat{\mathbb{Y}}_{\mathrm{npls}} &= \mathbb{X}_1 \widehat{\beta}_{\mathrm{npls}} = P_{S_q} \mathbb{Y}_1. \end{aligned}$$

Proof. The representations for $\mathbb{X}_{d+1}$ and $\mathbb{Y}_{d+1}$ follow straightforwardly from Lemma 3.2, (3.6) and the definition of $s_d = \mathbb{X}_d w_d$: Since, by Lemma 3.2, $Q_{s_{d+1}} Q_{s_j} = I - P_{s_{d+1}} - P_{s_j}$, the conclusion follows by using (3.6).

For the fitted values we take $\widehat{\beta}_{\text{npls}}$ from Table 3.1 to get,

$$\widehat{\mathbb{Y}}_{\text{npls}} = \mathbb{X}_1 \widehat{\beta}_{\text{npls}} = \mathbb{X}_1 W_q (L_q^T W_q)^{-1} M_q^T.$$

The covariance matrix identity is a consequence of (A.5) and (A.6):

$$S_{X_d, Y_d} = n^{-1} \mathbb{X}_d^T \mathbb{Y}_d = n^{-1} \mathbb{X}_1^T Q_{S_{d-1}} Q_{S_{d-1}} \mathbb{Y}_1 = n^{-1} \mathbb{X}_1^T Q_{S_{d-1}} \mathbb{Y}_1 = S_{X_d, Y_1}.$$

The representation of w_{d+1} follows similarly.

To show that $S_d = \mathbb{X}_1 W_d$, consider a typical diagonal element s_{k+1} of S_d,

$$\begin{aligned} s_{k+1} &= \mathbb{X}_{k+1} w_{k+1} \\ &= Q_{S_k} \mathbb{X}_1 w_{k+1}. \end{aligned}$$

It follows from (A.7) that $\mathbb{X}_1 w_{k+1}$ must fall into the orthogonal complement of the subspace spanned by the columns of S_k. Consequently, $s_{k+1} = Q_{S_k} \mathbb{X}_1 w_{k+1} = \mathbb{X}_1 w_{k+1}$. Thus, $S_d = \mathbb{X}_1 W_d$.

Let $D_d = \text{diag}(\|s_1\|^2, \ldots, \|s_d\|^2)$. As defined in Table 3.1, $l_d = \mathbb{X}_d^T s_d / \|s_d\|^2$. Substituting the form of $\mathbb{X}_d$ from (A.5), we get

$$l_d = \mathbb{X}_d^T s_d / \|s_d\|^2 = \mathbb{X}_1^T Q_{S_{d-1}} s_d / \|s_d\|^2 = \mathbb{X}_1^T s_d / \|s_d\|^2,$$

where the second equality follows from the first consequence of Lemma 3.2. This implies that $L_d = \mathbb{X}_1^T S_d D_d^{-1}$. Similarly, $M_d = \mathbb{Y}_1^T S_d D_d^{-1}$. Substituting these into $\widehat{\mathbb{Y}}_{\text{npls}}$ and using the fact that $S_q = \mathbb{X}_1 W_q$ we get

$$\begin{aligned} \widehat{\mathbb{Y}}_{\text{npls}} &= \mathbb{X}_1 W_q (L_q^T W_q)^{-1} M_q^T \\ &= \mathbb{X}_1 W_q (D_q^{-1} S_q^T \mathbb{X}_1 W_q)^{-1} D_q^{-1} S_q^T \mathbb{Y}_1 \\ &= S_q (S_q^T S_q)^{-1} S_q^T \mathbb{Y}_1 = P_{S_q} \mathbb{Y}_1. \end{aligned}$$

□

A.3.4 Lemma 3.4

Recall that V denotes the $p \times p_1$, $p_1 \leq p$, of eigenvectors of S_X with non-zero eigenvalues, and that $\mathbb{X}_1^T = V \mathbb{Z}_1^T$, where $\mathbb{Z}_1^T$ is an $p_1 \times n$ matrix that contains the coordinates of $\mathbb{X}_1^T$ in terms of the eigenvectors of S_X. Recall also that w_d^* and s_d^* denote the weights and scores that result from applying NIPALS to data $(\mathbb{Z}_1, \mathbb{Y}_1)$.

Lemma 3.4 states that, for $d = 1, \ldots, q$, (a) $s_d = s_d^*$ and (b) $w_d = Vw_d^*$. We see from the form of w_d that, for $d = 1, \ldots, q$, $w_d \in \text{span}(S_X)$,

$$
\begin{aligned}
w_{d+1} &= \ell_1(\mathbb{X}_1^T Q_{S_d} \mathbb{Y}_1 \mathbb{Y}_1^T Q_{S_d} \mathbb{X}_1) \\
&= \ell_1(V \mathbb{Z}_1^T Q_{S_d} \mathbb{Y}_1 \mathbb{Y}_1^T Q_{S_d} \mathbb{Z}_1 V^T) \\
&= V \ell_1(\mathbb{Z}_1^T Q_{S_d} \mathbb{Y}_1 \mathbb{Y}_1^T Q_{S_d} \mathbb{Z}_1).
\end{aligned}
$$

If conclusion (a) holds then $S_d = S_d^*$ and we have

$$
\begin{aligned}
w_{d+1} &= V \ell_1(\mathbb{Z}_1^T Q_{S_d^*} \mathbb{Y}_1 \mathbb{Y}_1^T Q_{S_d^*} \mathbb{Z}_1) \\
&= V w_d^*,
\end{aligned}
$$

and conclusion (b) follows.

We next demonstrate (a) by induction. For $d = 1$,

$$
\begin{aligned}
s_1 &= \mathbb{X}_1 \ell_1(\mathbb{X}_1^T \mathbb{Y}_1 \mathbb{Y}_1^T \mathbb{X}_1) \\
&= \mathbb{Z}_1 V^T \ell_1(V \mathbb{Z}_1^T \mathbb{Y}_1 \mathbb{Y}_1^T \mathbb{Z}_1 V^T) \\
&= \mathbb{Z}_1 \ell_1(\mathbb{Z}_1^T \mathbb{Y}_1 \mathbb{Y}_1^T \mathbb{Z}_1) \\
&= s_1^*.
\end{aligned}
$$

For $d = 2$

$$
\begin{aligned}
s_2 &= \mathbb{X}_1 \ell_1(\mathbb{X}_1^T Q_{s_1} \mathbb{Y}_1 \mathbb{Y}_1^T Q_{s_1} \mathbb{X}_1) \\
&= \mathbb{Z}_1 V^T \ell_1(V \mathbb{Z}_1^T Q_{s_1^*} \mathbb{Y}_1 \mathbb{Y}_1^T Q_{s_1^*} \mathbb{Z}_1 V^T) \\
&= \mathbb{Z}_1 \ell_1(\mathbb{Z}_1^T Q_{s_1^*} \mathbb{Y}_1 \mathbb{Y}_1^T Q_{s_1^*} \mathbb{Z}_1) \\
&= s_2^*.
\end{aligned}
$$

Under the induction hypothesis, assume that the conclusion holds for $d < q$. Then we have for the next term in the sequence

$$
\begin{aligned}
s_{d+1} &= \mathbb{X}_1 \ell_1(\mathbb{X}_1^T Q_{S_d} \mathbb{Y}_1 \mathbb{Y}_1^T Q_{S_d} \mathbb{X}_1) \\
&= \mathbb{Z}_1 V^T \ell_1(V \mathbb{Z}_1^T Q_{S_d^*} \mathbb{Y}_1 \mathbb{Y}_1^T Q_{S_d^*} \mathbb{Z}_1 V^T) \\
&= \mathbb{Z}_1 \ell_1(\mathbb{Z}_1^T Q_{S_d^*} \mathbb{Y}_1 \mathbb{Y}_1^T Q_{S_d^*} \mathbb{Z}_1) \\
&= s_{d+1}^*,
\end{aligned}
$$

where the second equality follows from the induction hypothesis.

A.3.5 Lemma 3.5

RESTATEMENT. *Let V denote a $p \times c$ matrix of rank c, let v denote a $p \times 1$ vector that is not contained in* span(V), *let Σ denote a $p \times p$ positive definite matrix and let $\Delta = Q^T_{V(\Sigma)}\Sigma Q_{V(\Sigma)}$. Then*

(a) $\Delta = Q^T_{V(\Sigma)}\Sigma = \Sigma Q_{V(\Sigma)}$

(b) $P_{(V,v)(\Sigma)} = P_{V(\Sigma)} + P_{Q_{V(\Sigma)}v(\Sigma)}$

(c) $Q_{V(\Sigma)}Q_{v(\Delta)} = Q_{(V,v)(\Sigma)}$

(d) $Q^T_{v(\Delta)}\Delta Q_{v(\Delta)} = Q^T_{(V,v)(\Sigma)}\Sigma Q_{(V,v)(\Sigma)}$.

Proof. Part (a) follows by straightforward algebra and its proof is omitted. For part (b) we have

$$\begin{aligned} P_{(V,v)(\Sigma)} &= \Sigma^{-1/2}P_{(\Sigma^{1/2}V,\Sigma^{1/2}v)}\Sigma^{1/2} \\ &= \Sigma^{-1/2}P_{\Sigma^{1/2}V}\Sigma^{1/2} + \Sigma^{-1/2}P_{Q_{\Sigma^{1/2}V}\Sigma^{1/2}v}\Sigma^{1/2} \\ &= P_{V(\Sigma)} + \Sigma^{-1/2}P_{Q_{\Sigma^{1/2}V}\Sigma^{1/2}v}\Sigma^{1/2}. \end{aligned}$$

Next, consider the second addend on the right hand side:

$$\begin{aligned} Q_{\Sigma^{1/2}V}\Sigma^{1/2} &= \Sigma^{1/2}Q_{V(\Sigma)} \\ P_{Q_{\Sigma^{1/2}V}\Sigma^{1/2}v} &= \Sigma^{1/2}Q_{V(\Sigma)}v\{v^TQ^T_{V(\Sigma)}\Sigma Q_{V(\Sigma)}v\}^{-1}v^TQ^T_{V(\Sigma)}\Sigma^{1/2} \\ \Sigma^{-1/2}P_{Q_{\Sigma^{1/2}V}\Sigma^{1/2}v}\Sigma^{1/2} &= Q_{V(\Sigma)}v\{v^TQ^T_{V(\Sigma)}\Sigma Q_{V(\Sigma)}v\}^{-1}v^TQ^T_{V(\Sigma)}\Sigma \\ &= P_{Q_{V(\Sigma)}v(\Sigma)}. \end{aligned} \tag{A.8}$$

This establishes part (b).

To show part (c), first write

$$\begin{aligned} Q_{v(\Delta)} &= I - v(v^T\Delta v)^{-1}v^T\Delta \\ &= I - v(v^TQ^T_{V(\Sigma)}\Sigma Q_{V(\Sigma)}v)^{-1}v^TQ^T_{V(\Sigma)}\Sigma Q_{V(\Sigma)} \\ &= I - v(v^TQ^T_{V(\Sigma)}\Sigma Q_{V(\Sigma)}v)^{-1}v^TQ^T_{V(\Sigma)}\Sigma, \end{aligned}$$

where the second equality follows by substituting for Δ and the third follows from part (a). Next, multiplying on the left by $Q_{V(\Sigma)}$ and using (A.8) we have

$$\begin{aligned} Q_{V(\Sigma)}Q_{v(\Delta)} &= Q_{V(\Sigma)} - Q_{V(\Sigma)}v(v^TQ^T_{V(\Sigma)}\Sigma Q_{V(\Sigma)}v)^{-1}v^TQ^T_{V(\Sigma)}\Sigma \\ &= Q_{V(\Sigma)} - P_{Q_{V(\Sigma)}v(\Sigma)} \\ &= I - P_{V(\Sigma)} - P_{Q_{V(\Sigma)}v(\Sigma)}. \end{aligned}$$

From part (b) we then have the desired result:

$$Q_{V(\Sigma)}Q_{v(\Delta)} = I - P_{(V,v)(\Sigma)} = Q_{(V,v)(\Sigma)}.$$

To show part (d) we begin by substituting for the middle Δ and then using part (c):

$$\begin{aligned} Q_{v(\Delta)}^T \Delta Q_{v(\Delta)} &= Q_{v(\Delta)}^T Q_{V(\Sigma)}^T \Sigma Q_{V(\Sigma)} Q_{v(\Delta)} \\ &= Q_{(V,v)(\Sigma)}^T \Sigma Q_{(V,v)(\Sigma)}. \end{aligned}$$

A.3.6 Lemma 3.6

RESTATEMENT. Let $\Sigma \in \mathbb{S}^p$ be positive definite and let $\mathcal{B}_j$ be a subspace of $\mathbb{R}^p$, $j = 1, 2$. Then

$$\mathcal{E}_\Sigma(\mathcal{B}_1 + \mathcal{B}_2) = \mathcal{E}_\Sigma(\mathcal{B}_1) + \mathcal{E}_\Sigma(\mathcal{B}_2).$$

Proof.

$$\mathcal{B}_1 + \mathcal{B}_2 \subseteq \mathcal{E}_\Sigma(\mathcal{B}_1) + \mathcal{E}_\Sigma(\mathcal{B}_2) \subseteq \mathcal{E}_\Sigma(\mathcal{B}_1 + \mathcal{B}_2). \tag{A.9}$$

The first containment seems clear, as $\mathcal{B}_j \subseteq \mathcal{E}_\Sigma(\mathcal{B}_j), j = 1, 2$. For the second containment, $\mathcal{E}_\Sigma(\mathcal{B}_1) \subseteq \mathcal{E}_\Sigma(\mathcal{B}_1 + \mathcal{B}_2)$ and $\mathcal{E}_\Sigma(\mathcal{B}_2) \subseteq \mathcal{E}_\Sigma(\mathcal{B}_1 + \mathcal{B}_2)$. The second containment follows since $\mathcal{E}_\Sigma(\mathcal{B}_1 + \mathcal{B}_2)$ is a subspace.

We next wish to show that $\mathcal{E}_\Sigma(\mathcal{B}_1) + \mathcal{E}_\Sigma(\mathcal{B}_2)$ reduces Σ. The envelopes $\mathcal{E}_\Sigma(\mathcal{B}_1)$ and $\mathcal{E}_\Sigma(\mathcal{B}_2)$ both reduce Σ and so by definition,

$$\Sigma\mathcal{E}_\Sigma(\mathcal{B}_j) \subseteq \mathcal{E}_\Sigma(\mathcal{B}_j),\ j = 1, 2.$$

In consequence

$$\begin{aligned} \Sigma\mathcal{E}_\Sigma(\mathcal{B}_1) + \Sigma\mathcal{E}_\Sigma(\mathcal{B}_2) &= \Sigma(\mathcal{E}_\Sigma(\mathcal{B}_1) + \mathcal{E}_\Sigma(\mathcal{B}_2)) \\ &\subseteq \mathcal{E}_\Sigma(\mathcal{B}_1) + \mathcal{E}_\Sigma(\mathcal{B}_2). \end{aligned}$$

It follows that $\mathcal{E}_\Sigma(\mathcal{B}_1) + \mathcal{E}_\Sigma(\mathcal{B}_2)$ is an invariant subspace of Σ. That it is a reducing subspace follows from the symmetry of Σ.

In short, $\mathcal{E}_\Sigma(\mathcal{B}_1) + \mathcal{E}_\Sigma(\mathcal{B}_2)$ is a reducing subspace of Σ that, by (A.9) contains $\mathcal{B}_1 + \mathcal{B}_2$. But $\mathcal{E}_\Sigma(\mathcal{B}_1 + \mathcal{B}_2)$ is by construction the smallest reducing subspace of Σ that contains $\mathcal{B}_1 + \mathcal{B}_2$. Thus,

$$\mathcal{E}_\Sigma(\mathcal{B}_1 + \mathcal{B}_2) \subseteq \mathcal{E}_\Sigma(\mathcal{B}_1) + \mathcal{E}_\Sigma(\mathcal{B}_2).$$

Together with (A.9) we have the desired conclusion. □

A.4 Proofs for Chapter 4

A.4.1 Proposition 4.1

The proof in this section was adapted from the unpublished notes of Cook, Helland, and Su (2013).

RESTATEMENT. Assume the regression structure given in (4.1) and (4.7) with p fixed. Then

(i) The PLS estimator $\widehat{\beta}_{\text{pls}}$ of β has the expansion

$$\sqrt{n}(\widehat{\beta}_{\text{pls}}-\beta) = \frac{1}{\delta\sqrt{n}}\sum_{i=1}^{n}\left\{(X_i-\mu_X)\epsilon_i + Q_{\Phi}(X_i-\mu_X)(X_i-\mu_X)^T\beta\right\}+O_p\left(\frac{1}{\sqrt{n}}\right),$$

where ϵ is the error for model (4.1).

(ii) $\sqrt{n}(\widehat{\beta}_{\text{pls}}-\beta)$ is asymptotically normal with mean 0 and variance

$$\text{avar}(\sqrt{n}\widehat{\beta}_{\text{pls}}) = \delta^{-2}\left\{\Sigma_X\sigma^2_{Y|X} + \text{var}(Q_{\Phi}(X-\mu_X)(X-\mu_X)^T\beta)\right\}.$$

(iii) If, in addition, $P_{\Phi}X$ is independent of $Q_{\Phi}X$ then

$$\text{avar}(\sqrt{n}\widehat{\beta}_{\text{pls}}) = \delta^{-1}\sigma^2_{Y|X}P_{\Phi} + \delta^{-2}\sigma^2_Y\Phi_0\Delta_0\Phi_0^T.$$

Proof. For notational convenience in this proof, we let $\sigma = \sigma_{X,Y}$, $\Sigma = \Sigma_X$, $\|\sigma\|^2 = \sigma^T\sigma$ and recall that P_σ is the projection onto span(σ). Without loss of generality we prove the proposition using the centered variables $x = X - \mu_X$ and $y = Y - \mu_Y$. Then

$$\widehat{\beta}_{\text{pls}} = \widehat{\sigma}(\widehat{\sigma}^T\widehat{\sigma})(\widehat{\sigma}^T S_X\widehat{\sigma})^{-1}.$$

We first expand $\widehat{\sigma}(\widehat{\sigma}^T\widehat{\sigma})$ and $\widehat{\sigma}^T S_X\widehat{\sigma}$. For this, we need the following expansions (see Cook and Setodji, 2003)

$$\begin{aligned}\sqrt{n}(\widehat{\sigma}-\sigma) &= n^{-\frac{1}{2}}\sum_{i=1}^{n}(x_iy_i-\sigma)+O_p(n^{-\frac{1}{2}}),\\ \sqrt{n}(S_X-\Sigma) &= n^{-\frac{1}{2}}\sum_{i=1}^{n}(x_ix_i^T-\Sigma)+O_p(n^{-\frac{1}{2}}).\end{aligned}$$

Step I: Expand $\widehat{\sigma}\|\widehat{\sigma}\|^2$.

$$\begin{aligned}\widehat{\sigma}(\widehat{\sigma}^T\widehat{\sigma}) &= (\widehat{\sigma}-\sigma+\sigma)(\widehat{\sigma}-\sigma+\sigma)^T(\widehat{\sigma}-\sigma+\sigma)\\ &= (\widehat{\sigma}-\sigma)\|\sigma\|^2+\sigma(\widehat{\sigma}-\sigma)^T\sigma+\sigma\sigma^T(\widehat{\sigma}-\sigma)+\sigma\|\sigma\|^2+O_p(n^{-1}),\end{aligned}$$

so

$$\begin{aligned}\sqrt{n}(\widehat{\sigma}\|\widehat{\sigma}\|^2-\sigma\|\sigma\|^2 &= \sqrt{n}\{(\widehat{\sigma}-\sigma)\|\sigma\|^2+\sigma\sigma^T(\widehat{\sigma}-\sigma)+\sigma\sigma^T(\widehat{\sigma}-\sigma)\}\\ &\quad+O_p(n^{-\frac{1}{2}}), \qquad\qquad (A.10)\\ &= \sqrt{n}(\widehat{\sigma}-\sigma)\|\sigma\|^2+2\sqrt{n}\sigma\sigma^T(\widehat{\sigma}-\sigma)+O_p(n^{-\frac{1}{2}}),\\ &= \|\sigma\|^2\sqrt{n}\{(\widehat{\sigma}-\sigma)+2P_\sigma(\widehat{\sigma}-\sigma)\}+O_p(n^{-\frac{1}{2}}),\\ &= \|\sigma\|\{I_p+2P_\sigma\}\sqrt{n}(\widehat{\sigma}-\sigma)+O_p(n^{-\frac{1}{2}}), \qquad (A.11)\\ &= \|\sigma\|\{I_p+2P_\sigma\}n^{-\frac{1}{2}}\sum_{i=1}^{n}(x_iy_i-\sigma)+O_p(n^{-\frac{1}{2}}).\end{aligned}$$

Step II. Expand $(\widehat{\sigma}^T S_X\widehat{\sigma})^{-1}$.

$$\begin{aligned}\sqrt{n}(\widehat{\sigma}^T S_X\widehat{\sigma}-\sigma^T\Sigma\sigma) &= \sqrt{n}\{(\widehat{\sigma}-\sigma+\sigma)^T(S_X-\Sigma+\Sigma)(\widehat{\sigma}-\sigma+\sigma)-\sigma^T\Sigma\sigma\}\\ &= \sqrt{n}(\widehat{\sigma}-\sigma)^T\Sigma\sigma+\sqrt{n}\sigma^T(S_X-\Sigma)\sigma+\sqrt{n}\sigma^T\Sigma(\widehat{\sigma}-\sigma)\\ &\quad+O_p(n^{-\frac{1}{2}}),\\ &= \sqrt{n}\sigma^T(S_X-\Sigma)\sigma+2\sqrt{n}\sigma^T\Sigma(\widehat{\sigma}-\sigma)+O_p(n^{-\frac{1}{2}}).\end{aligned}$$

Next, we derive a general result for inverse expansions. Let $\widehat{A}$, $A\in\mathbb{R}^{q\times q}$ with A nonsingular. Assume that $\sqrt{n}(\widehat{A}-A)$ converges in distribution at rate $\sqrt{n}$ and that $\widehat{A}^{-1}=A^{-1}+O_p(n^{-\frac{1}{2}})$. Then

$$\begin{aligned}\sqrt{n}(\widehat{A}\widehat{A}^{-1}-I) &= 0\Rightarrow\\ \sqrt{n}\{(\widehat{A}-A+A)(\widehat{A}^{-1}-A^{-1}+A^{-1})-I\} &= 0\Rightarrow\\ \sqrt{n}\{(\widehat{A}-A)(\widehat{A}^{-1}-A^{-1})+(\widehat{A}-A)A^{-1}+A(\widehat{A}^{-1}-A^{-1}) &= 0.\end{aligned}$$

Since $\sqrt{n}\{(\widehat{A}-A)(\widehat{A}^{-1}-A^{-1})=O_p(n^{-\frac{1}{2}})$, we have

$$\begin{aligned}\sqrt{n}\{(\widehat{A}-A)A^{-1}+A(\widehat{A}^{-1}-A^{-1})\} &= O_p(n^{-\frac{1}{2}})\Rightarrow\\ \sqrt{n}(\widehat{A}^{-1}-A^{-1}) &= -\sqrt{n}A^{-1}(\widehat{A}-A)A^{-1}+O_p(n^{-\frac{1}{2}}).\end{aligned}$$

Selecting $\widehat{A}=\widehat{\sigma}^T S_X\widehat{\sigma}$, we get

$$\begin{aligned}&\sqrt{n}\{(\widehat{\sigma}^T S_X\widehat{\sigma})^{-1}-(\sigma^T\Sigma\sigma)^{-1}\}=-\sqrt{n}(\sigma^T\Sigma\sigma)^{-2}(\widehat{A}-A)+O_p(n^{-\frac{1}{2}})\\ &= -\sqrt{n}(\sigma^T\Sigma\sigma)^{-2}\sigma^T(S_X-\Sigma)\sigma-2\sqrt{n}(\sigma^T\Sigma\sigma)^{-2}\sigma^T\Sigma(\widehat{\sigma}-\sigma)+O_p(n^{-\frac{1}{2}}).\end{aligned} \qquad (A.12)$$

Step III. Combining the previous results.

$$\begin{aligned}
\sqrt{n}(\widehat{\beta}_{\text{pls}} - \beta) &= \sqrt{n}\{\widehat{\sigma}\|\widehat{\sigma}\|^2(\widehat{\sigma}^T S_X \widehat{\sigma})^{-1} - \sigma\|\sigma\|^2(\sigma^T \Sigma \sigma)^{-1}\} \\
&= \sqrt{n}\{(\widehat{\sigma}\|\widehat{\sigma}\|^2 - \sigma\|\sigma\|^2)(\sigma^T \Sigma \sigma)^{-1}\} \\
&\quad + \sqrt{n}\sigma\|\sigma\|^2\{(\widehat{\sigma}^T S_X \widehat{\sigma})^{-1} - (\sigma^T \Sigma \sigma)^{-1}\} + O_p(n^{-\frac{1}{2}}) \\
&= (\sigma^T \Sigma \sigma)^{-1}\|\sigma\|^2\{I_p + 2P_\sigma\}\sqrt{n}(\widehat{\sigma} - \sigma) \ \text{(From (A.11))} \\
&\quad - \sigma\|\sigma\|^2(\sigma^T \Sigma \sigma)^{-2}\sqrt{n}\sigma^T(S_X - \Sigma)\sigma \ \text{(From (A.12))} \\
&\quad - 2\sigma\|\sigma\|^2(\sigma^T \Sigma \sigma)^{-2}\sqrt{n}\sigma^T\Sigma(\widehat{\sigma} - \sigma) + O_p(n^{-\frac{1}{2}}) \\
&= \left\{(\sigma^T \Sigma \sigma)^{-1}\|\sigma\|^2(I_p + 2P_\sigma) - 2\|\sigma\|^4(\sigma^T \Sigma \sigma)^{-2}P_\sigma\Sigma\right\}\sqrt{n}(\widehat{\sigma} - \sigma) \\
&\quad - \sigma\|\sigma\|^2(\sigma^T \Sigma \sigma)^{-2}\sqrt{n}\sigma^T(S_X - \Sigma)\sigma + O_p(n^{-\frac{1}{2}}).
\end{aligned}$$

Letting $\Phi = \sigma/\|\sigma\|$, we have

$$\begin{aligned}
\sqrt{n}(\widehat{\beta}_{\text{pls}} - \beta) &= \{(\Phi^T \Sigma \Phi)^{-1}(I_p + 2P_\Phi) - 2(\Phi^T \Sigma \Phi)^{-2}P_\Phi\Sigma\}\sqrt{n}(\widehat{\sigma} - \sigma) \\
&\quad -(\Phi^T \Sigma \Phi)^{-2}P_\Phi\sqrt{n}(S_X - \Sigma)\sigma + O_p(n^{-\frac{1}{2}}).
\end{aligned}$$

Step IV. Substitute the expansions for $\widehat{\sigma} - \sigma$ and $S_X - \Sigma$.

$$\begin{aligned}
\sqrt{n}(\widehat{\beta}_{\text{pls}} - \beta) &= n^{-\frac{1}{2}}(\Phi^T \Sigma \Phi)^{-1}\sum_{i=1}^{n}[\{I_p + 2P_\Phi - 2(\Phi^T \Sigma \Phi)^{-1}P_\Phi\Sigma\}(x_i y_i - \sigma) \\
&\quad - (\Phi^T \Sigma \Phi)^{-1}P_\Phi(x_i x_i^T - \Sigma)\sigma] + O_p(n^{-\frac{1}{2}}).
\end{aligned}$$

Since $q = 1$, $P_\Phi\Sigma = \Phi\Delta\Phi^T = \delta P_\Phi$, and

$$(\Phi^T \Sigma \Phi)^{-1} = \delta^{-1} \Rightarrow (\Phi^T \Sigma \Phi)^{-1}P_\Phi\Sigma = P_\Phi.$$

Then we have a representation as the sum of independent and identically distributed terms,

$$\sqrt{n}(\widehat{\beta}_{\text{pls}} - \beta) = n^{-\frac{1}{2}}\delta^{-1}\sum_{i=1}^{n}\{(x_i y_i - \sigma) - \delta^{-1}P_\Phi(x_i x_i^T - \Sigma)\sigma\} + O_p(n^{-\frac{1}{2}}).$$

Substituting $y_i = x_i^T\beta + \epsilon_i$, and using that

$$\begin{aligned}
\beta &= \Sigma^{-1}\sigma = \Phi\delta^{-1}\Phi^T\sigma = \Phi\delta^{-1}\|\sigma\| = \sigma\delta^{-1} \\
\Sigma\sigma/\delta &= \sigma,
\end{aligned}$$

we see that this expression is the same as conclusion (i) in the proposition.

We next need to calculate the variance of

$$R = \delta^{-1}\{(xy - \sigma) - \delta^{-1}P_{\Phi}(xx^T - \Sigma)\sigma\}.$$

Clearly, $\mathrm{E}(R) = 0$. Substituting $y = x^T\beta + \epsilon = x^T\sigma/\delta + \epsilon$,

$$R = \delta^{-1}\{(x\epsilon + xx^T\sigma\delta^{-1} - \sigma) - \delta^{-1}P_{\Phi}(xx^T - \Sigma)\sigma\},$$

where $\epsilon \perp\!\!\!\perp x$.

Step V. Study R.

$$R = \delta^{-1}(x\epsilon + xx^T\sigma\delta^{-1} - P_{\Phi}xx^T\sigma\delta^{-1} - \sigma + \delta^{-1}\Sigma\sigma).$$

But $\delta^{-1}\Sigma\sigma = P_{\Phi}\sigma = \sigma$. So we get

$$\begin{aligned} R &= \delta^{-1}(x\epsilon + xx^T\sigma\delta^{-1} - P_{\Phi}xx^T\sigma\delta^{-1}) \\ &= \delta^{-1}(x\epsilon + Q_{\Phi}xx^T\sigma\delta^{-1}) = \delta^{-1}(x\epsilon + Q_{\Phi}xx^T\beta). \end{aligned}$$

Since $\epsilon \perp\!\!\!\perp x$, the two terms in R are uncorrelated,

$$\mathrm{var}(R) = \delta^{-2}(\mathrm{var}(x\epsilon) + \mathrm{var}(Q_{\Phi}xx^T\beta)).$$

Next, $\mathrm{var}(x\epsilon) = \mathrm{var}(x)\mathrm{var}(\epsilon) = \Sigma\sigma^2_{Y|X}$, since $\mathrm{E}(x) = 0$, $\mathrm{E}(\epsilon) = 0$. In consequence,

$$\mathrm{var}(R) = \delta^{-2}\left(\Sigma\sigma^2_{Y|X} + \mathrm{var}(Q_{\Phi}xx^T\beta)\right),$$

which proves conclusion (ii) of the proposition.

To prove conclusion (iii) of the proposition, we first have

$$\mathrm{var}(Q_{\Phi}xx^T\sigma) = \mathrm{var}\{\Phi_0(\Phi_0^T x)(x^T\sigma)\} = \Phi_0\mathrm{var}(\Phi_0^T x \cdot x^T\sigma)\Phi_0^T.$$

Assuming that $\Phi_0^T x \perp\!\!\!\perp x^T\sigma$, we get

$$\begin{aligned} \mathrm{var}(Q_{\Phi}xx^T\sigma) &= \Phi_0\mathrm{var}(\Phi_0^T x)\mathrm{var}(\sigma^T x)\Phi_0^T \\ &= \Phi_0(\Phi_0^T\Sigma\Phi_0)\Phi_0^T\sigma^T\Sigma\sigma \\ &= \Phi_0\Delta_0\Phi_0^T\delta\|\sigma\|^2. \end{aligned}$$

Thus,

$$\begin{aligned} \mathrm{var}(R) &= \delta^{-2}(\Sigma\sigma^2_{Y|X} + \Phi_0\Delta_0\Phi_0^T\|\sigma\|^2\delta^{-1}) \\ &= \delta^{-2}\{\Phi\delta\Phi^T\sigma^2_{Y|X} + \Phi_0\Delta_0\Phi_0^T(\sigma^2_{Y|X} + \|\sigma\|^2\delta^{-1})\} \\ &= \delta^{-1}\Phi\Phi^T\sigma^2_{Y|X} + \Phi_0\Delta_0\Phi_0^T\delta^{-2}(\sigma^2_{Y|X} + \|\sigma\|^2\delta^{-1}) \\ &= \delta^{-1}\Phi\Phi^T\sigma^2_{Y|X} + \Phi_0\Delta_0\Phi_0^T\delta^{-2}\sigma^2_Y. \end{aligned}$$

Where the final step follows because

$$\sigma_Y^2 = \beta^T \Sigma \beta + \sigma_{Y|X}^2 = \delta^{-2} \sigma^T \Sigma \sigma + \sigma_{Y|X}^2 = \delta^{-1} \|\sigma\|^2 + \sigma_{Y|X}^2.$$

This then gives conclusion (iii) from Proposition 4.1:

$$\text{var}(R) = \delta^{-1} P_\Phi \sigma_{Y|X}^2 + \Phi_0 \Delta_0 \Phi_0^T \delta^{-2} \sigma_Y^2.$$

A.4.2 Notes on Corollary 4.1

RESTATEMENT.

Assume that the regression of $Y \in \mathbb{R}^1$ on $X \in \mathbb{R}^p$ follows model (4.1) with $\Delta_0 = \delta_0 I_{p-1}$ and that $C = (Y, X^T)^T$ correspondingly follows a multivariate normal distribution. Then

$$\text{avar}(\sqrt{n}\widehat{\beta}) = \sigma_{Y|X}^2 \delta^{-1} \Phi \Phi^T + \eta^2 \left(\eta^2 \delta_0 / \sigma_{Y|X}^2 + (\delta_0/\delta)(1 - \delta/\delta_0)^2 \right)^{-1} \Phi_0 \Phi_0^T$$

$$\text{avar}(\sqrt{n}\widehat{\beta}_{\text{pls}}) = \sigma_{Y|X}^2 \delta^{-1} \Phi \Phi^T + (\sigma_Y^2 \delta_0 / \delta^2) \Phi_0 \Phi_0^T.$$

$\text{avar}(\sqrt{n}\widehat{\beta})$ follows as a special case of Proposition 2.1 which was proved by Cook, Forzani, and Rothman (2013). Briefly, substituting the hypotheses of Proposition 4.1 into the conclusion of Proposition 2.1 we have

$$\begin{aligned} \text{avar}\{\sqrt{n}\text{vec}(\widehat{\beta})\} &= \Sigma_{Y|X} \otimes \Phi \Delta^{-1} \Phi^T + (\eta^T \otimes \Phi_0) M^\dagger (\eta \otimes \Phi_0^T), \\ &= (\sigma_{Y|X}^2/\delta) \Phi \Phi^T + \eta^2 \Phi_0 M^\dagger \Phi_0^T, \end{aligned}$$

where

$$\begin{aligned} M &= \eta \Sigma_{Y|X}^{-1} \eta^T \otimes \Delta_0 + \Delta \otimes \Delta_0^{-1} + \Delta^{-1} \otimes \Delta_0 - 2 I_q \otimes I_{p-q} \\ &= (\eta^2 \delta_0 / \sigma_{Y|X}^2) I_{p-1} + (\delta/\delta_0) I_{p-1} + (\delta_0/\delta) I_{p-1} - 2 I_{p-1} \\ &= \left\{ (\eta^2 \delta_0 / \sigma_{Y|X}^2) + (\delta_0/\delta)(1 - \delta/\delta_0)^2 \right\} I_{p-1}. \end{aligned}$$

Thus, we can use ordinary inverses and

$$\begin{aligned} \text{avar}\{\sqrt{n}\text{vec}(\widehat{\beta})\} &= (\sigma_{Y|X}^2/\delta) \Phi \Phi^T + \eta^2 \Phi_0 M^{-1} \Phi_0^T = (\sigma_{Y|X}^2/\delta) \Phi \Phi^T \\ &\quad + \eta^2 \{ (\eta^2 \delta_0 / \sigma_{Y|X}^2) + (\delta_0/\delta)(1 - \delta/\delta_0)^2 \}^{-1} \Phi_0 \Phi_0^T. \end{aligned}$$

Dividing $\eta^2 \{ (\eta^2 \delta_0 / \sigma_{Y|X}^2) + (\delta_0/\delta)(1 - \delta/\delta_0)^2 \}^{-1}$, the cost for the envelope estimator $\widehat{\beta}$, by the corresponding PLS cost $(\sigma_Y^2 \delta_0 / \delta^2)$ from $\text{avar}\{\sqrt{n}\text{vec}(\widehat{\beta}_{\text{pls}})\}$, which comes directly from Cook et al. (2013), and using the relationship $\sigma_Y^2 = \eta^2 \delta + \sigma_{Y|X}^2$ leads to cost ratio given below Corollary 4.1.

A.4.3 Form of Σ_X under compound symmetry

Under compound symmetry, $\text{var}(X_j) = \pi^2$ and $\text{cov}(X_j, X_k) = \rho$. Let 1_p denote the $p \times 1$ vector of ones. Then in matrix form

$$\begin{aligned}\Sigma_X &= \pi^2\{\rho 1_p 1_p^T + (1-\rho)I_p\} \\ &= \pi^2\{p\rho P_{1_p} + (1-\rho)(P_{1_p} + Q_{1_p})\} \\ &= \pi^2\{(1-\rho+p\rho)P_{1_p} + (1-\rho)Q_{1_p}\}.\end{aligned}$$

A.4.4 Theorem 4.1

RESTATEMENT. If model (4.1) holds, $\beta^T\Sigma_X\beta \asymp 1$ and $K_j(n,p)$, $j = 1,2$, converges to 0 as $n, p \to \infty$ then

$$D_N = O_p\left\{n^{-1/2} + K_1(n,p) + K_2^{1/2}(n,p)\right\}.$$

To ease notion, in this section we let $\sigma = \sigma_{X,Y}$, and we let

$$\begin{aligned}K_1(n,p) &= \frac{\text{tr}(\Delta_0)}{n\|\sigma\|^2} \\ K_2(n,p) &= \frac{\text{tr}(\Delta_0^2)}{n\|\sigma\|^4} \\ K_3(n,p) &= \frac{\text{tr}^{1/2}(\Delta_0^3)}{n\|\sigma\|^3} = \text{tr}^{1/2}(\Delta_\sigma^3)/n,\end{aligned}$$

where Δ_σ is the signal to noise ratio defined near (4.11).

The following proposition will be used to prove Theorem 4.1 by implication.

Proposition A.1. *If model (4.1) holds, $\beta^T\Sigma_X\beta \asymp 1$ and $K_j(n,p)$, $j = 1,2,3$, converges to 0 as $n, p \to \infty$ then*

$$\begin{aligned}D_N &= (\widehat{\beta}_{\text{pls}} - \beta_{\text{pls}})^T\omega_N \\ &= \left(\widehat{\sigma}^T\widehat{\sigma}(\widehat{\sigma}^T S_X\widehat{\sigma})^{-1}\widehat{\sigma}^T - \sigma^T\sigma(\sigma^T\Sigma_X\sigma)^{-1}\sigma^T\right)\omega_N \quad \text{(A.13)} \\ &= O_p\{n^{-1/2} + K_1(n,p) + K_2^{1/2}(n,p) + K_3(n,p)\}, \quad \text{(A.14)}\end{aligned}$$

where $\omega_N = X_N - \text{E}(X) \sim N(0, \Sigma_X)$ as defined near (4.8), (A.13) is a restatement of the definitions and (A.14) is what we need to demonstrate.

This proposition, which is a restatement of Theorem 1 in Cook and Forzani (2018), contains an extra addend $K_3(n,p)$ that does not appear in

Theorem 4.1. It turns out that the addend $K_3(n,p)$ is superfluous because the hypothesis of Theorem 4.1, $K_j(n,p) \to 0$ for $j = 1,2$, implies that $K_3(n,p) \to 0$:

$$K_3(n,p) \le (K_1(n,p)K_2(n,p))^{1/2} \le \frac{1}{\sqrt{2}}\left(K_1(n,p) + K_2(n,p)\right),$$

which establishes that K_3 is at most the order of $K_1 + K_2$. Nevertheless, to maintain a connection with the literature, we prove Proposition A.1, which then implies Theorem 4.1.

Proof. The proof of Proposition A.1 is facilitated by establishing key intermediate results, the first of which involves the asymptotic behavior of certain ratios:

Proposition A.2. *Assume the hypothesis of Theorem 4.1. Then*

$$\frac{\widehat{\sigma}^T S_X \widehat{\sigma}}{\sigma^T \Sigma_X \sigma} = 1 + O_p\{n^{-1/2} + K_1(n,p) + K_2(n,p) + K_3(n,p)\}. \quad \text{(A.15)}$$

$$\frac{\widehat{\sigma}^T \widehat{\sigma}}{\sigma^T \sigma} = 1 + O_p\{n^{-1/2} + K_1(n,p)\}. \quad \text{(A.16)}$$

$$\frac{\widehat{\sigma}^T \widehat{\sigma}}{\widehat{\sigma}^T S_X \widehat{\sigma}} = \frac{\sigma^T \sigma}{\sigma^T \Sigma_X \sigma} O_p(1). \quad \text{(A.17)}$$

Proof of Proposition A.2. Turning to the estimated terms in (A.13) and recalling that $\widehat{\sigma} = n^{-1}\mathbb{X}^T\mathbb{Y}$ and $S_X = n^{-1}\mathbb{X}^T\mathbb{X} \ge 0$, we have

$$\begin{aligned}
\widehat{\sigma}^T\widehat{\sigma} &= \frac{1}{n^2}\left\{\mathbb{Y}^T\mathbb{X}\mathbb{X}^T\mathbb{Y}\right\} \\
&= \frac{1}{n^2}\left\{(\mathbb{X}\beta_{\text{pls}} + \varepsilon)^T\mathbb{X}\mathbb{X}^T(\mathbb{X}\beta_{\text{pls}} + \varepsilon)\right\} \\
&= \frac{1}{n^2}\beta_{\text{pls}}^T W^2 \beta_{\text{pls}} + \frac{2}{n^2}\beta_{\text{pls}}^T W \mathbb{X}^T \varepsilon + \frac{1}{n^2}\varepsilon^T\mathbb{X}\mathbb{X}^T\varepsilon \qquad \text{(A.18)} \\
\widehat{\sigma}^T S_X \widehat{\sigma} &= \frac{1}{n^3}\beta_{\text{pls}}^T W^3 \beta_{\text{pls}} + \frac{2}{n^3}\beta_{\text{pls}}^T W^2 \mathbb{X}^T \varepsilon + \frac{1}{n^3}\varepsilon^T\mathbb{X}W\mathbb{X}^T\varepsilon, \qquad \text{(A.19)}
\end{aligned}$$

where $W \sim W_{n-1}(\Sigma_X)$. The justification below is intended to give only an indication of the steps necessary to obtain the result. Additional details are available from the Supplement to Cook and Forzani (2018).

Since we are going to need the expectation of powers of Wishart marices to determine the orders of (A.18) and (A.19), we begin with the following results from Letac and Massam (2004). An online Wishart moment calculator is available from Forzani, Percíncula, and Toledano (2022).

Lemma A.1. *Let $U \sim W_n(\Theta)$. Then*

$$
\begin{aligned}
E(U) &= n\Theta \\
E(U^2) &= n\Theta\mathrm{tr}(\Theta) + n(n+1)\Theta^2 \\
E(U^3) &= n\Theta\mathrm{tr}^2(\Theta) + n(n+1)(\Theta\mathrm{tr}(\Theta^2) + 2\Theta^2\mathrm{tr}(\Theta)) + n(n^2+3n+4)\Theta^3 \\
E(U^5) &= \left(n^5 + 10\,n^4 + 65\,n^3 + 160\,n^2 + 148\,n\right)\Theta^5 \\
&\quad + \left(4\left(n^4 + 6\,n^3 + 21\,n^2 + 20\,n\right)(\mathrm{tr}\,\Theta)\right)\Theta^4 \\
&\quad + \Big(6\left(n^3 + 3\,n^2 + 4\,n\right)(\mathrm{tr}\,\Theta)^2 \\
&\qquad + 3\left(n^4 + 5\,n^3 + 14\,n^2 + 12\,n\right)(\mathrm{tr}\,\Theta^2)\Big)\Theta^3 \\
&\quad + \Big\{4\left(n^2 + n\right)(\mathrm{tr}\,\Theta)^3 + 4\left(2\,n^3 + 5\,n^2 + 5\,n\right)(\mathrm{tr}\,\Theta^2)(\mathrm{tr}\,\Theta) \\
&\quad + 2\left(n^4 + 5\,n^3 + 14\,n^2 + 12\,n\right)(\mathrm{tr}\,\Theta^3)\Big\}\Theta^2 \\
&\quad + \Big\{n(\mathrm{tr}\,\Theta)^4 + 6\left(n^2 + n\right)(\mathrm{tr}\,\Theta^2)(\mathrm{tr}\,\Theta)^2 + \left(2\,n^3 + 5\,n^2 + 5\,n\right)(\mathrm{tr}\,\Theta^2)^2 \\
&\quad + 4\left(n^3 + 3\,n^2 + 4\,n\right)(\mathrm{tr}\,\Theta^3)(\mathrm{tr}\,\Theta) \\
&\quad + \left(n^4 + 6\,n^3 + 21\,n^2 + 20\,n\right)(\mathrm{tr}\,\Theta^4)\Big\}\Theta.
\end{aligned}
$$

The next three lemmas, each with its own PROOF section, give ingredients to establish (A.15)–(A.17) from (A.18) and (A.19).

Lemma A.2. *As $n, p \to \infty$ subject to the conditions of Theorem 4.1,*

$$
\frac{\varepsilon^T \mathbb{X}\mathbb{X}^T \varepsilon}{n^2 \sigma^T \sigma} = O_p(1)\left(\frac{1}{n} + \frac{\mathrm{tr}(\Delta_0)}{n\sigma^T\sigma}\right) \tag{A.20}
$$

$$
\frac{\varepsilon^T \mathbb{X} W \mathbb{X}^T \varepsilon}{n^3 \sigma^T \Sigma \sigma} = O_p(1)\left(\frac{1}{n} + \frac{\mathrm{tr}(\Delta_0^2)}{n\sigma^T\Sigma\sigma} + \frac{\mathrm{tr}^2(\Delta_0)}{n^2\sigma^T\Sigma\sigma}\right). \tag{A.21}
$$

PROOF OF LEMMA A.2. Since both quantities are positive we need only to compute their expectations using Lemma A.1 and then employ Markov's inequality. Recall from the preamble to Section 4.4 that ε is the $n \times 1$ vector with model errors ϵ_i as elements.

$$
\begin{aligned}
E(n^{-2}\varepsilon^T \mathbb{X}\mathbb{X}^T \varepsilon) &= \frac{\sigma^2_{Y|X}}{n^2} E(\mathrm{tr}(W)) = O(n^{-1})\mathrm{tr}(\Sigma_X) \\
E(n^{-3}\varepsilon^T \mathbb{X} W \mathbb{X}^T \varepsilon) &= \frac{\sigma^2_{Y|X}}{n^3} E(\mathrm{tr}(W^2)) = O(1)\left(\frac{\mathrm{tr}(\Sigma_X^2)}{n} + \frac{\mathrm{tr}^2(\Sigma_X)}{n^2}\right).
\end{aligned}
$$

Therefore, since $\sigma^T\Sigma_X\sigma \asymp (\sigma^T\sigma)^2$, $\delta \asymp \sigma^T\sigma$ and $\Sigma_X = \delta\Phi\Phi^T + \Phi_0\Delta_0\Phi^T$, we have

$$E\left(\frac{\varepsilon^T \mathbb{X}\mathbb{X}^T \varepsilon}{n^2 \sigma^T \sigma}\right) = O(1)\frac{\mathrm{tr}(\Sigma_X)}{n\sigma^T\sigma}$$
$$= O(1)\left(\frac{1}{n} + \frac{\mathrm{tr}(\Delta_0)}{n\sigma^T\sigma}\right)$$
$$= O(1)\left(\frac{1}{n} + K_1(n,p)\right)$$

$$E\left(\frac{\varepsilon^T \mathbb{X} W \mathbb{X}^T \varepsilon}{n^3 \sigma^T \Sigma_X \sigma}\right) = O(1)\frac{1}{(\sigma^T\sigma)^2}\left(\frac{\mathrm{tr}(\Sigma_X^2)}{n} + \frac{\mathrm{tr}^2(\Sigma_X)}{n^2}\right)$$
$$= O(1)\left(\frac{1}{n} + \frac{\mathrm{tr}(\Delta_0^2)}{n(\sigma^T\sigma)^2} + \frac{\mathrm{tr}^2(\Delta_0)}{n^2(\sigma^T\sigma)^2}\right)$$
$$= O(1)\left(\frac{1}{n} + K_2(n,p) + K_1^2(n,p)\right).$$

□

Lemma A.3. *As $n, p \to \infty$ subject to the conditions of Theorem 4.1,*

$$\mathrm{var}\left(\frac{\beta_{\mathrm{pls}}^T W \mathbb{X}^T \varepsilon}{n^2 \sigma^T \sigma}\right) = O(n^{-1}) \tag{A.22}$$

$$\mathrm{var}\left(\frac{\beta_{\mathrm{pls}}^T W^2 \mathbb{X}^T \varepsilon}{n^3 \sigma^T \Sigma_X \sigma}\right) = O(1)\left[\frac{1}{n} + K_2^2(n,p) + K_3^2(n,p)\right]. \tag{A.23}$$

Proof of Lemma A.3. Each term has expectation zero, so we compute their variances with the help of Lemma A.1.

$$\mathrm{var}\left(\frac{\beta_{\mathrm{pls}}^T W \mathbb{X}^T \varepsilon}{n^2 \sigma^T \sigma}\right) = \frac{\tau^2}{n^4(\sigma^T\sigma)^2}E(\beta_{\mathrm{pls}}^T W \mathbb{X}^T \mathbb{X} W \beta_{\mathrm{pls}}) = \frac{\tau^2}{n^4(\sigma^T\sigma)^2}E(\beta_{\mathrm{pls}}^T W^3 \beta_{\mathrm{pls}})$$
$$= O(1)\frac{1}{n^4(\sigma^T\sigma)^2}\left(n^3\beta_{\mathrm{pls}}^T\Sigma_X^3\beta_{\mathrm{pls}} + n^2\beta_{\mathrm{pls}}^T\Sigma_X^2\beta_{\mathrm{pls}}\mathrm{tr}(\Sigma_X)\right.$$
$$\left.+n^2\beta_{\mathrm{pls}}^T\Sigma_X\beta_{\mathrm{pls}}\mathrm{tr}(\Sigma_X^2) + n\beta_{\mathrm{pls}}^T\Sigma_X\beta_{\mathrm{pls}}\mathrm{tr}^2(\Sigma_X)\right).$$

Now, $\beta_{\mathrm{pls}}^T\Sigma_X^3\beta_{\mathrm{pls}} = \sigma^T\Sigma_X\sigma \asymp (\sigma^T\sigma)^2$ and therefore we have

$$\mathrm{var}\left(\frac{\beta_{\mathrm{pls}}^T W \mathbb{X}^T \varepsilon}{n^2 \sigma^T \sigma}\right) = O(1)\left(\frac{1}{n} + n^{-1}\{K_1(n,p) + K_2(n,p) + K_1^2(n,p)\}\right).$$

Conclusion (A.22) follows by hypothesis since the three terms involving K_j are all $o(1/n)$.

Conclusion (A.23) follows in a similar vein:

$$\begin{aligned}\operatorname{var}\left(\frac{\beta_{\text{pls}}^T W^2 \mathbb{X}^T \varepsilon}{n^2 \sigma^T \Sigma_X \sigma}\right) &= \frac{\sigma_{Y|X}^2}{n^6 (\sigma^T \Sigma_X \sigma)^2} E(\beta_{\text{pls}}^T W^2 \mathbb{X}^T \mathbb{X} W^2 \beta_{\text{pls}}) \\ &= \frac{\sigma_{Y|X}^2}{n^6 (\sigma^T \Sigma_X \sigma)^2} E(\beta_{\text{pls}}^T W^5 \beta_{\text{pls}}).\end{aligned}$$

$E(W^5)$ can now be evaluated using Lemma A.1 and the results simplified to yield (A.23). □

Lemma A.4. *As $n, p \to \infty$ subject to the conditions of Theorem 4.1,*

$$\frac{\beta_{\text{pls}}^T W^2 \beta_{\text{pls}}}{n^2 \sigma^T \sigma} - 1 = O_p\left(\frac{1}{\sqrt{n}} + K_1(n,p)\right) \tag{A.24}$$

$$\frac{\beta_{\text{pls}}^T W^3 \beta_{\text{pls}}}{n^3 \sigma^T \Sigma_X \sigma} - 1 = O_p\left(\frac{1}{\sqrt{n}} + K_1(n,p) + K_2(n,p)\right). \tag{A.25}$$

Proof. The proof follows the same logic as the proofs of the other lemmas. See Cook and Forzani (2018) for details. □

Lemmas A.2–A.4 give the orders of scaled versions of all six addends on the right hand sides of (A.18) and (A.19). These are next used to determine the orders (A.15)–(A.17). By combining (A.20), (A.22) and (A.24) we see that $\widehat{\sigma}^T \widehat{\sigma} / \|\sigma\|$ is of the order of

$$\left(\frac{1}{n} + K_1(n,p)\right) + \frac{1}{\sqrt{n}} + \left(\frac{1}{\sqrt{n}} + K_1(n,p)\right),$$

which has the order given in (A.16)

$$\left(\frac{1}{\sqrt{n}} + K_1(n,p)\right).$$

Combining (A.21), (A.23) and (A.25) and using that $\sigma^T \Sigma_X \sigma \asymp (\sigma^T \sigma)^2$, we find the order of $\widehat{\sigma}^T S_X \widehat{\sigma} / \sigma^T \Sigma_X \sigma$ is equal to the order of

$$\left(\frac{1}{n} + K_2(n,p) + K_1^2(n,p)\right) + \left(\frac{1}{\sqrt{n}} + K_2(n,p) + K_3(n,p)\right) + \left(\frac{1}{\sqrt{n}} + K_2(n,p) + K_1(n,p)\right).$$

Following the previous logic, we arrive at the stated order (A.15). (A.17) follows immediately from (A.16) and (A.15).

END PROOF OF PROPOSITION A.2. □

Continuing now with the proof of Proposition A.1, the next step is to rewrite (A.13) in a form that makes use of Proposition A.2. Recall that $\delta = \sigma^T \Sigma_X \sigma / \|\sigma\|^2$ is the eigenvalue of Σ_X that is associated with the basis vector $\Phi = \sigma/\|\sigma\|$ of the envelope $\mathcal{E}_{\Sigma_X}(\mathcal{B})$. Let $\widehat{\delta} = \widehat{\sigma}^T S_X \widehat{\sigma} / \widehat{\sigma}^T \widehat{\sigma}$. From (A.13) then, we need to find the order of

$$\begin{aligned} D_N &= (\widehat{\delta}^{-1}\widehat{\sigma} - \delta^{-1}\sigma)^T \omega_N \\ &= \widehat{\delta}^{-1}(\widehat{\sigma} - \sigma)^T \omega_N - \widehat{\delta}^{-1}(\widehat{\sigma}^T S_X \widehat{\sigma} - \sigma^T \Sigma_X \sigma)(\sigma^T \Sigma_X \sigma)^{-1} \sigma^T \omega_N \\ &\quad + (\widehat{\sigma}^T \widehat{\sigma} - \sigma^T \sigma)(\sigma^T \Sigma_X \sigma)^{-1} \sigma^T \omega_N. \end{aligned}$$

It follows from Equation (A.17) that $\widehat{\delta}^{-1}\delta = O_p(1)$. Consequently, multiplying the first two addends of D_N by $\delta\delta^{-1}$ we have

$$\begin{aligned} D_N = (\widehat{\delta}^{-1}\delta)\delta^{-1}(\widehat{\sigma} - \sigma)^T \omega_N - (\widehat{\delta}^{-1}\delta)\delta^{-1}(\widehat{\sigma}^T S_X \widehat{\sigma} - \sigma^T \Sigma_X \sigma)(\sigma^T \Sigma_X \sigma)^{-1} \sigma^T \omega_N \\ + (\widehat{\sigma}^T \widehat{\sigma} - \sigma^T \sigma)(\sigma^T \Sigma_X \sigma)^{-1} \sigma^T \omega_N. \end{aligned}$$

Therefore an order for D_N can be found by adding the orders of the following three terms.

$$\begin{aligned} I &= \delta^{-1}(\widehat{\sigma} - \sigma)^T \omega_N. \\ II &= \delta^{-1}(\widehat{\sigma}^T S_X \widehat{\sigma} - \sigma^T \Sigma_X \sigma)(\sigma^T \Sigma_X \sigma)^{-1} \sigma^T \omega_N. \\ III &= (\widehat{\sigma}^T \widehat{\sigma} - \sigma^T \sigma)(\sigma^T \Sigma_X \sigma)^{-1} \sigma^T \omega_N. \end{aligned}$$

The orders of these terms is as follows.

Term I.

Since $\mathrm{var}(\widehat{\sigma}) \asymp n^{-1}(\mathrm{var}(Y)\Sigma_X + \sigma\sigma^T)$ (Cook, Forzani, and Rothman, 2013) we have

$$\begin{aligned} \mathrm{var}(I) &= \delta^{-2} E((\widehat{\sigma} - \sigma)^T \Sigma_X (\widehat{\sigma} - \sigma)) \asymp \delta^{-2} \mathrm{tr}\{\mathrm{var}(\widehat{\sigma})\Sigma_X\} \\ &\asymp \delta^{-2} \frac{\mathrm{var}(Y)\mathrm{tr}(\Sigma_X^2) + \sigma^T \Sigma_X \sigma}{n} \\ &\asymp n^{-1}\delta^{-2}\mathrm{var}(Y)\left\{\delta^2 + \mathrm{tr}(\Delta_0^2)\right\} + n^{-1}\delta^{-2}\sigma^T \Sigma_X \sigma \\ &\asymp n^{-1} + K_2(n,p). \end{aligned}$$

Consequently, since $E(I) = 0$,

$$I = O_p\left(n^{-1/2} + K_2^{1/2}(n,p)\right).$$

Term II.

From conclusion (A.15) of Proposition A.2,

$$(\widehat{\sigma}^T S_X \widehat{\sigma} - \sigma^T \Sigma_X \sigma)(\sigma^T \Sigma_X \sigma)^{-1} = O_p\left(n^{-1/2} + K_1(n,p) + K_2(n,p) + K_3(n,p)\right)$$

and

$$\mathrm{var}(\delta^{-1}\sigma^T e_N) = \left(\delta^{-1}\right)^2 \sigma^T \Sigma_X \sigma = (\sigma^T\sigma)^2(\sigma^T \Sigma_X \sigma)^{-1} \asymp 1.$$

Therefore

$$II = O_p\left(n^{-1/2} + K_1(n,p) + K_2(n,p) + K_3(n,p)\right).$$

Term III.

It follows from conclusion (A.16) of Proposition A.2 that

$$(\widehat{\sigma}^T\widehat{\sigma} - \sigma^T\sigma)(\sigma^T\sigma)^{-1} = O\left(n^{-1/2} + K_1(n,p)\right)$$

and, from term II that $\mathrm{var}\left(\delta^{-1}\sigma^T w_N\right) \asymp 1$. Accordingly,

$$III = O\left(n^{-1/2} + K_1(n,p)\right).$$

Thus,

$$\begin{aligned} D_N &= I + II + III \\ &= O_p\left(n^{-1/2} + K_2^{1/2}(n,p) + K_1(n,p) + K_2(n,p) + K_3(n,p)\right) \\ &= O_p\left(n^{-1/2} + K_2^{1/2}(n,p) + K_1(n,p) + K_3(n,p)\right), \end{aligned}$$

which is the conclusion claimed.

A.4.5 Proposition 4.2

RESTATEMENT. Assume that the eigenvalues of Δ_0 are bounded and $\beta^T \Sigma_X \beta \asymp 1$. Then $\kappa(p) \asymp p$ and, if $\|\Sigma_{X,Y}\|^2 \to \infty$, then $\eta(p) \to \infty$, where $\|\cdot\|^2$ denotes the Euclidean norm.

Proof. First $\kappa(p) = \mathrm{tr}(\Delta_0) \asymp p$ is immediate since the eigenvalues of Δ_0 are bounded.

For the second conclusion, recall that in the multicomponent model $\Sigma_X = \Phi\Delta\Phi^T + \Phi_0\Delta_0\Phi_0$. We assume without loss of generality that $\Delta = \text{diag}(\delta_1, \ldots, \delta_q)$. The initial part of the proof is driven by the boundedness of $\beta^T\Sigma_X\beta$:

$$\begin{aligned}\beta^T\Sigma_X\beta &= \sigma_{X,Y}^T\Sigma_X^{-1}\sigma_{X,Y} \\ &= \sigma_{X,Y}^T\left\{\Phi\Delta^{-1}\Phi^T + \Phi_0\Delta_0^{-1}\Phi_0\right\}\sigma_{X,Y} \\ &= \sigma_{X,Y}^T\left\{\Phi\Delta^{-1}\Phi^T\right\}\sigma_{X,Y},\end{aligned}$$

where the last step follows because $\sigma_{X,Y}$ is contained in $\mathcal{E}_{\Sigma_X}(\mathcal{B})$, which has semi-orthogonal basis Φ. Let ϕ_i denote the i-th column of Φ and define

$$w_i = \frac{\sigma_{X,Y}^T P_{\phi_i}\sigma_{X,Y}}{\sigma_{X,Y}^T P_{\Phi}\sigma_{X,Y}}, \ i = 1, \ldots, q,$$

which are positive and sum to 1. Then

$$\begin{aligned}\beta^T\Sigma_X\beta &= \sigma_{X,Y}^T\left\{\Phi\Delta^{-1}\Phi^T\right\}\sigma_{X,Y} \\ &= \sum_{i=1}^{q} w_i(\|\sigma_{X,Y}\|^2/\delta_i).\end{aligned}$$

If the regression is abundant so $\eta(p) = \text{tr}(\Delta) \to \infty$ then $\|\sigma_{X,Y}\|^2 \to \infty$ since $\beta^T\Sigma_X\beta \asymp 1$. If $\|\sigma_{X,Y}\|^2 \to \infty$ then we must have $\delta_i \to \infty$ since again $\beta^T\Sigma_X\beta \asymp 1$. □

A.5 Proofs for Chapter 5

A.5.1 Proposition 5.1

RESTATEMENT. Define the subspaces $\mathcal{S} \subseteq \mathbb{R}^p$ and $\mathcal{R} \subseteq \mathbb{R}^r$. Then the two condition

$$\text{(a) } Q_{\mathcal{S}}X \perp\!\!\!\perp (Y, P_{\mathcal{S}}X) \text{ and (b) } Q_{\mathcal{R}}Y \perp\!\!\!\perp (X, P_{\mathcal{R}}Y)$$

hold if and only if the following two condition hold:

$$\text{(I) } Q_{\mathcal{R}}Y \perp\!\!\!\perp Q_{\mathcal{S}}X \text{ and (II) } (P_{\mathcal{R}}Y, P_{\mathcal{S}}X) \perp\!\!\!\perp (Q_{\mathcal{R}}Y, Q_{\mathcal{S}}X).$$

Proof. For notational convenience, let $M = (P_{\mathcal{R}}Y, P_{\mathcal{S}}X)$. We first show that conditions (a) and (b) are equivalent to the conditions

$$\text{(I}'\text{)}\ Q_{\mathcal{R}}Y \perp\!\!\!\perp Q_{\mathcal{S}}X \mid (P_{\mathcal{R}}Y, P_{\mathcal{S}}X) \text{ and (II) } (P_{\mathcal{R}}Y, P_{\mathcal{S}}X) \perp\!\!\!\perp (Q_{\mathcal{R}}Y, Q_{\mathcal{S}}X).$$

Assume that conditions (a) and (b) hold. Let $R \in \mathbb{R}^{r\times u}$ and $R_0 \in \mathbb{R}^{r\times r-u}$ be semi-orthogonal basis matrices for $\mathcal{R}$ and its orthogonal complement $\mathcal{R}^{\perp}$. Then (R, R_0) is an orthogonal matrix and condition (a) holds if and only if $Q_{\mathcal{S}}X \perp\!\!\!\perp (R^TY, R_0^TY, P_{\mathcal{S}}X)$. Consequently, condition (a) implies that (see (Cook, 1998, Proposition 4.6) for background on conditional independence.)

$$\begin{aligned} Q_{\mathcal{S}}X \perp\!\!\!\perp (P_{\mathcal{R}}Y, Q_{\mathcal{R}}Y, P_{\mathcal{S}}X) &\Rightarrow Q_{\mathcal{S}}X \perp\!\!\!\perp (M, Q_{\mathcal{R}}Y) \\ &\Rightarrow \text{(a1) } Q_{\mathcal{S}}X \perp\!\!\!\perp Q_{\mathcal{R}}Y \mid M \text{ and (a2) } Q_{\mathcal{S}}X \perp\!\!\!\perp M \\ &\Rightarrow \text{(a3) } Q_{\mathcal{S}}X \perp\!\!\!\perp M \mid Q_{\mathcal{R}}Y \text{ and (a4) } Q_{\mathcal{S}}X \perp\!\!\!\perp Q_{\mathcal{R}}Y. \end{aligned}$$

Similarly, condition (b) implies

$$\begin{aligned} Q_{\mathcal{R}}Y \perp\!\!\!\perp (P_{\mathcal{S}}X, Q_{\mathcal{S}}X, P_{\mathcal{R}}Y) &\Rightarrow Q_{\mathcal{R}}Y \perp\!\!\!\perp (M, Q_{\mathcal{S}}X) \\ &\Rightarrow \text{(b1) } Q_{\mathcal{S}}X \perp\!\!\!\perp Q_{\mathcal{R}}Y \mid M \text{ and (b2) } Q_{\mathcal{R}}Y \perp\!\!\!\perp M \\ &\Rightarrow \text{(b3) } Q_{\mathcal{R}}Y \perp\!\!\!\perp M \mid Q_{\mathcal{S}}X \text{ and (b4) } Q_{\mathcal{S}}X \perp\!\!\!\perp Q_{\mathcal{R}}Y. \end{aligned}$$

Condition (I$'$) follows immediately from either condition (a1) or condition (b1). Condition (II) is implied by (a3) and (b2). Thus condition (a) and (b) imply conditions (I$'$) and (II).

Assume that conditions (I$'$) and (II) hold. Then these imply that

$$\text{(I}'\text{)}\ Q_{\mathcal{R}}Y \perp\!\!\!\perp Q_{\mathcal{S}}X \mid M, \text{ (II1) } M \perp\!\!\!\perp Q_{\mathcal{R}}Y \text{ and (II2) } M \perp\!\!\!\perp Q_{\mathcal{S}}X.$$

(I$'$) and (II1) imply that $Q_{\mathcal{R}}Y \perp\!\!\!\perp (M, Q_{\mathcal{S}}X)$, while conditions (I$'$) and (II2) imply that $Q_{\mathcal{S}}X \perp\!\!\!\perp (M, Q_{\mathcal{R}}Y)$. Replacing M with its definition, these implications give

$$Q_{\mathcal{R}}Y \perp\!\!\!\perp (P_{\mathcal{S}}X, Q_{\mathcal{S}}X, P_{\mathcal{R}}Y) \text{ and } Q_{\mathcal{S}}X \perp\!\!\!\perp (P_{\mathcal{R}}Y, Q_{\mathcal{R}}Y, P_{\mathcal{S}}X).$$

The required condition (a) and (b) follow from here since if A is independent of B then A is independent of any non-stochastic function of B. That is, these statements imply that

$$Q_{\mathcal{R}}Y \perp\!\!\!\perp (P_{\mathcal{S}}X + Q_{\mathcal{S}}X, P_{\mathcal{R}}Y) \text{ and } Q_{\mathcal{S}}X \perp\!\!\!\perp (P_{\mathcal{R}}Y + Q_{\mathcal{R}}Y, P_{\mathcal{S}}X).$$

This established that conditions (I′) and (II) are equivalent to conditions (a) and (b).

To establish (a) and (b) with (I) and (II), first assume conditions (a) and (b). Then (I′) and (II) hold. But (II) implies that (I) holds if and only if (I′) holds. Next, assume that (I′) and (II) hold. Then (a) and (b) hold and, again, (II) implies that (I) holds if and only if (I′) holds.

□

A.5.2 Proof of Lemma 5.1

RESTATEMENT. Under the simultaneous envelope model (5.7), canonical correlation analysis can find at most d directions in the population, where $d = \operatorname{rank}(\Sigma_{X,Y})$ as defined in (5.5). Moreover, the directions are contained in the simultaneous envelope as

$$\operatorname{span}(a_1, \ldots, a_d) \subseteq \mathcal{E}_{\Sigma_X}(\mathcal{B}), \ \operatorname{span}(b_1, \ldots, b_d) \subseteq \mathcal{E}_{\Sigma_{Y|X}}(\mathcal{B}'). \tag{A.26}$$

Proof. Recall that the canonical correlation directions are the pairs of vectors $\{a_i, b_i\} = \{\Sigma_X^{-1/2} e_i, \Sigma_Y^{-1/2} f_i\}$, where $\{e_i, f_i\}$ is the i-th left-right eigenvector pair of the correlation matrix $\rho = \Sigma_X^{-1/2}\Sigma_{X,Y}\Sigma_Y^{-1/2}$, $i = 1, \ldots, d$, $d = \operatorname{rank}(\Sigma_{X,Y})$. Now, the conclusion follows for the a_i's because

$$\operatorname{span}(a_1, \ldots, a_d) = \operatorname{span}(\Sigma_X^{-1}\Sigma_{X,Y}\Sigma_Y^{-1/2}) \subseteq \operatorname{span}(\Phi) = \mathcal{E}_{\Sigma_X}(\mathcal{B}).$$

The conclusion for the b_i's follows similarly. □

A.5.3 Proof of Lemma 5.3

RESTATEMENT. Using the sample estimators from Lemma 5.2, the sample covariance matrix of the residuals from model (5.7),

$$S_{\text{res}} = n^{-1} \sum_{i=1}^{n} \{(Y_i - \bar{Y}) - G\widehat{\eta}^T W^T X_i\}\{(Y_i - \bar{Y}) - G\widehat{\eta}^T W^T X_i\}^T,$$

can be represented as

$$S_{\text{res}} = \widehat{\Sigma}_{Y|X} + P_G S_{Y|W^TX} Q_G + Q_G S_{Y|W^TX} P_G,$$

where $\widehat{\Sigma}_{Y|X}$ is as given in Lemma 5.2.

Proof. Substituting $\widehat{\eta} = S_{W^TX}^{-1} S_{W^TX,G^TY}$ and expanding, we have

$$\begin{aligned} S_{\text{res}} &= n^{-1}\sum_{i=1}^{n}\{(Y_i - \bar{Y}) - G\widehat{\eta}^T W^T X_i\}\{(Y_i - \bar{Y}) - G\widehat{\eta}^T W^T X_i\}^T \\ &= S_Y - G S_{G^TY,W^TX} S_{W^TY}^{-1} S_{W^TX,Y} \\ &\quad - S_{Y,W^TX} S_{W^TX}^{-1} S_{W^TX,G^TY} G^T + G S_{G^TY,W^TX} S_{W^TX}^{-1} S_{W^TX,G^TY} G^T. \end{aligned}$$

Let $M = S_{Y,W^TX} S_{W^TX}^{-1} S_{W^TX,Y}$ so that $S_Y - M = S_{Y|W^TX}$. Then

$$\begin{aligned} S_{\text{res}} &= S_Y - P_G M - M P_G + P_G M P_G \\ &= S_Y - P_G M Q_G - Q_G M P_G - P_G M P_G. \end{aligned}$$

We next expand

$$S_Y = P_G S_Y P_G + Q_G S_Y P_G + P_G S_Y Q_G + Q_G S_Y Q_G,$$

substitute into S_{res} and rearrange terms to get

$$\begin{aligned} S_{\text{res}} &= P_G(S_Y - M)P_G + Q_G(S_Y - M)P_G + P_G(S_Y - M)Q_G + Q_G S_Y Q_G \\ &= G S_{G^TY|W^TX} G^T + Q_G(S_Y - M)P_G + P_G(S_Y - M)Q_G + G_0^T S_{G_0^TY} G_0 \\ &= \widehat{\Sigma}_{Y|X} + Q_G S_{Y|W^TX} P_G + P_G S_{Y|W^TX} Q_G. \end{aligned}$$

□

A.5.4 Justification of the two-block algorithm, Section 5.4.3

Here we give a justification of the two-block algorithm as described in Table 5.2. We start with the data-based algorithm from Weglin (2000, Section 4.2, PLS-W2A) from which the population version will follow. Reall that $\mathbb{X}_{n\times p}$ and $\mathbb{Y}_{n\times r}$ are the matrices of centered predictor and response data.

1. $k \leftarrow 1$.
2. $\mathbb{X}^{(1)} \leftarrow \mathbb{X}$ and $\mathbb{Y}^{(1)} \leftarrow \mathbb{Y}$.
3. Compute the first singular vectors of $(\mathbb{X}^{(k)})^T\mathbb{Y}^{(k)}$, u_k left and v_k right.
4. $\xi_k \leftarrow \mathbb{X}^{(k)} u_k$ and $\omega_k \leftarrow \mathbb{Y}^{(k)} v_k$.
5. Regress $\mathbb{X}^{(k)}$ on ξ_k and $\mathbb{Y}^{(k)}$ on ω_k obtaining first order approximations $\widehat{\mathbb{X}}^{(k)} = \xi_k(\xi_k^T\xi_k)^{-1}\xi_k^T\mathbb{X}^{(k)}$ and $\widehat{\mathbb{Y}}^{(k)} = \omega_k(\omega_k^T\omega_k)^{-1}\omega_k^T\mathbb{Y}^{(k)}$.

6. Let $S_{X^{(k)}} = \mathbb{X}^{(k)T}\mathbb{X}^{(k)}/n$ and $S_{Y^{(k)}} = \mathbb{Y}^{(k)T}\mathbb{Y}^{(k)}/n$. Construct the residuals:

$$\begin{aligned}
\mathbb{X}^{(k+1)} \leftarrow \mathbb{X}^{(k)} - \widehat{\mathbb{X}}^{(k)} &= \mathbb{X}^{(k)} - \xi_k(\xi_k^T\xi_k)^{-1}\xi_k^T\mathbb{X}^{(k)} \\
&= \mathbb{X}^{(k)} - \mathbb{X}^{(k)}u_k(u_k^T\mathbb{X}^{(k)T}\mathbb{X}^{(k)}u_k)^{-1}u_k^T\mathbb{X}^{(k)T}\mathbb{X}^{(k)} \\
&= \mathbb{X}^{(k)}Q_{u_k(S_{X^{(k)}})} \\
\mathbb{Y}^{(k+1)} \leftarrow \mathbb{Y}^{(k)} - \widehat{\mathbb{Y}}^{(k)} &= \mathbb{Y}^{(k)} - \omega_k(\omega_k^T\omega_k)^{-1}\omega_k^T\mathbb{Y}^{(k)} \\
&= \mathbb{Y}^{(k)} - \mathbb{Y}^{(k)}v_k(v_k^T\mathbb{Y}^{(k)T}\mathbb{Y}^{(k)}v_k)^{-1}v_k^T\mathbb{Y}^{(k)T}\mathbb{Y}^{(k)} \\
&= \mathbb{Y}^{(k)}Q_{v_k(S_{Y^{(k)}})}.
\end{aligned}$$

7. If $(\mathbb{X}^{(k+1)})^T\mathbb{Y}^{(k+1)} = 0$ stop and k is the rank of the PLS model. Otherwise, $k \leftarrow k+1$ and go to step 3.

Now, if $k = 1$, u_1 and v_1 are the first left and right singular values of $S_{X,Y}$ from the initialization step in Table 5.2. Then $\xi_1 = \mathbb{X}^{(1)}u_1$, $\omega_1 = \mathbb{Y}^{(1)}v_1$ and

$$\begin{aligned}
\mathbb{X}^{(2)} &= \mathbb{X}^{(1)}Q_{u_1(S_{X^{(1)}})} \\
S_{X^{(2)}} &= (\mathbb{X}^{(2)})^T\mathbb{X}^{(2)}/n = Q_{u_1(S_{X^{(1)}})}^T S_{X^{(1)}} Q_{u_1(S_{X^{(1)}})} \\
\mathbb{Y}^{(2)} &= \mathbb{Y}^{(1)}Q_{v_1(S_{Y^{(1)}})} \\
S_{Y^{(2)}} &= (\mathbb{Y}^{(2)})^T\mathbb{Y}^{(2)}/n = Q_{v_1(S_{Y^{(1)}})}^T S_{Y^{(1)}} Q_{v_1(S_{Y^{(1)}})} \\
S_{X^{(2)},Y^{(2)}} &= (\mathbb{X}^{(2)})^T\mathbb{Y}^{(2)}/n = Q_{u_1(S_{X^{(1)}}))}^T S_{X^{(1)},Y^{(1)}} Q_{v_1(S_{Y^{(1)}})}.
\end{aligned} \tag{A.27}$$

Now, consider u_2 and v_2, the first left and right singular vectors of $S_{X^{(2)},Y^{(2)}}$, and

$$\begin{aligned}
\mathbb{X}^{(3)} &= \mathbb{X}^{(2)}Q_{u_2(S_{X^{(2)}})} \\
&= \mathbb{X}^{(1)}Q_{u_1(S_{X^{(1)}})}Q_{u_2(S_{X^{(2)}})} \\
&= \mathbb{X}^{(1)}Q_{(u_1,u_2)(S_{X^{(1)}})},
\end{aligned}$$

where the last step follows from the form of $S_{X^{(2)}}$ and Lemma 3.5. Similarly, $\mathbb{Y}^{(3)} = \mathbb{Y}^{(1)}Q_{(v_1,v_2)(S_{Y^{(1)}})}$.

We next compute first left and right singular vectors of

$$\begin{aligned}
S_{X^{(2)},Y^{(2)}} &= (\mathbb{X}^{(3)})^T\mathbb{Y}^{(3)}/n = n^{-1}Q_{(u_1,u_2)(S_{X^{(1)}})}^T(\mathbb{X}^{(1)})^T\mathbb{Y}^{(1)}Q_{(v_1,v_2)(S_{Y^{(1)}})} \\
&= Q_{(u_1,u_2)(S_{X^{(1)}})}^T S_{X,Y} Q_{(v_1,v_2)(S_{Y^{(1)}})}.
\end{aligned} \tag{A.28}$$

We seen then that when $S_X = S_{X^{(1)}}$, $S_{X,Y}S_{X^{(1)},Y^{(1)}}$ and $S_{Y^{(1)}}$ are replaced with their population versions, the initialization and steps $k = 1, 2$ of this algorithm correspond to the population version in Table 5.2. Subsequent steps can be justified in the same way by induction.

A.6 Proofs for Chapter 6

A.6.1 Justification for the equivalence of (6.5) and (6.6)

RESTATEMENT. (6.5),

$$(a)\; Y \perp\!\!\!\perp X_1 \mid P_{\mathcal{S}}X_1, X_2 \text{ and (b) } P_{\mathcal{S}}\mathrm{var}(X_1 \mid X_2)Q_{\mathcal{S}} = 0,$$

is equivalent to (6.6),

$$(a)\; R_{Y|2} \perp\!\!\!\perp R_{1|2} \mid P_{\mathcal{S}}R_{1|2}, X_2 \text{ and (b) } P_{\mathcal{S}}\mathrm{var}(R_{1|2} \mid X_2)Q_{\mathcal{S}} = 0.$$

The equivalence of (6.5) and (6.6) relies on Proposition 4.4 from Cook (1998), which states that for U, V and W random vectors, $U \perp\!\!\!\perp V \mid W$ if and only if $U \perp\!\!\!\perp (V, W) \mid W$. From this we have that (6.5a) holds if and only if

$$(Y, X_2) \perp\!\!\!\perp (X_1, X_2) \mid P_{\mathcal{S}}X_1, X_2.$$

Since (Y, X_2) is a one-to-one function of $(R_{Y|2}, X_2)$ and (X_1, X_2) is a one-to-one function of $(R_{1|2}, X_2)$, this last statement holds if and only if

$$(R_{Y|2}, X_2) \perp\!\!\!\perp (R_{1|2}, X_2) \mid P_{\mathcal{S}}X_1, X_2.$$

Applying Proposition 4.4 from Cook (1998) again, the last statement holds if and only if

$$R_{Y|2} \perp\!\!\!\perp R_{1|2} \mid P_{\mathcal{S}}X_1, X_2.$$

We obtain (6.6a) by applying the same logic to the conditioning argument, $(P_{\mathcal{S}}X_1, X_2)$, and to (6.5b).

A.6.2 Derivation of the MLEs for the partial envelope, Section 6.2.3

Recall from Section 6.2.3 that the partial envelope model is

$$\begin{aligned} X_2 &\sim N(0, \Sigma_2) \\ X_1 \mid X_2 &\sim N(\beta_{1|2}^T X_2, \Phi\Omega\Phi^T + \Phi_0\Omega_0\Phi_0^T) \\ Y \mid (X_1, X_2) &\sim N(\eta^T\Phi^T X_1 + \beta_2^T X_2, \sigma_{Y|X}^2), \end{aligned}$$

where Y, X_1, and X_2 are taken to be centered without loss of generality. Recall also that the corresponding log likelihood is

$$\begin{aligned}\log L &= -\frac{n}{2}\log|\Sigma_2| - \frac{1}{2}\mathrm{tr}\sum_{i=1}^{n} X_{2i}^T\Sigma_2^{-1}X_{2i} - \frac{n}{2}\log|\Omega| - \frac{n}{2}\log|\Omega_0| \\ &\quad -\frac{1}{2}\sum_{i=1}^{n}(X_{1i} - \beta_{1|2}^T X_{2i})^T(\Phi\Omega^{-1}\Phi^T + \Phi_0\Omega_0^{-1}\Phi_0^T)(X_{1i} - \beta_{1|2}^T X_{2i}) \\ &\quad -\frac{n}{2}\log\sigma_{Y|X}^2 - \frac{1}{2\sigma_{Y|X}^2}\sum_{i=1}^{n}(y_i - \eta^T\Phi^T X_{1i} - \beta_2^T X_{2i})^2.\end{aligned}$$

The MLE of $\beta_{1|2}$ is $\widehat{\beta}_{1|2} = S_2^{-1}S_{2,1}$ and then $R_{1|2i}$ is the i-th residual from the fit of X_1 on X_2,

$$\begin{aligned}R_{1|2i} &= X_{1i} - \widehat{\beta}_{1|2}^T X_{2i} \\ S_{1|2} &= n^{-1}\sum_{i=1}^{n} R_{1|2i}R_{1|2i}^T \\ &= S_1 - S_{1,2}S_2^{-1}S_{2,1}.\end{aligned}$$

The MLE of Σ_2 is S_2. Substituting these we get the partially maximized log likelihood

$$\begin{aligned}\log L_1 &= -\frac{n}{2}\log|S_2| - \frac{np_2}{2} - \frac{n}{2}\log|\Omega| - \frac{n}{2}\log|\Omega_0| \\ &\quad -\frac{1}{2}\sum_{i=1}^{n} R_{1|2i}^T\Phi\Omega^{-1}\Phi^T R_{1|2i} - \frac{1}{2}\sum_{i=1}^{n} R_{1|2i}^T\Phi_0\Omega_0^{-1}\Phi_0^T R_{1|2i} \\ &\quad -\frac{n}{2}\log\sigma_{Y|X}^2 - \frac{1}{2\sigma_{Y|X}^2}\sum_{i=1}^{n}(y_i - \eta^T\Phi^T X_{1i} - \beta_2^T X_{2i})^2 \\ &= -\frac{n}{2}\log|S_2| - \frac{np_2}{2} - \frac{n}{2}\log|\Omega| - \frac{n}{2}\log|\Omega_0| \\ &\quad -\frac{n}{2}\sum_{i=1}^{n}\mathrm{tr}\left\{S_{1|2}\Phi\Omega^{-1}\Phi^T\right\} - \frac{n}{2}\sum_{i=1}^{n}\mathrm{tr}\left\{S_{1|2}\Phi_0\Omega_0^{-1}\Phi_0^T\right\} \\ &\quad -\frac{n}{2}\log\sigma_{Y|X}^2 - \frac{1}{2\sigma_{Y|X}^2}\sum_{i=1}^{n}(y_i - \eta^T\Phi^T X_{1i} - \beta_2^T X_{2i})^2.\end{aligned}$$

From this we have the MLEs $\widehat{\Omega} = \Phi^T S_{1|2}\Phi$, $\widehat{\Omega}_0 = \Phi_0^T S_{1|2}\Phi_0$ and thus the next partially maximized log likelihood is

$$
\begin{aligned}
\log L_2 &= -\frac{n}{2}\log|S_2| - \frac{np_2}{2} \\
&\quad -\frac{n}{2}\log|\Phi^T S_{1|2}\Phi| - \frac{n}{2}\log|\Phi_0^T S_{1|2}\Phi_0| - \frac{np_1}{2} \\
&\quad -\frac{n}{2}\log\sigma^2_{Y|X} - \frac{1}{2\sigma^2_{Y|X}}\sum_{i=1}^{n}(y_i - \eta^T\Phi^T X_{1\,i} - \beta_2^T X_{2\,i})^2.
\end{aligned}
$$

The next step is to determine and substitute the values of η and β_2 that maximize the log likelihood with Φ and σ^2 held fixed. To facilitate this we orthogonalize the terms in the sum. To do this we use the coefficients $\widehat{\beta}_{1|2}\Phi$ from the OLS fit of $\Phi^T X_{1\,i}$ on $X_{2\,i}$. Let $SS = \sum_{i=1}^n (y_i - \eta^T\Phi^T X_{1\,i} - \beta^T X_{2\,i})^2$. Then write

$$
\begin{aligned}
SS &= \sum_{i=1}^{n}(y_i - \eta^T(\Phi^T X_{1i} - \Phi^T\widehat{\beta}_{1|2}^T X_{2\,i} + \Phi^T\widehat{\beta}_{1|2}^T X_{2\,i}) - \beta_2^T X_{2\,i})^2 \\
&= \sum_{i=1}^{n}(y_i - \eta^T\Phi^T(X_{1i} - \widehat{\beta}_{1|2}^T X_{2\,i}) - \eta^T\Phi^T\widehat{\beta}_{1|2}^T X_{2\,i} - \beta_2^T X_{2\,i})^2 \\
&= \sum_{i=1}^{n}(y_i - \eta^T\Phi^T R_{1|2\,i} - \beta_2^{*T} X_{2\,i})^2,
\end{aligned}
$$

where $R_{1|2i} = X_{1i} - \widehat{\beta}_{1|2}^T X_{2\,i}$ and $\beta_2^* = \beta_2 + \widehat{\beta}_{1|2}\Phi\eta$. Since $\sum_{i=1}^n R_{1|2i}X_{2i}^T = 0$, we can fit the two terms in the last sum separately, getting straightforwardly

$$
\begin{aligned}
n^{-1}\sum_{i=1}^{n}\Phi^T R_{1|2i}Y_i &= \Phi^T\{S_{1Y} - S_{1,2}S_2^{-1}S_{2,Y}\} \\
S_{1|2} &= S_1 - S_{1,2}S_2^{-1}S_{2,1} \\
\eta &= (\Phi^T S_{1|2}\Phi)^{-1}\Phi^T S_{R_{1|2},Y} \\
S_{R_{1|2},Y} &= S_{1Y} - S_{1,2}S_2^{-1}S_{2,Y} \\
\beta_2^* &= S_2^{-1}S_{2,Y}
\end{aligned}
$$

as the values of η and β_2^* that minimize the sum. Substituting these values and simplifying we get the next partially maximized log likelihood

$$
\begin{aligned}
\log L_3 &= -\frac{n}{2}\log|S_2| - \frac{np}{2} \\
&\quad -\frac{n}{2}\log|\Phi^T S_{1|2}\Phi| - \frac{n}{2}\log|\Phi_0^T S_{1|2}\Phi_0| \\
&\quad -\frac{n}{2}\log\sigma^2_{Y|X} - \frac{n}{2\sigma^2_{Y|X}}\{S_{Y|2} - S_{R_{1|2},Y}^T\Phi[\Phi^T S_{1|2}\Phi]^{-1}\Phi^T S_{R_{1|2},Y}\}.
\end{aligned}
$$

This is maximized at the value

$$\sigma^2_{Y|X} = S_{Y|2} - S^T_{R_{1|2},Y}\Phi[\Phi^T S_{1|2}\Phi]^{-1}\Phi^T S_{R_{1|2},Y},$$

giving the next partially maximized log likelihood

$$\begin{aligned}\log L_4 &= -\frac{n}{2}\log|S_2| - \frac{n(p+1)}{2} \\ &\quad -\frac{n}{2}\log|\Phi^T S_{1|2}\Phi| - \frac{n}{2}\log|\Phi_0^T S_{1|2}\Phi_0| \\ &\quad -\frac{n}{2}\log\left\{S_{Y|2} - S^T_{R_{1|2},Y}\Phi[\Phi^T S_{1|2}\Phi]^{-1}\Phi^T S_{R_{1|2},Y}\right\}.\end{aligned}$$

Writing the third and fifth terms together

$$\begin{aligned}\log L_4 &= -\frac{n}{2}\log|S_2| - \frac{n(p+1)}{2} - \frac{n}{2}\log|\Phi_0^T S_{1|2}\Phi_0| \\ &\quad -\frac{n}{2}\log|\Phi^T S_{1|2}\Phi|\left\{S_{Y|2} - S^T_{R_{1|2},Y}\Phi[\Phi^T S_{1|2}\Phi]^{-1}\Phi^T S_{R_{1|2},Y}\right\}.\end{aligned}$$

Let

$$T_4 = \log|\Phi^T S_{1|2}\Phi|\left\{S_{Y|2} - S^T_{R_{1|2},Y}\Phi[\Phi^T S_{1|2}\Phi]^{-1}\Phi^T S_{R_{1|2},Y}\right\}.$$

For clarity we use det rather than $|\cdot|$ to denote the determinant operator in the remainder of this proof. Then

$$\begin{aligned}T_4 &= \det\begin{pmatrix}\Phi^T S_{1|2}\Phi & \Phi^T S_{R_{1|2},Y} \\ S^T_{R_{1|2},Y}\Phi & S_{Y|2}\end{pmatrix} \\ &= \det\left\{\Phi^T S_{1|2}\Phi - \Phi^T S_{R_{1|2},Y}S^{-1}_{Y|2}S^T_{R_{1|2},Y}\Phi\right\} \\ &= \det\left\{\Phi^T[S_{1|2} - S_{R_{1|2},Y}S^{-1}_{Y|2}S^T_{R_{1|2},Y}]\Phi\right\}.\end{aligned}$$

Ignoring constant terms, the overall objective function to minimize becomes

$$F(\Phi) = \det\left\{\Phi^T[S_{1|2} - S_{R_{1|2},Y}S^{-1}_{Y|2}S^T_{R_{1|2},Y}]\Phi\right\} + \det\left(\Phi^T S^{-1}_{1|2}\Phi\right).$$

Since $R_{1|2}$ is uncorrelated in the sample with X_2, we have $S_{R_{1|2},Y} = S_{R_{1|2},R_{Y|2}}$ and so

$$F(\Phi) = \det\left\{\Phi^T[S_{1|2} - S_{R_{1|2},R_{Y|2}}S^{-1}_{Y|2}S^T_{R_{1|2}R_{Y|2}}]\Phi\right\} + \det\left(\Phi^T S^{-1}_{1|2}\Phi\right).$$

Thus this corresponds to defining the response as $R_{Y|2i}$ and predictor as $R_{1|2i}$ and then run the usual envelope or PLS for the regression of $R_{Y|2i}$ on $R_{1|2i}$. The asymptotic variances under predictor reduction should apply to this context.

Finally, recognizing that

$$S_{1|Y,2} = S_{1|2} - S_{R_{1|2},R_{Y|2}} S_{Y|2}^{-1} S_{R_{1|2}R_{Y|2}}^T$$

gives the objective function at (6.9). The estimators are obtained by piecing together the various parameter functions that maximize the log likelihood.

A.7 Proofs for Chapter 9

A.7.1 Proposition 9.1

RESTATEMENT. The following statements are equivalent.

(i) $Y \perp\!\!\!\perp \mathrm{E}(Y \mid X) \mid \alpha^T X$,

(ii) $\mathrm{cov}\{(Y, \mathrm{E}(Y \mid X)) \mid \alpha^T X\} = 0$,

(iii) $\mathrm{E}(Y \mid X)$ is a function of $\alpha^T X$.

Proof. That (i) implies (ii) and (iii) implies (i) are immediate because if $E(Y \mid X)$ is a function of $\alpha^T X$ then given $\alpha^T X$ the mean function is a constant. It remains to show that (ii) implies (iii). The idea is to show that (ii) implies that $\mathrm{var}[E(Y \mid X) \mid \alpha^T X] = 0$, from which it follows that $E(Y|X)$ is constant given $\alpha^T X$. From (ii),

$$E[YE(Y|X) \mid \alpha^T X] = E(Y \mid \alpha^T X) E\{E(Y \mid X) \mid \alpha^T X\}.$$

The left hand side is

$$\begin{aligned} E[YE(Y|X) \mid \alpha^T X] &= E[E\{YE(Y \mid X) \mid X\} \mid \alpha^T X] \\ &= E[E(Y \mid X)E(Y \mid X) \mid \alpha^T X] \\ &= E[\{E(Y \mid X)\}^2 \mid \alpha^T X]. \end{aligned}$$

The right hand side is

$$E(Y \mid \alpha^T X) E\{E(Y \mid X) \mid \alpha^T X\} = \{E[E(Y \mid X) \mid \alpha^T X]\}^2.$$

Consequently,

$$E[\{E(Y \mid X)\}^2 \mid \alpha^T X] = \{E[E(Y \mid X) \mid \alpha^T X]\}^2,$$

erom which it follows that $\mathrm{var}[E(Y \mid X) \mid \alpha^T X] = 0$. □

A.7.2 Proposition 9.2

RESTATEMENT. Assume that W satisfies the linearity condition relative to $\mathcal{S}$ under Definition 9.3. Then

1. $M = \Sigma_W \alpha(\alpha^T \Sigma_W \alpha)^{-1}$.

2. M^T is a generalized inverse of α.

3. αM^T is the orthogonal projection operator for $\mathcal{S}$ relative to the Σ_W inner product.

4. $$E(W \mid \alpha^T W) - E(W) = P^T_{\mathcal{S}(\Sigma_W)}(W - E(W)),$$
   where for completeness we have allowed $E(W)$ to be non-zero.

5. If $\mathcal{S}$ reduces Σ_W then $P_{\mathcal{S}(\Sigma_W)} = P_{\mathcal{S}}$.

Proof. To see conclusion 1, we have $E(W \mid \alpha^T W) = M\alpha^T W$. Multiplying both sides on the right by $W^T\alpha$ and taking expectations we get for the left hand side

$$\begin{aligned} E\{E(W \mid \alpha^T W)W^T\alpha\} &= E\{E(WW^T\alpha \mid \alpha^T W)\} \\ &= E(WW^T\alpha) \\ &= \Sigma_W \alpha, \end{aligned}$$

and for the right hand side

$$E(M\alpha^T WW^T\alpha) = M\alpha^T\Sigma_W\alpha.$$

Consequently,

$$\Sigma_W\alpha = M\alpha^T\Sigma_W\alpha$$

and conclusion 1 follows. Conclusions 2–4 follow straightforwardly from conclusion 1 and the definitions of a generalized inverse and projection. Conclusion 5 follows from Proposition 1.3. □

A.7.3 Proposition 9.3

RESTATEMENT.

Assume that the distribution of $X \mid Y$ satisfies the linearity condition relative to $\mathcal{S}_{Y|X}$ for each value of Y. Then the marginal distribution of X also satisfies the linearity condition relative to $\mathcal{S}_{Y|X}$.

Proof. Let α be a basis matrix for $\mathcal{S}_{Y|X}$. The random vector $X - E(X|Y = y)$ has mean 0 for each value y of Y. Then by hypothesis and Definition 9.3 we have for each value of Y

$$\mathrm{E}(X \mid \alpha^T X, Y = y) - \mathrm{E}(X \mid Y = y) = M_Y \alpha^T (X - \mathrm{E}(X \mid Y = y)),$$

where M_Y is now permitted to depend on the value of Y. This implies algebraically that

$$\mathrm{E}(X \mid \alpha^T X, Y = y) = M_Y \alpha^T X - M_Y \alpha^T \mathrm{E}(X \mid Y = y) + \mathrm{E}(X \mid Y = y).$$

Since α is a basis for $\mathcal{S}_{Y|X}$, the left hand side does not depend on the value of Y. This implies that $\mathrm{E}(X \mid \alpha^T X, Y = y) = \mathrm{E}(X \mid \alpha^T X)$ and thus that

$$\mathrm{E}(X \mid \alpha^T X) - \mathrm{E}(X \mid Y = y) = M_Y \alpha^T (X - \mathrm{E}(X \mid Y = y)).$$

Taking the expectation with respect to the marginal distribution of Y we have

$$\mathrm{E}(X \mid \alpha^T X) - \mu_X = \mathrm{E}(M_Y) \alpha^T X - \mathrm{E}(M_Y \alpha^T \mathrm{E}(X \mid Y)). \tag{A.29}$$

The left and right hand sides are a function of only $\alpha^T X$. Since the expectation of the left hand side in $\alpha^T X$ is 0, the expectation of the right hand side must also be 0,

$$\mathrm{E}(M_Y \alpha^T \mathrm{E}(X \mid Y)) = \mathrm{E}(M_Y) \alpha^T \mu_X.$$

Replacing the second term on the right hand side of (A.29) with $\mathrm{E}(M_Y)\alpha^T \mu_X$, it follows that

$$\mathrm{E}(X \mid \alpha^T X) - \mu_X = \mathrm{E}(M_Y) \alpha^T (X - \mu_X).$$

□

A.7.4 Proposition 9.5

RESTATEMENT. Assume the linearity condition for X relative to $\mathcal{S}_{\mathrm{E}(Y|X)}$. Then under the non-linear model (9.1) we have

$$\mathcal{E}_{\Sigma_X}(\mathcal{B}) \subseteq \mathcal{E}_{\Sigma_X}(\mathcal{S}_{\mathrm{E}(Y|X)}).$$

Moreover, if the single index model holds and $\Sigma_{X,Y} \neq 0$ then $\mathcal{E}_{\Sigma_X}(\mathcal{B}) = \mathcal{E}_{\Sigma_X}(\mathcal{S}_{\mathrm{E}(Y|X)})$.

Proof. We know from Proposition 9.4 and Corollary 9.2 that $\mathcal{B} \subseteq \mathcal{S}_{\mathrm{E}(Y|X)}$ with equality under the single index model (9.2) with $\Sigma_{X,Y} \neq 0$. Let A_1 be semi-orthogonal basis matrix for $\mathcal{B}$ and extent it so that (A_1, A_2) is a semi-orthogonal basis matrix for $\mathcal{S}_{\mathrm{E}(Y|X)}$. Then $\mathcal{S}_{\mathrm{E}(Y|X)} = \mathrm{span}((A_1, A_2)) = \mathrm{span}(A_1) + \mathrm{span}(A_2)$. Applying Proposition 1.7 with $M = \Sigma_X$, $S_1 = \mathrm{span}(A_1) = \mathcal{B}$ and $S_2 = \mathrm{span}(A_2)$, we have $\mathcal{S}_{\mathrm{E}(Y|X)} = \mathcal{E}_{\Sigma_X}(\mathcal{B}) + \mathcal{E}_{\Sigma_X}(\mathrm{span}(A_2)$, which implies the desired conclusion, $\mathcal{E}_{\Sigma_X}(\mathcal{B}) \subseteq \mathcal{E}_{\Sigma_X}(\mathcal{S}_{\mathrm{E}(Y|X)})$.

When the single index model holds, $\mathcal{B} = \mathcal{S}_{\mathrm{E}(Y|X)}$, which implies equality $\mathcal{S}_{\mathrm{E}(Y|X)} = \mathcal{E}_{\Sigma_X}(\mathcal{B})$.

□

A.8 Proofs for Chapter 10

A.8.1 Model from Section 10.2

Lemma A.5. *From (10.1) and (10.2), we have*

$$\begin{pmatrix} X \\ Y \\ \xi \\ \eta \end{pmatrix} \sim \begin{pmatrix} \mu_X \\ \mu_Y \\ \mu_\xi \\ \mu_\eta \end{pmatrix} + \epsilon_{X,Y,\xi,\eta}$$

where $\mathrm{var}(\epsilon_{X,Y,\xi,\eta}) = \Sigma_{(X,Y,\xi,\eta)}$ *with*

$$\Sigma_{(X,Y,\xi,\eta)} = \begin{pmatrix} \Sigma_{X|\xi} + \beta_{X|\xi}\beta_{X|\xi}^T\sigma_\xi^2 & \beta_{X|\xi}\beta_{Y|\eta}^T\sigma_{\xi,\eta} & \beta_{X|\xi}\sigma_\xi^2 & \beta_{X|\xi}\sigma_{\xi,\eta} \\ \cdots & \Sigma_{Y|\eta} + \beta_{Y|\eta}\beta_{Y|\eta}^T\sigma_\eta^2 & \beta_{Y|\eta}\sigma_{\xi,\eta} & \beta_{Y|\eta}\sigma_\eta^2 \\ \cdots & \cdots & \sigma_\xi^2 & \sigma_{\xi,\eta} \\ \cdots & \cdots & \cdots & \sigma_\eta^2 \end{pmatrix}. \tag{A.30}$$

Proof. The first two elements of the diagonal are direct consequence of the fact that $\text{cov}(Z) = E(\text{cov}(Z|H)) + \text{var}(E(Z|H))$. For $\Sigma_{X,\xi}$ and $\Sigma_{Y,\eta}$ we use the fact that $\beta_{X|\xi} = \sigma_\xi^{-2}\Sigma_{X,\xi}$ and $\beta_{Y|\eta} = \sigma_\eta^{-2}\Sigma_{Y,\eta}$. To compute $\Sigma_{Y,\eta}$ we use that

$$\begin{aligned}
\Sigma_{Y,\eta} &= E((Y-\mu_Y)(\eta-\mu_\eta)) \\
&= E_\eta[E((Y-\mu_Y)(\eta-\mu_\eta))|\eta] \\
&= \beta_{Y|\eta}E_\eta(\eta-\mu_\eta)^2 \\
&= \beta_{Y|\eta}\sigma_\eta^2.
\end{aligned}$$

Analogously for $\Sigma_{X,\xi}$. Now for $\Sigma_{X,Y}$ we use that

$$\begin{aligned}
\Sigma_{Y,X} &= E((Y-\mu_Y)(X-\mu_X)^T) \\
&= E_{\eta,\xi}[E_{Y,X|(\eta,\xi)}\{(Y-\mu_Y)(X-\mu_X)^T\}] \\
&= \beta_{Y|\eta}E_{\eta,\xi}(\eta-\mu_\eta)(\xi-\mu_\xi)\beta_{X|\xi}^T \\
&= \beta_{Y|\eta}\sigma_{\xi,\eta}\beta_{X|\xi}^T.
\end{aligned}$$

Now, for $\Sigma_{X\eta}$

$$\begin{aligned}
\Sigma_{X\eta} &= E((X-\mu_X)(\eta-\mu_\eta)) \\
&= E_{\eta,\xi}E((X-\mu_X)(\eta-\mu_\eta)|(\eta,\xi)) \\
&= \beta_{X|\xi}E_{\eta,\xi}E((\xi-\mu_\xi)(\eta-\mu_\eta)|(\eta,\xi)) \\
&= \beta_{X|\xi}\sigma_{\xi,\eta}.
\end{aligned}$$

The proof of $\Sigma_{Y,\xi}$ is analogous. □

A.8.2 Constraints in CB|SEM

Proposition A.3. *The SEM models under the marginal and regression constraints are*

- *Marginal constraints,*

$$\begin{pmatrix} X \\ Y \end{pmatrix} \sim \begin{pmatrix} \mu_X \\ \mu_Y \end{pmatrix} + \epsilon_{X,Y}$$

where

$$\text{var}(\epsilon_{X,Y}) := \Sigma_{(X,Y)} = \begin{pmatrix} D_{X|\xi} + c^2BB^T & BA^T \\ AB^T & D_{Y|\eta} + d^2AA^T \end{pmatrix}. \quad \text{(A.31)}$$

and $c^2d^2\text{cor}^2(\eta,\xi) = 1$.

- *Regression constraints*

$$\begin{pmatrix} X \\ Y \end{pmatrix} \sim \begin{pmatrix} \mu_X \\ \mu_Y \end{pmatrix} + \epsilon_{X,Y}$$

From (A.43)

$$\Sigma_{(X,Y)} = \begin{pmatrix} D_{X|\xi} + (B^T D_{X|\xi}^{-1} B)^{-1} BB^T/(\sigma_\xi^2 - 1) \\ AB^T \\ BA^T \\ D_{Y|\eta} + (A^T D_{Y|\eta}^{-1} A)^{-1} AA^T/(\sigma_\eta^2 - 1) \end{pmatrix}$$

where $\mathrm{cor}^2(\eta, \xi)(\sigma_\eta^2 - 1)^{-1}(A^T D_{Y|\eta}^{-1} A)^{-1}(\sigma_\xi^2 - 1)^{-1}(B^T D_{X|\xi}^{-1} B)^{-1} = 1.$

Proof. For (A.31) we only need to prove $\Sigma_X = D_{X|\xi} + BB^T c^2$ and $\Sigma_Y = D_{Y|\eta} + AA^T d^2$ with $c^2 d^2 \mathrm{cor}(\eta, \xi) = 1$.

Now, $\Sigma_X = D_{X|\xi} + \Sigma_{X,\xi}\Sigma_{\xi,X}$ and $\Sigma_Y = D_{Y|\eta} + \Sigma_{Y,\eta}\Sigma_{\eta,Y}$. And since $\Sigma_{Y,X} = AB^T = \Sigma_{Y,\eta}\mathrm{cor}(\eta, \xi)\Sigma_{\xi,X}$ we have that $\Sigma_{Y,\eta} = Ad$ and $\Sigma_{X,\xi} = Bc$ with $dc \times \mathrm{cor}(\eta, \xi) = 1$ from what follows (A.31). Then the conclusion for the regression constraints follows from the proof of Lemma A.6. □

A.8.3 Lemma 10.1

RESTATEMENT. For model presented in Section 10.2 and without imposing either the regression or marginal constraints, we have $E(Y \mid X) = \mu_Y + \beta_{Y|X}(X - \mu_X)$, where $\beta_{Y|X} = AB^T\Sigma_X^{-1}$, and $A \in \mathbb{R}^{r\times 1}$ and $B \in \mathbb{R}^{p\times 1}$ are such that $\Sigma_{Y,X} = AB^T$.

Proof. This is direct consequence of Lemma A.5 since $\Sigma_{Y,X} = \beta_{Y|\eta}\sigma_{\xi,\eta}\beta_{X|\xi}^T = \Sigma_{Y,\eta}\sigma_\eta^{-2}\sigma_{\xi,\eta}\sigma_\xi^{-2}\Sigma_{\xi,X}$ and therefore rank of $\Sigma_{Y,X}$ is one and the lemma follows. □

A.8.4 Proposition 10.1

RESTATEMENT. The parameters Σ_X, Σ_Y, $\Sigma_{Y,X} = AB^T$, $\beta_{Y|X}$, μ_Y and μ_X are identifiable in the reduced rank model of Lemma 10.1, but A and B are not. Additionally, under the regression constraints, the quantities $\sigma_{\xi,\eta}/(\sigma_\eta^2\sigma_\xi^2)$,

$cor\{E(\xi|X), E(\eta|Y)\}$, $E(\xi|X)$, $E(\eta|Y)$, $\Sigma_{X,\xi}$ and $\Sigma_{Y,\eta}$ are identifiable in the reflexive model except for sign, while $\mathrm{cor}(\eta,\xi)$, σ_ξ^2, σ_η^2, $\sigma_{\xi,\eta}$, $\beta_{X|\xi}$, $\beta_{Y|\eta}$, $\Sigma_{Y|\eta}$ and $\Sigma_{X|\xi}$ are not identifiable. Moreover,

$$|\mathrm{cor}\{E(\xi|X), E(\eta|Y)\}| = |\sigma_{\xi,\eta}|/(\sigma_\eta^2\sigma_\xi^2) = \mathrm{tr}^{1/2}(\beta_{X|Y}\beta_{Y|X}) \quad (A.32)$$
$$= \mathrm{tr}^{1/2}(\Sigma_{X,Y}\Sigma_Y^{-1}\Sigma_{Y,X}\Sigma_X^{-1}) \quad (A.33)$$

where $|\cdot|$ denotes the absolute value of its argument.

Proof. The first part follows from reduced rank model literature (see for example Cook, Forzani, and Zhang (2015)). Now, using (A.30) and the fact that $\mu_\eta = 0$,

$$E(\eta|Y) = \Sigma_{\eta,Y}\Sigma_Y^{-1}(Y - \mu_Y),$$

and therefore

$$\mathrm{var}(E(\eta|Y)) = \Sigma_{\eta,Y}\Sigma_Y^{-1}\Sigma_{\eta,Y}^T = 1, \quad (A.34)$$

where we use the hypothesis that $\mathrm{var}(E(\eta|Y)) = 1$ in the last equal. In the same way

$$\mathrm{var}(E(\xi|X)) = \Sigma_{\xi,X}\Sigma_X^{-1}\Sigma_{\xi,X}^T = 1. \quad (A.35)$$

Now, since

$$\Sigma_{Y,X} = AB^T = \Sigma_{Y,\eta}\sigma_\eta^{-2}\sigma_{\xi,\eta}\sigma_\xi^{-2}\Sigma_{\xi,X}, \quad (A.36)$$

we have for some m_η and m_ξ that

$$\Sigma_{Y,\eta} = Am_\eta, \quad (A.37)$$
$$\Sigma_{X,\xi} = Bm_\xi. \quad (A.38)$$

(A.36), (A.37), and (A.38) together makes $AB^T = Am_\eta\sigma_\eta^{-2}\sigma_{\xi,\eta}\sigma_\xi^{-2}m_\xi B^T$ and therefore

$$m_\eta\sigma_\eta^{-2}\sigma_{\xi,\eta}\sigma_\xi^{-2}m_\xi = 1. \quad (A.39)$$

Now, using (A.34) and (A.35) together with (A.37) and (A.38) we have that

$$m_\eta^2 = (A^T\Sigma_Y^{-1}A)^{-1} \quad (A.40)$$
$$m_\xi^2 = (B^T\Sigma_X^{-1}B)^{-1}. \quad (A.41)$$

Plugging this into (A.39),

$$\begin{aligned}\sigma_\eta^{-2}\sigma_{\xi,\eta}\sigma_\xi^{-2} &= (m_\xi m_\eta)^{-1}\\ &= (A^T\Sigma_Y^{-1}A)^{1/2}(B^T\Sigma_X^{-1}B)^{1/2}\\ &= (A^T\Sigma_Y^{-1}AB^T\Sigma_X^{-1}B)^{1/2}\\ &= \mathrm{tr}^{1/2}((AB^T)^T\Sigma_Y^{-1}AB^T\Sigma_X^{-1})\\ &= \mathrm{tr}^{1/2}(\beta_{Y|X}\beta_{X|Y}).\end{aligned}$$

Let us note that from (A.40) and (A.41), m_ξ^2 and m_η^2 are not unique, nevertheless $m_\eta^2 m_\xi^2$ is unique since any change of A and B should be such AB^T is the same. As a consequence $m_\eta m_\xi$ is unique except for a sign. And therefore $\sigma_\eta^{-2}\sigma_{\xi,\eta}\sigma_\xi^{-2}$ is unique except for a sign.

Now, coming back to equations (A.37) and (A.38) we have that

$$\begin{aligned}\Sigma_{Y,\eta} &= A(A^T\Sigma_Y^{-1}A)^{-1/2}\\ \Sigma_{X,\xi} &= B(B^T\Sigma_X^{-1}X)^{-1/2}\end{aligned}$$

and again are unique except by a sign.

Now, using again (A.30), the fact that $\mu_\eta = \mu_\xi = 0$ and $\mathrm{var}(E(\xi|X)) = \mathrm{var}(E(\eta|Y)) = 1$ we have

$$\begin{aligned}E(\eta|Y) &= \Sigma_{\eta,Y}\Sigma_Y^{-1}(Y-\mu_Y)\\ \sigma_\eta^2 &= E(\mathrm{var}(\eta|Y))+1\\ E(\xi|X) &= \Sigma_{\xi,X}\Sigma_X^{-1}(X-\mu_X)\\ \sigma_\xi^2 &= E(\mathrm{var}(\xi|X))+1.\end{aligned}$$

Replacing $\Sigma_{Y,X} = \Sigma_{Y,\eta}\sigma_\eta^{-2}\sigma_{\xi,\eta}\sigma_\xi^{-2}\Sigma_{\xi,X}$ we have

$$\begin{aligned}\mathrm{cov}\{E(\eta|Y),E(\xi|X)\} &= \Sigma_{\eta,Y}\Sigma_Y^{-1}\Sigma_{Y,X}\Sigma_X^{-1}\Sigma_{X,\xi}\\ &= \Sigma_{\eta,Y}\Sigma_Y^{-1}\Sigma_{Y,\eta}\sigma_\eta^{-2}\sigma_{\xi,\eta}\sigma_\xi^{-2}\Sigma_{\xi,X}\Sigma_X^{-1}\Sigma_{X,\xi}.\\ \mathrm{cor}\{E(\eta|Y),E(\xi|X)\} &= (\Sigma_{\eta,Y}\Sigma_Y^{-1}\Sigma_{Y,\eta})^{1/2}\sigma_\eta^{-2}\sigma_{\xi,\eta}\sigma_\xi^{-2}(\Sigma_{\xi,X}\Sigma_X^{-1}\Sigma_{X,\xi})^{1/2}\\ &= \sigma_\eta^{-2}\sigma_{\xi,\eta}\sigma_\xi^{-2}\\ &= (m_\eta m_\xi)^{-1}\\ &= (A^T\Sigma_Y^{-1}A)^{1/2}(B^T\Sigma_X^{-1}B)^{1/2}.\end{aligned}$$

Now, we will prove that σ_ξ and σ_η are not identifiable. For that, since σ_η and σ_ξ have to be greater than 1 and $\sigma_\eta\sigma_\xi$ should be constant we can change ξ by $C\xi$ and η by $C^{-1}\eta$ in such a way that $C\sigma_\xi > 1$ and $C^{-1}\sigma_\eta > 1$. We take any C such that $\sigma_\xi^{-1} < C < \sigma_\eta$ and none of the other parameters change. From this follows that $\sigma_{\xi,\eta}$ and $\beta_{X|\xi}$, $\beta_{Y|\eta}$ is not identifiable. It is left to prove that $\Sigma_{Y|\eta}$ and $\Sigma_{X|\xi}$ are not identifiable. For that we use $\Sigma_X = \Sigma_{X|\xi} + \Sigma_{X,\xi}\sigma_\xi^{-2}\Sigma_{\xi,X}$. If $\Sigma_{X|\xi}$ is identifiable, since Σ_X and $\Sigma_{X,\xi}$ are identifiable, σ_ξ is identifiable, which is a contradiction since we have already proven that it is not so. The same argument holds for $\Sigma_{Y|\eta}$.

□

Lemma A.6. *Under the reflexive model of Section 10.2 and the regression constraints,*

$$\begin{aligned}\Sigma_X &= \Sigma_{X|\xi} + \sigma_\xi^{-2}(B^T\Sigma_X^{-1}B)^{-1}BB^T &\quad (A.42)\\ &= \Sigma_{X|\xi} + \frac{1}{\sigma_\xi^2 - 1}(B^T\Sigma_{X|\xi}^{-1}B)^{-1}BB^T. &\quad (A.43)\\ \Sigma_Y &= \Sigma_{Y|\eta} + \sigma_\eta^{-2}(A^T\Sigma_Y^{-1}A)^{-1}AA^T &\quad (A.44)\\ &= \Sigma_{Y|\eta} + \frac{1}{\sigma_\eta^2 - 1}(A^T\Sigma_{Y|\eta}^{-1}A)^{-1}AA^T. &\quad (A.45)\end{aligned}$$

Proof. Expressions (A.42) and (A.44) are equivalent to (10.9) and (10.10) in Chapter 10.

By the covariance formula and the fact that by Proposition 10.1 we have $\Sigma_{X,\xi} = B(B^T\Sigma_X^{-1}B)^{-1/2}$ and $\Sigma_{Y,\eta} = A(A^T\Sigma_Y^{-1}A)^{-1/2}$ from where we get (A.42) and (A.44). Now, taking inverse and using the Woodbury inequality we have

$$B^T\Sigma_X^{-1}B = \sigma_\xi^2 \frac{B^T\Sigma_{X|\xi}^{-1}BB^T\Sigma_X^{-1}B}{\sigma_\xi^2 B^T\Sigma_X^{-1}B + B^T\Sigma_{X|\xi}^{-1}B}.$$

As a consequence

$$B^T\Sigma_X^{-1}B = B^T\Sigma_{X|\xi}^{-1}B\frac{\sigma_\xi^2 - 1}{\sigma_\xi^2}$$

and (A.43) follows replacing this into (A.42). The proof of (A.45) follows analogously. □

Proposition A.4. *Assume the regression constraints. If $\Sigma_{Y|\eta}$ and $\Sigma_{X|\xi}$ are identifiable then σ_ξ^2, σ_η^2, $|\sigma_{\xi,\eta}|$, $\beta_{X|\xi}$, $\beta_{Y|\eta}$, and* $\mathrm{cor}(\eta,\xi)$ *are identifiable, and*

$$\begin{aligned} \mathrm{cor}(\eta,\xi) &= \mathrm{cor}\{E(\eta|Y), E(\xi|X)\}\sigma_\xi\sigma_\eta \\ \sigma_\xi^2 &= \frac{H_\xi}{H_\xi - 1} \qquad \text{(A.46)} \\ \sigma_\eta^2 &= \frac{H_\eta}{H_\eta - 1}, \qquad \text{(A.47)} \end{aligned}$$

where $H_\xi = \Sigma_{\xi,X}\Sigma_{X|\xi}^{-1}\Sigma_{X,\xi}$ and $H_\eta = \Sigma_{\eta,Y}\Sigma_{Y|\eta}^{-1}\Sigma_{Y,\eta}$. Moreover, $\Sigma_{Y|\eta}$ and $\Sigma_{X|\xi}$ are identifiable if and only if σ_ξ^2, σ_η^2 are so.

Proof. By (A.42)–(A.45) and using Proposition 10.1 we have that σ_ξ and σ_η are identifiable and as a consequence the rest of the parameters are identifiable. Now, to prove (A.46) let us use the formula

$$\Sigma_X = \Sigma_{X|\xi} + \Sigma_{X,\xi}\sigma_\xi^{-2}\Sigma_{\xi,X}.$$

Using Woodbury inequality

$$\Sigma_X^{-1} = \Sigma_{X|\xi}^{-1} - \Sigma_{X|\xi}^{-1}\Sigma_{X,\xi}(\sigma_\xi^2 + \Sigma_{\xi,X}\Sigma_{X,\xi}^{-1}\Sigma_{X,\xi})^{-1}\Sigma_{\xi,X}\Sigma_{X|\xi}^{-1}. \qquad \text{(A.48)}$$

Using the fact that $\Sigma_{\xi,X}\Sigma_X^{-1}\Sigma_{X,\xi} = 1$ proven in Proposition 10.1 and multiplying to the left and to the right of (A.48) by $\Sigma_{\xi,X}$ and $\Sigma_{X,\xi}$ we have

$$\begin{aligned} 1 &= \Sigma_{\xi,X}\Sigma_{X|\xi}^{-1}\Sigma_{X,\xi} - \Sigma_{\xi,X}\Sigma_{X|\xi}^{-1}\Sigma_{X,\xi}(\sigma_\xi^2 + \Sigma_{\xi,X}\Sigma_{X|\xi}^{-1}\Sigma_{X,\xi})^{-1}\Sigma_{\xi,X}\Sigma_{X|\xi}^{-1}\Sigma_{X,\xi} \\ &= H_\xi - H_\xi(\sigma_\xi^2 + H_\xi)^{-1}H_\xi \\ &= \frac{H_\xi\sigma_\xi^2}{\sigma_\xi^2 + H_\xi} \end{aligned}$$

from where we get (A.46). Analogously we get (A.47). □

A.8.5 Proposition 10.2

Restatement. Under the regression constraints, (I) if $\Sigma_{X|\xi}$ contains an off-diagonal element that is known to be zero, say $(\Sigma_{X|\xi})_{ij} = 0$, and if $(B)_i(B)_j \neq 0$ then $\Sigma_{X|\xi}$ and σ_ξ^2 are identifiable. (II) if $\Sigma_{Y|\eta}$ contains an off-diagonal element that is known to be zero, say $(\Sigma_{Y|\eta})_{ij} = 0$, and if $(A)_i(A)_j \neq 0$ then $\Sigma_{Y|\eta}$ and σ_η^2 are identifiable.

Proof. Identification of $\Sigma_{X|\xi}$ means that if we have two different matrices Σ_X and $\tilde{\Sigma}_X$ satisfying (10.9) then we should conclude that $\Sigma_{X|\xi} = \tilde{\Sigma}_{X|\xi}$ and that $\sigma_\xi^2 = \tilde{\sigma}_\xi^2$. The same logic applies to $\Sigma_{Y|\eta}$ and σ_η^2. More specifically, from (10.9) the equality

$$\Sigma_{X|\xi} + \sigma_\xi^{-2}(B^T\Sigma_X^{-1}B)^{-1}BB^T = \tilde{\Sigma}_{X|\xi} + \tilde{\sigma}_\xi^{-2}(B^T\Sigma_X^{-1}B)^{-1}BB^T \quad \text{(A.49)}$$

must imply that $\Sigma_{X|\xi} = \tilde{\Sigma}_{X|\xi}$ and $\sigma_\xi^2 = \tilde{\sigma}_\xi^2$. If σ_ξ^2 is identifiable, so $\sigma_\xi^2 = \tilde{\sigma}_\xi^2$, then (A.49) implies $\Sigma_{X|\xi} = \tilde{\Sigma}_{X|\xi}$ since $(B^T\Sigma_X^{-1}B)^{-1}BB^T$ is identifiable. Similarly, if $\Sigma_{X|\xi}$ is identifiable, so $\Sigma_{X|\xi} = \tilde{\Sigma}_{X|\xi}$, then (A.49) implies that $\sigma_\xi^2 = \tilde{\sigma}_\xi^2$ and thus that σ_ξ^2 is identifiable.

Now, assume that elements (i, j) and (j, i) of $\Sigma_{X|\xi}$ and $\tilde{\Sigma}_{X|\xi}$ are known to be 0 and let e_k denote the $p \times 1$ vector with a 1 in position k and 0's elsewhere. Then multiplying (A.49) on the left by e_i and on the right by e_j gives

$$\sigma_\xi^{-2}(B^T\Sigma_X^{-1}B)^{-1}(B)_i(B)_j = \tilde{\sigma}_\xi^{-2}(B^T\Sigma_X^{-1}B)^{-1}(B)_i(B)_j$$

Since $(B^T\Sigma_X^{-1}B)^{-1}BB^T$ is identifiable and $(B)_i(B)_j \neq 0$, this implies that $\sigma_\xi^{-2} = \tilde{\sigma}_\xi^{-2}$ and thus σ_ξ^2 is identifiable, which implies that $\Sigma_{X|\xi}$ is identifiable.
□

A.8.6 Lemma 10.2

RESTATEMENT. In the SEM model, $\text{cor}(\xi, \eta)$ is unaffected by choice of constraint, $\sigma_\xi^2 = \sigma_\eta^2 = 1$ or $\text{var}\{E(\xi|X)\} = \text{var}\{E(\eta|Y)\} = 1$.

Proof. If we have $\sigma_\xi^2 = \sigma_\eta^2 = 1$ and we define $\tilde{\xi} = c_\xi \xi$, $\tilde{\eta} = c_\eta \xi$ with $c_\xi = [\text{var}\{E(\xi|X)\}]^{-1/2}$ (we can defined because $\text{var}E(\xi|X)$ is identifible) we have that $\text{cor}(\tilde{\eta}, \tilde{\xi}) = \text{cor}(\xi, \eta)$. Now, if $\text{var}\{E(\eta|Y)\} = \text{var}\{E(\xi|X)\} = 1$ and $\Sigma_{X|\xi}$ and $\Sigma_{Y|\eta}$ are identifiable we can identify σ_ξ and σ_η and we could define $\tilde{\xi} = c_\xi \xi$, $\tilde{\eta} = c_\eta \xi$ with $c_\xi = \{\sigma_\xi\}^{-1/2}$ and $c_\eta = \{\sigma_\eta\}^{-1/2}$ since c_ξ and c_η are identifiable.

□

A.8.7 Proof of Proposition 10.3

RESTATEMENT. Starting from the model that stems from (10.1) and (10.2),

$$\Psi = [\text{var}\{E(\xi \mid X)\}\text{var}\{E(\eta \mid Y)\}]^{1/2}\frac{\text{cor}(\xi,\eta)}{\sigma_\xi\sigma_\eta}.$$

Under either the regression constraints or the marginal constraints,

$$|\Psi| \leq |\text{cor}(\eta,\xi)|. \tag{A.50}$$

Proof. We start with

$$\begin{aligned}\text{cov}\{E(\xi \mid X), E(\eta \mid Y)\} &= \text{cov}(\Sigma_{\xi,X}\Sigma_X^{-1}X, \Sigma_{\eta,Y}\Sigma_Y^{-1}Y)\\ &= \Sigma_{\xi,X}\Sigma_X^{-1}\Sigma_{X,Y}\Sigma_Y^{-1}\Sigma_{Y,\eta}.\end{aligned}$$

From (A.30),

$$\begin{aligned}\Sigma_{X,Y} &= \beta_{X|\xi}\beta_{Y|\eta}^T\sigma_{\xi,\eta}\\ &= \Sigma_{X,\xi}\Sigma_{\eta,Y}\frac{\sigma_{\xi,\eta}}{\sigma_\xi^2\sigma_\eta^2}\\ &= \Sigma_{X,\xi}\Sigma_{\eta,Y}\frac{\text{cor}(\xi,\eta)}{\sigma_\xi\sigma_\eta}.\end{aligned}$$

Substituting we get

$$\text{cov}\{E(\xi \mid X), E(\eta \mid Y)\} = \Sigma_{\xi,X}\Sigma_X^{-1}\Sigma_{X,\xi}\Sigma_{\eta,Y}\Sigma_Y^{-1}\Sigma_{Y,\eta}\frac{\text{cor}(\xi,\eta)}{\sigma_\xi\sigma_\eta}.$$

As in the proof of Proposition 10.1, we using (A.30) and the fact that $\mu_\eta = 0$ to get, $E(\eta|Y) = \Sigma_{\eta,Y}\Sigma_Y^{-1}(Y - \mu_Y)$. Therefore $\text{var}(E(\eta|Y)) = \Sigma_{\eta,Y}\Sigma_Y^{-1}\Sigma_{\eta,Y}^T$. In the same way $\text{var}(E(\xi|X)) = \Sigma_{\xi,X}\Sigma_X^{-1}\Sigma_{\xi,X}^T$. This allows us to express

$$\text{cov}\{E(\xi \mid X), E(\eta \mid Y)\} = \text{var}\{E(\xi \mid X)\}\text{var}\{E(\eta \mid Y)\}\frac{\text{cor}(\xi,\eta)}{\sigma_\xi\sigma_\eta}.$$

In consequence, we get the first conclusion,

$$\begin{aligned}\Psi &= \text{cor}\{E(\xi \mid X), E(\eta \mid Y)\}\\ &= [\text{var}\{E(\xi \mid X)\}\text{var}\{E(\eta \mid Y)\}]^{1/2}\frac{\text{cor}(\xi,\eta)}{\sigma_\xi\sigma_\eta}.\end{aligned}$$

To demonstrate the inequality, we use that

$$\begin{aligned}\sigma_\xi^2 &= E\{\text{var}(\xi \mid X)\} + \text{var}\{E(\xi \mid X)\}\\ \sigma_\eta^2 &= E\{\text{var}(\eta \mid Y)\} + \text{var}\{E(\eta \mid Y)\}.\end{aligned}$$

Under the marginal constraints, $\sigma_\xi^2 = \sigma_\eta^2 = 1$ and so the variances of the regression functions must be less than 1: $\text{var}\{E(\xi \mid X)\} \le 1$ and $\text{var}\{E(\eta \mid Y)\} \le 1$, which implies that $|\Psi| \le |\text{cor}(\eta, \xi)|$. Under the regression constraints, the variances of the regression functions are equal to 1: $\text{var}\{E(\xi \mid X)\} = 1$ and $\text{var}\{E(\eta \mid Y)\} = 1$. This implies that $\sigma_\xi^2 \ge 1$ and $\sigma_\eta^2 \ge 1$. It again follows that $|\Psi| \le |\text{cor}(\eta, \xi)|$.

□

To perhaps aid intuition, we can see the result in another way by using the Woodbury identity to invert $\Sigma_X = \Sigma_{X,\xi}\Sigma_{\xi,X} + \Sigma_{X|\xi}$ we have

$$\begin{aligned}\text{var}\{E(\xi \mid X)\} = \Sigma_{\xi,X}\Sigma_X^{-1}\Sigma_{X,\xi} &= \Sigma_{\xi,X}(\Sigma_{X,\xi}\Sigma_{\xi,X} + \Sigma_{X|\xi})^{-1}\Sigma_{X,\xi} \\ &= \frac{\Sigma_{\xi,X}\Sigma_{X|\xi}^{-1}\Sigma_{X,\xi}}{1 + \Sigma_{\xi,X}\Sigma_{X|\xi}^{-1}\Sigma_{X,\xi}}.\end{aligned}$$

Similarly,

$$\text{var}\{E(\eta \mid Y)\} = \Sigma_{\eta,Y}\Sigma_Y^{-1}\Sigma_{Y,\eta} = \frac{\Sigma_{\eta,Y}\Sigma_{Y|\eta}^{-1}\Sigma_{Y,\eta}}{1 + \Sigma_{\eta,Y}\Sigma_{Y|\eta}^{-1}\Sigma_{Y,\eta}},$$

and thus

$$\Psi = \left[\frac{\Sigma_{\xi,X}\Sigma_{X|\xi}^{-1}\Sigma_{X,\xi}}{1 + \Sigma_{\xi,X}\Sigma_{X|\xi}^{-1}\Sigma_{X,\xi}} \frac{\Sigma_{\eta,Y}\Sigma_{Y|\eta}^{-1}\Sigma_{Y,\eta}}{1 + \Sigma_{\eta,Y}\Sigma_{Y|\eta}^{-1}\Sigma_{Y,\eta}}\right]^{1/2} \text{cor}(\xi, \eta).$$

This form shows that $|\Psi| \le |\text{cor}(\xi, \eta)|$ and provides a more detailed expression of their ratio.

A.9 Proofs for Chapter 11

A.9.1 Proposition 11.1

RESTATEMENT. For a single-response regression with $\Sigma_X > 0$ and dimension q envelope $\mathcal{E}_{\Sigma_X}(\mathcal{B})$, we have

1. $\text{span}(r_i) = \text{span}(w_{i+1})$, $i = 0, \ldots, q-1$

2. $\text{span}(p_i) = \text{span}(v_{i+1})$, $i = 0, \ldots, q-1$

3. $\text{span}(r_0, \ldots, r_{q-1}) = \text{span}(W_q) = \text{span}(V_q) = \mathcal{E}_{\Sigma_X}(\mathcal{B})$

4. $\beta_{\text{cg}} = \beta_{\text{npls}} = \beta_{\text{spls}}$

The proof of this proposition, which is by induction, relies in part on the following special case of a result by Elman (1994).

Lemma A.7. *In the context of the single-response linear model, the vectors generated by CGA satisfy*

(i) $r_k^T p_i = r_k^T r_i = 0$ *for* $i < k$

(ii) $p_k^T \Sigma_X p_i = 0$ *for* $i < k$

(iii) $\text{span}(r_0, \ldots, r_{k-1}) = \text{span}(p_0, \ldots, p_{k-1}) = \mathcal{K}_k(\Sigma_X, \sigma_{X,Y})$

Proof. See Elman (1994, Lemma 2.1). □

Recall from our discussion in Section 11.3.2 that $r_0 = p_0 = \sigma_{X,Y}$ and so $\text{span}(r_0) = \text{span}(w_1)$ and $\text{span}(p_0) = \text{span}(v_1)$. Recall that $q = \dim(\mathcal{E}_{\Sigma_X}(\mathcal{B}))$. For clarity of exposition, let β_j denote the population CGA iterate of β when $q = j$. Additionally, $\beta_1 = p_0(p_0^T \Sigma_X p_0)^{-1} p_0^T \sigma_{X,Y}$ and $\beta_1 = \beta_{\text{npls}} = \beta_{\text{spls}}$ when $q = 1$. We will also use going forward that $p_0^T \sigma_{X,Y} = r_0^T r_0$. The proposition therefore holds for $i = 0$.

For $i = 1$ we have

$$
\begin{aligned}
r_1 &= r_0 - \frac{r_0^T r_0}{r_0^T \Sigma_X r_0} \Sigma_X r_0 \\
&= \sigma_{X,Y} - \frac{\sigma_{X,Y}^T \sigma_{X,Y}}{\sigma_{X,Y}^T \Sigma_X \sigma_{X,Y}} \Sigma_X \sigma_{X,Y} \\
&= \sigma_{X,Y} - \Sigma_X \sigma_{X,Y} (\sigma_{X,Y}^T \Sigma_X \sigma_{X,Y})^{-1} \sigma_{X,Y}^T \sigma_{X,Y} \\
&= Q_{\sigma_{X,Y}(\Sigma_X)}^T \sigma_{X,Y} \\
\text{span}(r_1) &= \text{span}(w_2) \\
r_1^T r_0 &= \sigma_{X,Y}^T Q_{\sigma_{X,Y}(\Sigma_X)} \sigma_{X,Y} = 0 \\
r_1^T p_0 &= 0.
\end{aligned}
$$

Turning to p_1:

$$\begin{aligned}
p_1 &= r_1 + \frac{r_1^T r_1}{r_0^T r_0} p_0 \\
&= \sigma_{X,Y} - \frac{\sigma_{X,Y}^T \sigma_{X,Y}}{\sigma_{X,Y}^T \Sigma_X \sigma_{X,Y}} \Sigma_X \sigma_{X,Y} \\
&\quad + \frac{\sigma_{XY}^T Q_{\sigma_{X,Y}(\Sigma_X)} Q_{\sigma_{X,Y}(\Sigma_X)}^T \sigma_{X,Y}}{\sigma_{X,Y}^T \sigma_{X,Y}} \sigma_{X,Y} \\
&= \frac{\sigma_{X,Y}^T \sigma_{X,Y} \sigma_{X,Y}^T \Sigma_X^2 \sigma_{X,Y}}{(\sigma_{XY}^T \Sigma_X \sigma_{X,Y})^2} \\
&\quad \times \left(\sigma_{X,Y} - \Sigma_X \sigma_{X,Y} (\sigma_{X,Y}^T \Sigma_X^2 \sigma_{X,Y})^{-1} \sigma_{X,Y}^T \Sigma_X \sigma_{X,Y}\right) \\
&= \frac{\sigma_{X,Y}^T \sigma_{X,Y} \sigma_{X,Y}^T \Sigma_X^2 \sigma_{X,Y}}{(\sigma_{XY}^T \Sigma_X \sigma_{X,Y})^2} Q_{\Sigma_X \sigma_{X,Y}} \sigma_{X,Y} \propto v_2 = Q_{\Sigma_X \sigma_{X,Y}} \sigma_{X,Y} \\
\text{span}(p_1) &= \text{span}(v_2), \text{ (from Table 3.4)}.
\end{aligned}$$

Additionally,

$$p_0^T \Sigma_X p_1 = 0 \text{ since } \text{span}(p_0) = \text{span}(v_1), \text{span}(p_1) = \text{span}(v_2) \text{ and } v_1^T \Sigma_X v_2 = 0$$

$$\begin{aligned}
p_1^T \sigma_{X,Y} &= r_1^T \sigma_{X,Y} + \frac{r_1^T r_1}{r_0^T r_0} p_0^T \sigma_{X,Y} \\
&= r_1^T \sigma_{X,Y} + r_1^T r_1 \\
&= r_1^T r_1 \text{ since } r_1^T \sigma_{X,Y} = 0.
\end{aligned} \tag{A.51}$$

For the next regression coefficients,

$$\begin{aligned}
\beta_2 &= \beta_1 + \frac{r_1^T r_1}{p_1^T \Sigma_X p_1} p_1 \\
&= p_0 (p_0^T \Sigma_X p_0)^{-1} p_0^T \sigma_{X,Y} + p_1 (p_1^T \Sigma_X p_1)^{-1} p_1^T \sigma_{X,Y} \\
&= \beta_{\text{npls}} = \beta_{\text{spls}} \text{ when } q = 2,
\end{aligned}$$

where the second equality follows from (A.51) and we used in the last equality the fact that $p_0^T \Sigma_X p_1 = 0$.

Since the proposition holds for $q = 1, 2$, we now suppose that it holds for $q = 1, \ldots, k-1$. We next prove that it holds for $q = k$; that is, we show that $\text{span}(p_k) = \text{span}(v_{k+1})$, $\text{span}(r_k) = \text{span}(w_{k+1})$ and that $\beta_{k+1} = \beta_{\text{npls}} = \beta_{\text{spls}}$ when the number of PLS components is $k+1$.

For the induction hypotheses we have for $q = 1, \ldots, k$

1. $\text{span}\{r_0, \ldots, r_{q-1}\} = \text{span}\{p_0, \ldots, p_{q-1}\} = \text{span}\{v_1, \ldots, v_q\} = \text{span}\{w_1, \ldots, w_q\}$

2. For $i = 0, \dots, q-1$, $\text{span}(r_i) = \text{span}(w_{i+1})$, $\text{span}(p_i) = \text{span}(v_{i+1})$,

3. $\beta_q = \sum_{i=0}^{q-1} p_i(p_i^T \Sigma_X p_i^T)^{-1} \sigma_{X,Y} = \beta_{\text{npls}} = \beta_{\text{spls}}$,

4. $r_{q-1}^T r_i = 0$, for $i = 0, \dots, k-2$,

5. $p_{q-1}^T \Sigma_X p_i = 0$ for $i = 0, \dots, q-2$,

6. $r_{q-1}^T p_i = 0$ for $i = 0, \dots, q-2$,

7. $p_i^T \sigma_{X,Y} = r_i^T r_i$ for $i = 0, \dots, q-1$.

Items 4–6 are implied by Lemma A.7. To prove the proposition we first show that items 1 and 2 hold for $q = k+1$. To do this, we first show that $\{r_0, \dots, r_k\}$ is an orthogonal set. Although this is implied by Lemma A.7(i), we provide an alternate demonstration for completeness.

Claim: $\{r_0, \dots, r_k\}$ is an orthogonal set.

Proof. We know from induction hypothesis 4 that $\{r_0, \dots, r_{k-1}\}$ is an orthogonal set so we need to show that $r_k^T r_i = 0$ for $i = 0, \dots, k-1$. We know from item 5 that for $q = k$, $p_{k-1}^T \Sigma_X p_i = 0$ for $i = 0, \dots, k-2$. From hypothesis 1,

$$\text{span}\{r_0, \dots, r_{k-2}\} = \text{span}\{p_0, \dots, p_{k-2}\}$$

and consequently we have also that $p_{k-1}^T \Sigma_X r_i = 0$ for $i = 0, \dots, k-2$. By construction $r_k = r_{k-1} - \alpha_{k-1} \Sigma_X p_{k-1}$. Multiplying both sides by r_i we have

$$r_i^T r_k = r_i^T r_{k-1} - \alpha_{k-1} r_i^T \Sigma_X p_{k-1} = 0, \; i = 0, \dots, k-2.$$

It remains to show that $r_{k-1}^T r_k = 0$. By construction

$$\begin{aligned} r_{k-1}^T r_k &= r_{k-1}^T r_{k-1} - \frac{r_{k-1}^T r_{k-1}}{p_{k-1}^T \Sigma_X p_{k-1}} r_{k-1} \Sigma_X p_{k-1} \\ &= r_{k-1}^T r_{k-1} \left\{ 1 - \frac{r_{k-1} \Sigma_X p_{k-1}}{p_{k-1}^T \Sigma_X p_{k-1}} \right\} \\ &= 0, \end{aligned}$$

where the last equality follows because $r_{k-1}^T \Sigma_X p_{k-1} = p_{k-1}^T \Sigma_X p_{k-1}$. To see this we have by construction

$$p_{k-1} = r_{k-1} + B_{k-2} p_{k-2}.$$

Multiplying both sides by $p_{k-1}^T \Sigma_X$ and recognizing that $p_{k-1}^T \Sigma_X p_{k-2} = 0$ by induction hypothesis 5 gives the desired result. □

Claim: $\text{span}(r_k) = \text{span}(w_{k+1})$ *and* $\text{span}(p_k) = \text{span}(v_{k+1})$

Proof. The proof follows from Proposition 3.7 and Lemma A.7(iii).

In a bit more detail, we know from induction hypothesis 2 that for $q = 1, \ldots, k$ and $i = 0, \ldots, q-1$, $\text{span}(r_i) = \text{span}(w_{i+1})$. From Lemma A.7(iii)

$$\mathcal{E}_{\Sigma_X}(\mathcal{B}) = \text{span}\{w_1, \ldots, w_{k+1}\} = \text{span}\{r_0, \ldots, r_k\} = \mathcal{K}_{k+1}(\Sigma_X, \Sigma_{X,Y}).$$

Since r_k is orthogonal to $(r_0, \ldots, r_{k-1})$ and w_{k+1} is orthogonal to $(w_1, \ldots, w_k)$, we must have $\text{span}(r_k) = \text{span}(w_{k+1})$. The same rational can be used to show that $\text{span}(p_k) = \text{span}(v_{k+1})$ since both have $p_i^T \Sigma_X p_k = v_i^T \Sigma_X v_k = 0$. □

Claim: $\beta_{k+1} = \sum_{i=0}^{k} p_i (p_i^T \Sigma_X p_i^T)^{-1} \sigma_{X,Y} = \beta_{\text{npls}} = \beta_{\text{spls}}$

Proof.

$$\begin{aligned}
\beta_{k+1} &= \beta_k + \alpha_k p_k \\
&= \sum_{i=0}^{k-1} p_i (p_i^T \Sigma_X p_i)^{-1} p_i^T \sigma_{X,Y} + \frac{r_k^T r_k}{p_k^T \Sigma_X p_k} p_k \\
&= \sum_{i=0}^{k-1} p_i (p_i^T \Sigma_X p_i)^{-1} p_i^T \sigma_{X,Y} + p_k (p_k^T \Sigma_X p_k)^{-1} r_k^T r_k \\
&= \sum_{i=0}^{k-1} p_i (p_i^T \Sigma_X p_i)^{-1} p_i^T \sigma_{X,Y} + p_k (p_k^T \Sigma_X p_k)^{-1} p_k^T \sigma_{X,Y}.
\end{aligned}$$

From Table 11.1,

$$\begin{aligned}
p_k^T \sigma_{X,Y} &= r_k^T \sigma_{X,Y} + B_{k-1} p_{k-1}^T \sigma_{X,Y} \\
&= r_k^T r_k \frac{p_{k-1}^T \sigma_{X,Y}}{r_{k-1}^T r_{k-1}} \\
&= r_k^T r_k
\end{aligned}$$

since $p_{k-1}^T \sigma_{X,Y} / r_{k-1}^T r_{k-1} = 1$ by induction hypothesis 7. □

Bibliography

Adragni, K. and R. D. Cook (2009). Sufficient dimension reduction and prediction in regression. *Philosophical Transactions of the Royal Society A 367*, 4385–4405.

Adragni, K. P. (2009). Some basis functions for principal fitted components. http://userpages.umbc.edu/~kofi/reprints/BasisFunctions.pdf.

Akter, S., S. F. Wamba, and S. Dewan (2017). Why pls-sem is suitable for complex modelling? an empirical illustration in big data analytics quality. *Production Planning & Control 28*(11–12), 1011–1021.

Aldrich, J. (2005). Fisher and regression. *Statistical Science 20*(4), 401–417.

Allegrini, F. and A. C. Olivieri (2013). An integrated approach to the simultaneous selection of variables, mathematical pre-processing and calibration samples in partial least-squares multivariate calibration. *Talanta 115*, 755–760.

Allegrini, F. and A. C. Olivieri (2016). Sensitivity, prediction uncertainty, and detection limit for artificial neural network calibrations. *Analytical Chemistry 88*, 7807–7812.

AlShiab, M. S. I., H. A. N. Al-Malkawi, and A. Lahrech (2020). Revisiting the relationship between governance quality and economic growth. *International Journal of Economics and Financial Issues* 10(4), 54–63.

Anderson, T. W. (1951). Estimating linear restrictions on regression coefficients for multivariate normal distributions. *Ann. Math. Statist. 22*(3), 327–351.

Anderson, T. W. (1999). Asymptotic distribution of the reduced-rank regression estimator under general conditions. *Annals of Statistics 27*(4), 1141–1154.

Arvin, M., R. P. Pradhan, and M. S. Nair (2021). Are there links between institutional quality, government expenditure, tax revenue and economic growth? Evidence from low-income and lower middle-income countries. *Economic Analysis and Policy* 70(C), 468–489.

Asongu, S. A. and J. Nnanna (2019). Foreign aid, instability, and governance in Africa. *Politics & Policy* 47(4), 807—848.

Baffi, G., E. Martin, and J. Modrris, A (1999). Non-linear projection to latent structures revisited: the quadratic PLS algorithm. *Computers & Chemical Engineering 23*(3), 395–411.

Baffi, G., E. B. Martin, and A. J. Morris (1999). Non-linear projection to latent structures revisited (the neural network pls algorithm). *Computers & Chemical Engineering 23*(9), 1293–1307.

Barker, M. and W. Ravens (2003). Partial least squares for discrimination. *Journal of Chemometrics 17*, 166–173.

Basa, J., R. D. Cook, L. Forzani, and M. Marcos (2022, December). Asymptotic distribution of one-component PLS regression estimators. *Canadian Journal of Statistics Published online*, 23 December 2022, https://doi.org/10.1002/cjs.11755.

Berntsson, P. and S. Wold (1986). Comparison between x-ray crystallographic data and physicochemical parameters with respect to their information about the calcium channel antagonist activity of 4-phenyl-1,4-dihydropyridines. *Molecular Informatics 5*, 45–50.

Björck, Å. (1966). *Numerical Methods for Least Squares Problems.* SIAM.

Boggaard, C. and H. H. Thodberg (1992). Optimal minimal neural interpretation of spectra. *Analytical Chemistry 64*(5), 545–551.

Bollen, K. A. (1989). *Structural Equations with Latent Variables.* New York: Wiley.

Box, G. E. P. and D. R. Cox (1964). An analysis of transformations. *Journal of the Royal Statistical Society B 26*, 211–246.

Breiman, L. (2001). Statistical modeling: The two cultures (with comments and a rejoinder by the author). *Statistical Science 16*(3), 199–231.

Brereton, R. G. and G. R. Lloyd (2014). Partial least squares discriminant analysis: taking the magic away. *Journal of Chemometrics 28*(4), 213–225.

Bridgman, P. W. (1927). *The Logic of Modern Physics.* Interned Archive, https://archive.org/details/logicofmodernphy0000pwbr/page/8/mode/2up.

Bro, R. (1998). *Multi-way Analysis in the Food Industry, Model, Algorithms and Applications.* PhD. thesis, Doctoral Thesis, University of Amsterdam.

Bura, E., S. Duarte, and L. Forzani (2016). Sufficient reductions in regressions with exponential family inverse predictors. *Journal of the American Statistical Association 111*(515), 1313–1329.

Chakraborty, C. and Z. Su (2023). A comprehensive Bayesian framework for envelope models. *Journal of the American Statistical Association Published online*, 24 August 2023 https://doi.org/10.1080/01621459.2023.2250096.

Chiappini, F. A., H. Allegrini, H. C. Goicoechea, and A. C. Olivieri (2020). Sensitivity for multivariate calibration based on multilayer perceptron artificial neural networks. *Analytical Chemistry 92*, 12265–12272.

Chiappini, F. A., H. C. Goicoechea, and A. C. Olivieri (2020). Mvc1-gui: A matlab graphical user interface for first-order multivariate calibration. An upgrade including artificial neural networks modelling. *Chemometrics and Intelligent Laboratory Systems 206*, 104162.

Chiappini, F. A., F. Gutierrez, H. C. Goicoechea, and A. C. Olivieri (2021). Interference-free calibration with first-order instrumental data and multivariate curve resolution. When and why? *Analytica Chimica Acta 1161*, 338465.

Chiappini, F. A., C. M. Teglia, A. G. Forno, and H. C. Goicoechea (2020). Modelling of bioprocess non-linear fluorescence data for at-line prediction of etanercept based on artificial neural networks optimized by response surface methodology. *Talanta 210*, 120664.

Chiaromonte, F., R. D. Cook, and B. Li (2002). Sufficient dimension reduction in regressions with categorical predictors. *The Annals of Statistics 30*(2), 475–497.

Chun, H. and S. Keleş (2010). Sparse partial least squares regression for simultaneous dimension reduction and predictor selection. *Journal of the Royal Statistical Society B 72*(1), 3–25.

Conway, J. (1994). *A Course in Functional Analysis.* Graduate Texts in Mathematics. Springer New York.

Cook, R. D. (1994). Using dimension-reduction subspaces to identify important inputs in models of physical systems. In *Proceedings of the Section on Engineering and Physical Sciences*, pp. 18–25. Alexandria, VA: American Statistical Association.

Cook, R. D. (1998). *Regression Graphics: Ideas for Studying Regressions Through Graphics.* Wiley Series in Probability and Statistics. New York: Wiley.

Cook, R. D. (2000). SAVE: A method for dimension reduction and graphics in regression. *Communications in Statistics – Theory and Methods 29*(9–10), 2109–2121.

Cook, R. D. (2007). Fisher lecture: Dimension reduction in regression. *Statistical Science 22*(1), 1–26.

Cook, R. D. (2018). *An Introduction to Envelopes: Dimension Reduction for Efficient Estimation in Multivariate Statistics.* Wiley Series in Probability and Statistics. New York: Wiley.

Cook, R. D. (2020). Envelope methods. *Wiley Interdisciplinary Reviews: Computational Statistics 12*(2), e1484.

Cook, R. D. and L. Forzani (2008a). Covariance reducing models: An alternative to spectral modelling of covariance matrices. *Biometrika 95*(4), 799–812.

Cook, R. D. and L. Forzani (2008b). Principal fitted components for dimension reduction in regression. *Statistical Science 23*(4), 485–501.

Cook, R. D. and L. Forzani (2009). Likelihood-based sufficient dimension reduction. *Journal of the American Statistical Association 104*(485), 197–208.

Cook, R. D. and L. Forzani (2018). Big data and partial least squares prediction. *The Canadian Journal of Statistics/La Revue Canadienne de Statistique 47*(1), 62–78.

Cook, R. D. and L. Forzani (2019). Partial least squares prediction in high-dimensional regression. *Annals of Statistics 47*(2), 884–908.

Cook, R. D. and L. Forzani (2020). Envelopes: A new chapter in partial least squares regression. *Journal of Chemometrics 34*(10), e3294.

Cook, R. D. and L. Forzani (2021). PLS regression algorithms in the presence of nonlinearity. *Chemometrics and Intelligent Laboratory Systems 213*, 194307.

Cook, R. D. and L. Forzani (2023, November). On the role of partial least squares in path analysis for the social sciences. *Journal of Business Research 167*, 114132.

Cook, R. D., L. Forzani, and L. Liu (2023a, January). Envelopes for multivariate linear regression with linearly constrained coefficients. *Scandinavian Journal of Statistics Publlished online*, 12 October 2023 https://doi.org/10.1111/sjos.12690.

Cook, R. D., L. Forzani, and L. Liu (2023b). Partial least squares for simultaneous reduction of response and predictor vectors in regression. *Journal of Multivariate Analysis 196*, https://doi.org/10.1016/j.jmva.2023.105163.

Cook, R. D., L. Forzani, and A. J. Rothman (2012). Estimating sufficient reductions of the predictors in abundant high-dimensional regressions. *The Annals of Statistics 40*(1), 353–384.

Cook, R. D., L. Forzani, and A. J. Rothman (2013). Prediction in abundant high-dimensional linear regression. *Electronic Journal of Statistics 7*, 3059–3088.

Cook, R. D., L. Forzani, and Z. Su (2016, September). A note on fast envelope estimation. *Journal of Multivariate Analysis 150*, 42–54.

Cook, R. D., L. Forzani, and X. Zhang (2015). Envelopes and reduced-rank regression. *Biomerika 102*(2), 439–456.

Cook, R. D., I. S. Helland, and Z. Su (2013). Envelopes and partial least squares regression. *Journal of the Royal Statistical Society B 75*(5), 851–877.

Cook, R. D. and B. Li (2002). Dimension reduction for the conditional mean in regression. *Annals of Statistics 30*(2), 455–474.

Cook, R. D., B. Li, and F. Chiaromonte (2010). Envelope models for parsimonious and efficient multivariate linear regression. *Statistica Sinica 20*(3), 927–960.

Cook, R. D. and L. Ni (2005). Sufficient dimension reduction via inverse regression: a minimum discrepancy approach. *Journal of the American Statistical Association 100*(470), 410–428.

Cook, R. D. and C. M. Setodji (2003). A model-free test for reduced rank in multivariate regression. *Journal of the American Statistical Association 46*, 421–429.

Cook, R. D. and Z. Su (2013). Scaled envelopes: Scale-invariant and efficient estimation in multivariate linear regression. *Biometrika 100*(4), 939–954.

Cook, R. D. and Z. Su (2016). Scaled predictor envelopes and partial least-squares regression. *Technometrics 58*(2), 155–165.

Cook, R. D. and S. Weisberg (1982). *Residuals and Influence in Regression.* London: Chapman and Hall.

Cook, R. D. and S. Weisberg (1991). Sliced inverse regression for dimension reduction: Comment. *Journal of the American Statistical Association 86*(414), 328–332.

Cook, R. D. and S. Weisberg (1999). *Applied Regression Including Computing and Graphics* (1st ed.). New York: Wiley.

Cook, R. D. and X. Yin (2001). Special invited paper: Dimension reduction and visualization in discriminant analysis (with discussion). *Australian and New Zealand Journal of Statistics 43*, 147–199.

Cook, R. D. and X. Zhang (2015a). Foundations for envelope models and methods. *Journal of the American Statistical Association 110*(510), 599–611.

Cook, R. D. and X. Zhang (2015b). Simultaneous envelopes and multivariate linear regression. *Technometrics 57*(1), 11–25.

Cook, R. D. and X. Zhang (2016). Algorithms for envelope estimation. *Journal of Computational and Graphical Statistics 25*, 284–300.

Cook, R. D. and X. Zhang (2018). Fast envelope algorithms. *Statistica Sinica 28*, 1179–1197.

Cox, D. R. and N. Reid (1987). Parameter orthogonality and approximate conditional inference. *Journal of the Royal Statistical Society B 49*, 1–39.

de Jong, S. (1993). SIMPLS: An alternative approach to partial least squares regression. *Chemometrics and Intelligent Laboratory Systems 18*(3), 251–263.

Despagne, F. and D. Luc Massart (1998). Neural networks in multivariate calibration. *The Analyst 123*(11), 157R–178R.

Diaconis, P. and D. Freedman (1984). Asymptotics of graphical projection pursuit. *Annals of Statistics 12*(3), 793–815.

Dijkstra, T. (1983). Some comments on maximum likelihood and partial least squares methods. *Journal of Econometrics 22*, 67–90.

Dijkstra, T. (2010). Latent variables and indices: Herman Wold's basic design and partial least squares. In E. V. Vinzi, W. W. Chin, J. Henseler, and H. Wang (Eds.), *Handbook of Partial Least Squares*, Chapter 1, pp. 23–46. New York: Springer.

Dijkstra, T. and J. Henseler (2015a). Consistent and asymptotically normal pls estimators for linear structural equations. *Computational Statistics and Data Analysis 81*, 10–23.

Dijkstra, T. and J. Henseler (2015b). Consistent partial least squares path modeling. *MIS Quarterly 39*(2), 297–316.

Ding, S. and R. D. Cook (2018). Matrix-variate regressions and envelope models. *Journal of the Royal Statistical Society B 80*, 387–408.

Ding, S., Z. Su, G. Zhu, and L. Wang (2021). Envelope quantile regression. *Statistica Sinica 31*(1), 79–105.

Downey, G., R. Briandet, R. H. Wilson, and E. K. Kemsley (1997). Near- and mid-infrared spectroscopies in food authentication: Coffee varietal identification. *Journal of Agricultural and Food Chemistry 45*(11), 4357–4361.

Eaton, M. L. (1986). A characterization of spherical distributions. *Journal of Multivariate Analysis 20*, 260–271.

el Bouhaddani, S., H.-W. Un, C. Hayward, G. Jongbloed, and Houwing-Duistermaat (2018). Probabilistic partial least squares model: Identifiability, estimation and application. *Journal of Multivariate Analysis 167*, 331–346.

Elman, H. C. (1994). Iterative methods for linear systems. In J. Gilbert and D. Kershaw (Eds.), *Large-Scale Matrix Problems and the Numerical Solution of Partial Differential Equations*, pp. 69–117. Oxford, England: Clarendon Press.

Emara, N. and I. M. Chiu (2015). The impact of governance environment on economic growth: The case of Middle Eastern and North African countries. *Journal of Economics Library*, 3(1), 24–37.

Etiévant, L. and V. Viallon (2022). On some limitations of probabilistic models for dimension-reduction: illustration in the case of probabilistic formulations of partial least squares. *Statistica Neerlandica 76*(3), 331–346.

Evermann, J. and M. Rönkkö (2021). Recent developmets in PLS. *Communications of the Association for Information Systems 44*, 123–132.

Fawaz, F., A. Mnif, and A. Popiashvili (2021, June). Impact of governance on economic growth in developing countries: A case of HIDC vs. LIDC. *Journal of Social and Economic Development 23*(1), 44–58.

Fisher, R. A. (1922). On the mathematical foundations of theoretical statistics. *Philosophical Transactions of the Royal Society A: Mathematical, Physical and Engineering Sciences 222*(594–604), 309–368.

Forzani, L., C. A. Percíncula, and R. Toledano (2022). Wishart moments calculator. Technical report, https://antunescarles.github.io/wishart-moments-calculator/.

Forzani, L., D. Rodrigues, E. Smucler, and M. Sued (2019). Sufficient dimension reduction and prediction in regression: Asymptotic results. *Journal of Multivariate Analysis 171*, 339–349.

Forzani, L., D. Rodriguez, and M. Sued (2023). Asymptotic results for nonparametric regression estimators after sufficient dimension reduction estimation. Technical report, arxiv.org/abs/2306.10537.

Forzani, L. and Z. Su (2020). Envelopes for elliptical multivariate linear regression. *Statistica Sinica 31*(1), 301–332.

Frank, I. E. and J. H. Frideman (1993). A statistical view of some chemometrics regression tools. *Technometrics 35*(2), 102–246.

Frideman, J., T. Hastie, S. Rosset, R. Tibshirani, and J. Zhu (2004). Discussion of three papers on boosting. *The Annals of Statistics 32*(1), 102–107.

Gani, A. (2011). Governance and Growth in Developing Countries. *Journal of Economic Issues 45*(1), 19–40.

Garthwaite, P. H. (1994). An interpretation of partial least squares. *Journal of the American Statistical Association 89*, 122–127.

Geladi, P. (1988). Notes on the history and nature if partial least squares (PLS) modeling. *Journal of Chemometrics 2*, 231–246.

Goh, G. and D. K. Dey (2019). Asymptotic properties of marginal least-square estimator for ultrahigh-dimensional linear regression models with correlated errors. *The American Statistician 73*(1), 4–9.

Goicoechea, H. C. and A. C. Oliveri (1999). Enhanced synchronous spectrofluorometric determination of tetracycline in blood serum by chemometric analysis. comparison of partial least-squares and hybrid linear analysis calibrations. *Analytical Chemistry 71*, 4361–4368.

Goodhue, D., W. Lewis, and R. Thompson (2023). Comments on Evermann and Rönkkö: Recent developments in PLS. *Communications of the Association for Information Systems 52*, 751–755.

Green, P. and B. Silverman (1994). *Nonparametric regression and generalized linear models.* Boca Raton, FL: CRC Press.

Guide, J. B. and M. Ketokivi (2015, July). Notes from the editors: Redefining some methodological criteria for the journal. *Journal of Operations Management 37*, v–viii.

Haaland, D. M. and E. V. Thomas (1988). Partial least-squares methods for spectral analyses. 1. relation to other quantitative calibration methods and the extraction of qualitative information. *Analytical Chemistry 60*(11), 1193–1202.

Haenlein, M. and A. M. Kaplan (2004). A beginner's guide to partial least squares analysis. *Understanding Statistics 3*(4), 283–297.

Hall, P. and K. C. Li (1993). On almost linearity of low dimensional projections from high dimensional data. *Annals of Statistics 21*, 867–889.

Han, X., H. Khan, and J. Zhuang (2014). Do governance indicators explain development performance? A cross-country analysis. *Asian Development Bank Economics Working Paper Series*, No. 417.

Hand, D. J. (2006). Classifier technology and the illusion of progress. *Statistical Science 21*(1), 1–14.

Hand, D. J. and K. Yu (2001). Idiot's Bbayes–Not so stupid after all. *International Statistical Review 69*, 385–398.

Hawkins, D. M. and S. Weisberg (2017). Combining the box-cox power and generalised log transformations to accommodate nonpositive responses in linear and mixed-effects linear models. *South African Statistics Journal 51*, 317–328.

Hayami, K. (2020). Convergence of the conjugate gradient method on singular systems. Technical report, ArXiv: https://arxiv.org/abs/1809.00793.

Helland, I. S. (1990). Partial least squares regression and statistical models. *Scandinavian Journal of Statistics 17*(2), 97–114.

Helland, I. S. (1992). Maximum likelihood regression on relevant components. *Journal of the Royal Statistical Society B 54*(2), 637–647.

Helland, I. S., S. Sæbø, T. Almøy, and R. Rimal (2018). Model and estimators for partial least squares regression. *Journal of Chemometrics 32*(9), e3044.

Henderson, H. and S. Searle (1979). Vec and vech operators for matrices, with some uses in Jacobians and multivariate statistics. *The Canadian Journal of Statistics/La Revue Canadienne de Statistique 7*(1), 65–81.

Henseler, J., T. Dijkstra, M. Sarstedt, C. Ringle, A. Diamantopoulos, D. Straub, D. J. Ketchen Jr, J. F. Hair, T. Hult, and R. Calantone (2014). Common beliefs and reality about PLS: Comments on Rönkkö & Evermann (2013). *Organizational Research Methods 17*(2), 182–209.

Hentenes, M. R. and E. Stiefel (1952). Methods of conjugate gradients for solving linear systems. *Journal of Research of the National Bureau of Standards 49*(6), 409–436.

Hotelling, H. (1933). Analysis of a complex statistical variable into principal components. *Journal of Educational Psychology 24*(6), 417–441.

Indahl, U. (2005). A twist to partial least squares regression. *Journal of Chemometrics 19*(1), 32–44.

Izenman, J. A. (1975). Reduced-rank regression for the multivariate linear model. *Journal of Multivariate Analysis 5*(2), 248–264.

Jöreskog, K. G. (1970). A general method for estimating a linear structural equation model. *ETS Research Bulletin Series 1970*(2), i–41.

Kalivas, J. H. (1997). Two data sets of near infrared spectra. *Chemometrics and Intelligent Laboratory Systems 37*(2), 255–259.

Kenward, M. G. (1987). A method for comparing profiles of repeated measurements. *Journal of the Royal Statistical Society C 36*(3), 296–308.

Kettaneh-Wold, N. (1992). Analysis of mixture data with partial least squares. *Chemometrics and Intelligent Laboratory Systems 14*, 57–69.

Khare, K., S. Pal, and Z. Su (2016). A bayesian approach for envelope models. *Annals of Statistics 45*(1), 196–222.

Kruskal, J. B. (1983). *Statistical Data Analysis*, Volume 28 of *Proceedings of Symposia in Applied Mathematics*, Chapter 4, pp. 75–104. Providence, Rhode Island: American Mathematical Society.

Lavoie, F. B., K. Muteki, and R. Gosselin (2019). A novel robust nl-pls regression methodology. *Chemometrics and Intelligent Laboratory Systems 184*, 71–81.

Letac, G. and H. Massam (2004). All invariant moments of the wishart distribution. *Scandinavian Journal of Statistics 31*(2), 295–318.

Li, B. (2018). *Sufficient Dimension Reduction: Methods and Applications with R.* Chapman & Hall/CRC Monographs on Statistics and Applied Probability. Boca Raton, Florida: CRC Press.

Li, B., P. A. Hassel, A. J. Morris, and E. B. Martin (2005). A non-linear nested partial least squares algorithm. *Computational Statistics & Data Analysis 48*, 87–101.

Li, B. and S. Wang (2007). On directional regression for dimension reduction. *Journal of the American Statistical Association 102*, 997–1008.

Li, K. C. (1991). Sliced inverse regression for dimension reduction (with discussion). *Journal of the American Statistical Association 86*(414), 316–342.

Li, L., R. D. Cook, and C.-L. Tsai (2007). Partial inverse regression. *Biometrika 94*(3), 615–625.

Li, L. and X. Zhang (2017). Parsimonious tensor response regression. *Journal of the American Statistical Association 112*(519), 1131–1146.

Liland, K. H., M. Hø y, H. Martens, and S. Sæ bø (2013). Distribution based truncation for variable selection in subspace methods for multivariate regression. *Chemometrics and Intelligent Laboratory Systems 122*, 103–111.

Lindgren, F., P. Geladi, and S. Wold (1993). The kernel algorithm for pls. *Journal of Chemometrics 7*(1), 44–59.

Liu, Y. and W. Ravens (2007). PLS and dimension reduction for classification. *Computational Statistics 22*, 189–208.

Lohmöller, J.-B. (1989). *Latent Variable Path Modeling with Partial Least Squares.* New York: Springer.

Magnus, J. R. and H. Neudecker (1979). The commutation matrix: some properties and applications. *Annals of Statistics 7*(2), 381–394.

Manne, R. (1987). Analysis of two partial-least-squares algorithms for multivariate calibration. *Chemometrics and Intelligent Laboratory Systems 1*, 187–197.

Martens, H. and T. Næs (1989). *Multivariate Calibration.* New York: Wiley.

Martin, A. (2009). A comparison of nine PLS1 algorithms. *Journal of Chemometrics 23*(10), 518–529.

McIntosh, C. N., J. R. Edwards, and J. Antonakis (2014). Reflections on partial least squares path modeling. *Organizational Research Methods 17*(2), 210–251.

McLachlan, G. J. (2004). *Discriminant Analysis and Statistical Pattern Recognition.* New York, NY: Wiley.

Muirhead, R. J. (2005). *Aspects of Multivariate Statistical Theory.* New York: Wiley.

Nadler, B. and R. R. Coifman (2005). Partial least squares, Beer's law and the net analyte signal: Statistical modeling and analysis. *Journal of Chemometrics 19*, 435–54.

Næ s, T. and I. S. Helland (1993). Relevant components in regression. *Scandinavian Journal of Statistics 20*(3), 239–250.

Neudecker, H. and T. Wansbeek (1983). Some results on commutation matrices, with statistical applications. *Canadian Journal of Statistics 11*(3), 221–231.

Nguyen, D. V. and D. M. Rocke (2004). On partial least squares dimension reduction for microarray-based classification: A simulation study. *Computational Statistics and Data Analysis 46*(3), 407–425.

Olivieri, A. and G. M. Escandar (2014). *Practical three-way calibration.* Elsevier.

Olivieri, A. C. (2018). *Introduction to Multivariate Calibration: A Practical Approach.* New York, NY: Springer.

Osborne, B. G., T. Fearn, A. R. Miller, and S. Douglas (1984). Application of near infrared reflectance spectroscopy to compositional analysis of biscuits and biscuit doughs. *Journal of the Science of Food and Agriculture 35*(1), 99–105.

Pardoe, I., X. Yin, and R. D. Cook (2007). Graphical tools for quadratic discriminant analysis. *Technometrics 49*(2), 172–183.

Park, Y., Z. Su, and D. Chung (2022). Envelope-based partial partial least squares with application to cytokine-based biomarker analysis for COVID-19. *Statistics in Medicine 41*(23), 4578–4592.

Phatak, A. and F. de Hoog (2002). Exploring the connection between pls, Lanczos methods and conjugate gradients: Alternative proofs of some properties of pls. *Journal of Chemometrics 16*, 361–367.

R Core Team (2022). *A Language and Environment for Statistical Computing.* R Foundation for Statistical Computing, Vienna, Austria.

Radulović, M. (2020, April – J). The impact of institutional quality on economic growth: A comparative analysis of the Eu and non-Eu countries of Southeast Europe. *Economic Annals 65*(225), 163–182.

Rao, C. R. (1979). Separation theorems for singular values of matrices and their applications in multivariate analysis. *Journal of Multivariate Analysis 9*, 362–377.

Reinsel, G. C. and R. P. Velu (1998). *Multivariate Reduced-rank Regression: Theory and Applications.* New York: Springer.

Rekabdarkolaee, H. M., Q. Wang, Z. Naji, and M. Fluentes (2020). New parsimonious multivariate spatial model: Spatial envelope. *Statistica Sinica 30*(3), 1583–1604. https://arxiv.org/abs/1706.06703.

Rigdon (2012). Rethinking partial least squares path modeling: In praise of simple methods. *Long Range Planning 45*, 341–354.

Rigdon, E. E., J.-M. Becker, and M. Sarstedt (2019). Factor indeterminacy as metrological uncertainty: Implications for advancing psychological measurement. *Multivariate Behavioral Research 54*(3), 429–443.

Rimal, R., A. Trygve, and S. Sæbø (2019). Comparison of multi-response prediction methods. *Chemometrics and Intelligent Laboratory Systems 190*, 10–21.

Rimal, R., A. Trygve, and S. Sæbø (2020). Comparison of multi-response estimation methods. *Chemometrics and Intelligent Laboratory Systems 205*, 104093.

Rönkkö, M. and J. Evermann (2013). A critical examination of common beliefs about partial least squares path modeling. *Organizational Research Methods 16*(3), 425–448.

Rönkkö, M., C. N. McIntosh, J. Antonakis, and J. R. Edwards (2016a). *Appendix A – Analysis file for R (R Code for Rönkkö et al., 2016b).* https://www.researchgate.net/publication/304253732_Appendix_A_-_Analysis_file_for_R.

Rönkkö, M., C. N. McIntosh, J. Antonakis, and J. R. Edwards (2016b). Partial least squares path modeling: Time for some serious second thoughts. *Journal of Operations Management 47–48*(SC), 9–27.

Rosipal, R. (2011). Nonlinear partial least squares: An overview. In H. Lodhi and Y. Yamanishi (Eds.), *Chemoinformatics and Advanced Machine Learning Perspectives: Complex Computational Methods and Collaborative Techniques*, Chapter 9, pp. 169–189. Hershey, Pennsylvania: IGI Global.

Rosipal, R. and N. Krämer (2006). Overview and recent advances in partial least squares. In C. Saunders, M. Grobelnik, S. Gunn, and J. Shawe-Taylor (Eds.), *Subspace, Latent Structure and Feature Selection Techniques*, pp. 34–51. New York: Springer.

Rosseel, Y. (2012). lavaan: An R package for structural equation modeling. *Journal of Statistical Software 48*(2), 1–36.

Rothman, A. J., P. J. Bickel, E. Levina, and J. Zhu (2008). Sparse permutation invariant covaraince estimation. *Electronic Journal of Statistics 2*, 495–515.

Royston, P. and W. Sauerbrei (2008). *Multivariable Model-Building: A pragmatic approach to regression analysis based on fractional polynomials for modelling continuous variables.* New York: Wiley.

Russo, D. and K.-J. Stol (2023). Don't throw the baby out with the bathwater: Comments on "recent developments in PLS". *Communications of the Association for Information Systems 52*, 700–704.

Samadi, Y. S. and H. M. Herath (2021). Reduced-rank envelope vector autoregressive model. Technical report, School of Mathematical and Statistical Science, Southern Illinois University, Carbondale, IL.

Sarstedt, M., J. F. Hair, C. M. Ringle, K. O. Thiele, and S. P. Gudergan (2016). Estimation issues with pls and cbsem: Where the bias lies! *Journal of Business Research 69*(10), 3998 – 4010.

Schönemann, P. H. and M.-M. Wang (1972). Some new results on factor indeterminacy. *Psychometrika 37*(1), 61–91.

Shan, P., S. Peng, Y. Bi, L. Tang, C. Yang, Q. Xie, and C. Li (2014). Partial least squares–slice transform hybrid model for nonlinear calibration. *Chemometrics and Intelligent Laboratory Systems 138*, 72–83.

Shao, Y., R. D. Cook, and S. Weisberg (2007). Marginal tests with sliced average variance estimation. *Biometrika 94*(2), 285–296.

Shao, Y., R. D. Cook, and S. Weisberg (2009). Partial central subspace and sliced average variance estimation. *Journal of Statistical Planning and Inference 139*(3), 952–961.

Sharma, P. N., B. D. Liengaard, M. Sarstedt, J. F. Hair, and C. Ringle (2022). Extraordinary claims require extraordinary evidence: A comment on "recent developments in PLS". *Communications of the Association for Information Systems 52*, 739–742.

Simonoff, J. S. (1996). *Smoothing Methods in Statistics.* New York, NY: Springer.

Skagerberg, B., J. MacGregor, and C. Kiparissides (1992). Multivariate data analysis applied to low-density polyethylene reactors. *Chemometrics and Intelligent Laboratory Systems 14*, 341–356.

Small, C. G., J. Wang, and Z. Yang (2000). Eliminating multiple root problems in estimation (with discussion). *Statistical Science 15*(4), 313–341.

Stocchero, M. (2019). Iterative deflation algorithm, eigenvalue equations, and PLS2. *Chemometrics 33*(10), e3144.

Stocchero, M., M. de Nardi, and B. Scarpa (2020). An alternative point of view on PLS. *Chemometrics and Intelligent Laboratory Systems 222*, 104513.

Su, Z. and R. D. Cook (2011). Partial envelopes for efficient estimation in multivariate linear regression. *Biometrika 98*(1), 133–146.

Su, Z. and R. D. Cook (2012). Inner envelopes: Efficient estimation in multivariate linear regression. *Biometrika 99*(3), 687–702.

Su, Z. and R. D. Cook (2013). Estimation of multivariate means with heteroscedastic errors using envelope models. *Statistica Sinica 23*, 213–230.

Tapp, H. S., M. Defernez, and E. K. Kemsley (2003). Ftir spectroscopy and multivariate analysis can distinguish the geographic origin of extra virgin olive oils. *Journal of Agricultural and Food Chemistry 51*(21), 6110–6005.

Tenenhaus, M. and E. V. Vinzi (2005). Pls regression, pls path modeling, and generalized procrustean analysis: A combined approach for multiblock analysis. *Journal of Chemometrics 19*(3), 145–153.

Tibshirani, R. (1996). Regression shrinkage and selection via the lasso. *Journal of the Royal Statistical Society B 58*, 267–288.

Tipping, M. E. and C. M. Bishop (1999). Probabilistic principal component analysis. *Journal of the Royal Statistical Society B 61*(3), 611–622.

Vinzi, E. V., L. Trinchera, and S. Amato (2010). Pls path modeling: From foundations to recent developments and open issues for model assessment and improvement. In E. V. Vinzi, W. W. Chin, J. Henseler, and H. Wang (Eds.), *Handbook of Partial Least Squares*, Chapter 2, pp. 47–82. Berlin: Springer-Verlag.

vonRosen, D. (2018). *Bilinear Regression Analysis.* Number 220 in Lecture Notes in Statistics. Switzerland: Springer.

Wang, L. and S. Ding (2018). Vector autoregression and envelope model. *Stat 7*(1), e203.

Weglin, J. A. (2000). A survey of partial least squares (pls) methods, with emphasis on the two-block case. Technical Report 371, Department of Statistics, University of Washington.

Wen, X. and R. D. Cook (2007). Optimal sufficient dimension reduction in regressions with categorical predictors. *Journal of Statistical Planning and Inference 137*, 1961–1975.

Wold, H. (1966). Estimation of principal components and related models by iterative least squares. In P. R. Krishnaia (Ed.), *Multivariate Analysis*, pp. 392–420. New York: Academic Press.

Wold, H. (1975a). Path models with latent variables: The NIPALS approach. In H. M. Blalock, A. Aganbegian, F. M. Borodkin, R. Boudon, and V. Capecchi (Eds.), *Quantative Scoiology*, Chapter 11, pp. 307–357. London: Academic Press.

Wold, H. (1975b). Soft modelling by latent variables: the non-linear iterative partial least squares (NIPALS) approach. *Journal of Applied Probability 12*(S1), 117–142.

Wold, H. (1982). Soft modeling: the basic design and some extensions. In K. G. Jörgensen and H. Wold (Eds.), *Systems under indirect observation: Causality, structure, prediction*, Vol. 2, pp. 1–54. Amsterdam: NorthHolland.

Wold, S. (1992). Nonlinear partial least squares modeling ii. Spline inner relation. *Chemometrics and Intelligent Laboratory Systems 14*(1–3), 71–84.

Wold, S., N. Kettaneh, and K. Tjessem (1996). Hierarchial multiblock PLS models for easier model interpretation and as an alternative to variable selection. *Journal of Chemometrics 10*, 463–482.

Wold, S., H. Martens, and H. Wold (1983). The multivariate calibration problem in chemistry solved by the pls method. In A. Ruhe and B. Kågström (Eds.), *Proceedings of the Conference on Matrix Pencils, Lecture Notes in Mathematics*, Vol. 973, pp. 286–293. Heidelberg: Springer Verlag.

Wold, S., J. Trygg, A. Berglund, and H. Antti (2001). Some recent developments in PLS modeling. *Chemometrics and Intelligent Laboratory Systems 7*, 131–150.

Yeh, I. C. (2007). Modeling slump flow of concrete using second-order regressions and artiicial neural networks. *Cement and Concrete Composites 29*(6), 474–480.

Zhang, J. and X. Chen (2020). Principal envelope model. *Journal of Statistical Planning and Inference 206*, 249–262.

Zhang, X. and L. Li (2017). Tensor envelope partial least-squares regression. *Technometrics 59*(4), 426–436.

Zhang, X. and Q. Mai (2019). Efficient integration of sufficient dimension reduction and prediction in discriminant analysis. *Technometrics 61*(2), 259–272.

Zheng, W., X. Fu, and Y. Ying (2014). Spectroscopy-based food classification with extreme learning machine. *Chemometrics and Intelligent Laboratory Systems 139*, 42–47.

Zhu, G. and Z. Su (2020). Envelope-based sparse partial least squares. *Annals of Statistics 48*(1), 161–182.

Index

Printed in the United States
by Baker & Taylor Publisher Services

Printed in the United States
by Baker & Taylor Publisher Services